Environmental engineering in mines

Environmental engineering in mines

V.S.VUTUKURI
University of New South Wales, Australia
R.D.LAMA
Kembla Coal and Coke Pty Ltd, Australia

CAMBRIDGE UNIVERSITY PRESS
Cambridge
London New York New Rochelle
Melbourne Sydney

CAMBRIDGE UNIVERSITY PRESS
Cambridge, New York, Melbourne, Madrid, Cape Town, Singapore,
São Paulo, Delhi, Dubai, Tokyo, Mexico City

Cambridge University Press
The Edinburgh Building, Cambridge CB2 8RU, UK

Published in the United States of America by Cambridge University Press, New York

www.cambridge.org
Information on this title: www.cambridge.org/9780521157391

First published 1986
First paperback edition 2010

A catalogue record for this publication is available from the British Library

Library of Congress Cataloguing in Publication Data
Vutukuri, V. S., 1937-
Environmental engineering in mines.
Includes bibliographies.
1. Mine ventilation. 2. Mine safety. 3. Mining engineering.
4. Environmental engineering.
I. Lama, R. D., 1940- II. Title.
TN301. V88 1985 622'.4 85-12753

ISBN 978-0-521-24605-7 Hardback
ISBN 978-0-521-15739-1 Paperback

CONTENTS

Appendix II

Appendix III

Appendix IV

PREFACE

The primary purpose of this book is to present to students and practising engineers a comprehensive and modern overview of the subject of environmental engineering, as related to the mining of minerals, in a simple, clear and logical form in adequate depth. The book is designed as a ready reference to the fundamental principles involved in environmental engineering and common practices as followed in modern mines. The subject matter has been limited to methods of solving basic problems experienced in day to day operations in modern underground and opencut mines and in the planning of new mines.

The breadth of the subject of environmental engineering in mines is so large that it is almost impossible to detail all its facets in depth. Limitations on the size of the book have influenced the exclusion of certain aspects of environmental engineering, such as the effect on environment of surface mineral treatment plants, refuse dumps and tailings dams (which invariably form part of the mining operations). We have also excluded or only given a passing reference to those aspects of engineering, particularly equipment, on which information is normally available in handbooks of related fields such as civil, chemical, electrical and mechanical engineering.

It is hoped that the book will be useful to practising engineers and planners in the solution of environmental engineering related problems in mines. The understanding of the basic principles developed will stimulate interest in the subject.

We wish to gratefully acknowledge the many reseachers from whose works materials have been liberally drawn for the preparation of this book.

We are deeply indebted to Mr L. Griffiths, Consultant, Moss Vale and

Mr W. B. Bell, Research Fellow, University of New South Wales for reviewing some chapters of the manuscript and offering valuable comments.

Thanks are also due to Professor F. F. Roxborough and Professor L. J. Thomas of the University of New South Wales and Mr R. J. Fraser, Managing Director of Kembla Coal & Coke Pty Ltd for their encouragement and support in the preparation of the manuscript.

We wish to express our deep appreciation and profound gratitude to our families for their patience, support and understanding during the preparation of the book.

V. S. V. and R. D. L.

1

Introduction

The total environment in which mineral-producing activities take place is the result of the physical, regional and circumstantial conditions that surround and influence the extraction of minerals. The general environment results from and is structured about the unit operations involved in mining, namely, fragmentation, loading, haulage and support, and has four components: physical, process, regulatory and social. The physical environment is a function of the geological, hydrological, geographical and ecological nature of the physical space in which the mineral body occurs. The process environment refers to the products, sensations and risks that result from the unit operations to the physical environment. It includes gases, dust, radiation, heat, explosions, fires, water, noise, illumination and other health and safety elements that are functions of the mineral being mined, its state of occurrence and the available technology. The regulatory environment refers to the engineering systems, standards and regulations required to minimise the impact of the environment on mining personnel. It involves performance standards, enforcement procedures and compliance sanctions. The social environment refers to the physiological, psychological and physical factors that affect the performance and quality of life of the individual involved in, or affected by, the mining operations. It is a complex function of the impact of the other three environments on the individual personality.

The physical environment dictates the type of environmental elements encountered in the mining operations (gas, heat, water, etc.).

The process elements, i.e. gases, dust, radiation, etc., may be physiologically or psychologically damaging to mine personnel and may impede the production process. Some selected environmental elements, their sources, possible control practices and the property controlled are given below:

Element	Source(s)	Some common practice(s)	Property controlled
Gas	Natural, unit operations	Ventilation, infusion, scrubbers, drainage	Threshold concentration, emission rate
Dust	Natural, unit operations	Ventilation, water spray, collection	Concentration, generation
Radiation	Natural	Ventilation	Concentration, emission rate
Heat	Natural, unit operations	Ventilation, refrigeration	Temperature, flow rate
Water	Natural, unit operations	Collection, drainage, cementation, freezing, neutralisation	Quantity, flow rate
Explosions	Natural, unit operations	Ventilation, stone dusting, barriers	
Fires	Natural, unit operations	Ventilation, fire-resistant materials	
Noise	Unit operations	Modulation, isolation, protection	Level
Illumination	Natural, unit operations	Lighting	Level, glare

The general methodology followed in this book is to treat each element from the following aspects:

(1) hazards, physiological effects;
(2) occurrence and sources;
(3) detection, monitoring, sampling and measurement;
(4) control measures.

In the case of gases, dust, radiation and heat, control is usually accomplished by dilution and removal with artificially induced air currents. This subject has been dealt with separately (Chapter 2). Details regarding the other control measures are given in the respective chapters. Methane and its control in underground coal mines are dealt with separately because of its importance in the coal mining industry (Chapter 4). Aspects of ventilation planning along with computer monitoring which are common for all elements of mine environments are dealt with in Chapter 10. Computer monitoring of the environment has almost become a primary prerequisite for the introduction of efficient production techniques in very large mines.

Any engineer responsible for the provision and maintenance of a safe and healthy mine environment has to be thorough with the application of various regulations and standards of the state/country concerned.

2

Mine ventilation engineering

Ventilation is the process of conducting an adequate flow of pure fresh air along airways, working places and service points underground. The main purpose is to provide oxygen for mine workers; but it is also essential

(1) to dilute the concentration of the explosive and toxic gases, fumes and radon to environmentally safe levels and to remove from the mine;
(2) to dilute the concentration of the airborne dust to physiologically acceptable levels and to remove from the mine;
(3) to provide a thermally acceptable environment in which persons can work without undue discomfort or any danger of exhaustion from heat and to remove heat from the mine as may be necessary.

2.1 Ventilation practice in mines

2.1.1 General circulation of air in mines

To ensure adequate ventilation of a mine, provision is made for a suitable path or paths (normally termed airways) for the air to flow down the mine to the working places and suitable routes out of the mine when it has become unsuitable for further use.

The airways through which the fresh air passes on its way from the surface to the working places are termed intake or downcast airways, and generally consist of the main shaft or drift and transport roadways of the mine. The airways through which the air passes after having ventilated the working places of the mine are termed return or upcast airways. These consist either of airways and shafts or drifts specially excavated for the purpose, or are worked-out areas and old disused shafts or drifts.

Although a limited amount of air will normally flow through a mine without the use of a fan, or fans, the volume flow rate is usually insufficient for adequate ventilation and may vary. To ensure a constant and adequate supply of air, fans are used.

2.1.2 Basic principles of mine ventilation

Most mines use 'ascensional ventilation' of the working places, i.e. the fresh air is taken direct to the bottom sections of the working places by means of a downcast system and then ascends through the working places. This is generally more satisfactory than 'descensional ventilation', in which the fresh air descends through the working places and is then taken directly to the surface from the bottom sections by means of an upcast system.

The basic ventilation system desired for mines is given in Fig. 2.1.

Air always flows along the path of least resistance, but this may not be where it is required for use, i.e. working places. To prevent the air from returning directly to the surface, for example along the upper levels connecting the downcast and upcast shafts, ventilation doors or airlocks are installed in these levels. An airlock consists of two doors such that when one is opened, the other can be kept closed, thus preventing any airflow. If the roadway is only used occasionally, and leakage of some air can be accepted on these occasions, then a single door may be sufficient. Ventilation doors should be as airtight as possible and preferably self-closing.

If a small volume flow rate of air is required continuously in a given level, then an opening of required size, i.e. a regulator, is made in the

Fig. 2.1. Basic ventilation system desired for mines; D – door; R – regulator; 1, 2, 3 – working faces.

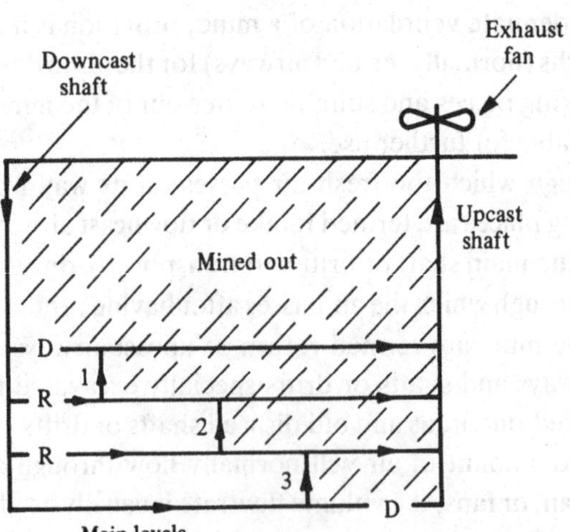

door or door frame. If such a regulator is to be installed across an airlock, it is best made in the form of a single ventilation duct of the required diameter running through both door frames.

Although the general airflow can be made to pass directly through the main working places, there will usually be several places where auxiliary ventilation will be required, e.g. development ends, hoist chambers, shaft loading stations, etc. To provide such ventilation requires an additional fan and a suitable route for the air. This route is generally a ventilation duct.

Ventilation of main working places

1. Stopes. A normal stope connects two levels, of which the lower usually acts as the intake airway and the upper as the return airway.

The distribution of air in a stope is controlled by closing the bottom and top of the stope, as far as possible, leaving openings only near the faces and other areas where persons may be working. This will direct most of the air towards the face. Fig. 2.2 gives an example of ventilation system for sublevel stoping.

2. Longwall faces. Typical ventilation schemes for longwall advance and retreat faces are given in Figs. 2.3 and 2.4 respectively.

Ventilation of dead ends

This includes the ventilation of development headings, raises and winzes, and refers particularly to the stage when these are being driven, but the principles are applicable also to places which remain dead ends after mining in them has stopped, e.g. a roadway off a main transport roadway leading to a loading station. For convenience, the term 'development end' will be used here.

Fig. 2.2. Ventilation system for sub-level stoping.

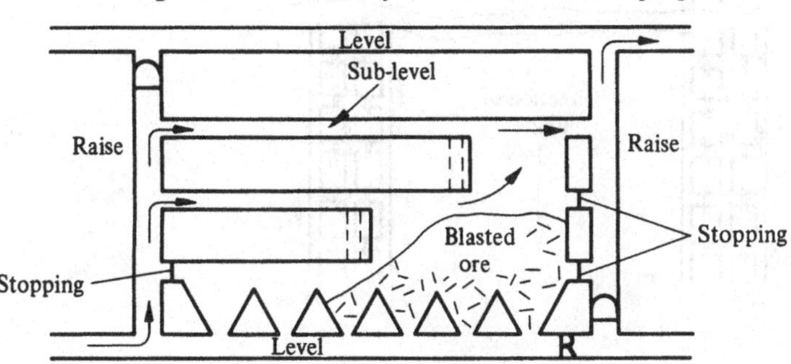

Fig. 2.3. Ventilation scheme for a longwall advance face.

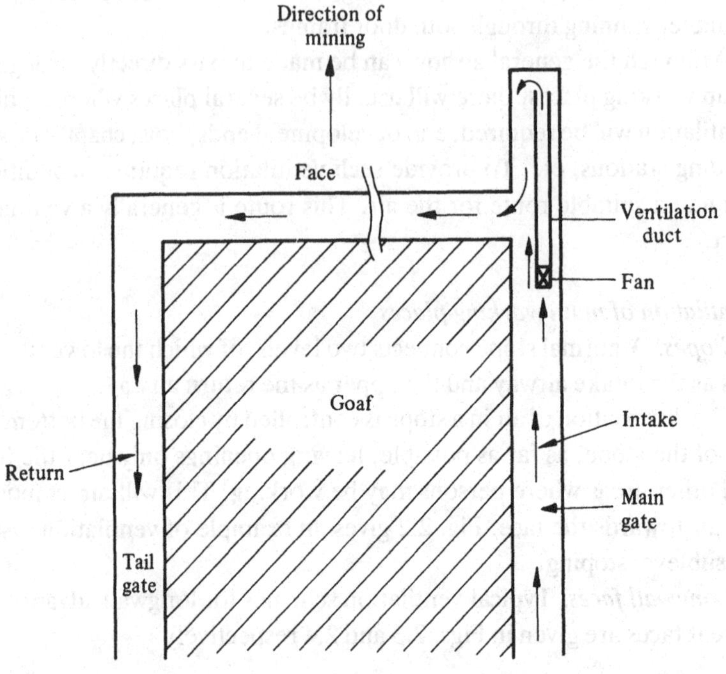

Fig. 2.4. Ventilation scheme for a longwall retreat face.

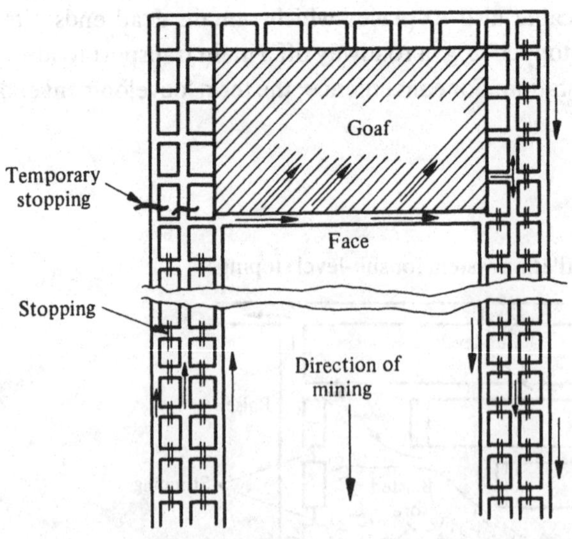

Development ends, etc. lead off an airway through which fresh air is flowing; some of this air must be used to ventilate the development end. It will not flow naturally into such ends, and must be induced to do so.

Three main systems can be used to ventilate the working face in a development end.

1. Forcing system (Fig. 2.5). The fresh air from the through airway is forced into the end through a ventilation duct by a fan, and this air returns via the development end.

The intake to the duct must extend well into the upstream portion of the fresh air stream (at least 5 m), otherwise there is a danger of the used air which flows out of the end entering the duct and thus being recirculated. The fan should be in the part of the duct which extends into the fresh air, so that it is accessible after blasting, when the roadway itself should not be entered. The intake and delivery ends of the duct should be fitted with wire screens to prevent any extraneous articles from being drawn into the fan and thereby damaging it or affecting its performance, and to prevent rock entering the duct during blasting. The duct should be kept in good condition and free from leaks, in order to deliver as much air as possible to the development end. The delivery end of the duct must be kept as close as possible to the working place. The end of the duct may be damaged by blasting if it is too close to the face. One method of avoiding this damage is to remove the last section or two of the ventilation duct before blasting and replace it at the time of loading, or to use a flexible retractable duct.

Fig. 2.5. Forcing system of ventilation.

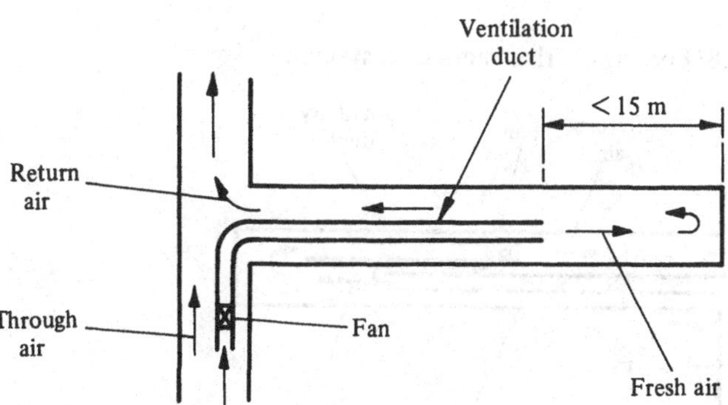

Fig. 2.6. Exhaust system of ventilation.

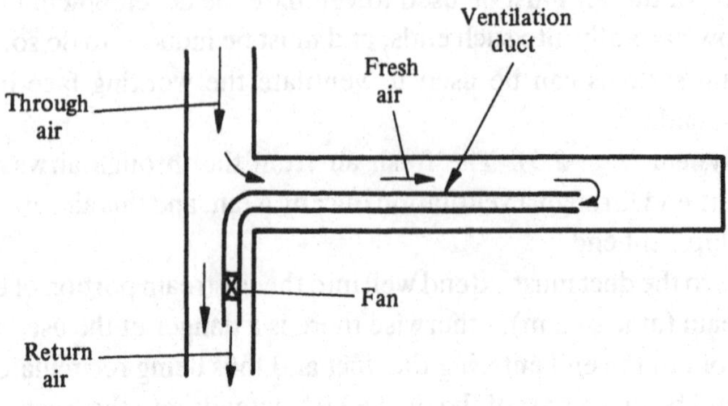

Fig. 2.7. Exhaust overlap ventilation system.

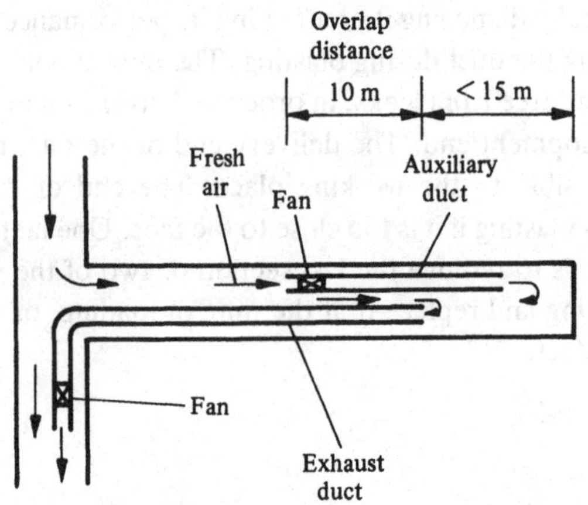

Fig. 2.8. Forcing overlap ventilation system.

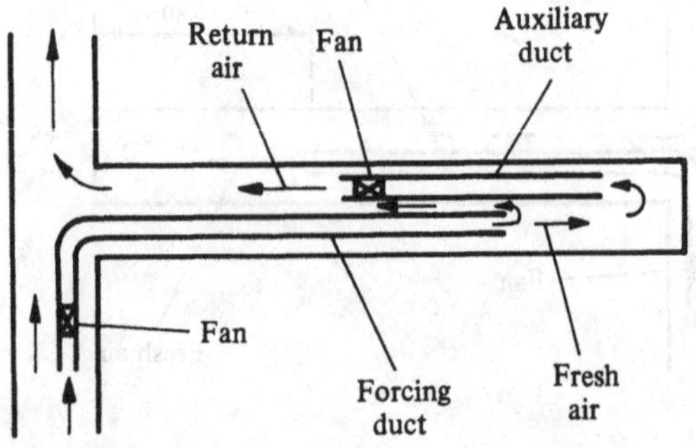

2. Exhaust system (Fig. 2.6). Air from near the working face of the end is exhausted through a ventilation duct, and is automatically replaced by fresh air drawn into and then along the roadway.

The requirements regarding the duct are much the same as above, but it must discharge the return air well beyond the entrance to the development end (at least 5 m), on the downstream side, to prevent it re-entering the end. This system is not commonly used because the fresh air drawn along the roadway will tend to enter the exhaust ventilation duct directly and very little of it will actually reach the face, as its velocity is low, and the intake of the return duct cannot always be kept right up to the face, owing to the danger of it being damaged by a blast.

3. Overlap systems. In exhaust overlap systems (Fig. 2.7), in addition to the ventilation duct exhausting air from the development end, there is an auxiliary duct which forces some fresh air from the roadway up to the face. The length of the auxiliary duct might be anything from 6 to 60 m.

The distance from the face to the intake end of the exhaust duct is generally kept within 15 m (but usually ~5 m) of the face. It is important that the exhaust duct and the auxiliary forcing duct 'overlap' by at least 5 m, and preferably more – a common length of overlap is 10 m. This is to ensure that the air entering the auxiliary forcing duct is fresh air coming up the roadway and is not air coming back from the face.

It is essential to ensure that the volume flow rate of air being exhausted from the face is at all times greater (2–4 times) than the volume flow rate being blown onto the face by the auxiliary duct, to prevent the short-circuiting of return air from the face. The drives of the fans must also be interlocked so that, if the main exhaust fan stops for any reason, the inbye forcing fan also stops. The reverse of the exhaust overlap system is the forcing overlap system shown in Fig. 2.8.

In coal mines, there are two face ventilation systems in use:

(1) use of a non-inflammable brattice;
(2) use of an auxiliary fan and a ventilation duct.

The exhaust system is the most common in coal mines. Often a non-inflammable brattice is used instead of an auxiliary duct to partition the development roadway into an intake and a return airflow.

Fig. 2.9 illustrates brattice ventilation systems for a typical four-heading development scheme. Fig. 2.10 gives a typical auxiliary fan layout for a seven-heading development scheme.

Fig. 2.9. Brattice ventilation systems for a four-heading development scheme (after Marshall and Allan, 1978).

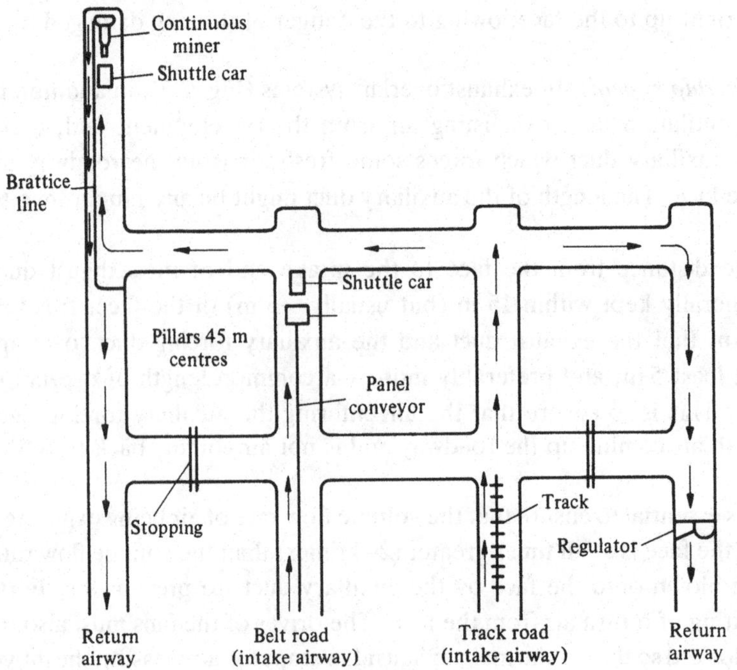

Sequence 1

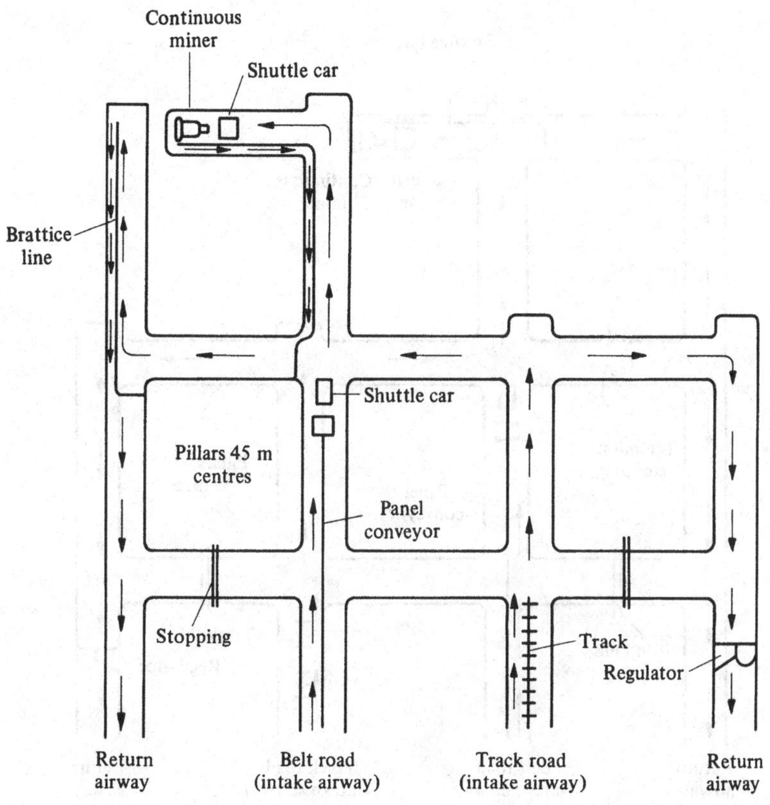

Sequence 2

(*Fig. continued*)

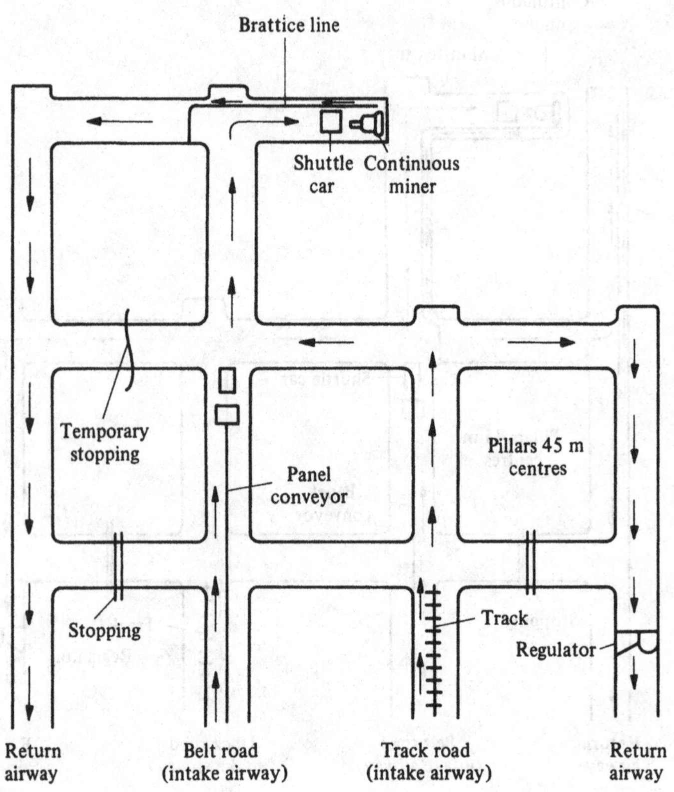
Brattice line
Shuttle car
Continuous miner
Temporary stopping
Panel conveyor
Pillars 45 m centres
Stopping
Track
Regulator
Return airway
Belt road (intake airway)
Track road (intake airway)
Return airway

Sequence 3

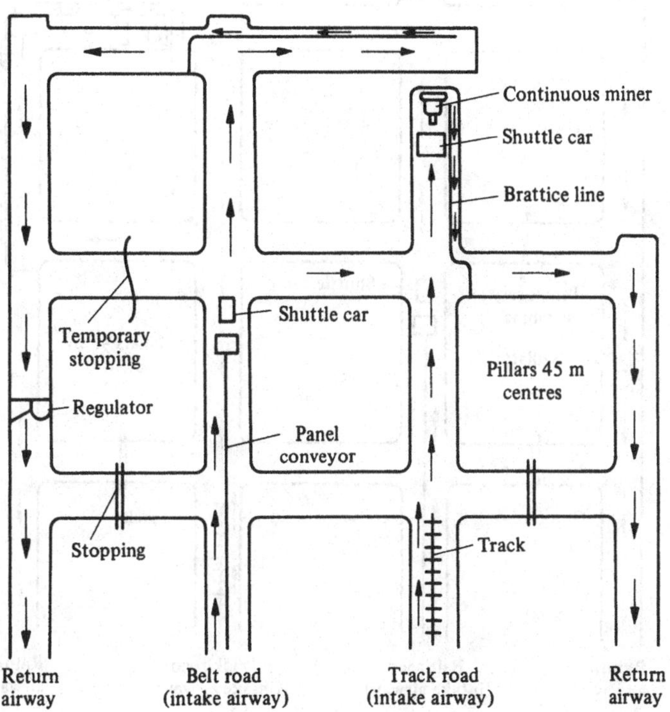

Sequence 4

(*Fig. continued*)

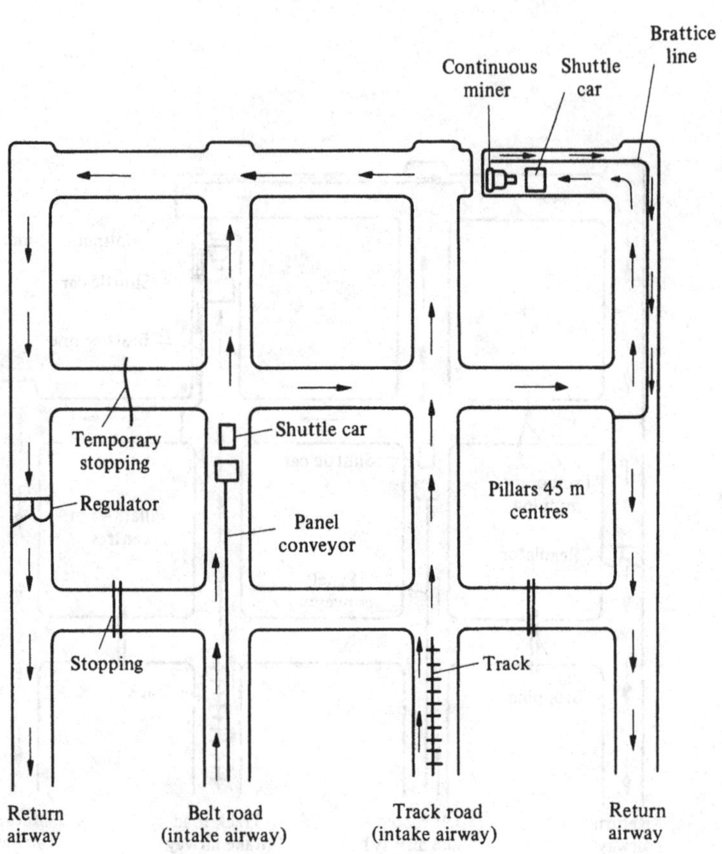
Brattice line
Continuous miner
Shuttle car
Temporary stopping
Shuttle car
Regulator
Panel conveyor
Pillars 45 m centres
Stopping
Track
Return airway
Belt road (intake airway)
Track road (intake airway)
Return airway

Sequence 5

Fig. 2.10. Auxiliary fan layout for a seven-heading development scheme (after Crous *et al.*, 1975).

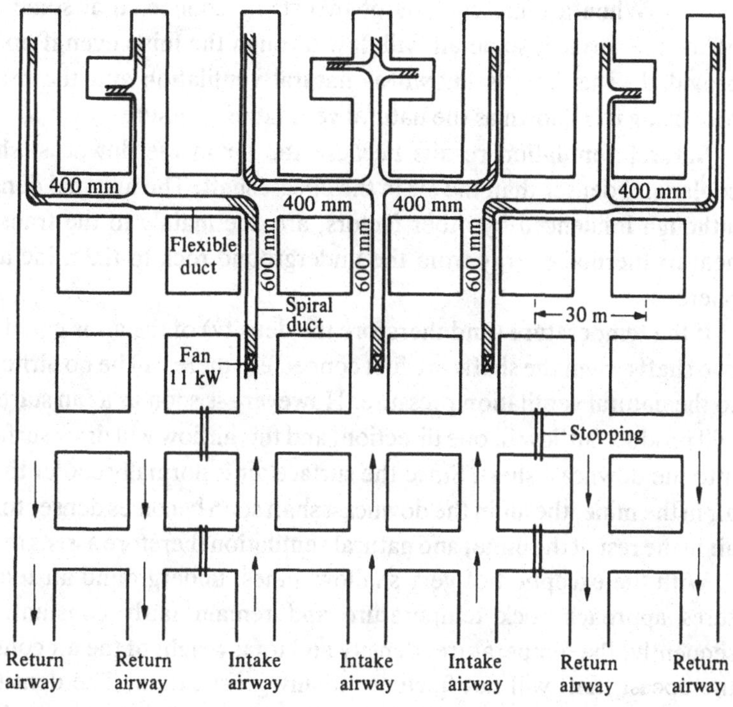

2.2 Forces causing airflow in mines

To cause airflow in mines, a difference in pressure between the intake and return shafts is essential. This pressure difference can be caused either by natural or mechanical ventilation.

2.2.1 Natural ventilation

When a mine consists of two shafts connected at some depth below the surface, some air will flow through the mine even if no fan is provided. This flow is known as natural ventilation and the pressure producing it is known as the natural ventilation pressure.

Natural ventilation results because the air in the downcast shaft is cooler and denser than the air in the upcast shaft. The unequal densities, although influenced by other factors, are due mainly to the transfer of heat or thermal energy from the underground rock to the mine atmosphere.

If the temperature (and therefore the density) of the air is equal in the two shafts when the shafts are first connected, there will be no airflow due to the natural ventilation pressure. However, as soon as a fan starts up it will produce airflow in one direction, and this airflow will draw surface air into the downcast shaft. Since the surface air is normally cooler than the air in the mine, the air in the downcast shaft soon becomes denser than the air in the rest of the mine, and natural ventilation therefore assists the fan.

With the exception of very shallow mines, underground air temperatures approach rock temperature and remain fairly constant. Consequently, the temperature, density and total weight of the air column in the upcast shaft will not fluctuate to any great extent. The density and total weight of the air column in the downcast shaft, however, is directly influenced by and fluctuates with the surface air temperature. Since a density change in either of the columns will cause a change in the natural ventilation pressure, this pressure difference may vary from a maximum in one direction to no difference to a maximum difference in the opposite direction, depending on whether the surface air temperature is lower than, equal to, or greater than the underground air temperature. This results in an unpredictable volume flow rate, which may vary from a maximum in one direction to some value in the opposite direction. In some instances, natural ventilation may maintain underground air quality standards but generally it alone is not a dependable means for continuous and effective ventilation and additional ventilation is provided by mechanical means, that is by the use of fans.

Determining the difference in density of two air columns of equal heights forms the basis for calculating natural ventilation pressures. The density difference multiplied by the length of the column and the gravitational constant gives the natural ventilation pressure.

2.2.2 *Mechanical ventilation*

A fan takes in air at one pressure and raises this air to a higher pressure, i.e. it increases the pressure of the air on its delivery side to a value above the pressure of the air on its intake side. Due to the pressure difference, the air is able to overcome the resistance of the system in which it is flowing.

Usually the main fan (or fans) is situated at the top of the upcast shaft or return opening. Its main advantages are:

(1) The working environment's pressure is slightly less than the outside atmospheric pressure, because the fan creates a negative pressure within the mine. When the fan stops, underground pressure builds up to atmospheric pressure and this increase in pressure slows down the liberation of strata gases and gas emissions from the goaf, and prolongs the time required for the gas to fill active workings.
(2) Through the use of discharge evasees, velocity pressure can be recovered enabling power savings.
(3) No heat is added to the air supply underground.

The disadvantages are:

(1) It can only be applied if all the return air flows in the upcast shaft.
(2) Foul air goes through the fan, making repair or maintenance difficult. Corrosive fumes and particles cause corrosion of the fan blades and reduce effective air passage area and can throw the fan out of balance.
(3) Natural ventilation pressure is reduced.
(4) It is difficult to hoist men and materials in the upcast shaft due to safety requirements.

In shallow mines having fractured ground, in areas of contiguous mining where there may be ground fractures into abandoned mines, and in mines with old workings outcropping on the surface, the most advantageous position for the fan is at the top of the downcast shaft or intake opening. Some of the disadvantages of this positioning are:

(1) access to the shaft is difficult, requiring special airlocks, etc.;
(2) it is difficult to transport men, material or ore in the shaft;
(3) heat is added to the downcast air.

From the thermodynamical point of view, the best location is at the bottom of the upcast shaft. Because the fan adds heat to the air in the upcast shaft, natural ventilation is increased. The other advantages are:

(1) no heat is added to the air breathed by miners;
(2) the downcast shaft is left clear for hoisting.

The disadvantages are:

(1) the fan handles foul air;
(2) it is not accessible during fires;
(3) any leakage across doors above the fan results in recirculation of foul air.

2.2.3 *Fundamentals of airflow*

Laminar and turbulent flow

In laminar flow the fluid particles move along smooth paths such that the movement of all particles on the same path is identical. This type of flow is associated with very low velocities and highly viscous fluids.

Turbulent flow is characterised by fluid particles moving in irregular paths and colliding with one another in a haphazard manner. Most practical flow problems of interest to the mining engineer are related to turbulent flow.

Analysis of airflow

Airflow is analysed as it passes through a fixed region in space defined by a control surface. The properties of the air are examined as it enters and leaves the volume bounded by the control surface, together with mass flow rates and energy transfer by work or heat. Three equations are obtained as follows.

1. The continuity equation. The mass flow rate entering the control surface must equal the mass flow rate leaving the control surface, i.e.

$$M_1 = M_2$$

i.e.
$$w_1 A_1 V_1 = w_2 A_2 V_2 \tag{2.1}$$

or if air is of constant density,

$$Q_1 = Q_2$$
$$A_1 V_1 = A_2 V_2. \tag{2.2}$$

In these equations,

M = mass flow rate (kg/s);
w = air density (kg/m^3);
A = flow area measured normal to the direction of flow (m^2);
V = flow velocity (m/s); and
Q = volume flow rate (m^3/s).

2. *The energy equation.* Air in motion possesses energy by virtue of the pressure under which it exists, its velocity and its height above some datum level of potential energy. These three forms of energy – pressure, kinetic and potential – may be expressed per unit weight of air concerned. The result is the pressure, kinetic and potential head. If a mass of air is subjected to a pressure P (Pa), moving with a velocity V (m/s) and having its centre of mass at a height Z (m) above a datum of potential energy, then

$$\text{the total head } (m) = \frac{P}{wg} + \frac{V^2}{2g} + Z \qquad (2.3)$$

where g = gravitational acceleration (m/s^2); and
w = air density (kg/m^3).

Equation (2.3) in terms of pressure (i.e. energy per unit volume) is

$$\text{the total pressure (Pa)} = P + \frac{wV^2}{2} + Zwg. \qquad (2.4)$$

The energy equation, also known as Bernoulli's equation, states that in the case of an incompressible and inviscous fluid the energy entering the control surface must equal the energy leaving it. Expressed as energy per unit volume or pressure,

$$P_1 + \frac{wV_1^2}{2} + Z_1wg = P_2 + \frac{wV_2^2}{2} + Z_2wg. \qquad (2.5)$$

Changes in density of air up to 5% can be considered as incompressible from a practical point of view. This equation can then be applied to airflow at atmospheric pressure, without unduly great errors, when height changes are limited to 500 m and velocity changes to 100 m/s. If the limits of incompressibility are exceeded, the application of the thermodynamic laws is required to describe flow phenomena.

In the case of air which is viscous, equation (2.5) may be modified to account for losses due to various causes* by adding the total pressure losses to the downstream side of the equation as follows:

$$P_1 + \frac{wV_1^2}{2} + Z_1wg = P_2 + \frac{wV_2^2}{2} + Z_2wg + p. \qquad (2.6)$$

where p = loss in total pressure between points 1 and 2.

* Viscous resistance of air (laminar flow), the frictional resistance of the control surface, e.g. the duct wall in the case of airflow in ducts (turbulent flow), and non-uniformities of the control surface, e.g. changes in area, bends, etc. Losses associated with non-uniformities are minor.

If a device such as a fan which increases the total pressure is incorporated between points 1 and 2, equation (2.6) can be modified further by adding the total pressure rise to the pressure on the upstream side of the equation to give:

$$P_1 + \frac{wV_1^2}{2} + Z_1wg + p_f = P_2 + \frac{wV_2^2}{2} + Z_2wg + p \qquad (2.7)$$

where p_f = total pressure rise through the device.
3. *The momentum equation.* The sum of the external forces acting on the control surface in a given direction is equal to the rate of change of air momentum in that direction.

Airflow in airways
Airflow through an airway results from difference of pressure between the higher pressure at the inflow end and the lower pressure at the outflow end.
Pressure loss in laminar flow. In laminar flow the pressure loss is caused by the viscous resistance of air as shear stresses are generated when adjacent layers of air pass over each other. These stresses cause a uniform loss of pressure in the direction of flow.

The loss of pressure in circular airways may be estimated by Hagen–Poiseuille's equation.

$$p = \frac{32\mu LV}{D^2} \qquad (2.8)$$

where p = pressure loss (Pa);
μ = dynamic viscosity of air (Pa.s);
L = airway length (m);
V = mean air velocity (m/s); and
D = airway diameter (m).

It is clear from the above equation that the pressure loss is proportional to the mean air velocity (V).
Pressure loss in turbulent flow. In flow of this nature the pressure loss is associated with the shear stresses caused by the frictional resistance of the airway wall.

The loss of pressure in circular airways may be estimated by Darcy's equation.

$$p = \frac{4fwLV^2}{2D} \qquad (2.9)$$

where p = pressure loss (Pa);

f = an experimentally determined friction factor (dimensionless);
w = air density (kg/m^3);
L = airway length (m);
V = mean air velocity (m/s); and
D = airway diameter (m).

It is clear from the above equation that the pressure loss is proportional to the square of the mean air velocity V^2.

For non-circular airways equation (2.9) can also be used if the diameter D is replaced by the equivalent diameter, defined as

$$D_e = \frac{4A}{C}$$

where D_e = equivalent diameter (m);
A = cross-sectional area of the airway (m^2); and
C = perimeter of the airway in contact with air flowing (m).

Equation (2.9) then becomes

$$p = \frac{4fwLV^2C}{2 \times 4A}. \qquad (2.10)$$

Reynolds number. The Reynolds number is defined by

$$R_e = \frac{wVD}{\mu} \quad \text{or} \quad \frac{VD}{\nu} \qquad (2.11)$$

where R_e = the Reynolds number (dimensionless);
w = air density (kg/m^3);
V = mean air velocity (m/s);
D = airway diameter (m);
μ = dynamic viscosity of air (Pa.s); and
ν = kinematic viscosity of air (m^2/s).

The value of the Reynolds number determines whether laminar or turbulent flow prevails. If $R_e < 2000$ the flow is laminar and if $R_e > 2300$ the flow is turbulent. Between these values is a critical zone in which the flow is transitional.

Relationship between Darcy's friction factor and the Reynolds number

Laminar flow: It is convenient to compare the pressure loss given by Hagen–Poiseuille's and Darcy's equations, i.e.

$$p = \frac{32\mu LV}{D^2} = \frac{4fwLV^2}{2D}.$$

Therefore

$$f = \frac{16\mu}{wVD}$$

$$= \frac{16}{R_e}. \qquad (2.12)$$

Turbulent flow: In turbulent flow f is a function of the Reynolds number and of the relative roughness of the airway. The relative roughness is the ratio of the average height of the surface irregularities (ε) to the airway diameter (D).

Moody's diagram: The relationship between f, R_e and ε/D is given in Fig. 2.11. The graphs in Fig. 2.11 enable pressure loss to be calculated for shafts, raises and lined roadways.

2.2.4 *Atkinson's equation*

In mine ventilation, an alternative friction factor K is used instead of f. Atkinson arrived at this factor by assuming the pressure loss to be proportional to the airway length, the perimeter, the square of the

Fig. 2.11. Chart for friction factors (Moody's diagram) (after Massey, 1978). w = air density (kg/m^3); V = mean air velocity (m/s); D = airway diameter (m); μ = dynamic viscosity of air (Pa·s); ν = kinematic viscosity of air (m^2/s); ε = average height of the surface irregularities.

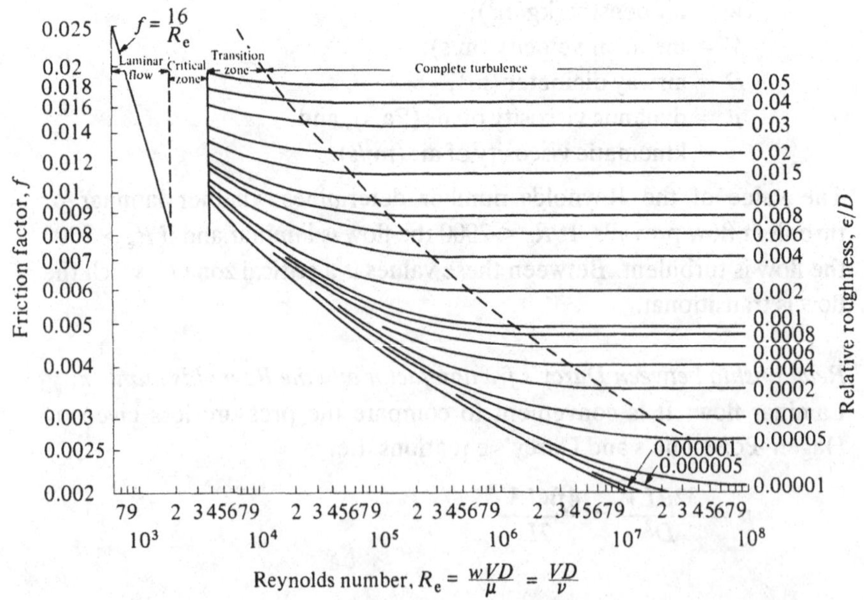

velocity and inversely proportional to the cross-sectional area. Thus,

$$p = \frac{KLCV^2}{A}$$

$$= \frac{KLCQ^2}{A^3} \text{ since } V = \frac{Q}{A} \tag{2.13}$$

where p = pressure loss (Pa);
K = friction factor (Ns^2/m^4);
L = airway length (m);
C = airway perimeter (m);
V = airflow velocity (m/s);
A = airway cross-sectional area (m^2); and
Q = volume flow rate (m^3/s).

Atkinson ignored the density effects in deriving equation (2.13) and it is therefore valid only for a standard air density of 1.2 kg/m^3.

The form of the equation used at any density is

$$p = \frac{KLCV^2}{A} \times \frac{w}{1.2}$$

$$= \frac{KLCQ^2}{A^3} \times \frac{w}{1.2} \tag{2.14}$$

where w = air density (kg/m^3).

Equating p from equations (2.10) and (2.14), the relationship between K and f is obtained as follows:

$$\frac{4fwLV^2C}{2 \times 4A} = \frac{KLCV^2}{A} \times \frac{w}{1.2}$$

$$\frac{f}{2} = \frac{K}{1.2}$$

or

$$K = 0.6f. \tag{2.15}$$

For a given airway, KLC/A^3 is constant and is substituted by the term R, the resistance of the airway. In that case,

$$p = RQ^2 \times \frac{W}{1.2}. \tag{2.16}$$

The unit of R is Ns^2/m^8.

Friction factor, K

The value of the friction factor, K, normally lies between 0.01 and 0.02 Ns^2/m^4. Wherever possible values obtained from pressure–volume flow rate surveys should be used. Some values are given in Tables 2.1 and 2.2.

Table 2.1. *Bureau of Mines schedule of friction factors for mine airways (converted to SI units using McElroy, 1935)*

		Values of K (Ns^2/m^4)											
					Sinuous or curved								
		Straight			Slightly			Moderately			High-degree		
Type of airway	Irregularities of surfaces, areas and alignment	Clean (basic values)	Slightly ob-structed	Moderately ob-structed	Clean	Slightly ob-structed	Moderately ob-structed	Clean	Slightly ob-structed	Moderately ob-structed	Clean	Slightly ob-structed	Moderately ob-structed
Smooth-lined	Minimum	0.0019	0.0028	0.0046	0.0037	0.0046	0.0065	0.0046	0.0056	0.0074	0.0065	0.0074	0.0093
	Average	0.0028	0.0037	0.0056	0.0046	0.0056	0.0074	0.0056	0.0065	0.0083	0.0074	0.0083	0.0102
	Maximum	0.0037	0.0046	0.0065	0.0056	0.0065	0.0083	0.0065	0.0074	0.0093	0.0083	0.0093	0.0111
Sedimentary (or coal)	Minimum	0.0056	0.0065	0.0083	0.0074	0.0083	0.0102	0.0083	0.0093	0.0111	0.0102	0.0111	0.0130
	Average	0.0102	0.0111	0.0130	0.0121	0.0130	0.0148	0.0130	0.0139	0.0158	0.0148	0.0158	0.0176
	Maximum	0.0130	0.0139	0.0158	0.0148	0.0158	0.0176	0.0158	0.0176	0.0186	0.0176	0.0186	0.0204
Timbered (1.5 m centres)	Minimum	0.0148	0.0158	0.0176	0.0167	0.0176	0.0195	0.0176	0.0186	0.0204	0.0195	0.0204	0.0223
	Average	0.0176	0.0186	0.0204	0.0195	0.0204	0.0223	0.0204	0.0213	0.0232	0.0223	0.0232	0.0250
	Maximum	0.0195	0.0204	0.0223	0.0213	0.0223	0.0241	0.0223	0.0232	0.0250	0.0241	0.0250	0.0269
Igneous rock	Minimum	0.0167	0.0176	0.0195	0.0186	0.0195	0.0213	0.0195	0.0204	0.0223	0.0213	0.0223	0.0241
	Average	0.0269	0.0278	0.0297	0.0288	0.0297	0.0306	0.0297	0.0306	0.0325	0.0315	0.0325	0.0362
	Maximum	0.0362	0.0371	0.0390	0.0380	0.0390	0.0408	0.0390	0.0399	0.0417	0.0408	0.0417	0.0436

All values of K are for air of standard density of 1.2 kg/m^3.
Note: The values of the friction factor K have been increased to take into account small losses of pressure due to changes in areas, changes in direction and obstructions. For separate calculations of large losses, see the section 'Losses of pressure due to causes other than viscous and frictional resistance', p. 26.

Table 2.2. *National Coal Board (UK) schedule of friction factors (K, Ns^2/m^4) for circular airways*[a] *(after NCB, 1979)*

Shafts	K (Ns^2/m^4)
Smooth, concrete-lined, unobstructed	0.0030
Brick-lined, unobstructed	0.0037
Smooth, concrete-lined, with rope guides and pipe ranges on buntons	0.0065
Brick-lined, with rope guides and pipe ranges on buntons	0.0074
Tubbing-lined, with no guides or cages	0.0139
Timber-lined, no middle buntons	0.0167
Brick-lined, two lines of side buntons, without tie girders	0.0176
Brick-lined, two lines of side buntons, one tie girder to each bunton	0.0223
Timber-lined, with middle buntons	0.0223
Steel-arched roadways	
Smooth concrete all round	0.0037
Concrete slabs or timber lagging between flanges all round	0.0074
Concrete slabs or timber lagging between flanges to spring	0.0093
Lagged behind arches, good condition	0.0121
Rough conditions with irregular roof, sides and floor	0.0158
Rectangular roadways	
Smooth, concrete lined	0.0037
Girders on brick or concrete walls	0.0093
Unlined airways with uniform sides	0.0121
Unlined airways, irregular conditions	0.0158
Girders or bars on timber props	0.0186

[a] For airways of other shapes, the value of the friction factor from the table is to be multiplied by the shape factor below:

Shape of roadway or shaft	Factor
Circular	1.00
Arched, straight legs	1.08
Arched, splayed legs	1.09
Square	1.11
Rectangular	
width : height = 1.5 : 1	1.13
= 2.0 : 1	1.16
= 3.0 : 1	1.23

Losses of pressure due to causes other than viscous and frictional resistances

When an airway changes its direction, cross-sectional area or has an obstruction placed in it there will be a loss of pressure due to the disturbance in flow. This is normally expressed by

$$p = \lambda \frac{wV^2}{2} \qquad (2.17)$$

where p = pressure loss (Pa);
λ = experimentally determined loss factor for the item concerned, and is regarded as having no strong dependence on the Reynolds number (dimensionless);
w = air density (kg/m^3); and
V = airflow velocity (m/s).

Alternatively, these pressure losses may be expressed in terms of the length of straight airway l, usually expressed as a multiple n of the airway diameter (D), to give the equivalent pressure loss. This method is approximate.

Some typical values of λ and $l/D = n$ for airways are given in Table 2.3.

Obstructions in airways retard the flow of air in two ways. Firstly, the obstruction causes a reduction in the free cross-sectional area and consequently an increase in air velocity which causes an increase in frictional resistance. But secondly, and usually more seriously, the flow pattern of the air is disrupted by the presence of the obstruction and this causes eddy currents which result in a loss of pressure. The extent of the disruption of the flow pattern is dependent not only upon the size of the obstruction, but also on its shape.

The loss factor λ can be reduced by using the following methods:

(1) At the bends by rounding off sharp corners or inserting guide vanes.

(2) At points of enlargements or contractions by limiting the maximum angles in diverging or converging passages to about 7° including angle.

(3) At splits or junctions by rounding off sharp corners; rounding off junctions; chamfering internal sharp edges; placing guide vanes; building smooth linings.

(4) At the intersection of vertical shaft and horizontal off-take by chamfering edges and inserting a cone (Fig. 2.12).

(5) For shaft buntons by streamlining, i.e. rounding the forward part facing the airflow and placing a sharp point on the rear side

Table 2.3. *Typical values of λ and n for various types of bends, area changes and entrances*

Description	λ	n (i.e. l/D)
Bends		
(i) Short radius		
22.5°	0.2	9
45.0°	0.4	18
90.0°	1.0	45
(ii) Long radius		
22.5°	0.1	5
45.0°	0.2	9
90.0°	0.4	18
Sudden enlargement[a]		
Inlet diameter : outlet diameter		
1:2	0.60	27
1:3	0.80	36
1:5 and over	1.00	45
Sudden contraction		
2:1	0.35	16
3:1	0.45	20
5 and over:1	0.50	22
Entrances[b]		
Plain end	0.9	
Flanged end	0.5	
60° cone into end	0.2	

[a] These loss factors are based on the velocity pressure in the narrow section.
[b] These values are based on the velocity pressure just inside the airway.

Fig. 2.12. Inserting a cone and chamfering the edges at the intersection of vertical shaft and horizontal off-take (after Anon., 1978).

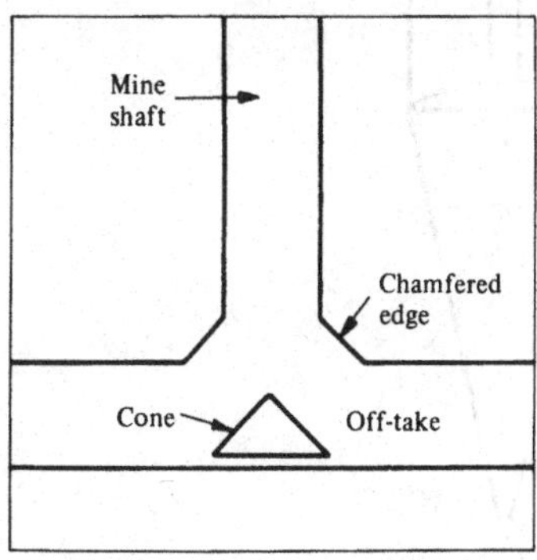

(Fig. 2.13). Squashed pipe or hexagonal buntons are used in new shafts.

(6) For cages by fitting them with annular fairing on top and with close-fitting panel doors.

Air power

When air over an area A (m^2) moves a distance l (m) in time t (s) against a pressure difference p (Pa), the work done is given by pAl (J), while the power expended is given by pAl/t (W). But Al/t is the volume of air moved in unit time, i.e. Q (m^3/s), and thus

$$\text{air power} = pQ. \tag{2.18}$$

Atkinson's equation (equation 2.13) may be substituted in the above equation:

$$\text{air power} = pQ = \frac{KLCQ^2}{A^3} \times Q = \frac{KLCQ^3}{A^3}. \tag{2.19}$$

In a given airway, K, L, C and A are constants, and therefore

$$\text{air power} \propto Q^3.$$

Fig. 2.13. Streamlined buntons.

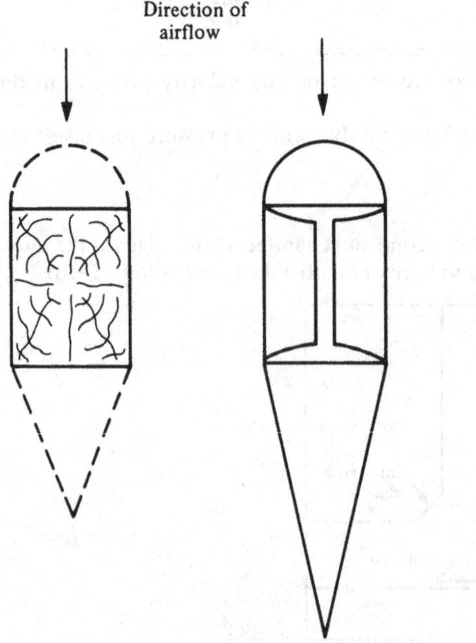

2.3 The ventilation resistance of the workings and the distribution of the mine air

Usually there are three basic combinations of mine airways: (1) series; (2) parallel; (3) compound.

2.3.1 Series circuits

When airways are connected in series (Fig. 2.14), the volume flow rate, Q, is the same in each branch; therefore

$$p_1 = R_1Q^2$$
$$p_2 = R_2Q^2$$
$$p_3 = R_3Q^2.$$

where p_1, p_2, p_3 = pressure drop in airways 1, 2 and 3; and
R_1, R_2, R_3 = the resistance of airways 1, 2 and 3.

The sum of these equations is the total pressure drop, p, for the system:

$$p = p_1 + p_2 + p_3 = (R_1 + R_2 + R_3)Q^2.$$

If the equivalent (total) resistance for the complete series circuit is R_s, then

$$p = R_sQ^2.$$

Hence $R_s = R_1 + R_2 + R_3$. In general, for b branches connected in series

$$R_s = \sum_{i=1}^{b} R_i. \tag{2.20}$$

2.3.2 Parallel circuits

Parallel connections between airways (also called splits) occur in a ventilation network where the airways branch at one point and rejoin at another. In this case, the pressure drop, p, is common to each parallel airway (Fig. 2.15); therefore

$$p = R_1Q_1^2 = R_2Q_2^2 = R_3Q_3^2$$

Fig. 2.14. Airways in series. Q = volume flow rate; p_1, p_2, p_3 = pressure drop in airways 1, 2 and 3; R_1, R_2, R_3 = resistance in airways 1, 2 and 3; p = total pressure drop.

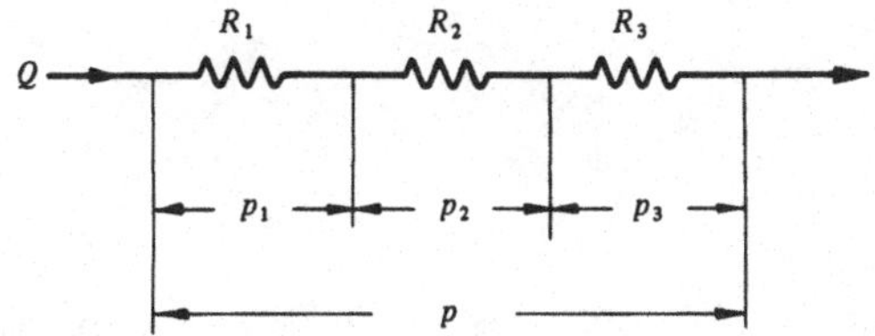

or

$$Q_1 = \sqrt{\frac{p}{R_1}}$$

$$Q_2 = \sqrt{\frac{p}{R_2}}$$

$$Q_3 = \sqrt{\frac{p}{R_3}}$$

Adding these three equations gives

$$Q = Q_1 + Q_2 + Q_3 = \sqrt{p}\left[\frac{1}{\sqrt{R_1}} + \frac{1}{\sqrt{R_2}} + \frac{1}{\sqrt{R_3}}\right].$$

Now, if the equivalent (total) resistance for the parallel circuit is R_p, then

$$Q = \sqrt{\frac{p}{R_p}}.$$

Hence

$$\frac{1}{\sqrt{R_p}} = \frac{1}{\sqrt{R_1}} + \frac{1}{\sqrt{R_2}} + \frac{1}{\sqrt{R_3}}.$$

In general, for b branches connected in parallel,

$$\frac{1}{\sqrt{R_p}} = \sum_{i=1}^{b} \frac{1}{\sqrt{R_i}}. \tag{2.21}$$

Since $p = R_1Q_1^2 = R_2Q_2^2 = R_3Q_3^2$

$$Q_1 \propto \frac{1}{\sqrt{R_1}},$$

$$Q_2 \propto \frac{1}{\sqrt{R_2}},$$

$$Q_3 \propto \frac{1}{\sqrt{R_3}},$$

$$Q \propto \frac{1}{\sqrt{R_p}} \propto \left[\frac{1}{\sqrt{R_1}} + \frac{1}{\sqrt{R_2}} + \frac{1}{\sqrt{R_3}}\right].$$

Therefore

$$Q_1 = Q\left[\sqrt{\frac{R_p}{R_1}}\right]$$

$$Q_2 = Q\left[\sqrt{\frac{R_p}{R_2}}\right]$$

$$Q_3 = Q\left[\sqrt{\frac{R_p}{R_3}}\right]. \tag{2.22}$$

The above equations give the volume flow rates distributed through airways connected in parallel depending on their resistances. If these volume flow rates differ from the requirements, one of the following three methods is used to redistribute the volume flow rate:

(1) By increasing the resistance of one airway by building in it a regulator or regulators. This causes higher volume flow rates into the other splits.
(2) By increasing volume flow rate in an airway by installing a booster fan in it.
(3) By reducing the resistance of the airway in which the volume flow rate has to be increased.

With a regulator for varying the volume flow rate, it must be remembered that the regulator increases the resistance not only of the airway in which it is built but also of the mine as a whole. The number of regulators should therefore be kept to a minimum. The increase in the volume flow rate into the other airways is much less than the reduction of the volume flow rate in the regulated airway and there is thus an overall reduction of the volume flow rate into the mine.

Installing a booster fan in one split increases the total volume flow rate entering the mine. The booster does, however, to some degree reduce the volume flow rates in the other splits. But the reduction in the other splits is less than the increase in the boosted split.

Varying the volume flow rate, by reducing the resistance of the airway in which the volume flow rate is to be increased, is done by increasing its cross-section, repairing it, cleaning it up, changing the timbering, supplying two or more paths for the air in parallel, etc. All these operations are expensive and are rarely justified for general ventilation control; but these are sometimes used in major airways such as shafts to increase the total volume flow rate through the mine.

Fig. 2.15. Airways in parallel. R_1, R_2, R_3 = resistance in airways 1, 2 and 3; Q_1, Q_2, Q_3 = volume flow rate in airways 1, 2 and 3; Q = total volume flow rate; p = total pressure drop.

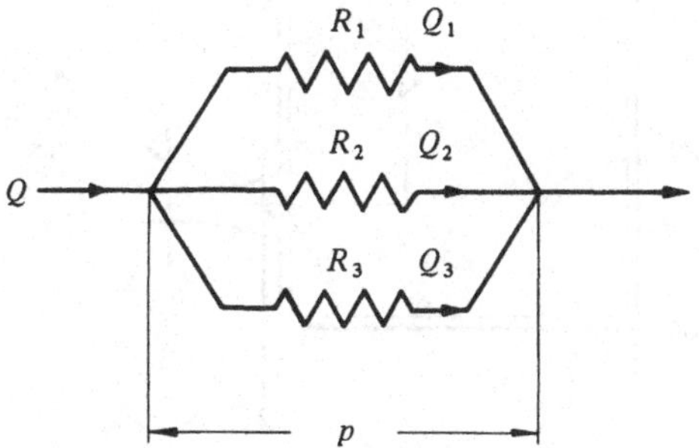

Calculation of size of the regulator

A regulator works as a sharp-edged orifice (Fig. 2.16). In this case, volume flow rate Q is given by

$$Q = A_2V_2$$

Since potential energy = velocity energy,

$$Z_1wg = \frac{wV_2^2}{2}$$

or

$$V_2 = \sqrt{(2Z_1g)}.$$

The ideal $Q = A_2\sqrt{(2Z_1g)}$.

Q = volume flow rate through the orifice, m^3/s;
A_2 = cross-sectional area of the orifice, m^2;
V_2 = velocity of airflow at the orifice, m/s;
Z_1 = height of air column above the centre of the orifice, m;
w = density of air, kg/m^3; and
g = gravitational acceleration, 9.81 m/s^2.

But the actual Q is less owing to losses and the contraction of the airstream after exit from the orifice. The coefficient of discharge, C_d, allows for both effects. Thus

$$Q = C_dA_2\sqrt{(2Z_1g)}. \tag{2.23}$$

The value of C_d is approximately 0.64. In other terms,

$$Q = 0.64A\sqrt{(2Hg)}$$

Fig. 2.16. Analysis of a sharp-edged orifice. C_d = coefficient of discharge; V = flow velocity (m/s); A = cross-sectional area (m^2).

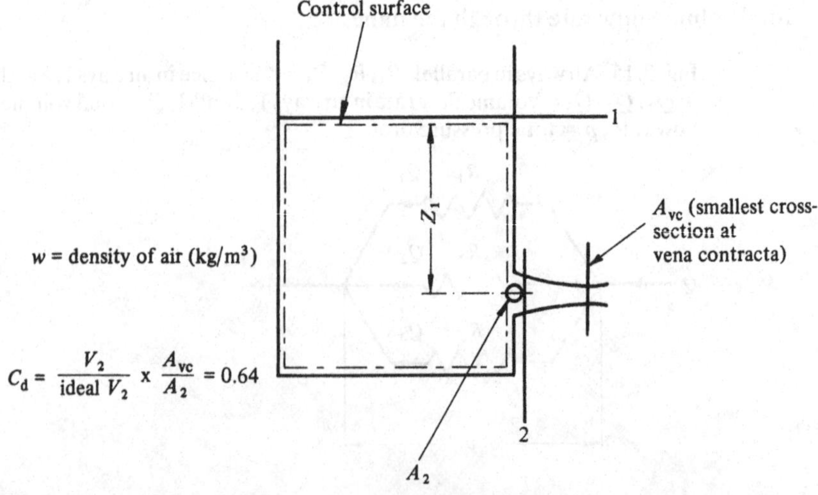

where Q = the volume flow rate, m^3/s;
A = the area of the regulator, m^2; and
H = the pressure producing flow, m of air column.

Therefore

$$A = \frac{Q}{0.64\sqrt{(2Hg)}}$$

If the pressure, p, is in Pa, then

$$H = \frac{p}{wg}$$

where w = density of air, kg/m^3.

Therefore

$$A = \frac{Q}{0.64\sqrt{\left(\frac{2pg}{wg}\right)}}$$

$$= 1.105Q\sqrt{\left(\frac{w}{p}\right)}. \tag{2.24}$$

The pressure absorbed by a regulator containing an opening of A (m^2) can be determined from

$$p = \frac{1.105^2Q^2w}{A^2} = \frac{1.221Q^2w}{A^2} \tag{2.25}$$

2.3.3 *Compound circuits*

A characteristic feature of compound circuits is the presence of one or more diagonal connections. Fig. 2.17 shows simple and complex

Fig. 2.17. Diagonal ventilation network: (*a*) simple, (*b*) complex.

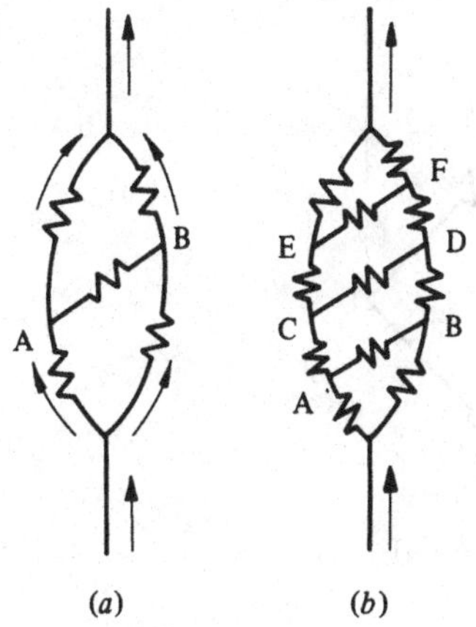

diagonal ventilation networks. Along the diagonals, depending on the pressure in the different airways, the air may flow in a reverse direction or not at all. Compound circuits are difficult to resolve into a series/parallel configuration and are therefore solved by other methods.

2.3.4 *Ventilation network analysis*

Fundamentals

A ventilation network can be divided into branches, junctions and meshes which are defined as follows:

A 'branch' is an airway between two junctions.

A 'junction' is a point where two or more airways meet.

A 'mesh' is a closed path traversed through a network of airways.

Kirchhoff's electrical laws are applicable to a ventilation network as follows:

First law. The total mass flow rate of air entering a junction is equal to the total mass flow rate of air leaving it. Where changes in density are small or all measurements have been corrected to a standard density, the total volume flow rate of air entering a junction is equal to the total volume flow rate of air leaving it.

Second law. The algebraic sum of the pressure drops in a mesh is zero. If any fan and/or natural ventilation pressure is present, the algebraic sum of all pressure drops around any closed mesh, less any fan and natural ventilation pressure, is equal to zero.

Fig. 2.18. Simple ventilation network. R_1, R_2, R_3, R_4, R_5 = resistance in airways 1–5; Q_1, Q_2, Q_3 = volume flow rate in airways; p = total pressure drop.

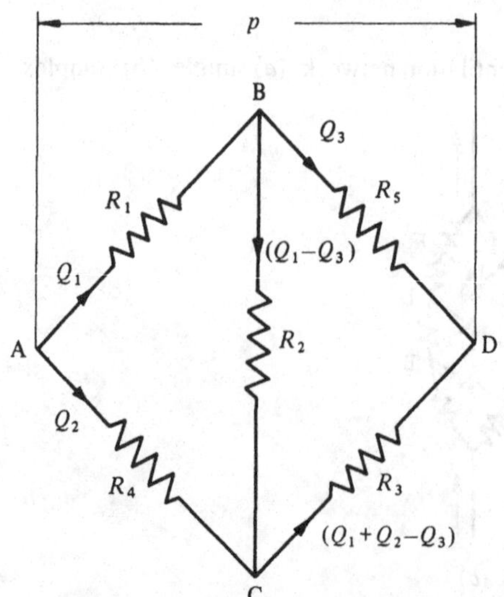

Direct analysis

This analysis involves the use of Kirchhoff's laws. If a network contains b branches and j junctions, b equations have to be solved to determine all the volume flow rates. From the first law, j equations can be written. However, since the volume flow rates at the jth junction are already defined by the volume flow rates at other junctions, only $(j-1)$ of the j equations are independent. Therefore, $b-(j-1)$ equations remain to be written and are obtained by choosing a minimum of $(b-j+1)$ meshes and using the second law for each of them.

Consider a simple ventilation network given in Fig. 2.18. Let the pressure applied across AD be p; the problem is to determine the volume flow rate distribution. The resistances of each of the airways are given in Fig. 2.18.

Kirchhoff's first law is applied by first writing down the volume flow rates in all branches. Then by applying Kirchhoff's second law and Atkinson's equation, i.e. $p = RQ^2$, the following are obtained:

For mesh ABC

$$R_1Q_1^2 + R_2(Q_1 - Q_3)^2 - R_4Q_2^2 = 0.$$

For mesh ABD

$$R_1Q_1^2 + R_5Q_3^2 = p.$$

For mesh BDC

$$R_5Q_3^2 - R_3(Q_1 + Q_2 - Q_3)^2 - R_2(Q_1 - Q_3)^2 = 0.$$

These are three equations for the three unknowns Q_1, Q_2 and Q_3. Though these equations are of a simple algebraic form, there is no straightforward rigorous solution. When rigorous methods fail, often numerical methods are applicable. A particularly useful method of this type is that of successive approximation.

The Hardy Cross method of analysis. Consider a volume flow rate of air, Q, passing through an airway of resistance R and obeying Atkinson's equation, i.e. $p = RQ^2$ where p is the pressure drop along the airway (Fig. 2.19).

To determine the true volume flow rate, Q, first an estimated value, Q_a, is assumed. If ΔQ is the error involved in the assumed volume flow rate, then

$$Q = Q_a + \Delta Q.$$

Similarly, Δp is the corresponding error in the pressure drop, p.

Calculation of ΔQ: In Fig. 2.19, the slope of the curve in the region of Q and Q_a is given approximately by $\Delta p/\Delta Q$. In the limit this becomes $\mathrm{d}p/\mathrm{d}Q$. Differentiating the equation $p = RQ^2$ gives

$$\frac{dp}{dQ} = 2RQ \text{ or } 2RQ_a$$

at the assumed volume flow rate. Therefore,

$$\frac{\Delta p}{\Delta Q} = 2RQ_a \text{ approximately,}$$

or

$$\Delta Q = \frac{\Delta p}{2RQ_a}.$$

But

$$\Delta p = RQ^2 - RQ_a^2,$$

giving

$$\Delta Q = \frac{RQ^2 - RQ_a^2}{2RQ_a}. \qquad (2.26)$$

If a series of b branches is now considered, forming a closed mesh within a network, equation (2.26) gives a composite value of volume flow rate correction, ΔQ_m, known as the mesh correction factor:

$$\Delta Q_m = \frac{\sum_{i=1}^{b} (R_i Q_i^2 - R_i Q_{i,a}^2)/b}{\sum_{i=1}^{b} 2R_i Q_{i,a}/b}. \qquad (2.27)$$

Fig. 2.19. The Hardy Cross method of analysis. R = resistance; Q_a = estimated volume flow rate.

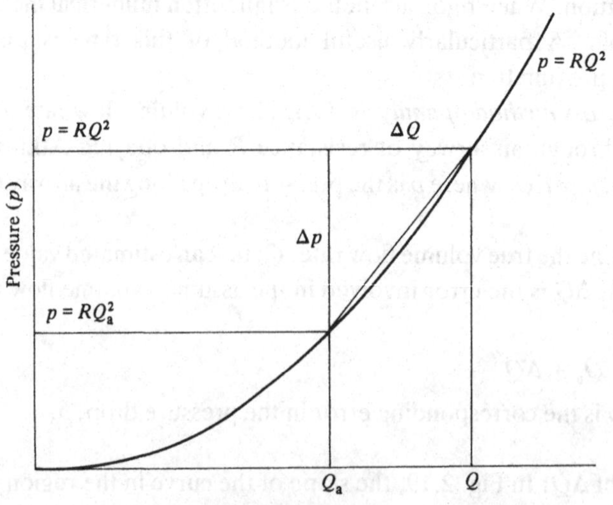

From Kirchhoff's second law

$$\sum_{i=1}^{b} R_i Q_i^2 = 0$$

and equation (2.27) reduces to

$$\Delta Q_{\mathrm{m}} = \frac{-\sum_{i=1}^{b} R_i Q_{i,\mathrm{a}}^2}{2\sum_{i=1}^{b} R_i Q_{i,\mathrm{a}}} \tag{2.28}$$

where ΔQ_{m} = mesh correction factor, $\mathrm{m^3/s}$;
b = number of branches in the mesh;
R_i = resistance of branch i, $\mathrm{Ns^2/m^8}$;
Q_i = true volume flow rate in branch i, $\mathrm{m^3/s}$; and
$Q_{i,\mathrm{a}}$ = assumed volume flow rate in branch i, $\mathrm{m^3/s}$.

This equation can be rewritten as follows to take into account the sign of pressure drop – the pressure drop is positive in the direction of flow:

$$\Delta Q_{\mathrm{m}} = -\frac{\sum_{i=1}^{b} R_i Q_{i,\mathrm{a}} |Q_{i,\mathrm{a}}|}{2\sum_{i=1}^{b} R_i |Q_{i,\mathrm{a}}|}$$

where $|Q_{i,\mathrm{a}}|$ is the absolute value of $Q_{i,\mathrm{a}}$.

Fans or natural ventilation pressures (nvp)* in the mesh and the slope of the fan characteristic are included in the above equation as follows:

$$\Delta Q_{\mathrm{m}} = \frac{-\left[\sum_{i=1}^{b} (R_i Q_{i,\mathrm{a}} |Q_{i,\mathrm{a}}| - p_{\mathrm{f},i}) - (nvp)_{\mathrm{m}}\right]}{\sum_{i=1}^{b} (2R_i |Q_{i,\mathrm{a}}| - S_{\mathrm{f},i})} \tag{2.29}$$

where $p_{\mathrm{f},i}$ and $S_{\mathrm{f},i}$ are the pressure and slope of the characteristic of a fan in branch i at a volume flow rate of $Q_{i,\mathrm{a}}$. This technique is applied repeatedly until all the mesh correction factors ΔQ_{m} approach the required accuracy. In summary, the procedure is as follows:

(1) Estimate the volume flow rates through each branch of the network and the pressures developed by the fans. The estimated volume flow rates should obey Kirchhoff's first law at each junction.

* It is assumed that natural ventilation pressures are independent of volume flow rate.

(2) Examine the network and decide upon a pattern of closed meshes. It is important to see that all branches of the network are represented in the mesh pattern. No mesh should contain more than one high-resistance branch and such a branch should not appear in more than one mesh. The minimum number of meshes is given by

(number of branches − number of junctions + 1).

(3) For each mesh, calculate the mesh correction factor using equation (2.29).

(4) Correct the volume flow rate in each branch.

(5) Repeat steps (3) and (4) until all values of ΔQ_m are below a prescribed level.

Digital computer. The application of iterative techniques to even simple ventilation networks involves a considerable amount of arithmetic. For practical mine systems, iterative methods become impractical if the repetitive calculations are to be carried out by manual means. Programmable pocket calculators assist in the calculation of mesh correction factors and the application of these to branch volume flow rates. However, the number of iterations which may be required, together with the organisational procedures of mesh selection and network updating, still involve a great amount of work.

The modern approach is to automate fully the complete process by utilising large and fast computers. Several computer programs have been developed for this purpose and the availability of such programs reduces the manual work to that of specifying the data which describes the network.

One such program is PENVEN Denver Technical Support Centre mine ventilation simulator for use on the CDC CYBER 74-28/16 computer system (Anderson and Dvorkin, 1978).

A program listing (in BASIC) for microcomputers is given in Appendix I.

2.3.5 *Analogue computer*

The ventilation network analogue computer consists of several variable resistors which are connected in the same fashion as the branch airways in the ventilation network. The fans in the network are simulated by applying suitable voltages corresponding to the pressures generated by them at the corresponding positions in the circuit.

The law of airflow is given by the equation $p = RQ^2$ whereas that of the electrical current flow, i, through a resistance, r, is given by $V = ri$, where

V is the voltage.* From these two equations it can be seen that the airway resistance can be represented by the electrical resistance as follows:

$$r = K(RQ)$$

where K = constant of proportionality.

Since r is a product involving Q, which is to be determined, the correct values of r cannot be directly set on the computer. First, a certain value of Q is assumed (the nearer this value is to the true value, the fewer the number of successive approximations) and the value of r on the resistors is successively adjusted according to the following iterative equation until successive values of r become almost constant and the correct value of Q is reached:

$$r_{m+1} = \tfrac{1}{2}(r_m + KRQ_{m+1})$$

where m = the number of the iteration.

A voltmeter graduated in units of air pressure and an ammeter calibrated in units of volume flow rate are used to measure respectively the pressure across the fans and the volume flow rate in each resistor or branch.

2.4 Mine fans and their operation

2.4.1 Fan pressure, power and efficiency – fan characteristic curves

In mines, fans are used to induce airflow continuously at moderate pressures. In the calculations of operation of a fan, it is usually assumed that air is incompressible, i.e. the volume flow rate of air through a fan remains constant between intake and discharge. Fan pressure is expressed in terms of pressure rise through the fan as follows:

(1) fan total pressure is the difference between the total pressures at the fan discharge and intake;
(2) fan static pressure is the fan total pressure minus the fan velocity pressure; and
(3) fan velocity pressure is the velocity pressure corresponding to the average velocity at the fan discharge. (The average velocity is calculated by dividing the volume flow rate of air by the area of the fan discharge orifice.)

Fan static pressure is often regarded as the useful fan pressure.

The total pressure developed by a fan depends on the volume flow rate through it. When intake and discharge are fully open, the fan total

* Electronic devices are available which permit V to be proportional to i^2. In that case, values of R are fed to the analogue to obtain the values of Q directly.

pressure is equal to the fan velocity pressure, and the fan static pressure is zero. As the volume flow rate decreases, the fan velocity pressure decreases and the fan total pressure (and consequently the fan static pressure) increases. Theoretically, the fan total pressure should reach a maximum value when intake and/or discharge are closed, i.e. when there is no airflow through it. Owing to friction and shock losses, the fan total pressure reaches a maximum value at a point in the range of the volume flow rate depending on the particular design of fan. A graph of fan pressure against volume flow rate of air is known as the fan characteristic and is determined by testing. The shape of the characteristic curve varies with the type of fan. A typical fan characteristic curve is given in Fig. 2.20.

The total power output of a fan (air power (total)) is calculated by using fan total pressure in equation (2.18) (i.e. air power = pQ). When fan static pressure is used, the resulting power is known as air power (static).

Fan efficiency is the ratio of output power to mechanical input power and is usually expressed as a percentage:

$$\text{fan total efficiency, } \eta = \frac{\text{air power (total)}}{\text{measured fan input power}} \times 100\,\%$$

$$\text{fan static efficiency, } \eta_s = \frac{\text{air power (static)}}{\text{measured fan input power}} \times 100\,\%.$$

The measured fan input may be that absorbed by the impeller only, or may include the power absorbed by bearings and auxiliary drive components according to the design and use of the fan.

Fig. 2.20. Fan characteristic curves.

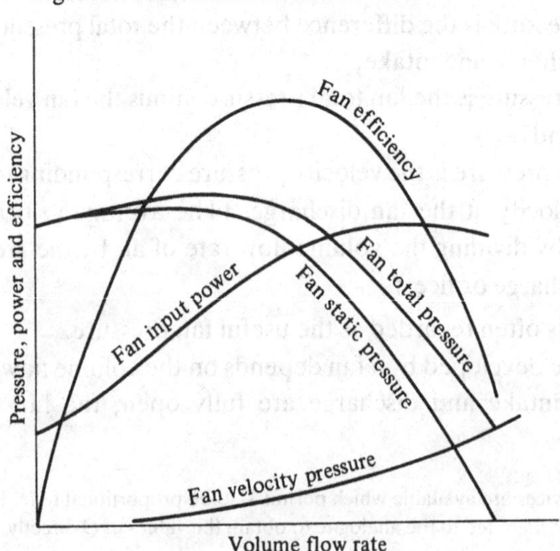

The overall efficiency is the ratio of output power, i.e. air power, to input power to electric motor expressed as a percentage:

$$\text{overall efficiency} = \frac{\text{air power}}{\text{electrical input power}} \times 100\,\%.$$

The electrical input power in watts is calculated from the following formulas:

For single-phase motors – watts = volts × amps × power factor.
For three-phase motors – watts = $\sqrt{3}$ × volts × amps × power factor.

2.4.2 *Fan laws*

From the fan laws it is possible to calculate characteristic curves for a fan at different speeds and air densities once the curve at any one speed and density is known, and also to determine curves for similar fans of different sizes.

The fan laws for a range of geometrically similar fans, and for a particular point of operation on the pressure/volume flow rate characteristic are:

$$\text{Volume flow rate, } Q \propto nD^3 \qquad (2.30)$$

$$\text{Fan pressure, } p \propto n^2D^2w \qquad (2.31)$$

$$\text{Fan power, } W \propto n^3D^5w \qquad (2.32)$$

where n = speed of rotation;
D = impeller diameter; and
w = air density.

Fan efficiency, η, remains constant but its position will move to the new volume flow rate line.

2.4.3 *Performance coefficients*

Non-dimensional performance coefficients are commonly used in expressing fan performance and are given below:

$$\text{Volume flow rate coefficient, } \phi = \frac{Q}{\left(\frac{\pi D^2}{4}\right)u} \qquad (2.33)$$

$$\text{Pressure coefficient, } \psi = \frac{p}{\frac{1}{2}wu^2} \qquad (2.34)$$

$$\text{Power coefficient, } \lambda = \frac{\phi\psi}{\eta} \qquad (2.35)$$

where u = impeller peripheral velocity = πDn.

Specific speed

Specific speed n_s is the speed at which a fan (of unspecified diameter) would rotate to give unit volume flow rate and unit fan pressure. This applies to all geometrically similar fans, and is derived from the fan laws by eliminating the diameter term, thus:

$$p \propto n^2 D^2 w$$

from which

$$D \propto \frac{\sqrt{p}}{n\sqrt{w}}$$

$$Q \propto nD^3$$

$$\propto \frac{np^{3/2}}{n^3 w^{3/2}}$$

$$\propto \frac{p^{3/2}}{n^2 w^{3/2}}$$

$$n \propto \frac{p^{3/4}}{Q^{1/2} w^{3/4}}$$

and substituting n_s when $Q = 1$ and $p = 1$, the air density remaining the same

$$\frac{n_s}{n} = \frac{Q^{1/2}}{p^{3/4}}$$

or

$$n_s = \frac{nQ^{1/2}}{p^{3/4}}. \tag{2.36}$$

Fig. 2.21. Construction of a centrifugal fan.

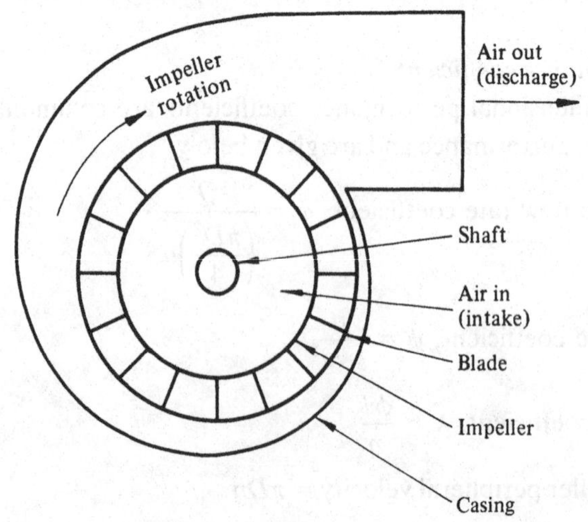

The performance coefficients and specific speed are almost always based on fan performance under standard air conditions.

2.4.4 *Centrifugal or radial flow fans*

A centrifugal fan consists of an impeller running in a casing which has a spirally shaped contour (Fig. 2.21). The air is drawn into the open end(s) of the impeller in an axial direction and is thrown out at the periphery, the impeller rotation being towards the casing discharge. If the air can enter on both sides of the impeller, it is called a double-inlet fan; if air entry is from one side only, it is called a single-inlet fan. Three main forms of blade are common (Fig. 2.22).

(1) Backward bladed, in which the blade tips incline away from the direction of rotation, and the blade angle, β, is less than 90°.
(2) Radially bladed, where the blade tips are radial, that is, $\beta = 90°$.
(3) Forward bladed, where the blade tips incline towards the direction of rotation, and β is greater than 90°.

The pressure developed increases with increase of blade angle, and for an impeller of given proportions, the volume flow rate also tends to increase.

Typical pressure/volume flow rate and power/volume flow rate characteristic curves for each type of centrifugal fan are given in Fig. 2.23. In this figure, the non-dimensional volume flow rate (ϕ), pressure (ψ) and power (λ) coefficients are used to provide realistic comparisons.

The power characteristic of a radially bladed fan is almost a straight line passing above the origin and showing an increase of power with increasing volume flow rate. For the backward bladed fan, the power characteristic is similar but tends to flatten with increasing volume flow rates and, as the maximum volume flow rate is approached, the power consumed may become constant or even decrease. This characteristic is said to be non-overloading. The power characteristic of the forward bladed fan steepens at high volume flow rates, and so is overloading.

Fig. 2.22. Blade forms of fans. β = blade angle.

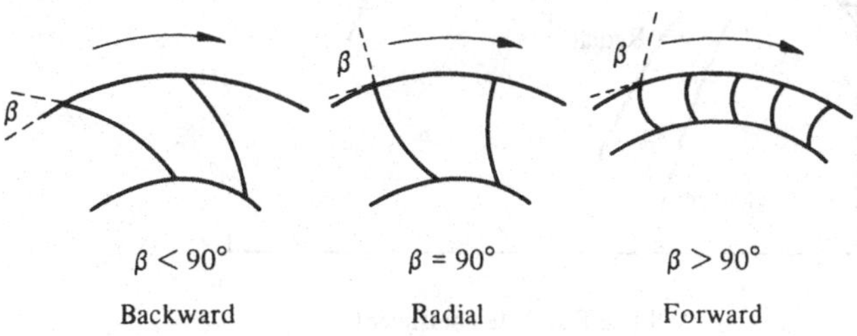

With backward bladed fans, the maximum power required is generally a little in excess of the power absorbed by the fan when working at maximum efficiency. Consequently, an electric motor able to cope with all operating conditions may economically be selected. In the case of forward bladed (and also radially bladed) fans the electric motor is usually selected for a particular duty with a margin of 25–30 % rather than for maximum possible fan power. In order to avoid the possibility of over-loading, ammeters are fitted to an electric motor control gear to give an indication of the running conditions.

Fig. 2.23. Centrifugal fan characteristics (after Osborne, 1966).

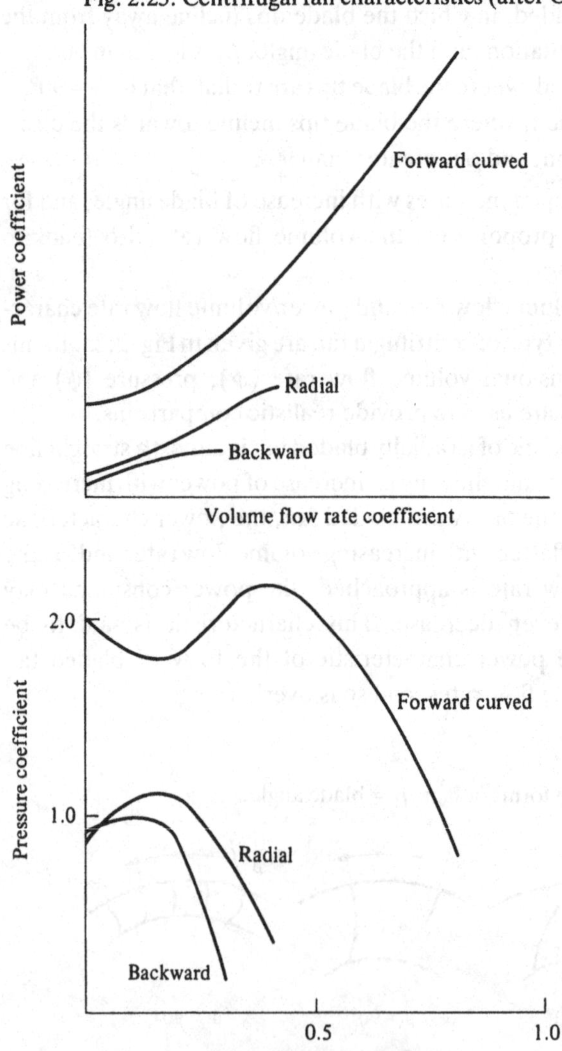

2.4.5 *Axial flow fans*

In an axial flow fan (Fig. 2.24), the tips of the impeller blades, which are commonly of aerofoil section, run with a fine clearance in a cylindrical section of the casing. Air approaches the impeller in an axial direction and leaves with a rotational component due to work done by the impeller torque. Thus the absolute velocity of the air leaving the impeller is higher than the axial velocity, with the result that some of the total pressure developed by the impeller does not appear as useful fan total pressure. To achieve the maximum amount of useful pressure, guide vanes are built downstream of the impeller which remove the rotational component, thus slowing down the air and converting some of the excess velocity pressure to more useful static pressure.

Some fans have two or three impellers, one behind the other, each with its own set of guide vanes. They are called two-stage or three-stage fans. In a contra-rotating fan there are two impellers but these rotate in opposite directions without the use of guide vanes.

The blades can be fixed at a certain angle, i.e. the blade air-cutting angle, they can have adjustable pitch, which means that the angle of the blades can be changed by undoing the nuts and lock screws at their bases, or they can be of the variable pitch type, which means that the angle of all the blades can be altered by operating a suitable mechanism while the fan is in motion. The casing consists of three parts – a cylindrical section round the impeller, an inlet cone to reduce the entrance losses, and a diffuser to reduce shock losses by allowing the air to slow down gradually. The fan

Fig. 2.24. Construction of an axial flow fan.

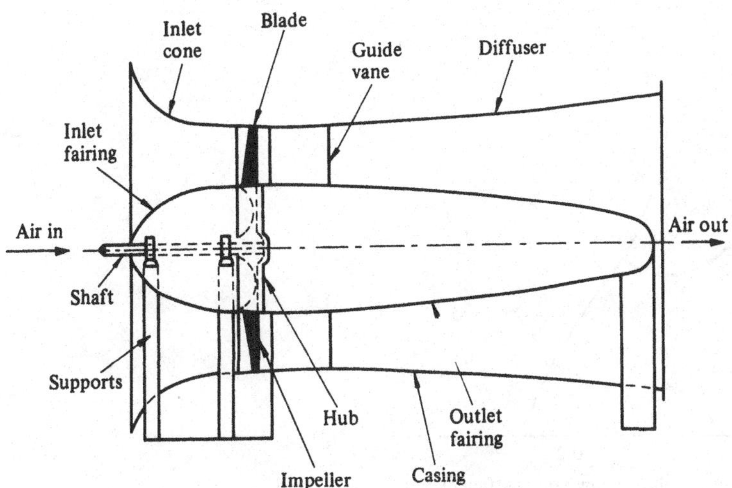

has a short, snub-nosed inlet fairing on the upstream side of the hub and a long, pointed outlet fairing on the downstream side to ensure smooth flow of the air through it. This streamlining of the hub improves the efficiency of the fan.

Typical characteristic curves for an axial flow fan are given in Fig. 2.25. The pressure/volume flow rate characteristic exhibits a region of discontinuity corresponding to stalling conditions on the blade aerofoils. It is wise not to operate fans in this region, or at lower volume flow rates.

Axial flow fans do not develop as much pressure as centrifugal fans of the same impeller diameter and speed.

Fig. 2.25. Axial flow fan characteristics (after Osborne, 1966).

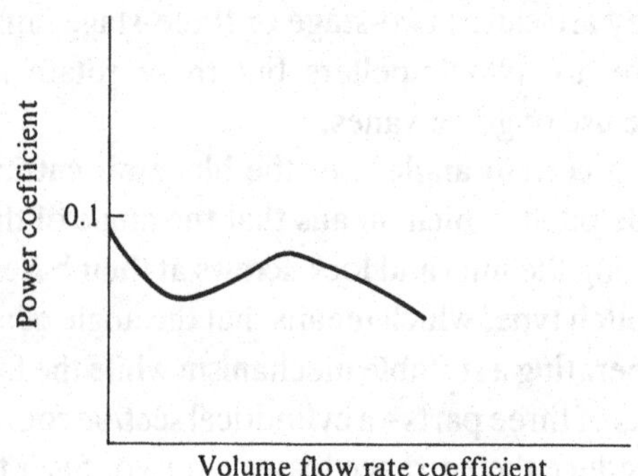

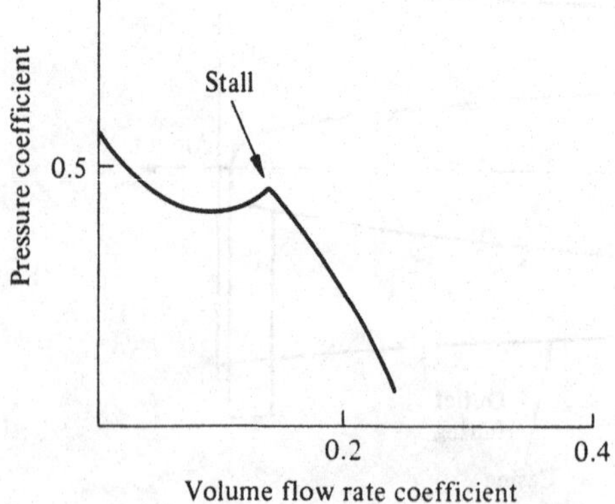

Evasee or diffuser

This is an expanding piece of duct work attached to the discharge end of the fan casing. In this, some of the velocity pressure is recovered by converting it to static pressure. The efficiency (i.e. the increase in static pressure divided by the decrease in velocity pressure) of an evasee decreases as the divergence angle increases. In practice, divergence angles are near $3\frac{1}{2}°$.

2.4.6 *Comparison of fans*

The performance coefficients of some of the main types of fans are given in Table 2.4. Fig. 2.26 gives specific speed versus efficiency for various types of fans. Curves (a) to (e) are for a set of centrifugal fans with backward curved blades and they have higher efficiencies compared to the centrifugal fans with radial blades and forward curved blades (curves j, k and l). Axial flow fans have high efficiencies as well as high specific speeds (curves f, g and h).

2.4.7 *Fan and system*

The purpose of a fan is to move air at the required volume flow rate, and to supply the total pressure loss (including the discharge velocity pressure) in the system to which it is applied. Thus selection of the fan

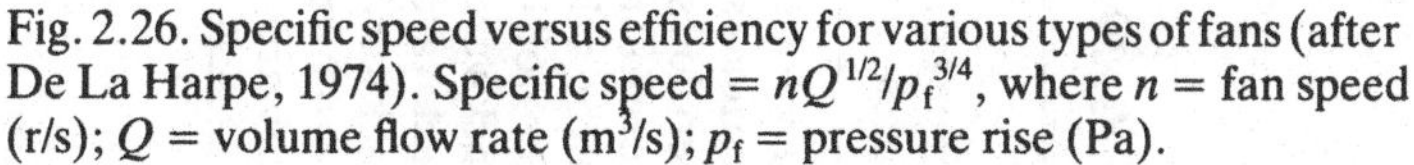
Fig. 2.26. Specific speed versus efficiency for various types of fans (after De La Harpe, 1974). Specific speed = $nQ^{1/2}/p_f^{3/4}$, where n = fan speed (r/s); Q = volume flow rate (m^3/s); p_f = pressure rise (Pa).

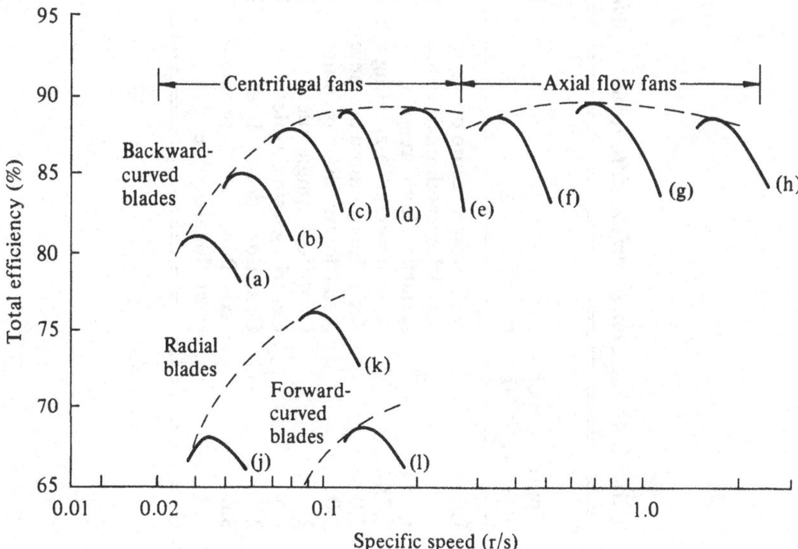

Table 2.4. *Fan characteristic data (after Rowlands, 1979)*

Approx. upper pressure (kPa)	Fan type	Blade type and stages	Pressure coefficient (ψ)	Volume flow rate coefficient (ϕ)	Static efficiency (%)	Total efficiency (%)
1.0	Axial flow medium pressure	Aerofoil single stage	0.38	0.28	82	86
1.0	Axial flow medium pressure	Laminar single stage	0.37	0.28	77	80
1.25	Axial flow high pressure	Aerofoil single stage	0.52	0.27	82	85
1.75	Axial flow contra-rotating	Aerofoil two stage	0.51	0.23	80	84
2.0	Axial flow medium pressure	Turbo axial single stage	0.65	0.27	80	82
2.0	Axial flow medium pressure	Aerofoil two stage	0.76	0.28	82	86
2.5	Centrifugal single inlet	Backward laminar	0.78	0.20	74	78
2.5	Centrifugal single inlet	Backward aerofoil	0.66	0.19	83	86
2.5	Centrifugal double inlet	Backward aerofoil	0.81	0.38	82	84
3.0	Axial flow medium pressure	Aerofoil three stage	1.12	0.28	83	88
3.0	Centrifugal double inlet	Forward	2.11	0.54	70	74

should be made on the basis of fan total pressure. However, fan manufacturers give details of performances in terms of fan static pressure under standard air conditions and it is therefore customary to select fans on this basis, thus:

> fan total pressure = system (total) pressure loss; or
> fan static pressure + fan velocity pressure = system (total) pressure loss; or
> fan static pressure = system (total) pressure loss − fan velocity pressure.

The system pressure loss/volume flow rate relationship is defined by equation (2.16). The fan pressure/volume flow rate relationship is defined by the fan characteristic. If these curves are plotted on a single graph (Fig. 2.27), the intersection will give the point of operation. The fan power and efficiency corresponding to the operating volume flow rate may be read off.

The total pressure loss and volume flow rate requirements for the system should be calculated as accurately as possible. If the true system characteristic is OA (Fig. 2.28) and a volume flow rate of Q_1 against pressure p_1 is required, but volume flow rate Q_2 is specified to the fan supplier, then the fan supplier will select a fan that will be stable and operate at high efficiency on a system characteristic curve OB. When it is installed, however, the fan will operate at the point of intersection of the fan characteristic curve and the system characteristic curve OA. The operation will be unstable and inefficient and it will be difficult, if not

Fig. 2.27. Fan and system characteristics.

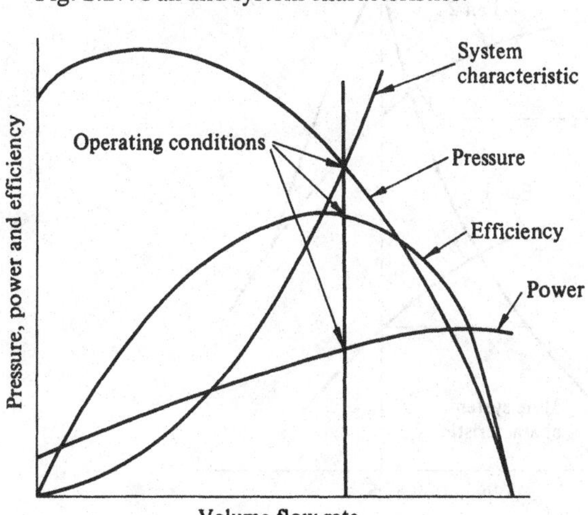

Fig. 2.28. Selection of a fan.

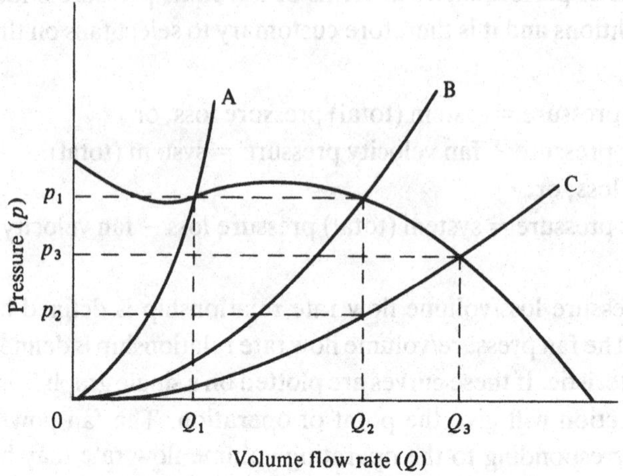

Fig. 2.29. The combined characteristic for two fans in series. OR = resistance curve; X, Y = operating points for fans 1 and 2, respectively; Z = operating point for fans 1 and 2 in series.

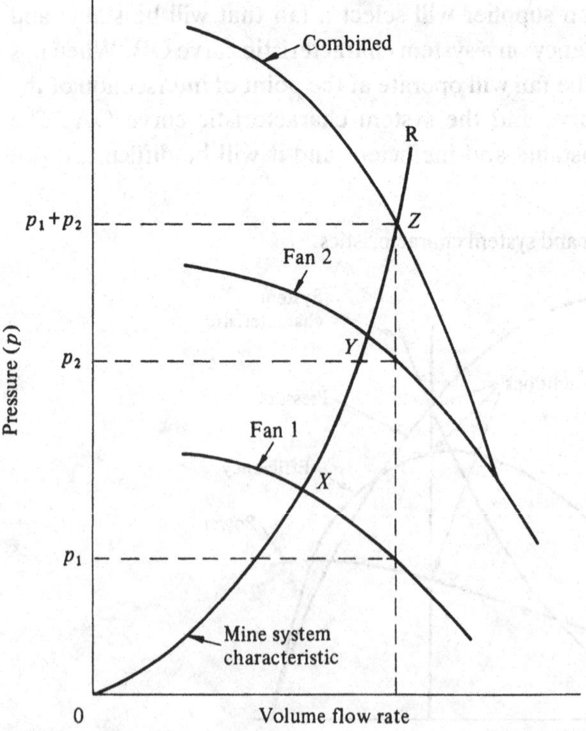

impossible, to solve the problem unless some air can be shortcircuited between the fan discharge and its inlet.

On the other hand, if a volume flow rate Q_2 is required at a pressure loss p_2, but a pressure loss p_1 is specified for that volume flow rate, then the fan will operate at the intersection of the fan characteristic curve and the system characteristic curve OC, delivering a volume flow rate Q_3 against pressure p_3. Again, the operation of the fan will be inefficient and possibly unstable. In addition, there would be a strong possibility of overloading the motor. The fan could be brought back to operate stably at volume flow rate Q_2 by installing extra resistance in the system but power will be wasted in overcoming this resistance.

2.4.8 *Fans in series*

When two (or more) fans are connected in series, the volume flow rate through each fan will be the same whilst the overall total pressure will be the sum of the individual fan total pressures. The combined characteristic is drawn by adding the fan total pressures at each volume flow rate for the two fans (Fig. 2.29). If fans 1 and 2 are installed in a mine system with resistance curve OR, then each fan on its own will operate at point X for fan 1 and point Y for fan 2. The two fans in series will operate at point Z.

2.4.9 *Fans in parallel*

When two (or more) fans are connected in parallel, the total pressure difference will be the same across each fan. The total volume flow rate will be the sum of the individual volume flow rates flowing through each fan at the same fan total pressure. The combined characteristic is drawn by adding the volume flow rate of each fan at the same fan total pressure for the two fans (Fig. 2.30). If fans 1 and 2 are installed in a mine system with resistance curve OR, then each fan on its own will operate at point X for fan 1 and point Y for fan 2. The two fans in parallel will operate at point Z.

In the case of fans which have pronounced stalls (such as axial flow or forward curved centrifugal fans), as many as three different volume flow rates are possible at certain values of fan total pressure. For two such identical fans in parallel, as many as six volume flow rates are thus possible (Fig. 2.31). If the mine system has a resistance curve OR_1 then the operation will be stable whether one or both fans are running, as the points of intersection of the system resistance curve with both fan curves will be single points (a) and (b). If the system has a resistance curve OR_2,

Fig. 2.30. The combined fan characteristic for two fans in parallel. OR = resistance curve; X, Y = operating points for fans 1 and 2, respectively; Z = operating point for fans 1 and 2 in parallel.

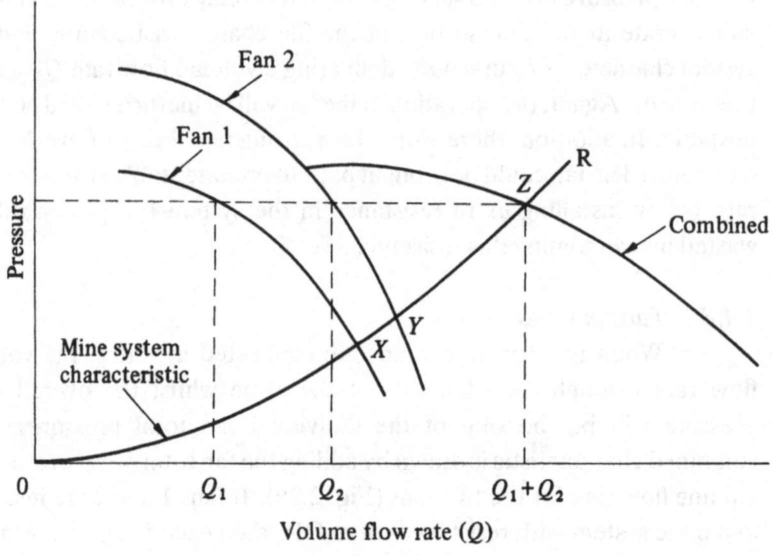

Fig. 2.31. The combined characteristic for two fans which have pronounced stall in parallel. OR_1 and OR_2 = resistance curves. See text for explanation of (a) to (f).

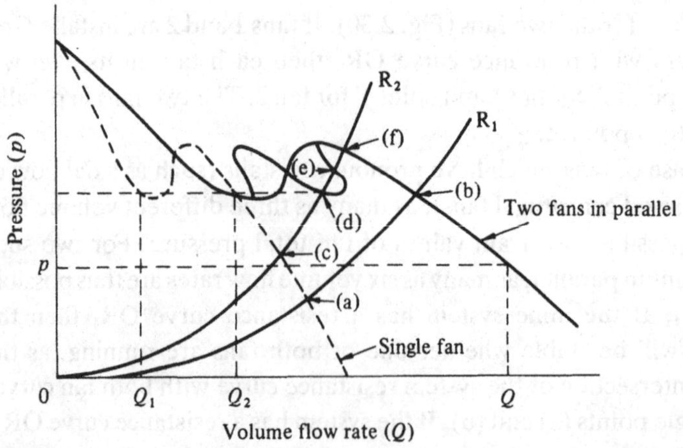

however, then the operation of one fan only will be stable at point (c), but the operation will be unstable when both fans are running, since the system resistance curve OR_2 will now intersect the combined curve at three points, namely, at (d), (e) and (f). It is impossible to forecast the actual working conditions and the fans may settle down at any single point or oscillate between all three. The desired operating point would be (f), when both fans share the load equally. If the point of operation is (e), the load will be shared unequally, fan 1 being underloaded whilst fan 2 is overloaded. (Fan 1 also operates in the unstable stall zone.) If the point of operation is (d), the difference is even greater, fan 2 being grossly overloaded. (In this case also, fan 1 operates in the unstable stall zone.)

If trouble is encountered on existing installations, it may be possible to solve the problem by adding resistance, possibly in the form of variable dampers, separately to each fan. The effect of this is to give a characteristic with a reduced 'hump'.

If fans in parallel are started one at a time, it is better to have automatic backdraught shutters fitted to prevent air recirculating through the fans not already running. This avoids reverse rotation of the impellers which could lead to heavy starting currents and mechanical shock on starting.

2.4.10 Fans partially in parallel

Where fans work in parallel through airways which are connected only at some distance from the fans (Fig. 2.32), they are said to be partially in parallel. In this case, first residual fan pressure curves are drawn for each fan and its airway, namely, $F_A - R_A$ and $F_B - R_B$

Fig. 2.32. Fans partially in parallel. R_A, R_B, R_c = resistance of airways A, B and C.

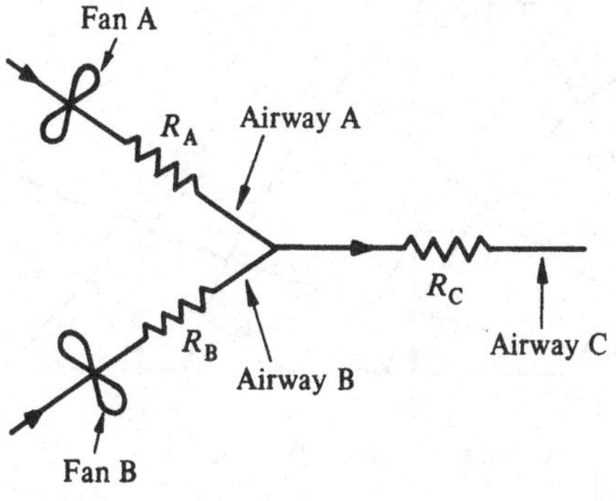

Fan A
R_A
Airway A
R_C
Airway C
R_B
Airway B
Fan B

(Fig. 2.33). These two residual curves are then used to draw a combined curve for the whole system, namely, $(F_A - R_A) + (F_B - R_B)$. The characteristic of the rest of the mine, i.e. R_C is then plotted and the intersection indicates the pressure available for ventilating the rest of the mine and the total volume flow rate handled. The volume flow rates and pressures of the two fans can be read off as shown in Fig. 2.33.

2.4.11 *Fans partially in series*

Fans are said to be partially in series when most of the volume flow rate from one fan passes through the other, but some volume flow rate is added or removed between the two positions (Fig. 2.34). In this case, first the residual system curve, i.e. $(R_B - F_B)$, is drawn for the airway B and the fan in it, i.e. fan B (Fig. 2.35). This curve and the curve for airway A, i.e. R_A, are then used to draw a combined curve $(R_B - F_B) + R_A$. The residual fan pressure curve for fan F_C and airway R_C, i.e. $F_C - R_C$, is then drawn. The intersection of the curves $(R_B - F_B) + R_A$ and $F_C - R_C$ gives Q_C. The procedure for determining the airflows through airways A and B and the operating points of fans B and C is shown in Fig. 2.35.

Fig. 2.33. The combined characteristic for two fans partially in parallel. See text for explanation of symbols.

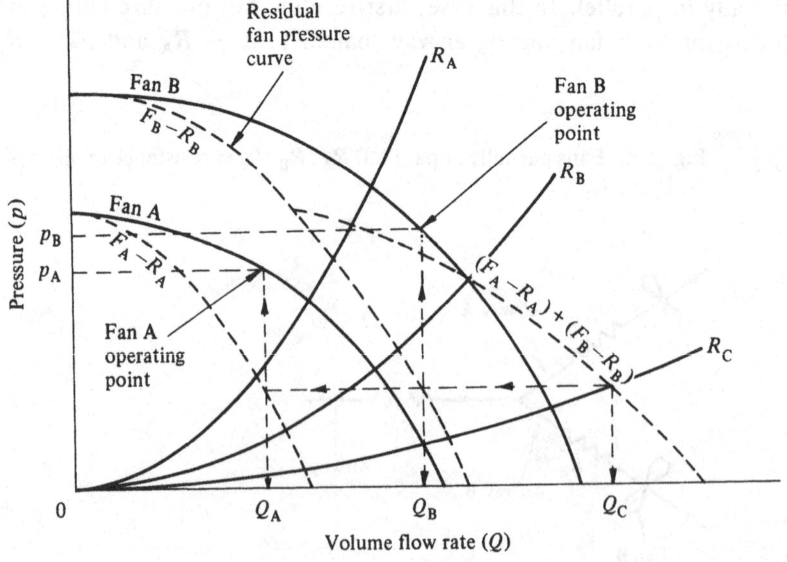

Fig. 2.34. Fans partially in series. R_A, R_B, R_C = resistance of airways A, B and C.

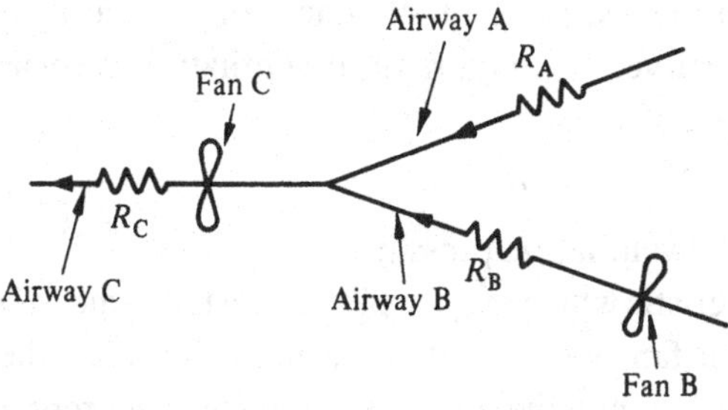

Fig. 2.35. The combined characteristic for two fans partially in series. See text for explanation of symbols.

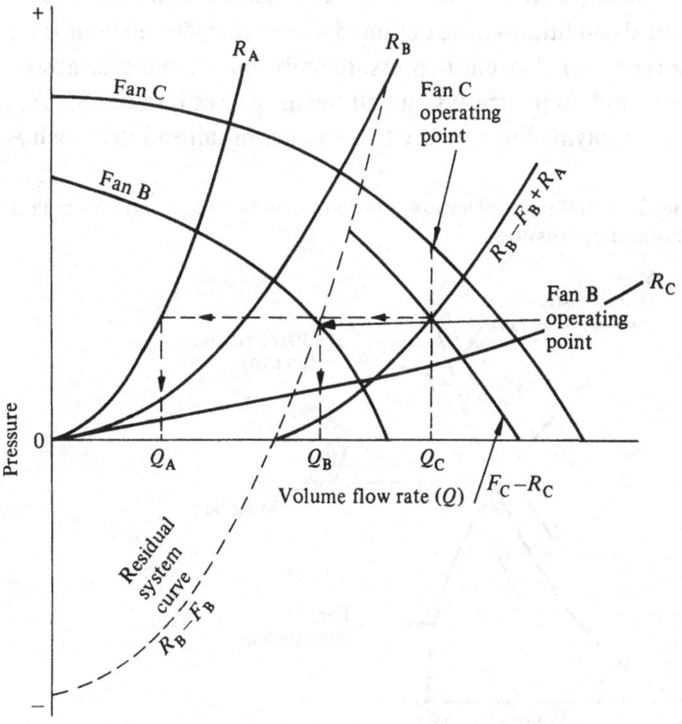

2.4.12 Natural ventilation in series with fans

Natural ventilation may increase or decrease the resistance which the fan is operating against, and influences the operating point of the fan characteristic curve. Usually, natural ventilation is regarded as creating a pressure, so that:

$$\text{fan pressure} = RQ^2 - nvp$$

where *nvp* is the natural ventilation pressure.

This gives a characteristic which may be plotted on the same axes as the fan characteristic. The fan operates at some point between the intersections of its characteristic and the two mine characteristics representing the extreme conditions of natural ventilation (Fig. 2.36).

2.4.13 Booster fans

Sometimes it is necessary to install a booster fan in a branch which may be in parallel with another branch (Fig. 2.37). This will increase the airflow in branch A but will reduce the airflow in the other branch B because a larger volume flow rate of air will be carried by the shafts and main airways, causing an increased pressure drop in these and leaving a smaller portion of the surface fan pressure for airflow in branch B. The critical condition will be obtained when the airflow in branch B just ceases (or reverses). The entire pressure of the main fan is then absorbed in the shafts and main airways and all the air passes into branch A, the booster fan supplying the pressure to circulate this air round branch A.

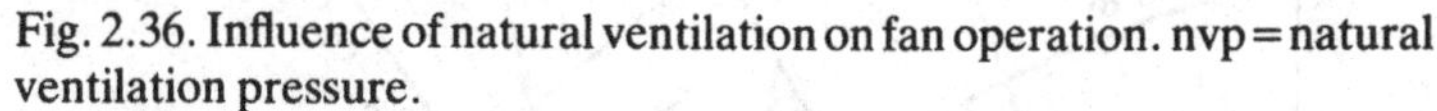
Fig. 2.36. Influence of natural ventilation on fan operation. nvp = natural ventilation pressure.

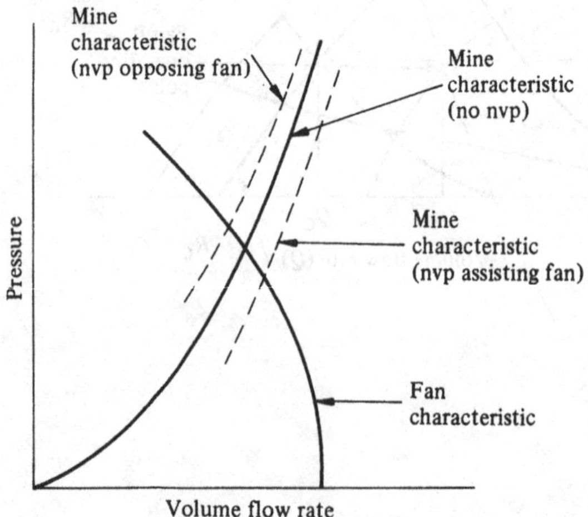

If p_m is the pressure developed by the main fan and p_A is the pressure developed by the booster fan in branch A where R_t and R_A are the resistances of the shafts and main airways and branch A, respectively, then for the critical condition when reversal commences in branch B, in which no air now passes,

$$p_m = R_t Q^2$$

where Q is the volume flow rate passing in the shafts and main airways. But all the volume flow rate now passes into branch A, so that

$$p_A = R_A Q^2$$

or

$$p_A = R_A \frac{p_m}{R_t}.$$

This gives the maximum permissible pressure of the booster fan; the flow in branch B will cease if the pressure reaches this value and will be reversed if it is exceeded. Reversed flow will mean recirculation round the two branches.

Examination of the above equation shows that if the resistance of the shafts and main airways, i.e. R_t, is high in relation to the branch resistance, i.e. R_A, and if the pressure due to the surface fan, i.e. p_m, is low, the critical pressure for the booster fan is low. Long, high-resistance main airways are often the source of trouble when booster fans are required. The increase in the total flow is small and the booster fan is merely changing the distribution. This is not the case, however, when a booster fan of large capacity is placed in series with the surface fan in the main airway, outside all the branches. The volume flow rate is then increased in all branches.

Fig. 2.37. Booster fan in a parallel circuit. R_A, R_B = resistance of airway branches A and B; p_m = pressure developed by the main fan.

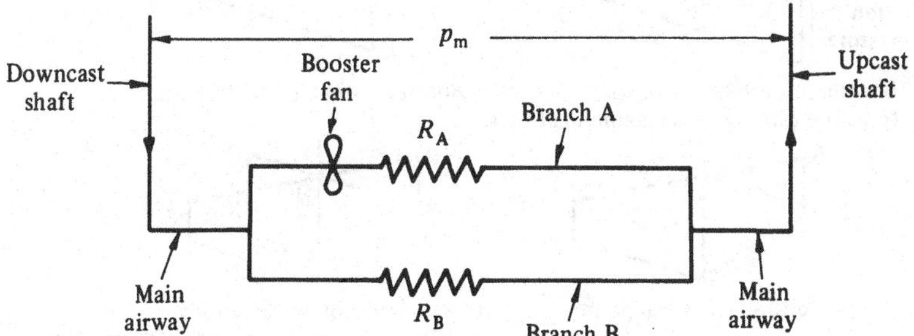

Siting of a booster fan

A booster fan is usually sited in the return airway to avoid the coal transport system which is usually in the intake airway. The position of the fan in the return airway is chosen carefully so that there is not appreciably higher pressure in the return airway than in the intake airway at any place where leakage between them can occur, resulting in unacceptable recirculation.

Wherever the booster fan is sited, the total ventilating pressure acting on the boosted split is the sum of the mine ventilation pressure (Fig. 2.38, AB) at the entrance to the split and the booster fan pressure (CD). If the ventilating pressure in the district is plotted against the distance inbye, the gradient of the curve along any portion of the airway is determined by the local airway resistance. For different positions of the booster fan the pressure difference between the intake airway and the return airway can be found by introducing the booster fan pressure into the pressure/distance curve at the corresponding position (Fig. 2.38).

The neutral point (N) is the position of the booster fan at which the pressure difference between the intake airway and the return airway is equal to the booster fan pressure. If the fan is sited inbye of this point the pressure in the return airway is higher than that in the intake airway and consequently if there is a leakage path some recirculation will occur. As the site of the fan is moved progressively outbye from the neutral point, the pressure difference between the intake and the return airways is increased, resulting in increased leakage flow in any leakage path there.

Fig. 2.38. Effect of booster fan site on ventilating pressure. AB = mine ventilation pressure; CD = booster fan pressure; N = neutral point.

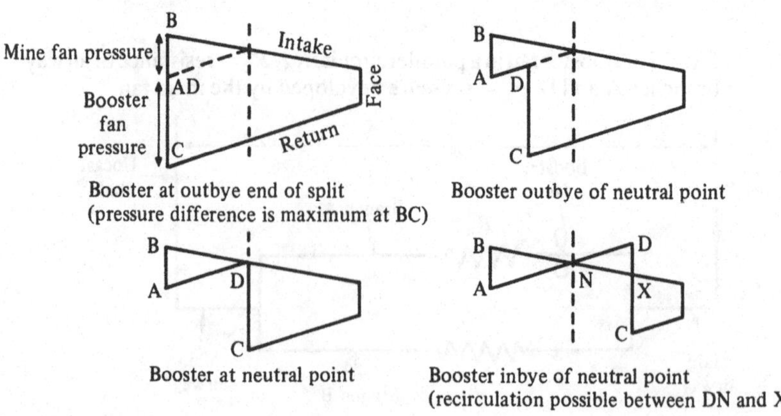

2.4.14 Reversal of airflow

In the event of a fire in or near the downcast shaft, it is necessary or desirable to reverse the direction of airflow through a mine. If axial flow fans are in use, reversal of impeller rotation will give reversal of airflow. The performance will generally be less favourable under reverse flow conditions.

If radial flow fans are in use, reversal of rotation of the impeller does not result in reversal of direction of airflow. When using these fans, flow reversal is achieved by means of a system of doors and ducting. Such an arrangement is shown diagrammatically in Fig. 2.39.

2.4.15 Regulation of volume flow rate

It is often necessary to vary the output of a fan during operation. To do this, an artificial change has to be made either to the fan characteristic or to the system.

(1) Varying the speed. The fan characteristic changes by changing the speed of the fan and can be predicted by the fan laws. Speed variation can be obtained by using a variable-speed electric motor as a driving unit. Hydraulic as well as power couplings and V-belt or gear drive may also provide this facility, but operating efficiencies may not be consistently high. This method is very effective and provides the most efficient means of volume flow rate regulation. However, initial cost is normally very high.

(2) Using variable inlet guide vanes. These guide vanes pre-rotate the air as it enters the fan impeller, in the same direction of rotation. This has the effect of reducing the amount of work done on the air by the impeller and each guide vane setting results in a different fan characteristic.

(3) Using adjustable impellers. Fans are available with adjustable impellers, for example:
   (a) axial flow fans with variable or adjustable pitch blades;
   (b) centrifugal fans with retracting or variable-angle blade tips;
   (c) centrifugal fans with variable-width impellers.

Fig. 2.39. Reversing circuit for a centrifugal fan.

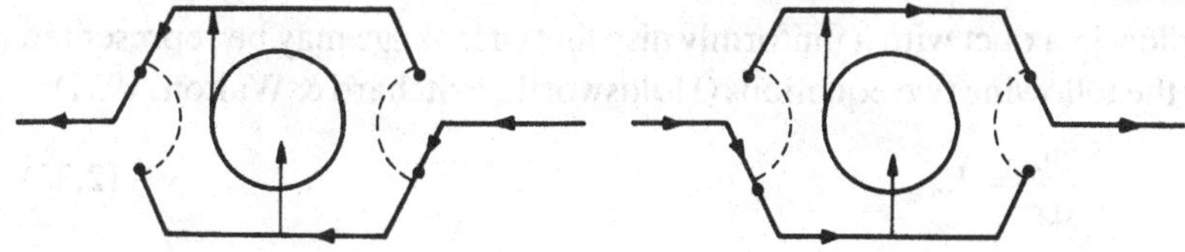

These adjustable impellers give volume flow rate regulation by variation of the fan characteristics.

(4) Using an adjustable damper. This increases the total system resistance, thus varying the operating point on the fan characteristic. Although of relatively low cost initially, it is not a very efficient method of regulation where a wide range of volume flow rate is required.

The method chosen will generally be a compromise between capital cost and running cost, although ease of installation and operation may be significant factors in some instances, as may be the possibility of automatic control.

2.5 Auxiliary ventilation calculations

Auxiliary ventilation relies upon fabricated ducting for the passage of either intake or return air. Because of the nature of the ducting and the numerous joints, complete elimination of leakage from or into the ducting is impossible in practice.

2.5.1 Airflow analysis in leaky ducts

Three methods are available for analysing the airflow problems in a leaky duct:

(1) assuming a uniformly distributed leakage and analysing the leaky duct mathematically;
(2) assuming a number of discrete leakage paths and treating the leaky duct, i.e. duct and leakage paths, as a ventilation network;
(3) assuming a number of discrete leakage paths and treating the airflow through the leaky duct as a series–parallel combination of airflows along the duct and through leakage paths.

1. Mathematical analysis of airflow in ducts with a uniformly distributed leakage

A schematic diagram for this method is shown in Fig. 2.40. In a leaky duct, both pressure and volume flow rate vary along the length. Two constants are required to define the duct characteristics, namely, the resistance coefficient of the duct and resistance coefficient of the leakage paths. The airflow in a duct with a uniformly distributed leakage may be represented by the following two equations (Holdsworth, Pritchard & Walton, 1951):

$$\frac{\mathrm{d}p}{\mathrm{d}x} = R_{\mathrm{d}}Q^2 \qquad (2.37)$$

and

$$\frac{dQ}{dx} = \sqrt{\frac{p}{R_l}} \tag{2.38}$$

where p = pressure at a distance x along the duct;
Q = volume flow rate at a distance x along the duct;
R_d = resistance coefficient of the duct;
R_l = resistance coefficient of the leakage paths; and
x = distance along the duct measured in the direction of increasing volume flow rate and pressure (p and Q are regarded as positive for the cases of both forcing and exhausting systems).

The validity of these two equations depends on the following reasonable assumptions:

(i) air density is constant;
(ii) the leakage is not influenced by the velocity pressure;
(iii) pressure losses at ends are neglected;

Fig. 2.40. Schematic diagram for mathematical analysis of a duct with a uniformly distributed leakage. Q_f – fan volume flow rate; p_f – fan pressure; R_d – duct resistance coefficient per unit length; Q_0 – delivered volume flow rate; $l_1, l_2 \ldots$ – duct leakage per unit length; $p_1, p_2 \ldots$ – pressure loss per unit length; n – number of leakage paths.

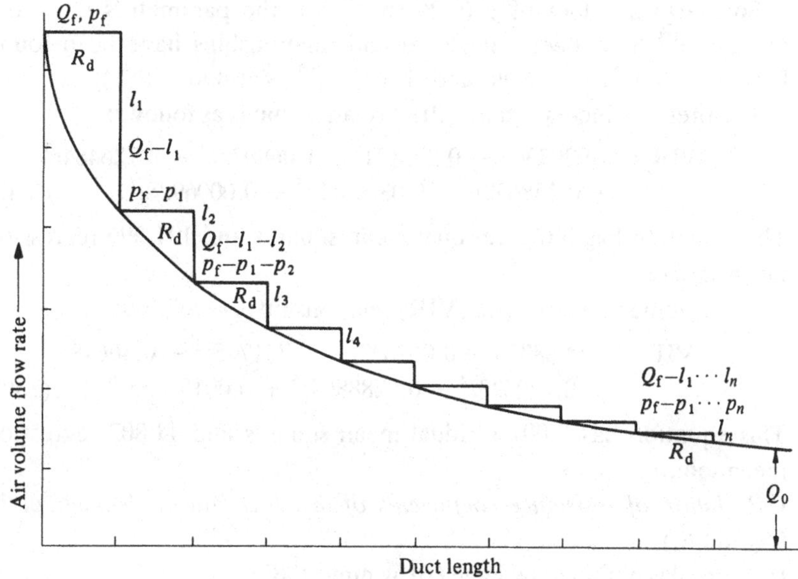

(iv) ducting is of uniform diameter and quality;
(v) velocity pressure in the air leaving the fan is small compared with total pressure.

By eliminating Q from equations (2.37) and (2.38), the following equation is obtained:

$$\left(\frac{\mathrm{d}p}{\mathrm{d}x}\right)^{1/2} \times \frac{\mathrm{d}^2 p}{\mathrm{d}x^2} = \left[\frac{R_\mathrm{d}}{R_\mathrm{l}}\right]^{1/2} \times 2 \times \left[\frac{\mathrm{d}p}{\mathrm{d}x}\right] \times p^{1/2}. \tag{2.39}$$

By taking boundary conditions, i.e. $p = 0$, $Q = Q_0$ = volume flow rate at the face, $x = 0$, and by introducing the dimensionless parameters y and t

$$y = \left(\frac{2}{R^{1/2} R_\mathrm{d}}\right)^{2/3} \frac{p}{Q_0^2} \tag{2.40}$$

$$t = \left[2\left(\frac{R_\mathrm{d}}{R_\mathrm{l}}\right)^{1/2}\right]^{2/3} x \tag{2.41}$$

the following equation is obtained after integration:

$$t = \int_0^y \frac{\mathrm{d}y}{(1 + y^{3/2})^{2/3}}. \tag{2.42}$$

The pressure increase ratio, defined as the ratio of pressure required to deliver Q_0 under leaky conditions to pressure required to deliver Q_0 under leakless conditions, is equal to y/t. The volume increase ratio, defined as the ratio of volume flow rate handled by the fan to volume flow rate reaching the face, is equal to $(1 + y^{3/2})^{1/3}$.

For various values of y (0.25 to 50.00), the parameters t, y/t and $(1 + y^{3/2})^{1/3}$ have been calculated and relationships have been found between t and y/t as well as t and $(1 + y^{3/2})^{1/3}$ (Vutukuri, 1983).

The pressure increase ratio (PIR) relationship is as follows:

$$\begin{aligned}\mathrm{PIR} = {} & 1.0203344 - 0.268171t + 1.068933t^2 - 0.726484t^3 \\ & + 0.258632t^4 - 0.0309553t^5 + 0.0006928t^6.\end{aligned} \tag{2.43}$$

This equation has 0.002 residual mean squares and 121.596 regression mean squares.

The volume increase ratio (VIR) relationship is as follows:

$$\begin{aligned}\mathrm{VIR} = {} & 1.0088958 - 0.062632t + 0.731745t^2 - 0.466485t^3 \\ & + 0.180633t^4 - 0.0288809t^5 + 0.00155255t^6.\end{aligned} \tag{2.44}$$

This equation has 0.001 residual mean squares and 44.802 regression mean squares.

Calculation of resistance coefficients of the duct and the leakage paths (R_d *and* R_l)

(i) Formulas obtained by Page (Browning, 1983).

$$R_\mathrm{d} = \frac{10(p_1 - p_2)}{L}\left(\frac{500}{2Q_1 + 3Q_2}\right)^2 \tag{2.45}$$

$$R_l = \frac{1000}{\left[\dfrac{150(Q_1 - Q_2)(p_1 - p_2)}{L(p_1^{3/2} - p_2^{3/2})}\right]^2} \tag{2.46}$$

where R_d = resistance coefficient of the duct, Ns^2/m^8 per 100 m;
R_l = resistance coefficient of the leakage paths, Ns^2/m^8 per 100 m;
p_1 = upstream pressure, kPa;
p_2 = downstream pressure, kPa;
Q_1 = volume flow rate at p_1, m^3/s;
Q_2 = volume flow rate at p_2, m^3/s; and
L = distance between p_1 and p_2, m.

Providing the fractional leakage is less than 15 % and the upstream pressure is at least 2.5 times the downstream pressure, these formulas are accurate to within 3 %.

At the discharge end of the duct, $p_2 = 0$ and $Q_2 = Q_0$. In that case

$$R_d = \frac{10p_1}{L}\left(\frac{500}{2Q_1 + 3Q_0}\right)^2 \tag{2.47}$$

$$R_l = \frac{1000}{\left[\dfrac{150(Q_1 - Q_0)}{Lp_1^{1/2}}\right]^2}. \tag{2.48}$$

(ii) Formula obtained by Woronin (Strzeminski, 1975).

$$R_l = \frac{R_d\left(\dfrac{L}{100}\right)^3}{\left\{3\left[\left(\dfrac{Q_1}{Q_0}\right)^{1/2} - 1\right]\right\}^2}. \tag{2.49}$$

(iii) Formula obtained by Vutukuri (1983).

By combining equations (2.41) and (2.44), the following equation is obtained:

$$\frac{Q_1}{Q_0} = 1.0088958 - 0.062632t + 0.731745t^2 - 0.466485t^3$$
$$+ 0.180633t^4 - 0.0288809t^5 + 0.00155255t^6 \tag{2.50}$$

where

$$t = \left[2\left(\frac{R_d}{R_l}\right)^{1/2}\right]^{2/3}\left(\frac{L}{100}\right).$$

(iv) Formula obtained by Browning (1983).

$$\frac{Q_1}{Q_0} = 1.02 + 0.0185\left(\frac{L^2}{10R_l}\right)^{1/2}\left(\frac{R_dL}{100}\right)^{1/2} + 0.00015\,\frac{L^2}{10R_l}\left(\frac{R_dL}{100}\right). \tag{2.51}$$

To use equations (2.47) to (2.51), downstream pressure p_2 must be zero, i.e. at the duct discharge. Equations (2.47), (2.48), (2.49) and (2.51) can be solved directly. Equation (2.50) has to be solved by the midpoint method. The range of application of equation (2.50) is up to leakages of 85 % and equation (2.51) 80 %.

To determine the resistance coefficients of the duct and leakage paths, the following are measured:

Q_1 = the volume flow rate at the fan end, m³/s;
Q_0 = the volume flow rate at the delivery end, m³/s;
p_1 = the pressure at the fan end, Pa; and
L = the duct length, m.

The resistance coefficient of the duct, R_d, is calculated from equation (2.47).

The resistance coefficient of the leakage paths, R_l, is calculated from equations (2.49) or (2.50) or (2.51). Equations (2.50) and (2.51) are more accurate than equation (2.49).

Design of a simple auxiliary ventilation system

The calculations usually involved are:

(1) To find the minimum ventilation pressure and volume flow rate to give the required volume flow rate at the face of the heading. The following values are required and will be known or selected, depending on the system and the heading: R_d, R_l, L and Q_0. Then the following are calculated:
   (i) t – from equation (2.41);
   (ii) VIR – from equation (2.44);
   (iii) PIR – from equation (2.43);
   (iv) the pressure required to deliver Q_0 under leakless conditions – from $p = R_d Q_0^2 \left(\frac{\text{length of duct}}{100}\right)$;
   (v) the fan volume flow rate, $Q_1 = (\text{VIR})Q_0$;
   (vi) the fan pressure, $p_1 = (\text{PIR})p$.

(2) To determine the fan operating point and the delivered volume flow rate at the face for a given duct and a fan. The following are required: R_d, R_l, L and the fan p–Q characteristic curve. The calculations are done as follows:
   (i) First Q_0 is assumed.
   (ii) Q_1 and p_1 are then calculated as above.
   (iii) Then the effective resistance of the system, R_e, is calculated from

$$R_e = \frac{p_1}{Q_1^2}.$$

(iv) The system resistance curve is then plotted on the fan curve. Where the resistance curve intersects the fan curve, the actual volume flow rate Q_1' and fan pressure p_1' are read off. (The fan p–Q curve is usually represented by a second-degree polynomial $p = aQ^2 + bQ + c$, where a, b and c are constants, and the duct resistance curve by $p = R_e Q^2$. The point of intersection can also be determined mathematically.) The actual delivered volume flow rate Q_0' is calculated using:

$$Q_0' = Q_1'/\text{VIR}.$$

(3) To find the inlet and delivered volume flow rates when a fixed pressure is applied (e.g. across doors). The following are required: R_d, R_l, L and pressure available. The calculations are done as follows:

(i) First Q_0 is assumed.

(ii) Q_1 and p_1 are then calculated as above.

(iii) If the fixed pressure is p_1' the inlet volume flow rate, Q_1', is obtained from:

$$Q_1' = Q_1\left(\frac{p_1'}{p_1}\right)^{1/2}.$$

(iv) The actual delivered volume flow rate Q_2' is calculated using:

$$Q_0' = Q_1'/\text{VIR}.$$

2. Network analysis of leaky duct

Fig. 2.41 gives a schematic diagram of this method of analysis. Although any network analysis program can be used for calculations of the type discussed in the earlier section (p. 38), this method is best used for analysing a duct system with multiple fans placed along the duct line. Recirculation can be noticed if it exists. This analysis requires a computer.

3. Treatment of the airflow through a leaky duct as a series–parallel combination of airflows along the duct and through leakage paths (Vutukuri, 1983)

Fig. 2.42 gives a schematic diagram of this method of analysis. This figure is drawn in a different way to that of Fig. 2.41 but both represent the same.

The air enters the duct at (0) and leaves it at $(n + 1)$; $1, 2, 3, \ldots, n$ are joints, and leakage occurs at these points. Depending upon the type of joints, leakage paths have some resistance and this is represented by the resistance coefficient of the leakage paths. Assuming 0–1 = 1–2 = 2–3 = 3–4 = ... = $n - (n + 1)$, the resistance coefficient of all duct sections (R_d) is the same. Assuming all joints are made and maintained in the same

Fig. 2.41. Schematic diagram for ventilation network analysis. Q_f – fan volume flow rate; p_f – fan pressure; R_e – duct effective resistance; R_d – duct resistance coefficient per unit length; Q_0 – delivered volume flow rate; $l_1, l_2 \ldots$ – duct leakage per unit length; $p_1, p_2 \ldots$ – pressure loss per unit length; n – number of leakage paths; 1, 2, 3, . . . n are inside the duct, 1′, 2′, 3′, . . . n′ are outside the duct.

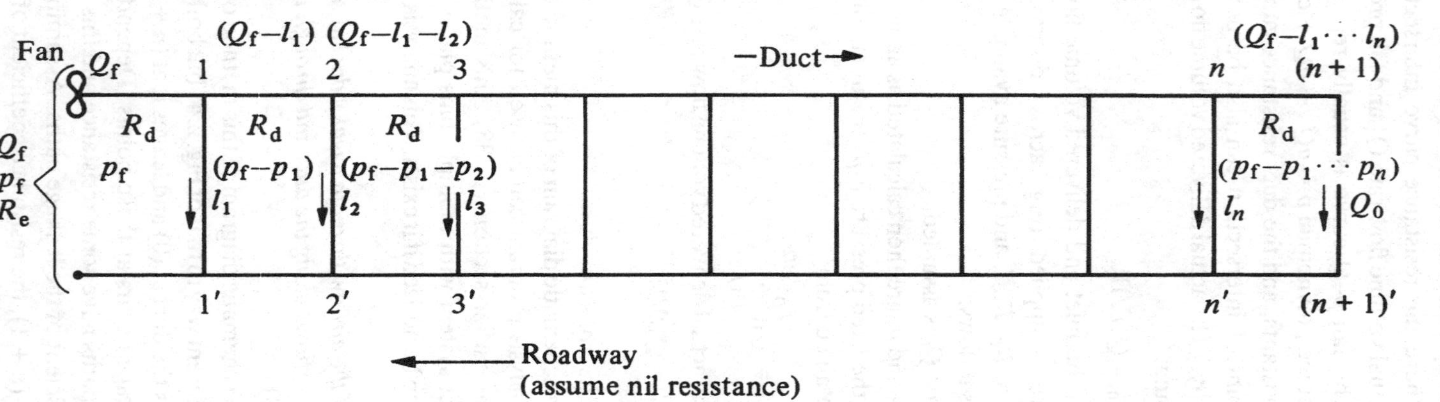

fashion, the resistance coefficient of the leakage paths at all the joints (R_l) is also the same. The pressure outside the duct is assumed to be the same along the airway. Then

$$\frac{1}{\sqrt{R_{n-(n+1)}}} = \frac{1}{\sqrt{R_d}} + \frac{1}{\sqrt{R_l}}$$

$$\sqrt{R_{n-(n+1)}} = \frac{\sqrt{R_d R_l}}{\sqrt{R_l} + \sqrt{R_d}}$$

$$R_{n-(n+1)} = \frac{R_d R_l}{(\sqrt{R_l} + \sqrt{R_d})^2}.$$

Similarly

$$\frac{1}{\sqrt{R_{(1-n)-n}}} = \frac{1}{\sqrt{R_d + \dfrac{R_d R_l}{(\sqrt{R_l} + \sqrt{R_d})^2}}} + \frac{1}{\sqrt{R_l}}.$$

From this, $R_{(1-n)-n}$ can be determined. By following the same steps successively, $R_{0-(n+1)}$ can also be determined.

If p_f is known, Q entering the duct at (0), i.e. Q_{0-1}, can be calculated as follows:

$$Q_{0-1} = \sqrt{\frac{p_f}{R_{0-(n+1)}}}.$$

If a fan is installed at (0), then the duct resistance curve is plotted on the fan p–Q curve and the point of intersection of these two curves gives the operating point of the fan, i.e. Q_{0-1} and p_f.

Fig. 2.42. Schematic diagram of a series–parallel combination of airflows along the duct and through leakage paths. Q_f – fan volume flow rate; p_f – fan pressure; R_e – duct effective resistance; R_d – duct resistance coefficient per unit length; R_l – resistance coefficient of leakage paths per unit length; $l_1, l_2 \ldots$ – duct leakage per unit length; $p_1, p_2 \ldots$ – pressure loss per unit length; n – number of leakage paths; 1, 2, 3, . . . n are leakage points along the duct.

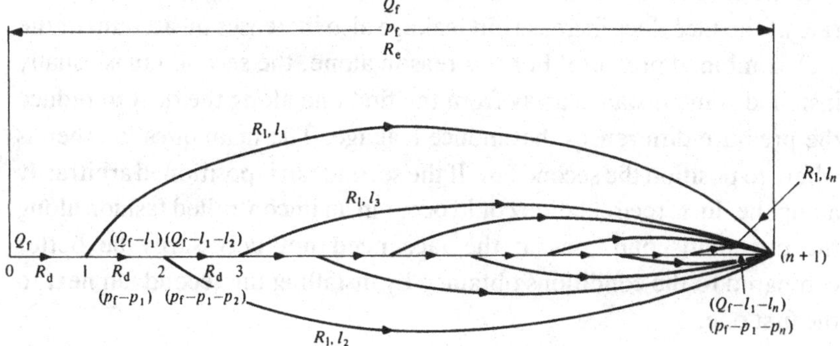

Q_{0-1} splits into Q_{1-2} and $Q_{1-(n+1)}$.

$Q_{0-1} = Q_{1-(n+1)} + Q_{1-2}$.

Since 1–2–3–4– . . . –$(n + 1)$ and 1–$(n + 1)$ are in parallel

$R_1 Q^2_{1-(n+1)} = R_{1-(n+1)} Q^2_{1-2}$.

From these two equations

$$Q_{1-(n+1)} = \frac{Q_{0-1}\sqrt{\dfrac{R_{1-(n+1)}}{R_1}}}{\left[1 + \sqrt{\dfrac{R_{1-(n+1)}}{R_1}}\right]}$$

and $Q_{1-2} = Q_{0-1} - Q_{1-(n+1)}$.

In a similar fashion Q_{2-3}, $Q_{2-(n+1)}$, Q_{3-4}, $Q_{3-(n+1)}$, $Q_{n-(n+1)}$ through the duct and $Q_{n-(n+1)}$ through the leakage paths can be calculated.

This type of calculation is best done by computers. The main advantage to this method is that the effects of bends, air coolers, dust control equipment, etc. along the duct line can be taken into account in the design of an auxiliary ventilation system.

Computer programs for methods 1 and 3 are given in Appendix II.

2.5.2 *Principles of design of auxiliary ventilation systems for long drivages*

Owing to limitations on the size of fans for underground use and for economic reasons, multiple fans are selected for long drivages. When a ventilation duct in a development end is extended, as the face advances the air volume flow rate reaching the face gradually decreases owing to both the increasing resistance of the duct and the increasing leakage. When the face air volume flow rate decreases to the minimum acceptable value, another fan is installed in the duct. The question then arises of whether the second fan should be installed next to the existing fan at the one end of the duct or some distance away from the first fan along the duct.

If the second fan is installed next to the first one, the air volume flow rate at the face does increase but leakage also increases on account of the high combined pressure. For this reason alone, the second fan is usually installed some distance away from the first one along the duct to reduce the pressure differences that induce leakage. The main question then is where to position the second fan. If the second fan is positioned arbitrarily along the duct, recirculation could occur in an uncontrolled fashion along the duct and conditions at the face need not necessarily be better compared to the conditions obtained by installing the second fan next to the first one.

In very long ducts, multiple fans may have to be used. The important principle to follow in the design of auxiliary ventilation systems using multiple fans spaced in series in long ducts is: each of the fans must overcome the resistance of motion arising over the sector of ventilation duct line between it and the fan immediately following it.

For a constant length duct, the design problem may be approached in two ways:

(1) by determining the optimum locations for fans of given characteristics;
(2) by deciding the locations for the fans and then choosing the fans with the required characteristics.

In normal development work, as the face advances, the duct is extended. In this case, controllable pitch axivane fans are the best to use. The fans are installed at the calculated optimal locations for a particular total duct length as required and as the duct is extended the pitch of the blades of the fans is adjusted to control recirculation. A detailed worked example has been given elsewhere (Vutukuri, 1983).

If fixed pitch axivane fans are installed, as required, at their optimal locations, there would be some recirculation. If unacceptable recirculation takes place, the fan locations can be shifted by some distance towards the intake, away from the face, to prevent recirculation. A detailed worked example has been given elsewhere.* The alternative is to place a variable area, conical restricting fitting on the outlet from the ventilation ducting to increase the resistance of this sector to prevent recirculation.

2.5.3 *Effects of variables in auxiliary ventilation*

The detailed effects of variables in auxiliary ventilation installations – fan characteristic curve, duct diameter, standard of installation and duct length – on the effective resistance of the duct and the volume flow rate of air at the face have been given elsewhere (Vutukuri, 1984) and the most important conclusions are given below.

(1) The effective resistance of a duct initially decreases as its diameter increases and then tends to level off.
(2) The effective resistance of a duct initially increases as its length increases and then tends to level off: the extent to which this happens is a function of the standard of installation.

* Vutukuri, V. S. (1983). Principles of design of auxiliary ventilation systems for long drivages. Unpublished report.

(3) For short duct lengths the effect of standard of installation on the effective resistance of a duct is not significant, but for long lengths the effective resistance increases as the standard of installation improves.
(4) The volume flow rate of air at the face decreases as the duct length increases; at long lengths it remains more or less constant.
(5) For a given fan the volume flow rate of air at the face increases as the duct diameter increases: the choice of different sizes of duct for a given fan produces a wide range of possible volume flow rates of air at the face.
(6) If the duct is long, the choice of fan does not greatly affect the volume flow rate of air at the face.

2.6 Economic size of airways

An airway is considered as a shaft or tunnel excavated specifically for ventilation. The size of a mine airway for passing a specific volume flow rate of air is selected to minimise the total ventilation cost which consists of:

(1) capital cost for airway development (capital cost of the ventilating machinery may also be added to this cost); and
(2) power cost of ventilation (maintenance cost of the ventilating plant may also be added to this cost).

In this discussion these costs are considered on the basis of equivalent annual costs.

Capital cost for airway development

The cost function takes various forms but the following is considered here:

$$\text{Cost} = C_d \times A \times L \text{ (in dollars)}$$

where C_d = development cost, \$/m^3;
A = cross-sectional area of the airway, m^2; and
L = length of the airway, m.

C_d depends on whether the airway is lined or unlined, equipped or unequipped, ground conditions difficult or easy, etc.

Power cost of ventilation

$$\text{Power in the air} = p \times Q \text{ (watts)}$$

where p = ventilating pressure required, Pa; and
Q = volume flow rate, m^3/s.

Since

$$p = \frac{KLCQ^2w}{1.2A^3}$$

$$\text{the power in the air} = \frac{KLCQ^3w}{1.2A^3}$$

where K = friction factor for the airway, Ns^2/m^4;
C = perimeter of the airway, m; and
w = density of air flowing, kg/m^3.

If the combined fan, transmission and motor efficiency is η (%) and the power cost is C_p (\$/W/annum), the annual power cost of ventilation becomes

$$\frac{KLCQ^3w}{1.2A^3} \times \frac{100}{\eta} \times C_p\ (\$).$$

Equivalent annual cost

The capital cost can be considered as a loan which is repaid with interest by annual payments over the life of the airway. The annual payments then become equivalent annual costs for the capital cost (expenditure).

$$\text{Annual payments (costs)} = \text{capital cost} \times \left[\frac{i/100}{1 - \left(1 + \frac{i}{100}\right)^{-n}}\right]$$

where i = interest rate per year, %; and
n = life of the airway, years.

Optimisation

As the airway size increases, the annual development cost (curve 1, Fig. 2.43) increases while the annual power cost (curve 2) decreases for a given air volume flow rate. The two curves are added (curve 3) to obtain the optimum airway size, point A, where the total cost is at a minimum.

Mathematical formulation

Total annual cost,

$$C_t = \frac{100KLCQ^3wC_p}{1.2A^3\eta} + C_dAL\left[\frac{i/100}{1 - \left(1 + \frac{i}{100}\right)^{-n}}\right]$$

For a 1 m length square airway with a as the side of airway in m, the above equation becomes

$$C_t = \frac{100K \times 1 \times 4a \times Q^3 w C_p}{1.2a^6\eta} + C_d a^2 \times 1 \left[\frac{i/100}{1 - \left(1 + \frac{i}{100}\right)^{-n}}\right]$$

Differentiating C_t with respect to a and equating dC_t/da to zero results in an equation from which the economic airway size can be determined.

$$\frac{dC_t}{da} = \left(\frac{400}{1.2}\right) \times \frac{(-5) \times KQ^3 w C_p}{a^6\eta} + 2C_d a \left[\frac{i/100}{1 - \left(1 + \frac{i}{100}\right)^{-n}}\right]$$

$$a = \left[\frac{1000}{1.2} \times \frac{KQ^3 w C_p}{\eta C_d \times C_c}\right]^{1/7}$$

where

$$C_c = \left[\frac{i/100}{1 - \left(1 + \frac{i}{100}\right)^{-n}}\right].$$

In the case of circular airway, the above equation becomes:

$$a = \left[\frac{64000}{1.2 \times \pi^3} \times \frac{KQ^3 w C_p}{\eta C_d C_c}\right]^{1/7}$$

where a is the diameter of the airway.

Fig. 2.43. Size of airway versus costs, 1, annual development cost; 2, annual power cost; 3, annual total costs. Point A is where the total cost is at a minimum.

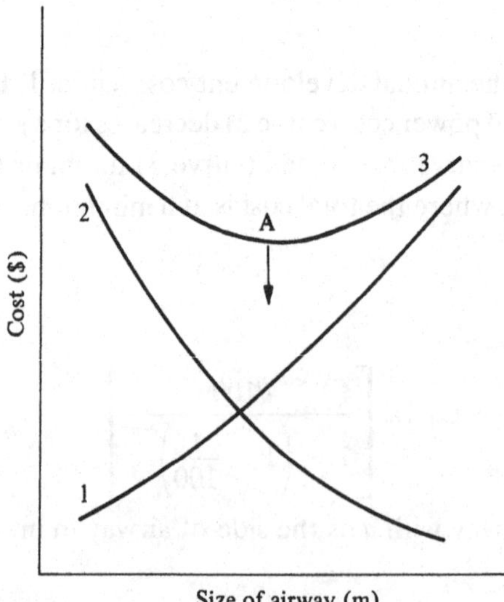

2.7 Ventilation surveys

The two most important types of surveys usually carried out are: (1) pressure surveys and (2) volume flow rate surveys. The pressure survey is done to measure the pressure distribution and the volume flow rate survey is done to measure the volume flow rate distribution and to locate leakages. Measurements of volume flow rate, when used in conjunction with observations of pressure loss (i.e. pressure surveys) along various sections of the airway, enable the resistance of each section to be determined.

The surveys should be carried out on a shift when there is little or no activity in the mine, for example at the weekend, so that accurate and consistent readings are obtained. During normal working shifts, readings are affected by the opening and closing of doors and by the movement of vehicles in shafts and roadways.

2.7.1 Equipment

Pressure measurement

The instruments for measuring ventilation pressures (i.e. pressure differences) are of two types, namely, manometers and barometers.

Manometers. The simplest manometer is a vertical U-tube half filled with paraffin or similar liquid (because this gives a better shape of meniscus and a rather more open scale of pascals than water) (Fig. 2.44). The two legs of the U-tube are joined by plastic tubes to the points for which the pressure difference is required; after connecting, the liquid rises in the leg which is connected to the point with the lower pressure, and falls in the other leg. The difference in liquid levels in the two legs is the pressure difference, calculated as follows:

$$p = P_1 - P_2 = Hwg$$

Fig. 2.44. A manometer. P_1, P_2 = pressure at points 1 and 2 in the ventilation system; H = vertical difference in levels.

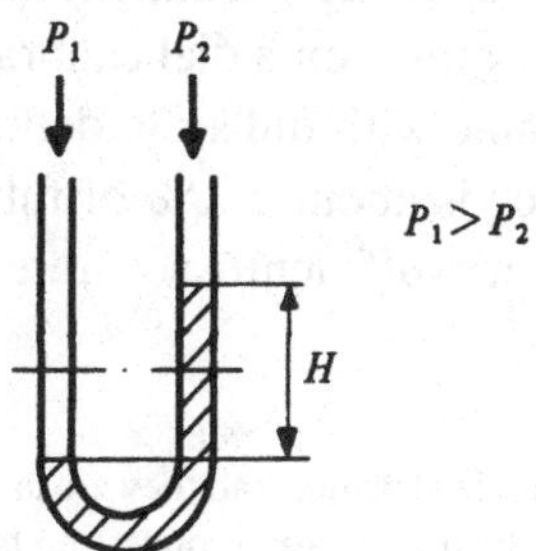

where p = pressure difference, Pa;
H = vertical difference in levels, m;
w = density of the liquid, kg/m^3; and
g = gravitational acceleration, m/s^2.

The scale is usually graduated to give directly pressure in pascals and it should normally be read to 10 Pa.

For measuring small pressure differences, inclined manometers are used where the measurement precision is magnified by the slope ratio, i.e. hypotenuse/vertical.

Barometers. Barometers measure the absolute pressure of the air at the site of the instruments. These are of two types, namely, mercury barometers and aneroid barometers (altimeters*).

In mercury barometers, a mercury column is balanced against the air pressure, but as they are not easily portable, aneroid (i.e. without liquid) barometers are more commonly used for pressure surveys.

Aneroid barometers consist of a vacuum chamber with corrugated elastic diaphragms which are held from collapsing by a spring. As the external pressure varies, the distance between the diaphragms changes. This change in distance, after magnification, is transmitted to a pointer that moves over a graduated scale. Many instruments are temperature-compensated and can be read to 5 Pa for a range of 20 kPa.

The characteristics of aneroid barometers change with time and they should be calibrated at least once a year with additional calibration against a mercury barometer before a major survey. Calibration curves should be obtained both for increasing pressures and decreasing pressures so that an appropriate one is used for correction to obtain the true pressure.

Magnehelic manometer. This instrument is similar to an aneroid barometer and consists of a diaphragm across which the pressure difference is applied, causing it to move against a spring. At the end of the diaphragm there is a magnet. Adjacent to the magnet there is a helix which is free to turn so as to maintain the minimum air gap between itself and the magnet. The position of the helix is indicated by a pointer on a dial calibrated in pressure units. The instruments are available with full scale deflections ranging from 100 up to 4000 Pa. Accuracy is about ±2 % of full-scale deflection, so the lower range instruments are sufficiently accurate for all underground work.

* In the case of altimeters, readings are in metres elevation and they are later converted to pressures in Pa by multiplying by the elevation conversion factor.

Volume flow rate measurement

For calculating volume flow rates, air velocities and cross-sectional areas of airways are measured. Tracer gas techniques are used for leakage surveys.

Air velocity. Air velocity measuring instruments are usually classified as low velocity, medium velocity and high velocity, the approximate limits being 0.5 and 10 m/s.

For low-velocity airflow measurements, smoke clouds are used. The velocity is measured by timing the movement of a puff of chemical smoke over a known distance, e.g. 5 m, with a stop watch. The smoke is produced by forcing air from an aspirator bulb through a glass tube containing a smoke-generating reagent (usually titanium tetrachloride).

For medium-velocity airflow measurements, two types of instruments, the vane anemometer and the velometer, are usually employed. The vane anemometer is a small windmill; the air turns the rotor at a speed proportional to the air velocity and this rotation operates through a gear train to a counting system. The anemometer is operated over a certain time with a stop watch and the velocity is determined from the total revolutions divided by the time.

The velometer is a direct-reading velocity meter. Air enters a port on one side, deflecting a spring-controlled vane and leaves from the other side. The instrument is designed in such a manner that the deflection is directly proportional to the air velocity. The vane movement is transmitted to a pointer moving over a scale.

For high-velocity airflow measurements, both high-velocity anemometers and velometers may be used although a pitot tube along with a differential pressure gauge capable of reading to 2 Pa is usually used. The construction of a pitot tube is shown in Fig. 2.45. It has two concentric tubes, one within the other, with the end bent at right angles to the shaft. The inner tube is open on the end and receives the total pressure of the airflow (i.e. static plus velocity) whereas the outer tube is closed at the end and receives the static pressure only of the airflow through a number of pin holes set back from the end. The tip is tapered or made hemispherical to reduce interference with the airflow. The other end of the pitot tube is provided with two fittings, one for each tube. After facing the tapered end of the pitot tube directly into and parallel to the airflow, hoses from both the tubes are connected to the pressure gauge (e.g. inclined manometer) and the difference between the total pressure and the static pressure gives the velocity pressure. The velocity is then calculated from the following equation:

$$v = \sqrt{\frac{2p}{w}}$$

where v = air velocity, m/s;
p = velocity pressure, Pa; and
w = air density, kg/m^3.

Windmill-type digital instruments are available with a velocity range of 0.3 to 40 m/s.

Cross-sectional area of airway. The cross-sectional area of an airway is determined at the point where the velocity of airflow is measured to calculate the volume flow rate. It is usually possible to calculate the cross-sectional area as the sum of component areas of regular shapes. If the shape is too irregular, a tape is stretched diagonally across the airway and offsets to the roof, floor and sides are measured at right angles to the tape. The measurements are plotted and the cross-sectional area determined. Photographic methods of area measurement are also available.

Tracer gas techniques. For the measurement of both leakage and recirculation airflows, smoke cloud and anemometer are inadequate. Sulphur hexafluoride (SF_6) tracer gas has been used for such surveys.

The tracer gas, which is non-toxic and odourless, is released from a pressurised bottle and air samples are taken with syringe bottles. Gas concentrations as low as 10^{-9} are determined accurately with a portable gas chromatograph.

There are two techniques, namely, pulse release and steady release. In the pulse-release technique, gas is released over a short time interval in a

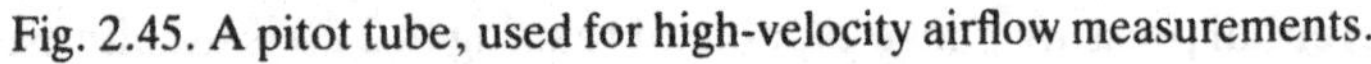
Fig. 2.45. A pitot tube, used for high-velocity airflow measurements.

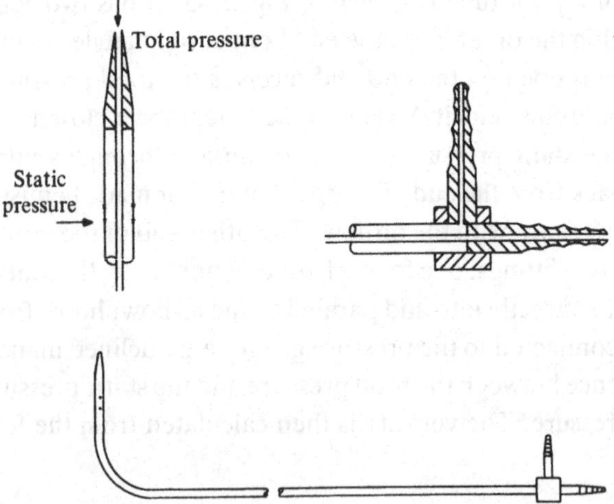

donor airway, say from 2 to 10 minutes. Samples are taken at 2 to 15 minute intervals, depending on the suspected gas concentration, at the point of investigation in the receiver airway until the gas concentration becomes too small to measure. The volume of gas released is determined from the mass released (loss in mass of the gas container) and the air temperature and pressure in the airway at the point under investigation. The volume of tracer gas passing the sampling point is estimated from the equation below:

$$V_{SF_6} = Q_{air} \int c\,dt$$

where V_{SF_6} = volume of SF_6, m^3;
Q_{air} = air volume flow rate at the sampling point, m^3/s; and
c = concentration of SF_6 at time t.

Integration of $c\,dt$ is done by simply multiplying C_{avg}, the average SF_6 concentration in those sample bottles that contain SF_6, by the total time T during which measurable amounts of SF_6 were found at the point of investigation. This total time T is the sampling interval multiplied by the number of samples containing measurable concentrations of SF_6. The above equation therefore reduces to

$$V_{SF_6} = Q_{air} C_{avg} T.$$

The amount of air leakage can be calculated as follows:

$$\text{leakage rate} = \frac{V_{SF_6}}{V_{SF_6\,\text{released}}} \times \text{volume flow rate of air at the location of release.}$$

In the steady-release technique, the tracer gas is released at a steady rate for a number of hours in the donor airway and samples are taken for a shift or so in the receiver airway. The volume flow rate of tracer gas released is obtained from the mass flow rate, the temperature and pressure in the receiver airway. The volume flow rate of tracer gas in the receiver airway is determined from the concentration on a volume basis when this has reached a steady state, and the volume flow rate of air in the receiver airway.

The amount of air leakage can be calculated as follows:

$$\text{leakage rate} = \left[\frac{\text{volume flow rate of tracer gas in the receiver airway}}{\text{volume flow rate of tracer gas released}}\right] \times \text{volume flow rate of air at the location of release.}$$

2.7.2 *Pressure surveys*

Pressure surveys are conducted to determine the pressure losses in various segments of a ventilation system, to determine existing conditions and to obtain information upon which to plan for the future. The resistances of the airways may be calculated if the air volume flow rates also are determined. The value of the friction factor may be determined if the airway dimensions, i.e. the cross-section details and the length and the air density,* are determined.

Two methods are commonly used, the trailing hose method and the barometric method.

Trailing hose method

In this method, two airtight hoses or tubes (example – 6 mm internal diameter polythene tubes) from the ends of the section of the airway are connected to a manometer (either inclined or Magnehelic) to measure directly the difference between the pressures at the two ends. Since the differences in static pressures are usually measured, the ends of the tubes are held or placed normal to the direction of airflow. Alternatively, the outer tubes of the pitot tubes are connected to the ends of trailing tubes.

This method is simple, requires a minimum of time, and can be used during the normal shift. The main advantage is that the pressure loss can be read directly from the manometer. Its main disadvantages are:

(1) Long lengths of tubing are difficult to install and to check for airtightness. The maximum length is probably 50 m.
(2) The tubes may get blocked by kinking or by water droplets inside them.
(3) When long lengths are used, response to pressure changes may become sluggish.
(4) Errors are introduced by differences in temperature and humidity between the tube and the airway.
(5) Any errors made during the survey are cumulative.

Barometric method

In conducting a pressure survey, either of two methods may be used, both requiring two barometers. In the first, both barometers are taken underground and readings are taken simultaneously at adjacent stations. Two

* To determine the air density, wet- and dry-bulb temperatures, usually measured by a whirling hygrometer, and the absolute pressure at the point of measurement are determined.

operators are required. In the second, one barometer is taken underground while the second barometer is kept at the surface or at some base point underground, and readings of both are taken at a prearranged time schedule. If a recording barometer is available for the base, only one operator is required. Temperatures and elevations are recorded at each station.

The calculation procedures are as follows:

(1) calculate barometric pressures from barometer readings, applying instrument corrections, correcting for time changes, etc.;
(2) calculate the required psychrometric properties;
(3) calculate the pressure loss or energy loss values;
(4) calculate resistances or friction factors.

The pressure loss can be calculated from the following equation:

$$p = -(P_2 - P_1) - g \int w \mathrm{d}Z$$

where p = frictional pressure loss, Pa;
P = barometric pressure, Pa (subscript 1 refers to the upstream end and 2 to the downstream end of the airway);
g = gravitational acceleration, m/s^2;
w = air density, kg/m^3; and
Z = elevation, m (measured above some convenient datum).

$(P_2 - P_1)$ is the actual pressure increase, while $-g \int w \mathrm{d}Z$ is the theoretical pressure increase.

Density, w, can be calculated from the basic data of barometric pressure, wet-bulb and dry-bulb temperatures using the standard psychrometric equation as follows:

$$w = \frac{P - 0.378P_{\mathrm{w}}}{287(t_{\mathrm{db}} + 273.15)}$$

where w = true density of the air–water vapour mixture, kg/m^3;
P = barometric pressure, Pa;
P_{w} = vapour pressure in the air, Pa, and is equal to

$$610.5 \exp\left(\frac{17.27t_{\mathrm{wb}}}{t_{\mathrm{wb}} + 237.3}\right) - 0.000644P(t_{\mathrm{db}} - t_{\mathrm{wb}});$$

t_{db} = dry-bulb temperature, °C; and
t_{wb} = wet-bulb temperature, °C.

The density can be calculated at the two ends of the airway, but how it varies at points in-between is not known. The most common assumption is that the density varies linearly with elevation Z. Then

$$\int w\mathrm{d}Z = \frac{(w_1 + w_2)(Z_2 - Z_1)}{2}.$$

The pressure loss can also be calculated from the following equation:

$$p = F \times w_m$$

where p = frictional pressure loss, Pa;
F = frictional energy loss, J/kg; and
w_m = mean air density, kg/m^3.

The frictional energy loss is calculated from the steady flow equation as follows (Hemp, 1982):

$$F = -\int_1^2 v\mathrm{d}P - g(Z_2 - Z_1) - \tfrac{1}{2}g(r_1 + r_2)(Z_2 - Z_1)$$

where F = frictional energy loss in the airway, J/kg (expressed per unit mass of dry air);
v = specific volume, m^3/kg (expressed per unit mass of dry air);
P = barometric pressure, Pa;
Z = height above datum, m;
g = gravitational acceleration, m/s^2; and
r = moisture content, kg/kg (expressed per unit mass of dry air).

Now, assuming that the flow along the airway is polytropic, i.e. Pv^k = constant, it can be shown that the polytropic index

$$k = \frac{\log (P_2/P_1)}{\log (v_1/v_2)}$$

and

$$\int_1^2 v\mathrm{d}P = \frac{k}{k-1}(P_2v_2 - P_1v_1).$$

The specific volume, v, is calculated from the following equation:

$$v = \frac{287(t_{db} + 273.15)}{(P - P_w)}.$$

where t_{db} = dry-bulb temperature, °C;
P = barometric pressure, Pa; and
P_w = vapour pressure in the air, Pa.

Moisture content, r, is calculated from the following equation:

$$r = \frac{0.622P_w}{P - P_w}.$$

When a pressure survey is made with altimeters, the following equation can be used for the calculation of pressure loss (Anderson and Nugent, 1977):

$$p_t = \{[(E_2 - E_1)(DR)] - [(A_2 - A_1) - (B_2 - B_1)]\} \times \frac{1}{608.998} \log^{-1}\left[1.476 - \frac{(A_2 - \text{offset}) + (A_1 - \text{offset})}{38151.67}\right] + \frac{V_2^2 - V_1^2}{414.557}$$

where p_t = total pressure loss between two pressure stations, Pa;
E_1, E_2 = elevations at two pressure stations, m;
DR = density ratio, density of mine air to density of dry air at 10 °C;
A_1, A_2 = roving altimeter readings at two pressure stations, m of air;
B_1, B_2 = base altimeter readings taken at the same time as roving altimeter readings, m of air;
V_1, V_2 = air velocity at pressure stations, m/s; and
offset = the number of m of air the altimeter scale zero differs from zero elevation.

Full-volume flow rate and reduced-volume flow rate survey method. In this method, barometric pressures are measured at the two ends of an airway, both during the normal or full-flow condition and during a period when the flow is reduced. The change in flow rate might be accomplished by changing door settings or stopping fans not in the airway. Then

$$P_{F_1} - P_{F_2} = RQ_F^2 \quad \text{and} \quad P_{R_1} - P_{R_2} = RQ_R^2$$

so that

$$(P_{F_1} - P_{F_2}) - (P_{R_1} - P_{R_2}) = (P_{F_1} - P_{R_1}) - (P_{F_2} - P_{R_2}) = R(Q_F^2 - Q_R^2).$$

The subscripts F and R refer to full and reduced flow conditions. Subscripts 1 and 2 refer to the two ends of an airway.

This equation can be solved for R. The main advantage of this method is that the barometers need only indicate changes in pressure accurately.

The barometric method is suitable for performing an extended survey over a mine or part of a mine because it is quicker and requires much less effort than does the trailing hose method and because the errors are not cumulative.

2.7.3 *Volume flow rate surveys*

Volume flow rate surveys are conducted to determine the volume flow rates of air being circulated through the mine workings. Some of the reasons for conducting these surveys are:

(1) The survey enables air leakage to be located, and gives information on how to take measures for reducing such leakage losses.
(2) Air volume flow rate and pressure loss measurements along various sections of the mine enable the calculation of resistance for each section.
(3) When static pressures are measured during a pressure survey, volume flow rate measurements are necessary to calculate total pressures.

Volume flow rate measurements involve the measurement of air velocities and cross-sectional areas of airways and should, if possible, be taken where the airway is straight and the cross-sectional area is regular for 20 m or so (five or six airway widths) on either side of the actual point of measurement.

Velocity measurement

Smoke cloud method. First, a known distance, say 5 m, is measured along the airway. Two persons are required, one with the smoke tube at the upstream point and the other with a stop watch at the downstream point. Smoke clouds are released at quarter points and the times they take to travel the measured distance are recorded. The results will be on average about 10 % high and should be corrected accordingly.

Anemometer method. For ordinary rough measurements, the anemometer is held in the hand. For precise work, it should be mounted on a rod of sufficient length so that it can be held at least 1.5 m away from the operator to reduce the effect of hand, arm and body on the airflow.

The plane of measurement should be at right angles to the axis of the airway. For airways that are horizontal or less steep than 1 in 2, either a rigidly attached or a freely suspended anemometer may be used. For airways steeper than 1 in 2, the anemometer should be attached rigidly to the rod because of the yaw effect if the instrument is freely suspended.

There are three main methods of measuring air velocity, namely, spot readings, traversing and division.

Spot readings: A velocity reading is taken at the centre of the airway. The corrected reading is multiplied by a constant to give the average velocity of flow. The value of this constant should be determined by comparison with a traverse reading and it varies between 0.7 and 0.9.

Traversing: This is the most commonly used method. Continuous traverses are made by moving the anemometer at a uniform speed (say 0.3 m/s) through a plane at right angles to the axis of the airway. The path of the traverse should comprise either vertical or horizontal runs connected by semicircular loops. Traverses should start alternately from one end to the other end. Each traverse should not take less than a minute. It is necessary to take at least four acceptable readings and then to average them. Any freak readings (due perhaps to the opening of doors) are to be ignored. All acceptable readings must lie within ±5 % of their mean.

Division: Where precise measurements are required, such as for fan tests, this method is used. In this method the cross-sectional area of the airway is divided into a number of equal areas (>16) of a similar shape to that of the airway and the velocity is measured at the centre of each area. The average of the corrected velocities in each of these areas is the mean air velocity at the test section.

Where a freely suspended (i.e. vertical) anemometer is used in an airway inclined at 1 in 3 or more, the observed reading needs to be multiplied by a correction factor as follows:

Gradient	Correction factor
1 in 3	1.02
1 in 2.5	1.04
1 in 2	1.06

At gradients steeper than 1 in 2, the correction would be excessive, and therefore freely suspended instruments must not be used. At gradients less steep than 1 in 3, no correction is required.

Anemometers give a velocity somewhere within ±10 % of the true velocity. Calibrations are in the form of tables of plus or minus corrections to be made to the observed velocities.

Anemometer readings are virtually independent of the air densities encountered in underground work. When a correction is necessary, the following procedure is used:

(1) Multiply the observed velocity by $(w_1/w_0)^{1/2}$ where w_1 = actual density and w_0 = calibration density.

(2) Apply the correction to the velocity obtained under (1).

(3) Multiply this corrected velocity by $(w_0/w_1)^{1/2}$ to obtain the true velocity.

Velometer method. As the velometer is a direct-reading velocity meter, measurements are carried out at specific points. For routine measurements, the method of spot readings is used. For precise measurements, the method of division is used.

Pitot tube method. For volume flow rate measurements in ducts where the air velocities are high, the pitot tube is most commonly used. It can be inserted through a small hole in the duct, thereby permitting velocity pressures to be determined by a pressure gauge. For routine measurements, spot readings are used. For precise measurements, the method of division is used.

Fig. 2.46 shows a rectangular duct divided into 16 areas; the velocity is measured at the positions shown. A number of holes are necessary in the duct opposite the row of rectangles to be measured. These holes can be plugged when not used.

A circular duct is normally divided into five concentric circles, and measurements are made at the centre of each area along the horizontal and vertical diameters. The radii for the equal areas are $\sqrt{0.2}$, $\sqrt{0.4}$, $\sqrt{0.6}$, $\sqrt{0.8}$ and $\sqrt{1.0}$ of the radius of the duct. The instrument locations are at $\sqrt{0.1}$, $\sqrt{0.3}$, $\sqrt{0.5}$, $\sqrt{0.7}$ and $\sqrt{0.9}$ or 0.32, 0.55, 0.71, 0.84 and 0.95 of the duct radius. Fig. 2.47 shows the positions of points where the velocity is

Fig. 2.46. Measuring positions in a rectangular duct when using a pitot tube. (*a*) Division of a rectangular duct into 16 areas of the same geometric shape; '+' indicates position where measurements are taken. (*b*) Arrangement of measuring positions in area (top left-hand corner).

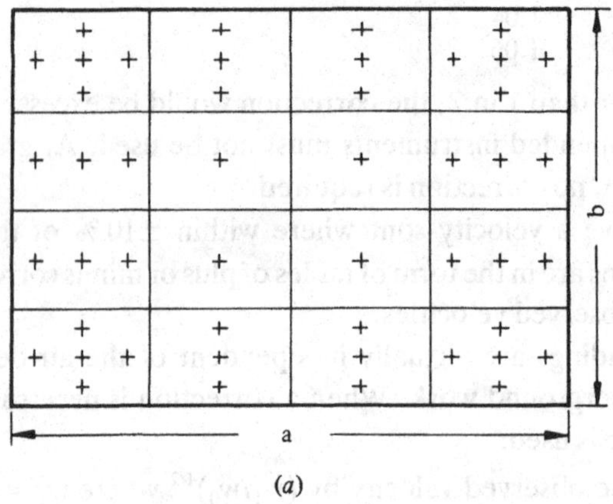

(*a*)

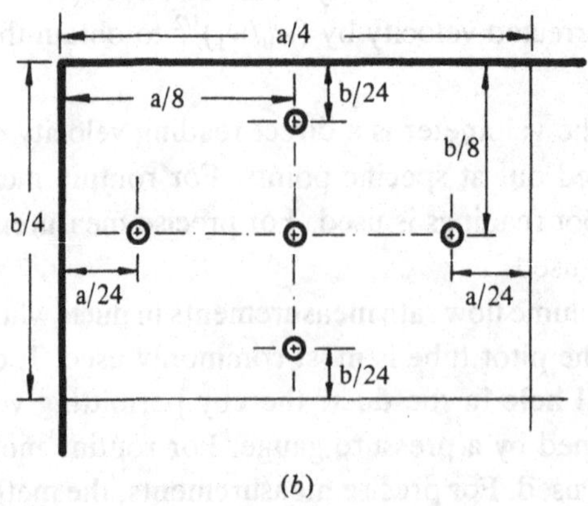

(*b*)

measured. Four holes are necessary in the horizontal and vertical diameters of the circular section to make the measurements.

When determining the average of several readings made with a pitot tube to calculate average velocity, the direct average of the velocity pressures obtained cannot be used because velocity pressures vary as the square of the velocities; the velocities are calculated for all the pitot tube readings first and then averaged.

In the case of a rectangular duct, velocities are first calculated from velocity pressures at all the positions shown in Fig. 2.46. The average velocity in each sub-section is then calculated by taking the average of the velocities at the positions in that sub-section. The average velocity of the airflow in the duct is calculated by the average of the sub-section average velocities.

In the case of a circular duct, after calculating the velocities from velocity pressures at all positions, the average velocity of the airflow in the duct is obtained from the average of all the position velocities.

For accurate results, the pitot tube head must be aligned with the direction of the air current. Correct alignment can be found by yawing the pitot head and noting the position at which the maximum reading occurs.

Fig. 2.47. Measuring positions in a circular duct when using a pitot tube.

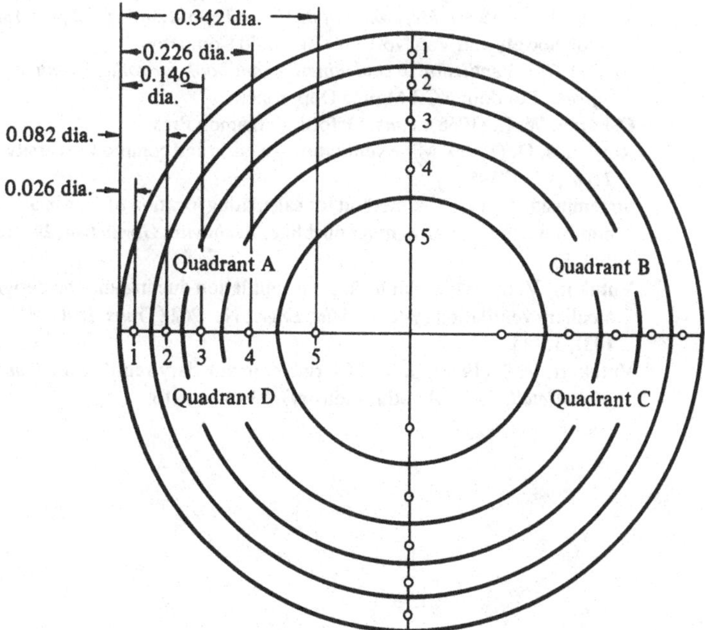

2.8 References

Anderson, T. C. and Dvorkin, D. (1978). *Mine ventilation digital simulation and analysis capabilities at MESA's Denver Technical Support Centre.* US Mining Enforcement and Safety Administration Inf. Rept 1066.

Anderson, T. C. and Nugent, J. W. (1977). *Mine ventilation pressure differentials with a programmable hand calculator.* US Mining Enforcement and Safety Administration Inf. Rept 1050.

Anon. (1978). More air for miners underground. *CSIRO Industrial Res. News*, Nov., No. 131.

Browning, E. J. (1983). An approximate method for auxiliary ventilation calculations. *Min. Engr.*, No. 264 (*Trans. Instn Min. Engrs*, **143**), 129–34.

Crous, S. J., Haven, Y., Roos, C. J., Swanepoel, A. J., Thorp, N. and Van Dyk, J. H. (1975). Review of ventilation in South African coal mines. In *International mine ventilation congress*, Johannesburg, ed. R. Hemp and F. H. Lancaster, pp. 187–94. Johannesburg: The Mine Ventilation Society of South Africa.

De La Harpe, J. H. (1974). Basic fan engineering. In *The ventilation of South African gold mines*, pp. 375–97. Johannesburg: The Mine Ventilation Society of South Africa.

Hemp, R. (1982). Pressure surveys. In *Environmental engineering in South African mines*, pp. 123–53. Johannesburg: The Mine Ventilation Society of South Africa.

Holdsworth, J. F., Pritchard, F. W. and Walton, W. H. (1951). Fluid flow in ducts with a uniformly distributed leakage. *Br. J. Appl. Phys.*, **2**, 321–4.

McElroy, G. E. (1935). *Engineering factors in the ventilation of metal mines.* USBM Bull. 385. Washington: US Bureau of Mines.

Marshall, P. and Allan, B. S. (1978). *Dust control in bord and pillar workings. Second Australian pneumoconiosis conference*, Sydney, Paper D.3, 18 pp.

Massey, B. S. (1970). *Mechanics of fluids*, 2nd edn. London: English Language Book Society and Van Nostrand Reinhold Company.

NCB (1979). *Ventilation in coal mines: a handbook for colliery ventilation officers.* London: NCB Mining Department.

Osborne, W. C. (1966). *Fans.* Oxford: Pergamon Press.

Rowlands, D. (1979). Mine ventilation – a basic and common necessity. *Min. Tech.*, **61**, 171–5.

Strzeminski, J. (1975). A method for calculating location of fans in a non-branched non-airtight air duct line. *Archiwum Gornictwa*, **20**, No. 2, 143–53.

Vutukuri, V. S. (1983). Air leakage in ventilation ducting and the design of auxiliary ventilation systems. *Min. Engr.*, No. 262 (*Trans. Instn Min. Engrs*, **143**), 37–43.

Vutukuri, V. S. (1984). Study of variables in auxiliary ventilation. *Trans. Instn Min. Metall.* (Sect. A: Min. Industry), **93**, A10–A16.

3

Mine gases and their control

3.1 Mine air

Normal dry air contains 20.93 % oxygen, 79.04 % nitrogen (including argon and rare gases) and 0.03 % carbon dioxide by volume. The total amount of other gases is less than 0.01 %. In normal atmospheric air, moisture is also always present, from a fraction of 1 % up to a maximum of about 6 %.

In addition to oxygen, nitrogen and carbon dioxide, the underground atmosphere contains small amounts of other gases, namely, methane, carbon monoxide, nitrogen oxides, unburned hydrocarbons, partially oxidised hydrocarbons (aldehydes), ammonia, hydrogen sulphide and sulphur dioxide, even under normal conditions. The major characteristics of some of these gases are given in Table 3.1.

3.2 Hazards

3.2.1 Toxic hazard

Toxic gases are those, if breathed in sufficient concentrations for a sufficient time, will seriously disable and possibly kill a person. They are discussed in the following groups depending on their mode of action on the body:

(1) asphyxiant gases;
(2) irritant gases;
(3) poisonous gases.

Asphyxiant gases

These are of two types: simple and chemical. Simple asphyxiant gases are those which have no specific toxic effect, but act by excluding oxygen from the lungs. Carbon dioxide and methane are of this type. The effect is proportional to the extent to which they decrease the partial pressure of

Table 3.1. *Characteristics of mine gases*

Name	Symbol	Specific gravity (air = 1)	Other physical properties	Harmful effects	Source	Detection	Max. allowed conc. (%)	Fatal point (%)
Oxygen	O	1.1056	Odourless, colourless, tasteless	Non-toxic	Normal air	Breathing, flame safety lamp, detector tube, electrochemical detector, para-magnetic method	19.0 (minimum)	6
Nitrogen	N	0.9673	Odourless, colourless, tasteless	Asphyxiating (due to oxygen deficiency)	Normal air, strata	Extinguishes flame safety lamp	80	—
Carbon dioxide	CO_2	1.5291	Odourless, colourless, slight acid taste	Asphyxiating (due to oxygen deficiency)	Breathing, strata, fire, explosion, blasting, diesel engine, any combustion	Breathing, flame safety lamp, detector tube, infra-red analyser	TLV-TWA–0.5 TLV-STEL–1.5	18
Methane	CH_4	0.5545	Odourless, colourless, tasteless	Asphyxiating (due to oxygen deficiency), explosive	Strata, blasting, diesel engine, organic decay	Flame safety lamp, detector tube, optical detector, detector using thermal conductivity, detector using catalytic combustion, detector using infra-red absorption	1.0 or 1.25	(5.3–14 explosive)
Carbon monoxide	CO	0.9672	Odourless, colourless, tasteless	Asphyxiating (chemical), explosive	Blasting, diesel engine, fire, explosion, any	Canaries, detector tube, catalytic combustion	TLV-TWA–0.005 TLW-STEL–	0.03 (12.5–74 explosive)

					incomplete combustion	detector, electrochemical detector, infra-red analyser	0.04	
Nitrogen oxides	NO_2	1.5895	Faint smell, colourless, relatively insoluble in water	Irritant	Blasting, diesel engine, incomplete combustion	Odour, colour, detector tube, electrochemical detector	TLV-C– 0.005	0.005
	NO	1.04	Distinct smell, reddish brown colour, very soluble in water	Irritant, poisonous			TLV-TWA– 0.0025 TLV-C– 0.0035	
Hydrogen sulphide	H_2S	1.1912	Rotten egg odour, colourless, acid taste	Irritant; also poisonous; explosive	Strata water, strata gas, blasting	Odour, detector tube, electro-chemical detector	TLV-TWA– 0.001 TLV-STEL– 0.0015	0.1 (4–44 explosive)
Sulphur dioxide	SO_2	2.2636	Irritating odour, colourless, acid taste	Irritant, poisonous	Diesel engine, blasting of certain sulphide ores, combustion of sulphide ore, fire	Odour, detector tube	TLV-C– 0.0005	0.1

Firedamp is CH_4 + air (explosive).
Blackdamp is $CO_2 + N_2$ + air (asphyxiant).
Afterdamp is $CO + CO_2 + CH_4 + O_2 + N_2 + H_2$ (follows fire or explosion; asphyxiant–chemical).
Whitedamp is CO + air (asphyxiant–chemical).
Stinkdamp is H_2S + air (irritant as well as poisonous).
Bottom gas is $CO_2 + CH_4$ (asphyxiant as well as explosive).

oxygen in the air. Oxygen may be decreased to two-thirds of its normal percentage in air before appreciable symptoms develop, i.e. the concentration of a simple asphyxiant should be 33 % in a mixture of air and gas. When the concentration of the simple asphyxiant reaches 50 %, marked symptoms are produced. A concentration of 75 % is fatal in a matter of minutes.

The first symptoms produced by asphyxiant gases are fast respiration and hunger for air. Mental alertness is decreased and muscular coordination becomes worse. Later, judgement becomes faulty and all sensations are depressed. Emotional instability often results and fatigue occurs rapidly. With time, there may be nausea and vomiting, prostration, i.e. laying flat on ground, loss of consciousness, and finally, convulsions, deep coma and death.

Carbon monoxide is a chemical asphyxiant. Its affinity for haemoglobin is about 210 times that of oxygen. By combining with the haemoglobin, it temporarily deprives the oxygen-carrying capacity of the latter, i.e. the effect on the body is therefore predominantly one of asphyxia. The presence of carboxyhaemoglobin in the blood also interferes with the dissolution of the remaining haemoglobin. The erythrocytes (red blood cells) do not appear to be damaged in any way by saturation with the gas and can readily resume their function as oxygen carriers.

The concentration of carbon monoxide in the blood depends on the concentration of carbon monoxide in air, the length of exposure and the

Fig. 3.1. The effect of various concentrations of carbon monoxide in air on the percentage of carboxyhaemoglobin present in persons engaged in light activity (after Spencer and Lawther, 1972).

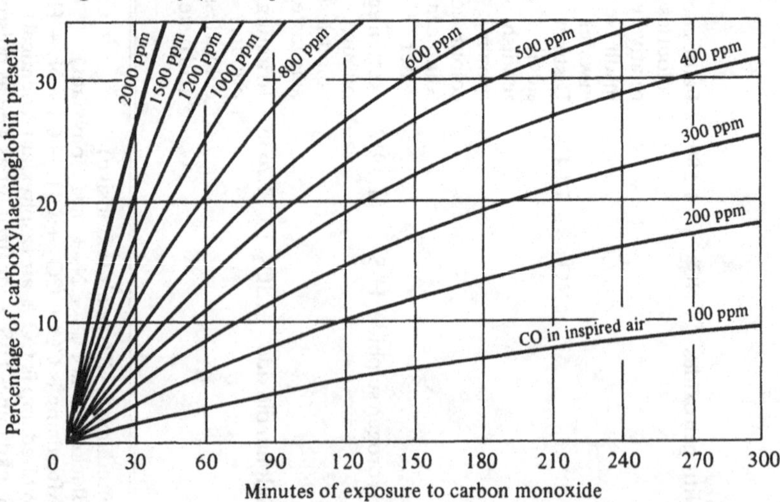

type of activity undertaken during exposure. An individual's response will vary depending upon age and general physical condition. The effects of various concentrations of carbon monoxide in air on the percentage of carboxyhaemoglobin for persons engaged in light activity are given in Fig. 3.1. The following equation can also be used to calculate approximately the percentage of carboxyhaemoglobin present on exposure to small concentrations of carbon monoxide over short periods (Spencer and Lawther, 1972):

$$b = \frac{4ate}{100} \tag{3.1}$$

where b = carboxyhaemoglobin content in blood, %;
a = concentration of carbon monoxide in air, ppm;
t = time of exposure, hours; and
e = a factor depending upon the type of activity undertaken during exposure, 1 for rest, 2 for walking, 3 for working.

The effects of various concentrations of carboxyhaemoglobin in blood are given below:

Below 20 %	No symptoms
20 to 30 %	Throbbing headache
30 to 50 %	Dizziness, nausea, muscular weakness, danger of collapse
50 % and above	Unconciousness and death

Irritant gases

These gases induce inflammation in tissues such as the skin, conjunctiva of the eyes, the membranes of the respiratory tract when they come into contact with them. If the gases are not sufficiently soluble to be absorbed into the moist tissues of the upper respiratory tract, they enter the lungs, causing inflammation leading to pulmonary oedema, i.e. exudation of the fluid in the lungs which may lead to suffocation. Nitric oxide, nitrogen dioxide, hydrogen sulphide, partially oxidised hydrocarbons (aldehydes) and sulphur dioxide are the most common irritant gases found in mines. In the case of nitrogen dioxide, inflammation of the lungs can occur and is of the greatest concern to health.

Poisonous gases

These gases destroy the tissues with which they come into contact. Nitric oxide, hydrogen sulphide and sulphur dioxide are not only irritant but also poisonous.

Threshold limit values (TLVs)
These values refer to concentrations of toxic agents in the air and represent conditions to which workers may be repeatedly exposed day after day without adverse effects. There are three categories of TLVs:

(1) time-weighted average (TLV-TWA);
(2) short-term exposure limit (TLV-STEL);
(3) ceiling (TLV-C).

The time-weighted average is based on a conventional 8-hour work shift or 40-hour work week and allows concentrations above the TLVs for short periods if the time-weighted average is not exceeded. The TLV-TWA is commonly referred to as the TLV. The short-term exposure limit is the maximum concentration for exposures up to 15 minutes without adverse effects provided that (i) no more than four exposures per day are permitted, with at least 60 minutes between exposures and (ii) the daily TLV-TWA is not exceeded. The ceiling is the concentration that should not be exceeded even for a moment.

The American Conference of Governmental Industrial Hygienists has published TLVs for a large number of industrial materials (ACGIH, 1979).

3.2.2 Explosion hazard

Some gases are explosive. For example, a methane–air mixture is explosive when the methane concentration in air is between 5.3 and 14 %. Below and above these figures, methane will burn if ignited. The other explosive gases found in mines are:

Carbon monoxide	12.5 to 74 % explosive range
Hydrogen sulphide	4 to 44 % explosive range

3.3 Occurrence and control of mine gases

Harmful gases in mines commonly occur on the following occasions:

(1) blasting;
(2) using diesel engines;
(3) mining coal;
(4) underground explosions and fires;
(5) oxidation of timber, loose coal, pyrites, etc.;
(6) drilling through into stagnant water.

The first two are discussed separately. The third is discussed in Chapter 4. Explosions and fires in mines generally produce, in addition to carbon dioxide, large amounts of carbon monoxide, owing to incomplete combustion. Oxygen deficiency can also occur. These topics are discussed in Chapters 8 and 9.

Decaying timber slowly combines with oxygen, giving off carbon dioxide often through the agency of fungi. Under wet conditions, some methane may also be formed. Loose coal, pyrites, etc. also oxidise. In places where there is no air circulation, oxygen deficiency may develop. In extreme cases, there may not be sufficient oxygen to support life.

Drilling or 'holing' through into old workings containing water often results in the liberation of hydrogen sulphide accumulated in the stagnant water by action on pyrites.

The techniques selected to keep gases within acceptable levels depend on the source of the gas and nature of its occurrence. The general techniques used in the prevention of exposure of persons to mine gases are as follows:

(1) prevention of formation of gases;
(2) prevention of exposure of persons;
(3) dilution of gases;
(4) removal of gases.

3.4 Detection and monitoring of mine gases

Mine regulations in various countries require frequent testing for the presence of toxic as well as explosive gases in air to determine whether or not the atmosphere complies with the standards laid down for safety and health. For rapid, on-the-spot approximate determinations, hand-held portable instruments are used. For the continuous monitoring of mine gases, portable instruments are mounted at strategic locations, usually sounding an alarm when the pre-set threshold of the gas is exceeded. (For example, continuous miners, used extensively with the bord and pillar method, are fitted with methane monitors. If the methane concentration reaches 1 %, a warning light operates on the control panel alerting the operator that methane is present in small quantities. When it reaches 2 %, the monitor disconnects the electric power to the miner, and prevents resetting until the concentration is reduced to an acceptable level.) Also, stationary instrumentation can be used with a telemetry system to transmit the information from several stations to the surface or to a central point underground to enable mine management to take appropriate action without delay. (In some coal mines using the longwall method, the methane-monitoring system is programmed so that when the methane concentration exceeds the statutory limit, the power supply disconnects in the area concerned.) Complete and exact determinations of the nature and amount of gases present are made by the laboratory analysis of samples collected underground. In recent years, devices that can be worn by a miner have been developed for personal monitoring and dosimetry.

3.4.1 Sampling mine air for laboratory analysis

The sampling consists of collecting a representative portion of the mine air in containers and may be done by one of four methods, namely, water displacement, air displacement, evacuation and compression.

Water displacement method

The bottle is filled with water and the water emptied out at the place to be sampled. Air replaces the water in the bottle and the bottle is corked. Some gases which are soluble in water dissolve in the drops left behind in the bottle and this may affect the results. For routine sampling of carbon dioxide, however, this method is quite suitable.

Air displacement method

The air in a dry bottle is displaced by pumping the air from the place being sampled into the bottle by means of bellows or a pump. The volume of air pumped in should be about 10 times the capacity of the bottle to make sure of complete replacement of the air originally in the bottle, although there is always some doubt as to whether or not the original air in the bottle is completely replaced.

Evacuation

The sampling bottle is fitted with a glass stopcock and evacuated with a suction pump on the surface. The sample is taken by opening the stopcock. The bottle may be checked with a vacuum gauge just before the sample is taken to ensure that the proper vacuum still exists. Bottles with a nominal capacity of 250 cm^3 are normally used but these are large and cumbersome and limit the sampling capacity. Evacuated glass vials of 10 cm^3 capacity stoppered by 6 mm thick butyl rubber septums (commonly used for routine blood sampling) have been tested by the US Bureau of Mines and the US Mining Enforcement and Safety Administration as sample collectors. They have given results comparable with those obtained with conventional bottle samplers of 250 cm^3 nominal capacity.

Compression

A small, double-acting pump is used to compress the air being sampled into a small steel cylinder at high pressure. This not only increases the volume sampled but also facilitates the injection of the sample into the equipment used for analysis. The pump can be fitted with a rubber tube to assist when traversing the roadway to obtain a mean sample. A number of these cylinders can easily be carried in a pocket.

The above methods take an instantaneous or near instantaneous sample of the air, i.e. a spot sample. To obtain the time-weighted average concentration, a number of samples has to be taken over a period of time. To overcome the disadvantages of spot sampling, the US Bureau of Mines has developed a mechanically driven sequential sampler that can automatically take air samples for 1 to 8 days. The sampler (Fig. 3.2) consists of a spiral of plastic tubing wound around a cylinder, which is tilted so that one end of the tube is submerged in a pan of water. The other end of the cylinder is connected to a spring-wound motor which rotates the cylinder at set intervals of once a minute to once an hour. A sample is taken every revolution. As the cylinder rotates, the low end of the tubing comes out of the water for half a revolution of the cylinder. As the end of the tube submerges again, a plug of air is trapped in the tube between two plugs of water, and successive air samples are automatically spiralled upward in the tubing as the cylinder rotates. Clamps are used to prevent the samples from mixing when the tubing is taken from the mine to the laboratory for analysis. A syringe is used to draw samples from the plastic tube. The samples should be analysed as quickly as is practicable otherwise some gases may be lost by diffusion from the plastic tubing. If necessary, glass tubing can be used instead and the samples transferred to plastic tubing in the laboratory, where they can be removed by a syringe.

Fig. 3.2. Sequential air sampler (after Anon., 1977).

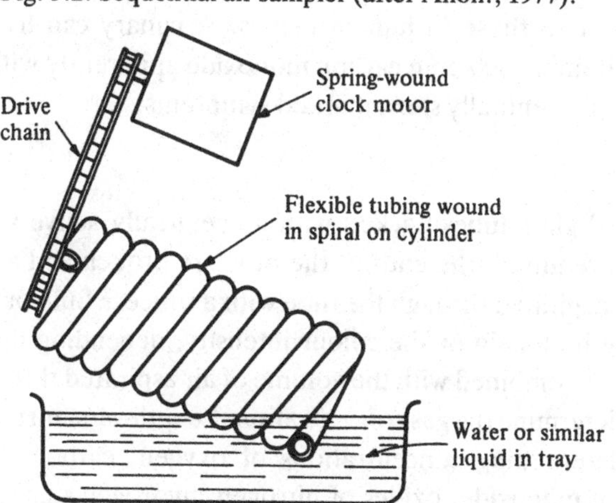

3.4.2 *Instruments for detection and monitoring*

Flame safety lamp

The flame safety lamp is the oldest type of detector still used for determining methane concentrations. The flame, when appropriately adjusted, shows a faint blue cap if methane is present. The ratio of the height of the gas cap from the base of the flame to the overall width of the flame is used to measure the percentage of methane in air. The conventional lamp shows concentrations of 1.25 to 5 %. Safety lamps using the Garforth principle in which the samples are injected into the lamp can be used to detect methane concentrations of up to 20 %. A trained and experienced person is required to use the flame safety lamp for measuring the methane concentrations. Where there is permanent lighting, it cannot be used.

The flame safety lamp can also be used for detecting oxygen deficiency as well as for determining concentrations of carbon dioxide approximately. The lamp extinguishes in the range of 14 to 17 % oxygen and at about 3 % carbon dioxide.

Canaries

Canaries are still used in some coal mines for detecting carbon monoxide and are a reliable method of warning of high concentrations. A canary's blood attains equilibrium a great deal faster than that of humans. The concentration of carboxyhaemoglobin in the blood levels off in about 20 minutes, whereas the corresponding period for a resting man exceeds 6 hours. When air contains 2500 ppm of carbon monoxide, a canary will fall off its perch in about 4 minutes; a resting man will probably feel no effects in 20 minutes. At low concentrations, however, it takes approximately two or three times as much carbon monoxide to produce symptoms in a canary equivalent to those in human beings. A canary can live in an atmosphere containing 500 ppm carbon monoxide apparently without ill effects; a man will eventually show marked symptoms.

Detector tubes

These are sealed glass tubes packed with a chemically active granular material. When required, the ends of the tube are broken and a known volume of air is aspirated through the tube with a squeeze bulb or special pump. The length of stain or the colour intensity, depending upon the type of tube used, combined with the volume of air aspirated through the tube is used to determine the gas concentration. Length-of-stain tubes are available for determining concentrations of oxygen, carbon dioxide, methane, carbon monoxide, oxides of nitrogen, hydrogen sulphide and sulphur dioxide. For carbon monoxide, colour-change tubes are also available. Each manufacturer has a method or calibration scale for use with their detector tubes.

Detector tubes for oxides of nitrogen usually measure either NO or NO_x, i.e. NO + NO_2. The NO_2 detector tube has an oxidising material at one end. If the sample is aspirated from that end, the NO will be oxidised and so the tube reads total oxides (NO_x). Therefore, by using two tubes, one aspirated in each direction, both NO and NO_x can be assessed.

Filament instruments

There are two types of instruments: (1) thermal conductivity instruments and (2) catalytic combustion instruments.

Thermal conductivity instruments. The thermal conductivity of a sample of air containing a gas, for example methane, differs from that of air or other reference gas. An electrically heated wire immersed in the sample cools at a rate different from its cooling rate in the reference gas. This principle is used for measuring the gas concentration by forming a Wheatstone bridge circuit in which two physically similar wires serve as detector elements, one immersed in the sample to be tested and the other in the reference gas. With the reference gas element resistance held constant, the change of the resistance of the element in the sample unbalances the bridge in proportion to the concentration of the gas present in the sample. Handheld detectors are available for methane.

Catalytic combustion instruments. These instruments are used for measuring the concentration of combustible gas such as methane in air. They also use two heated elements in a Wheatstone bridge circuit. In this case, however, the elements are coated. One element has a platinum or palladium compound to catalyse the combustion of combustible gases; the other, the comparison element, is coated with an inert compound to inhibit combustion. In these instruments, both elements are immersed in the combustible gas–air mixture. Combustion takes place on the surface of the active element with resultant heating which increases its resistance, giving rise to an out-of-balance current in the bridge which is calibrated in terms of the percentage of combustible gas. These instruments become insensitive when exposed to high concentrations of combustible gas. Instruments are available specifically for methane. However, above about 10 % methane, or in oxygen-deficient atmospheres, the instrument gives lower readings. Handheld, continuous miner-mounted (monitors), portable as well as stationary laboratory instruments are available for methane, using the principle of catalytic combustion.

Paramagnetic instruments
Portable as well as stationary laboratory instruments using the paramagnetic property are available for oxygen. Air is drawn past a magnetic field, disturbing the magnetic flux while an indicator shows the amount of oxygen present.

Optical instruments
There are three types of instruments in this category:

(1) infra-red radiation;
(2) light emission from chemical reaction, i.e. chemiluminescence;
(3) interference fringes.

Infra-red radiation instruments. These instruments use the basic principle of a source emitting infra-red radiation through a sample cell, through which the air sample flows, to an infra-red radiation detector. When a gas which absorbs infra-red radiation is present in the sample, the radiation reaching the detector (and thus its output) is reduced. At present, portable as well as stationary laboratory instruments are used for methane, carbon monoxide and carbon dioxide. They can provide digital readout, continuous monitoring and alarms at predetermined set points. Personal alarms are available for carbon monoxide, using solid-state (semiconductor) detectors.

Instruments using chemiluminescence. At present, instruments utilising chemiluminescence are available for laboratory analysis of NO, NO_2 and ozone. In an NO/NO_2 instrument with an NO_2 to NO converter, NO reacts with ozone, producing NO_2 and excited oxygen, which then decays producing light. The amount of light produced is proportional to the NO concentration. These instruments produce ozone and use high-voltage photodetectors.

Instruments using interference fringes. These instruments are based on the difference in the index of refraction between two gases. Interference fringes result when light from a source is split into paths of slightly unequal length and then recombined. If the sample cell is included in one of the light paths, changes in the refractive index will alter the optical path length and alter the position of the fringe. This movement can be used as a measure of the refractive index change. Handheld portable instruments using this principle, in which the fringes are observed on a scale, are available which can be used for the detection of combustible gases, carbon dioxide and so on.

Gas chromatography
In the gas chromatographic method, a slug of air is passed through a long tube by a carrier gas, usually helium. The tube contains absorptive

materials, the effect of which is to allow different gases in the sample to travel the length of tube at different rates. The rate at which each gas passes through the tube identifies the gas. The quantity is measured by a suitable non-specific detector such as a hydrogen flame ionisation detector, a thermal conductivity detector, a flame photometric detector or a radioactive ionisation detector.

Gas chromatography is the primary method for laboratory analysis. Up to eight components can be separated and measured. It requires about 7 minutes to detect normal constituents in coal mines (oxygen, nitrogen, carbon dioxide and methane) and an additional 3 minutes to detect carbon monoxide.

Electron-capture instruments

As the air or gas passes into the chamber containing the α-source it gets ionised, giving a steady background ion current. Gases of interest are converted into small particulate or aerosol form by bubbling through a reagent. When these particles pass through the chamber, they collect ions and reduce the current. The reduction in ion current is proportional to the concentration of the gas present and an electrometer amplifier converts this to a usable readout. Portable instruments are available for the detection of NO_2, ammonia, chlorine or HCl.

Electrochemical instruments

The basic instrument has a cell with two electrodes immersed in an electrolyte. When the sample gas reacts chemically with the electrolyte, or electrolytically at the electrode–electrolyte interface, the electrical characteristics of the cell are altered and a current or voltage is derived which can be related to the gas concentration. Portable as well as on-line electrochemical instruments are available for oxygen, carbon monoxide, hydrogen sulphide, sulphur dioxide and oxides of nitrogen. For oxygen, personal indicators also are available. Electrochemical cells wear out and a replacement is required every three to six months of continuous use.

Catalytic oxidation instruments

Portable instruments are available for detecting carbon monoxide. In these instruments, the air sample is passed through a bed of Hopcalite (a mixture of cupric and manganese oxides) which oxidises carbon monoxide to carbon dioxide, producing an amount of heat proportional to the amount of carbon monoxide present. A thermocouple or thermistor measures this heat and converts it to a voltage, which is read out on a sensitive voltmeter calibrated to read directly in parts per million of carbon monoxide. Since Hopcalite is affected by moisture, the sample must be passed through a drier before it is drawn through the Hopcalite. One disadvantage of this instrument is that it requires replacement of Hopcalite at infrequent intervals.

3.5 Gases from blasting and their control

3.5.1 Gases from blasting

Gases resulting from blasting are principally carbon dioxide, nitrogen and steam. However, toxic gases, including carbon monoxide and nitrogen oxides, also result from any blasting. Toxic gases are also called fumes in the explosives industry. In addition to fumes, smoke also results, which is composed mainly of steam and the solid products of combustion. Although smoke is non-toxic, excessive exposure to smoke, especially that produced by dynamite, can cause severe headaches and should be avoided. The main constituents of blasting fumes and their quantities depend on the chemical and physical properties of the explosives and the conditions of use. The factors that increase fumes include poor explosive formulation, inadequate priming, insufficient water resistance, lack of confinement, inadequate burden, reactivity of the explosive with the rock (appreciable quantities of hydrogen sulphide may be produced on blasting in heavily pyritic rock), and incomplete detonation. Depending on the type of explosive and the conditions under which it is detonated, 100 kg of explosives can produce approximately the following amounts of gases:

Carbon dioxide	10 to 27 m^3
Carbon monoxide	1.2 to 4.0 m^3
Nitrogen oxides	0.6 to 4.4 m^3
Ammonia	0.03 to 0.3 m^3

3.5.2 Control of gases from blasting

As the formation of blasting fumes cannot be prevented, every effort must be made to see that miners are not exposed to it. Various regulations exist to prevent the exposure of miners to the products of blasting, e.g. blasting must be done at the end of each shift, at set times, so that all other personnel can move beforehand to fresh air; a certain time must elapse for the fumes to clear before anyone re-enters the working places; and adequate ventilation must be provided.

The system of ventilation is as important as the volume flow rate of air. In a study conducted by Nicholas and Wall (1971), exhausting ventilation has been found to be the best. Forcing ventilation has been found to be the worst. There has been little improvement with the forcing overlap system. In the case of coal mine headings which are gassy, a main exhausting ventilation with a forcing overlap, i.e. an exhausting overlap system, may be more suitable for keeping the face of the heading free from methane and blasting fumes.

Observations in tunnels indicate that fumes extend to a distance of about 20 m from the face (Szechy, 1976). Hence the volume flow rate of fresh air required can be approximately calculated using the following relationship:

$$Q = \frac{20AN}{t} \tag{3.2}$$

where Q = volume flow rate of fresh air required, m^3/s;
A = cross-sectional area of the tunnel, m^2;
N = number of times air is to be exchanged, around five; and
t = time within which the fumes should be diluted, s.

It is also important to see that the air velocity in the tunnel is within the range of 0.7 to 1.0 m/s for rapid diffusion.

The nitrous fumes from blasting can be reduced considerably by the spray from a water blast, which dissolves all of them fairly readily except NO. Some of the fumes can be trapped under the broken rock and might be liberated when the rock is loaded. Proper watering down will remove most of these gases.

Up to about 100 kg of explosives are used for a blast in a large development end in a mine. When such an end is worked on the multi-shift blast system, the ventilation air sometimes has to pass through other areas in which people are working. In such cases both the fumes and the dust have to be removed from the air by means of a fume–dust filter (Rabson, Quillium and Goldblatt, 1960). The air from the development end is drawn slowly (<15 m/minute) through a 1 m thick bed of exfoliated vermiculite thoroughly sprayed with a solution of potassium permanganate (5 % by weight) and sodium carbonate (5 % by weight) in water. The NO is oxidised by the potassium permanganate and the NO_2 thus formed is absorbed by the sodium carbonate solution. The air then passes through a dust filter.

To minimise the formation of toxic fumes, it is advisable to carry out the following practices. The largest diameter cartridges of explosive consistent with the work to be done should be used. Explosives that have obviously deteriorated or are damaged should not be used. The explosives should not be removed from the cartridges. Overcharging of the holes should not be done. Paper or other combustible material should not be added to the charge. All conditions that may cause burning rather than detonation should be avoided. In wet holes, explosives having adequate water resistance should be used. Firing should take place as soon as practicable after loading. The charge should be confined with incombustible stemming.

3.6 Gases from diesel engines and their control

3.6.1 Exhaust emission from diesel engines

Diesel fuel consists of 85–86 % carbon, 13–14 % hydrogen and 0.05–0.7 % sulphur, all by mass. For every kg of diesel fuel, 15 kg of air are required for complete combustion, giving about 6.4 m^3 of exhaust products. Under an ideal fuel–air mixture and oxidation, the exhaust would consist of approximately 73 % nitrogen, 13 % carbon dioxide and 13 % water by volume. The remaining 1 % of exhaust emissions produced under imperfect or actual engine conditions consist of unburned hydrocarbons,* partially oxidised hydrocarbons (includes aldehydes responsible for odour), carbon monoxide, smoke (particulates),† oxides of nitrogen and sulphur dioxide.

Typical concentrations of the principal components present in the exhaust emissions of a diesel engine are as follows:

	Range	Mean
Carbon dioxide	9.0 to 13.8 %	12.5 % by volume
Sulphur dioxide	Very small	0.02 % by volume
Hydrocarbons	0.01 to 2.5 %	1.14 % by volume
Carbon monoxide	0.01 to 0.36 %	0.04 % by volume
Particulates	33 to 200 mg/m^3	
Oxides of nitrogen (usually 90 % are NO)	0.007 to 0.33 %	0.0735 % by volume

Factors affecting the exhaust emission characteristics

Several factors affect emissions, including the size of the engine, design and operational factors. The combustion chamber configuration, the duty cycle and the altitude of the mine are also important parameters. The

* Noted for their nuisance value, haze and fogging potential and surface contamination.

† Smoke generally means the visible choking fumes as they issue from the exhaust pipe. Particulates are the same materials after they have equilibrated with the ambient air and have absorbed any other toxic material from the cooled exhaust. They are essentially carbon at high loads, but they contain a significant percentage of hydrocarbon material at low and medium loads. Particulates constitute an irritation and odour problem and are of greatest concern to health. They can contribute to health effects by carrying the irritants to the deeper, more sensitive recesses of the lung, where they can act to initiate, promote, or accelerate emphysematous, fibrotic or carcinogenic processes.

quality of the fuel, the use of fuel additives, the composition of the intake air and the type of maintenance influence exhaust emissions, acting independently or in combination.

Diesel exhaust emissions are highly dependent upon the ratio of fuel to air. An increase in the amount of fuel injected increases the power but more oxides of nitrogen, carbon monoxide and smoke are produced. By limiting the amount of fuel injected into the engine, i.e. derating the engine, these emissions can be reduced below a given limit. This, however, reduces the power output of the engine.

Indirect-injection engines* produce significantly less oxides of nitrogen (about 50 % of that of the direct-injection engines) by lowering peak combustion temperatures as well as the incomplete combustion products by improving fuel spray distribution in the pre-chamber and increasing turbulence in the main chamber. Retarding fuel injection reduces oxides of nitrogen but increases slightly incomplete combustion products. Turbocharging† improves the input, raises the combustion temperature and enhances combustion, with the result that the power is increased. As the turbocharging pressure increases, the emission of incomplete combustion products drops. Turbocharging pressure has little influence on the emission of NO but increases the NO_2 emission. Two-cycle engines produce more hydrocarbons and smoke than four-cycle engines. Both engines produce about the same amounts of carbon monoxide and oxides of nitrogen. The amount of sulphur dioxide produced is directly proportional to the sulphur content in the fuel.

As the methane content of the intake air increases, the emission of carbon monoxide increases and at the same time the production of NO is slightly reduced. With respect to NO_2 emission, it decreases in direct-injection engines and increases in indirect-injection engines as the methane content increases. Even after the increase, the indirect-injection engines produce lower NO_2 emissions than direct-injection engines.

* Two systems of fuel injection are common for diesels, the direct-injection (DI) and indirect-injection (IDI) systems. In the DI system, the fuel is injected at high pressure into the cylinder. The intake air moves in a 'squish' (radial flow) motion as the piston rises in the compression stroke. In the IDI system, there is a pre-chamber (communicating with the main chamber) and the fuel is injected into it, where combustion takes place. Fuel-injection pressures are lower and turbulent air motion is much more violent in the IDI case.

† Diesel engines can be either naturally aspirated or turbocharged. Turbocharging can increase the engine output by up to 100 %. A turbocharger is a compressor, powered by the exhaust gas energy, which pressurises the intake air. This results in a larger mass of air in the cylinder and permits a proportional increase in the injected fuel volume.

The concentration of all the undesirable emissions increases with load. In general, nitrogen oxides form when high combustion temperatures are reached, whereas this condition produces smaller quantities of other pollutants. If the engine is not properly maintained, the engine produces a lot of incomplete combustion products, i.e. smoke, aldehydes, hydrocarbons and carbon monoxide, but the levels of oxides of nitrogen decrease.

Health hazards from diesel emissions

The exhaust components of greatest health impact are identified as oxides of nitrogen and particulates (Table 3.2). For adequate protection of workers' health, concentrations of toxic agents in diesel exhaust have to be kept below their respective TLVs (Table 3.1).

3.6.2 *Control of exhaust emission*

For each kW delivered, a typical diesel engine produces about 0.0006 m^3/s of exhaust. The toxicants present in typical exhaust fumes are at concentrations well above TLVs. Control is effected by adding diluent air in quantities sufficient to lower the concentration of the toxicants in the resultant mine air, or by reducing the offensive toxicants from the exhaust to safe levels. Dilution by ventilation is the simpler, and, in most cases, the more practical of the two.

Table 3.2. *Health severity score of diesel emission components*

Diesel component	Health severity score
CO_2	1
SO_2	1 (2)
Hydrocarbons	1 (carcinogenic potential)
Aldehydes	1
CO	1 (2)
Particulates	3 (carcinogenic potential)
NO	1
NO_2	3 (2) (1)

Key: 3 – greatest concern to health;
2 – lesser health threat;
1 – least harmful substances;
( ) – represents minority viewpoints.

Dilution by ventilation

The minimum ventilation requirement depends upon the engines, but some generalisation can be made. The ventilation requirement is basically a function of the engine power. For example, concentrations of the principal pollutants present in the exhaust of a diesel engine are as follows:

Carbon dioxide	100 000 ppm
Carbon monoxide	2 000 ppm
Nitric oxide	900 ppm
Nitrogen dioxide	100 ppm
Sulphur dioxide	215 ppm

The air volume flow rate required to reduce the concentration of any gas to acceptable levels can be calculated as follows:

$$Q = \frac{c_E \times q}{(c_{TLV} - c_a)} \tag{3.3}$$

where Q = volume flow rate of air required in m^3/s per kW;
c_E = concentration of gas in exhaust, ppm;
q = exhaust quantity, m^3/kW;
c_{TLV} = TLV of gas, ppm; and
c_a = ambient concentration of gas in normal air, ppm.

The ambient concentration of carbon dioxide and carbon monoxide in fresh air are 2000 ppm and 10 ppm, respectively. To dilute the exhaust carbon dioxide to the acceptable level of 5000 ppm, the volume flow rate of air required will be

$$\frac{100000 \times 0.0006}{5000 - 2000} = 0.02\ m^3/s \text{ per kW}.$$

In the same way, 0.03 m^3/s per kW are required to dilute the exhaust carbon monoxide to an acceptable level; nitric oxide dilution requires 0.0216 m^3/s per kW, nitrogen dioxide dilution requires 0.012 m^3/s per kW; and sulphur dioxide dilution requires 0.0258 m^3/s per kW. These requirements are not additive. Therefore, only the highest one need be met, i.e. 0.03 m^3/s per kW.

Exhaust emissions as released from the tailpipe require dilution in some manner to protect the machine operator. This dilution may be accomplished by discharging the exhaust into the engine cooling fan blast, or by pumping dilution air into the exhaust gas in some kind of mixing chamber. Fig. 3.3 shows a fume diluter. It has a dilution ratio of about 10:1 and the diluted gases are directed away from the driver. No appreciable quantities of noxious gases can be detected at a distance of 1.2

to 1.5 m from the point of emission. It also collects exhaust particulate matter but by cleaning it regularly, engine back-pressure can be minimised.

Fan-powered fume diluters in association with a thermal gradient in the drift have also been used in recent years to direct the diluted exhaust away from the face.

Engine emission reduction

Although ventilation can be used to dilute toxicants to safe levels from virtually any initial concentration in the exhaust, the hazard from toxicants can be reduced by limiting the toxicant emissions into the mine atmosphere. There are several possibilities for controlling diesel emissions at the source. One or more of the emission control systems have helped to reduce toxic emissions but oxides of nitrogen and particulates do present problems.

Exhaust recirculation. Controlled recirculation of a fraction of the exhaust (20 %) into the engine intake has been found to be effective in reducing by a half the amount of nitrogen oxides produced. This, however, does present problems by way of increased production of incomplete combustion products. It is also less desirable since smoke also recirculates and fouls the intake tract. This controlled recirculation is justified if the engine is equipped with both a catalytic purifier and a water scrubber.

Water scrubber. Water scrubbers (Fig. 3.4) trap a portion of the particulates, hydrocarbons, aldehydes, sulphur and nitrogen dioxides. Because sulphur and nitrogen dioxides form acids, stainless steel construction and buffering chemicals* are often recommended to reduce corrosion. Carbon monoxide and nitric oxide are not very soluble in water and are not absorbed in any significant quantities. These scrubbers should have an adequate capacity of water (about 3.4 litres/kW) so that the replenishment or topping of water is required only once in a working shift. The typical retention of various pollutants in a water scrubber are as follows:

Particulates	30 %
Hydrocarbons	20 to 25 %
Aldehydes	50 to 85 %
Sulphur dioxide	50 to 80 %
Nitrogen dioxide	20 %

* A solution consisting of 10 % sodium sulphite and 0.5 % hydroquinone in water may be used instead of water. This also improves retention of the aldehydes.

Fig. 3.3. General arrangement of diesel vehicle exhaust fume diluter (after Fermor, 1975).

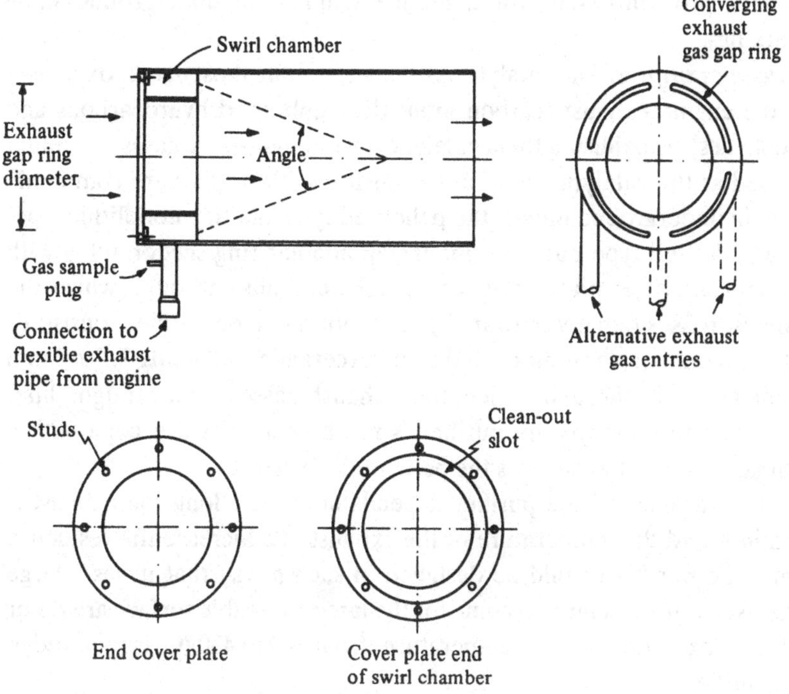

Fig. 3.4. General arrangement of a two-cell water scrubber for an engine with a power range of 50–100 kW (after Fermor, 1975).

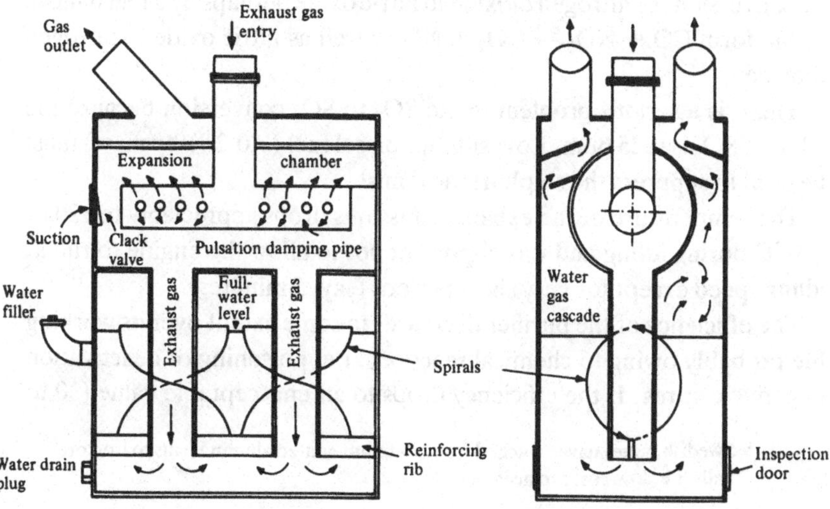

The exhaust leaving the scrubber is water-saturated and could contain water droplets carried out of the scrubber. This water could create fogging problems that would reduce visibility in underground mine workings.

Catalytic purifiers. The catalytic purifiers eliminate most of the toxic gases in the engine exhaust (carbon monoxide, unburned hydrocarbons and aldehydes) by oxidising them catalytically into non-toxic gases. Platinum is used as the catalytic agent in the purifiers. Two types are commonly used in underground mines: the pelletised type and the monolithic type. The pelletised-type purifier contains an annular ring section filled with 6.4 mm diameter platinum-covered alumina balls* through which the exhaust passes radially outward. The monolithic-type purifier consists of a honeycomb of platinum-coated alumina ceramic containing 2.5 to 3 mm diameter cells through which the exhaust passes in a straight line. Although the honeycomb purifier is much smaller it can cope with a similar volume of exhaust as the pelletised purifier.

The efficiency of the purifier depends upon how long the exhaust is within it and the temperature of the exhaust. To increase the residence time, the purifier should be designed in such a way that it has a large cross-section and a large volume for the largest possible surface area. For efficient oxidation, a high temperature, from 260 to 430 °C, is required in the purifier.

The catalytic purifiers convert carbon monoxide to carbon dioxide (50 % conversion at 260 °C and 85 % conversion at 450 °C) and oxidise most of the hydrocarbons (90 %) and aldehydes (85 %). They do not significantly alter smoke or nitrogen dioxide concentrations. There could be conversions of nitrogen dioxide to nitric oxide perhaps by a mechanism of the form $CO + NO_2 \rightarrow CO_2 + NO$ as well as nitric oxide to nitrogen dioxide.

There is a serious problem in the SO_2 to SO_3 conversion because the TLV of SO_3 is 0.25 ppm. Low-sulphur diesel fuel (<0.2 % by mass) must be used to suppress the sulphuric acid mist.

The temperature of the exhaust emissions drops appreciably to 120 or 150 °C during idling and it is important not to allow the engine to run at idling speed except for very short periods (say, 5 minutes).

The efficiency of the purifier decreases to some extent over its working life probably owing to chemical reactions, i.e. poisoning or deactivation of active centres. If the efficiency drops to an unacceptable value (30 to

* Swedish mines have switched from platinum-coated alumina balls to iron ore balls, i.e. low-cost replacement.

50 %), the catalyst can be regenerated either thermally or chemically. The effective life of the catalyst varies widely and 10 000 hours may be taken as a guide.

Exhaust filters. Stainless steel–fibreglass filters have high efficiencies for trapping diesel exhaust particulates (80 to 90+ %). Engine back-pressure has been found to be acceptable with these filters.

Water injection or water/diesel fuel emulsions. A reduction in the emission of nitric oxide can be achieved by these means. Water contamination in the engine will eventually cause serious maintenance problems. With emulsions, stability is also of concern.

3.6.3 Maintenance of diesel equipment

Maintenance of diesel equipment or in many cases the lack of it is a matter of major concern in mines. Due attention should be paid to:

(1) air intake system – flame trap, air filter;
(2) exhaust system – water scrubber, catalytic purifier, exhaust filter;
(3) engine – valve clearance, stall speed rpm, full-speed no-load rpm, fuel injector, oil filter;
(4) fuel system – clean fuel, fuel pre-cleaner and filter.

It is a statutory requirement in mines that all diesel engines should have the raw engine emission tested regularly at a given interval (e.g. every 28 days) and from this regular testing it is possible to note, on a progressive basis, any increase in the various pollutants from the engine and suggest a cause. For example, a sudden increase in carbon monoxide would suggest either an intake air blockage or a problem with the injectors. An increase in carbon monoxide combined with black smoke would suggest air intake troubles. An increase in carbon monoxide with an increase in aldehydes would suggest that the injectors need to be examined. An increase in oxides of nitrogen would suggest that an examination of the fuel injection timing system is needed.

3.6.4 Regulations governing the use of diesel engines underground

Regulations promulgated by legislative authorities limit the concentrations of certain constituents in the raw exhaust and mine air and define calculation methods to determine the minimum ventilation requirements which must be supplied to the machine when operating.

Diesel engines, before they can be accepted for use in underground mines, must be certified by the testing authority of the country. The Mine Safety and Health Administration requirements for the use of diesel equipment in underground mines in the USA are given in Table 3.3.

Table 3.3. *Mine Safety and Health Administration requirements for diesel equipment*[a]

	Maximum allowable fuel/air ratio	Undiluted exhaust gas		Diluted exhaust gas				Quality of ventilating air[b]					Permissible gases in working environment			
	CO	CO	NO_x	CO	CO_2	NO_x	Aldehydes	CO	CO_2	NO_x	O_2	Inflammable gas	CO	CO_2	NO_2	O_2
Vehicle type																
Permissible diesel locomotives used in coal mines (gassy and dusty)	≤0.25	≤0.25	—	≤0.01	—	≤0.0025[c]	≤0.001[d]	—	—	—	≥20	≤0.25	≤0.01	≤0.5	≤0.0025	≥20
Mobile diesel-powered equipment for non-coal mines	≤0.25	≤0.25	—	≤0.01	—	≤0.0025[c]	≤0.001[d]	—	≤0.5	—	≥20	≤0.25[f]	≤0.01	≤0.5	≤0.0025	≥20
Mobile diesel-powered transportation equipment for gassy, non-coal mines and tunnels	≤0.30	≤0.30	≤0.20[e]	≤0.01	≤0.5	≤0.0025[e]	≤0.001[d]	≤0.005	≤0.25	≤0.00125[e]	≥20	—	—	—	—	—

[a] Values in per cent by volume.
[b] Values apply to air current before exhaust gases from diesel equipment are added.
[c] Values calculated as equivalent nitrogen peroxide.
[d] Values calculated as equivalent formaldehyde.
[e] Values calculated as equivalent nitrogen dioxide, NO_2.
[f] Value for flammable gas.

To take into account the effects of the methane in intake air on the concentration of harmful gases emitted in the exhaust fumes, regulations have been revised in the Federal Republic of Germany in 1976 as follows:

	Permissible limits		
CH_4 (vol. %)	CO (ppm)	NO_x (ppm)	NO_2 (ppm)
0	500	750	75
1.0	1200	1000	200
1.5	1800	1000	200

3.7 References

ACGIH. (1979). *Threshold limit values for chemical substances and physical agents in the workroom environment with intended changes for 1979.* Cincinnati, Ohio, USA: American Conference of Governmental Industrial Hygienists.

Anon. (1977). Archimedes screw samples air automatically. *Coal Age*, **82**, 140, 143.

Fermor, R. B. (1975). A study of the factors present in planning the ventilation needs of diesel vehicle operation mines in Quebec Province, Canada. In *International mine ventilation congress*, Johannesburg, ed. R. Hemp and F. H. Lancaster, pp. 437–42. Johannesburg: The Mine Ventilation Society of South Africa.

Nicholas, E. J. M. and Wall, R. (1971). Atmospheric pollution in headings following shotfiring. *Min. Engr.*, No. 128, 509–18.

Rabson, S. R., Quillium, J. H. and Goldblatt, E. (1960). The elimination of nitrous fumes from blasting gases. *J. S. Afr. Inst. Min. Met.*, **61**, 152–82.

Spencer, T. D. and Lawther, P. J. (1972). Mine gases. In *Medicine in the mining industries*, ed. J. M. Rogan, pp. 224–36. London: Heinemann.

Szechy, K. (1966). *The art of tunnelling*. Budapest: Akademiai Kiado.

4

Methane and its control in underground coal mines

Methane or firedamp, as it is called in many coalfields, is the result of a transformation of the material forming the source for the development of coal seams. During the formation of coal, methane is liberated in the later stages only, and because of increased depths and pressures it remains associated with the coal. When the seams are uplifted, methane is likely to escape into the atmosphere. In areas close to the outcrops and major tension faults methane escapes into the atmosphere. So where the distance from the outcrop is large and the depth of coal seams is high, large quantities of methane can be expected to be present in coal seams.

As soon as an excavation is made in such a seam, methane gas is liberated into the excavations. The amount of gas liberated and its rate of emission depends upon several factors which are discussed later. It is, however, important to remember that because of tectonic stresses to which the coal seam may have been subjected during the course of its history and the effect of certain other factors, methane gas may also be associated with some other gases. These may include nitrogen, carbon dioxide, higher hydrocarbons, hydrogen sulphide, small quantities of hydrogen and carbon monoxide. Most of these gases are in traces except for nitrogen and carbon dioxide which may have been intruded into the coal seam as a result of tectonic activity. Higher hydrocarbons may infiltrate into the coal seams from surrounding rocks containing oil-bearing shales and sands.

4.1 Presence of methane in coal seams

Methane is present in coal seams in two forms: (1) adsorbed and (2) free. Free gas is present in the pores and open cracks and forms only a very small portion (~10 % or less) of the total gas present in the coal seam. Most of the gas is present in an adsorbed state on the internal surface of the coal in a monomolecular layer. The high gas adsorption capacity of coal is due to the very large internal surface area of coal (~200 m^2/g). The internal surface area of coal-bearing strata such as sandstones and shales is much smaller and as such the amount of adsorbed gas in those rocks is small.

The amount of free gas present is related to the porosity of the rock and depends upon the gas pressure and temperature. It can be calculated using the relationship

$$Q = \frac{273\eta p}{T p_0} \tag{4.1}$$

where Q = free gas, m^3/t;
η = total porosity, m^3/t;
p = absolute gas pressure, kPa;
p_0 = atmospheric pressure, 101.3 kPa; and
T = absolute strata temperature in degrees Kelvin.

Thus if the total porosity of a material, coal or sandstone is 6 %, then the amount of gas present at 2500 kPa and 293 K will be of the order of 1.38 m^3/t. The porosity of coal is generally of the order of 5 to 10 %, that of sandstones 2 to 25 % and that of shales 5 to 15 %. As such even at comparatively high pressures, the amount of free gas present in coal and the surrounding rocks is small.

Adsorbed methane gas in coal at pressures normally encountered in coal mines can be adequately described by Langmuir's equation of the type

$$Q = \frac{abp}{1 + bp} \tag{4.2}$$

where Q = quantity of methane gas adsorbed at a given pressure p, m^3/t;
a = Langmuir's constant representing quantity of methane gas adsorbed at pressure $\rightarrow \infty$, m^3/t; and
b = Langmuir's constant with dimensions of $(1/p)$.

The constants a and b are very much dependent upon temperature, moisture content, rank of coal, etc. High-rank coal can adsorb much more gas. For example, at 2000 kPa the amount of methane adsorbed in coal with 20 % volatile matter may be about 13 m^3/t but for coal with 10 % volatile matter, this value may rise to 22 m^3/t.

Fig. 4.1 gives some idea of the adsorption capacity of coal as a function of gas pressure and volatile matter. However, it must be pointed out that these curves should not be used in any gas estimations and are only given as a guide.

Adsorption capacity is highly dependent upon the moisture content of coal and can be described by

$$\frac{Q_w}{Q_d} = \frac{1}{1 + 0.31w} \tag{4.3}$$

where Q_d = gas content of dry coal, m^3/t;
Q_w = gas content of moist coal, m^3/t; and
w = moisture content, %.

When the moisture content reaches 4 %, saturation occurs and no further drop in gas adsorption takes place. Thus the limiting effect of moisture is given by

$$Q_w = 0.446Q_d. \tag{4.4}$$

Increase in temperature decreases the adsorption capacity by about 0.8 % per degree for bituminous coals and 0.6 % per degree for anthracite coals.

The gas adsorption capacity of coal for carbon dioxide is almost 2.5 to 3 times that of methane. When a mixture of methane and carbon dioxide is present in a coal seam, then methane will be liberated first followed by carbon dioxide. As such the liberated gas will be richer in methane in the earlier stages and richer in carbon dioxide in the later stages. The equilibrium may be reached in a few minutes or may take days, depending upon the type of coal, the partial pressures of the two gases and the amount of coal present.

Fig. 4.1. Adsorption isotherms for dry coal containing from 5 to 30 % volatiles, temperature 303 K (after Boxho *et al.*, 1980).

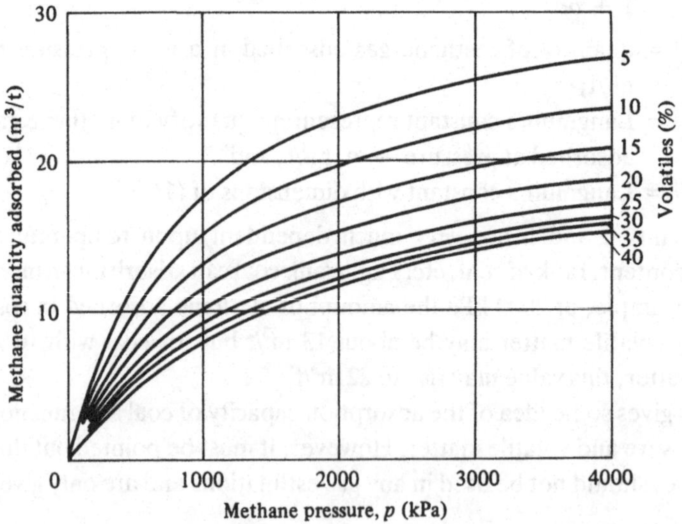

4.2 Emission of gas from coal

The adsorption of gas in coal is a reversible phenomenon. When a coal sample is exposed to the atmosphere, gas escapes into the atmosphere. The law which gas desorption from coal obeys is not very clear, but it can be argued that it could possibly follow the laws of diffusion, where gas from the coal matrix desorbs and diffuses into the cracks and, through a network of cracks, flows out into the excavation. If the unit particles forming the coal are assumed to be spherical (an approximation), then the amount of gas Q desorbed in a given time t can be given by (Boxho *et al.*, 1980)

$$Q_t = Q_\infty \sqrt{1 - e^{-4\pi^2 (D/d^2)t}} \tag{4.5}$$

where Q_t = gas desorbed, m^3 in given time, t;
Q_∞ = total gas desorbed after time $t = \infty$, i.e. the gas present initially in coal sample;
t = time, seconds;
D = diffusion coefficient, cm^2/s; and
d = equivalent particle diameter, cm; $d = 6V/A_0$ (V = particle volume and A_0 = surface area).

The diffusion coefficient depends upon temperature, initial and final gas content and the nature of the gas. The value of D for methane is about 10^{-10} cm^2/s. For small values of $t[t < (d^2/\pi D)]$, equation (4.5) can be rewritten as

$$Q_t = Q_\infty \left(\frac{12}{d} \sqrt{\frac{Dt}{\pi}} - \frac{12Dt}{d^2} \right)$$

or

$$\frac{Q_t}{Q_\infty} = \frac{12}{d} \sqrt{\frac{Dt}{\pi}} - \frac{12Dt}{d^2}. \tag{4.6}$$

Since D and d are constants for a given sample, one can state that

$$\frac{Q_t}{Q_\infty} = k_1 \sqrt{t}. \tag{4.7}$$

In other words, for a short period, the rate of desorption is proportional to the square root of time. This expression has been used in gas estimation (Kissell, McCulloch and Elder, 1973). However, if the time elapsed is large, this expression breaks down. Experiments conducted in the laboratory have shown that this square root law does not hold good particularly for broken coal samples (Lama and Bartosiewicz, 1983).

Certain researchers have put forward a gas emission relationship of the type (Airey, 1968)

$$Q_t = Q_\infty[1 - e^{(-(t/t_0))^n}] \quad (4.8)$$

where t_0 = time for desorption of 63 % of the total gas; and
n = 0.33 for bituminous coal and 0.5 for anthracite coal.

Equation (4.7) has been commonly used to estimate lost gas in the first few minutes of desorption of gas although in many cases it may not be applicable (Lama and Bartosiewicz, 1983). From equations (4.6) and (4.7)

$$k_1 = \frac{12}{d}\sqrt{\frac{D}{\pi}} - \frac{12D\sqrt{t}}{d^2} \quad (4.9)$$

Since D is very small and d is comparatively large, the second term of equation (4.9) can be neglected. Hence

$$k_1 = \frac{12}{d}\sqrt{\frac{D}{\pi}}$$

or

$$\frac{D}{d^2} = \frac{\pi k_1^2}{12^2}. \quad (4.10)$$

Therefore from the initial desorption rate, the value of k_1 can be determined, which can help assess the value of D/d^2.

4.3 Gas content of coal seams

Several methods have been developed to estimate the gas content of coal seams. Basically, all these methods can be grouped into two categories: (1) direct methods and (2) indirect methods.

4.3.1 Direct methods

The most commonly used direct method (also called the US Bureau of Mines' method) consists of obtaining a core sample from the seam by drilling and enclosing it as soon as possible in a bomb and measuring the gas released with time (Fig. 4.2) (Kissell *et al.*, 1973). To minimise time lost, wire line core barrels are used. Gas emission measurement is continued using the water displacement method until the gas emitted from the sample is less than 0.05 cm^3/g per day for 5 consecutive days. Testing can be conducted by taking a part of or the full length of the core. Taking a full length of the core is preferable since variations in gas content of the coal seam can occur due to the variability in the quality of coal. After desorption has been completed, the core is taken out of the bomb and its mass is determined. The core is then crushed in a sealed bomb using steel rods or steel balls and the gas released is measured. The results of gas emission are plotted on graph paper to calculate gas lost during the transfer of the sample from seam to bomb (Fig. 4.3) assuming that the release of gas follows the square root law (equation 4.7). The total gas content of the core Q is obtained using the relationship

$$Q = Q_1 + Q_2 + Q_3 \tag{4.11}$$

where Q_1 = gas lost between drilling of the core and transfer of the core from the drill hole to the bomb;

Q_2 = gas liberated from the core after placing it in the bomb; and

Q_3 = gas liberated when the coal sample is crushed, most frequently called the rest gas.

In estimating the amount of gas lost, lost time t has to be established accurately. It can be assumed that gas emission will start only when the gas pressure in the core is equal to or greater than the hydraulic head. Therefore, to calculate at what stage gas liberation from the core starts, an estimate of gas pressure is essential. In general it can be said that gas pressure in the coal seam is about 0.6 of the hydraulic head at that depth. In exceptional cases this value may be equal to 0.9 of the hydraulic head, but this will be rather rare. Using this estimate, the time when desorption would have possibly started can be determined and the value of Q_1 can be estimated.

Fig. 4.2. US Bureau of Mines method: sample container. Gas emission is measured by displacement of water.

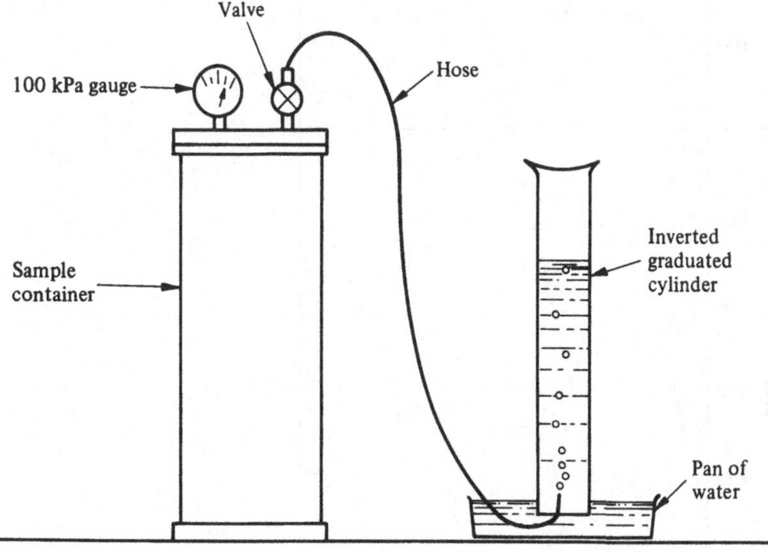

When drilling horizontal holes underground, the lost time should be taken from the start of drilling of that section of the core which is to be included in the bomb for gas estimation, because in such a case there is no hydraulic head on the core to stop desorption of gas.

Some investigators believe that the determination of lost gas is not essential because of the overall inaccuracy of the system, and take the lost gas equal to 10 % of $(Q_2 + Q_3)$. It is also argued that in the estimation of gas that will be emitted into the mine, the value of Q_3 should not be taken into account.

Estimation of gas content underground

Because of the variability of the gas content of coal in different sections of a coal seam and difficulties of drilling underground to depths where the

Fig. 4.3. Estimation of gas lost during the transfer of a sample from the seam to the bomb. Core from Vesta No. 5 site, Pittsburgh coal bed.

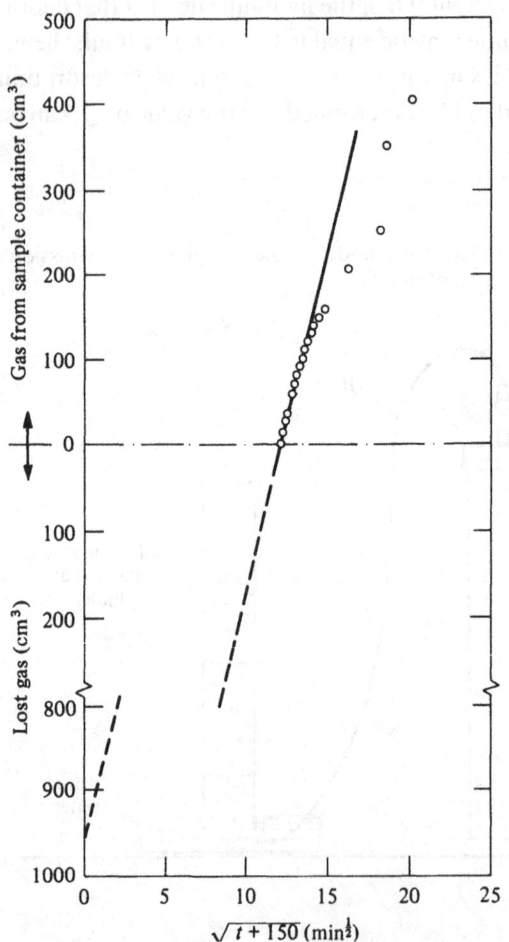

gas content is not disturbed as a result of the excavation and obtaining a full section of the coal seam, probabilistic methods have been adopted. A large number of coal samples (a minimum of 20 but preferably 50) is taken from the coal face, with the mass of each sample in the range of 25 to 40 g, and each is sealed individually in glass bottles (Creedy, 1980). These samples are then crushed in a sealed bomb and the gas content of each sample determined. The results are plotted on probability paper. The upper 90 percentile of the log-normal distribution of the measurement set is taken as the gas content *in situ* of the coal seam. When measuring gas content of highly gassy seams, the upper 95 percentile is taken to give better results and it is preferable to take smaller sized samples (20 to 30 g) (Lama, 1984*a*).

4.3.2 *Indirect methods*

Several indirect methods have been used for estimating the gas content of coal seams. One of the simplest is that suggested by Kim (1977). According to her

$$Q = \frac{(100 - \%\ \text{moisture} - \%\ \text{ash})}{100}(0.75) \times \left[k(0.096h)^n - 0.14\left(\frac{1.8h}{100} + 1\right)\right] \tag{4.12}$$

where Q = gas content, m^3/t;
h = depth, m; and
k and n = constants.

The value of k varies between 5.7 and 20 and n varies between 0.31 and 0.12. Fig. 4.4 gives the relationship between k, n and the ratio of fixed carbon to volatile matter of coal on an ash-free dry basis. The equation

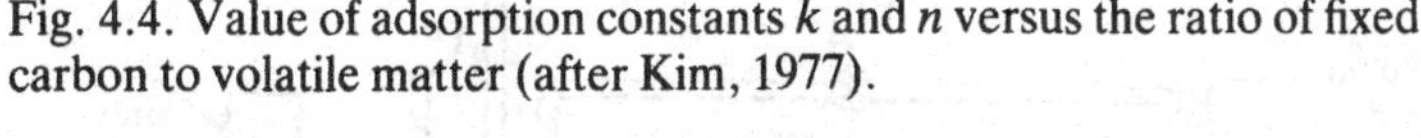

Fig. 4.4. Value of adsorption constants k and n versus the ratio of fixed carbon to volatile matter (after Kim, 1977).

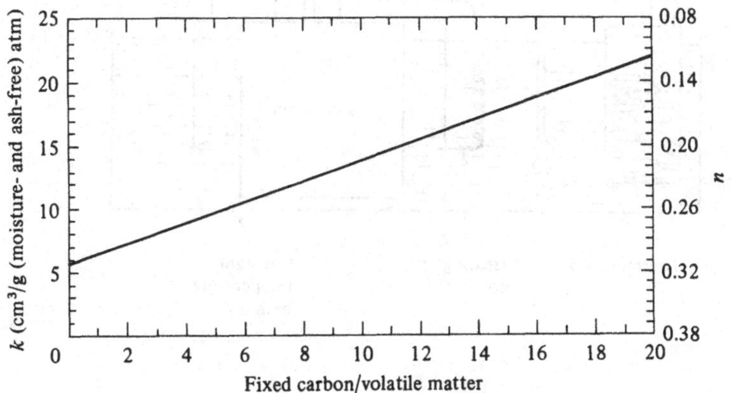

seems to give good results for higher depths (higher pressures) but breaks down at lower depths.

A more common method is based upon high-pressure adsorption of gas by coal. The technique applied consists of subjecting the sample to gas pressures and measuring the gas uptake of the sample at a given pressure until equilibrium is reached. The amount of gas adsorbed is calculated. The pressure is then varied and the volume of gas adsorbed until equilibrium is reached is recalculated. By varying the pressure a complete gas adsorption curve can be obtained.

The method used for calculating the amount of gas adsorbed could either be volumetric or gravimetric.

In the volumetric method, a known mass of coal sample is enclosed in a bomb of known volume and the bomb is evacuated. The bomb is then connected to a large chamber with a known value of gas pressure (Fig. 4.5). The line connecting the chamber and the sample bomb is opened and the pressure is monitored until there is no change in pressure. The amount of gas adsorbed is then calculated using the following equation:

$$Q_{\mathrm{p}} = \frac{1}{w}\left[p_{\mathrm{c}}V_2 - p\left(V_2 + V_1 - \frac{w}{\rho_{\mathrm{c}}}\right)\right] \qquad (4.13)$$

where Q_p = quantity of gas adsorbed at pressure p, $\mathrm{m^3/t}$;
V_1 = volume of bomb containing the sample, $\mathrm{cm^3}$;
V_2 = volume of chamber containing gas, $\mathrm{cm^3}$;

Fig. 4.5. Arrangement for volumetric measurement of gas adsorption.

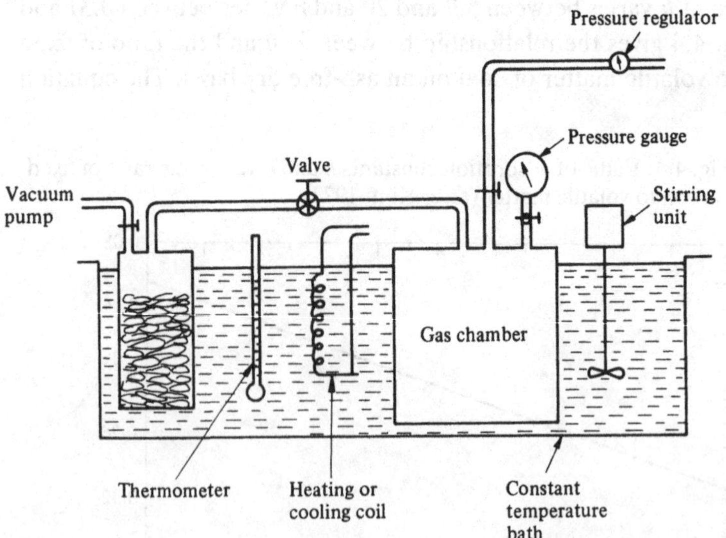

p_c = pressure of gas in the chamber;
p = pressure of gas in the chamber and bomb after equilibrium is reached;
w = mass of coal sample, g; and
ρ_c = helium density of coal, g/cm³.

It is essential that the diameter of the connecting pipe is small, otherwise the volume of this pipe must be taken into account. The temperature of the bath must be kept constant.

A schematic drawing of the equipment for the gravimetric method is given in Fig. 4.6. This system allows up to six samples of different coal to be tested simultaneously. As in the volumetric method, the sample bombs are evacuated and mass determined before and after introducing the coal samples into them. Gas is then introduced at constant pressure and the mass of the bombs determined until there is no change in mass at a given pressure. The volume of gas adsorbed at a given pressure is calculated using the following equation (Lama and Bartosiewicz, 1983):

$$Q_p = \frac{1}{w_2 - w_1}\left[\left(\frac{w_3 - w_2}{\rho_g}\right) - \frac{V_1}{V_0}\left(V_0 - \frac{w_2 - w_1}{\rho_c}\right)\right] \tag{4.14}$$

where Q_p = gas adsorbed at a given equilibrium pressure p, m³/t;
w_1 = mass of the evacuated bomb, g;
w_2 = mass of the coal sample plus bomb when evacuated, g;

Fig. 4.6. High-pressure adsorption system. 1–6, primary supply system; 7–12, pressure release valves; 13–18, isolation valves; A–F, sample bombs (after Lama and Bartosiewicz, 1983).

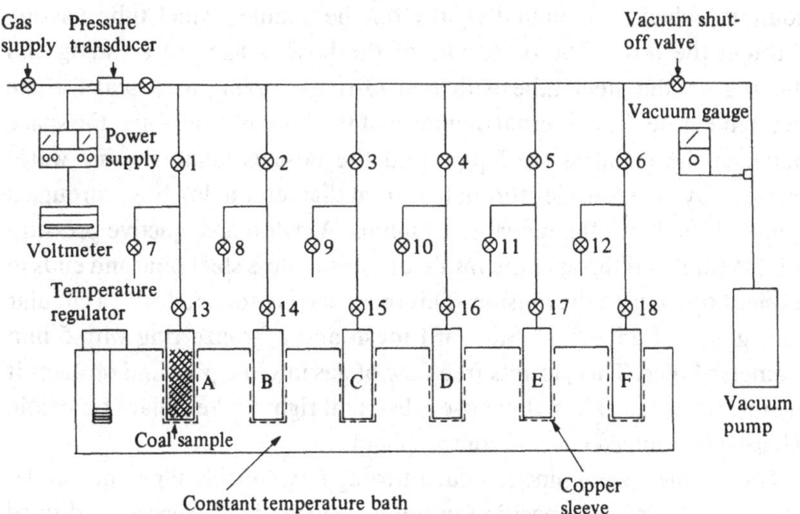

w_3 = mass of the bomb containing coal sample when saturated with gas at pressure p, g;
V_0 = free volume of the bomb when empty at NTP, cm^3;
V_1 = volume of gas at NTP contained in bomb when pressurised to pressure p, cm^3;
ρ_g = density of gas, g/cm^3; and
ρ_c = density of coal, g/cm^3.

When pressures are not high, volume V_1 can be calculated assuming ideal gas conditions. However, the adsorption of gas on the surface of the bomb can introduce serious errors. As such it is always advisable to obtain calibration curves for each bomb to obtain V_1 values for the pressure range at which tests are to be conducted.

The density of gas and coal must be accurately determined. An error of 1.5 % in the density of coal could result in an error of 5 % in calculated value of adsorbed gas at high pressures but at lower pressures the error is small (<0.2 %).

4.4 Measurement of gas pressure

To use high-pressure gas adsorption measurements, knowledge of the in-situ gas pressure is essential. Gas pressure measurements can be conducted in horizontal bore holes in coal seams when accessible or in vertical bore holes where the seam is intersected by vertical bore holes. The method consists of sealing a section of the bore hole, using inflatable packers. An arrangement using triple packers in a horizontal bore hole is given in Fig. 4.7. The packers consist of an inflatable hose with a stainless steel tube passing through it. One end of the hose is fixed to a screwed coupling which is mounted rigidly on the stainless steel tube passing through the hose. The other end of the hose is fixed to a sliding end moving on the steel tube with two O-rings serving as a seal for the inflatable hose. The internal diameter of the hose is 22 mm and the space between the stainless steel pipe and the hose is inflated using water pumped by an intensifier through a 3 mm diameter nylon hose through a nipple drilled into the movable coupling. A nylon hose (active pressure line) is threaded through the inside of the stainless steel pipe and ends in a spigot opening to the outside. This open end is protected with a circular wire gauze of 25 British Standard mesh and a bronze ring with 5 mm diameter holes. This permits free flow of gas into the pipe and protects it against blockage. The nylon hose is brought right to the collar of the hole where it is connected to the control panel.

The stainless steel pipe is joined through extendable pipe and can be placed at any desired position in the measuring hole. Second and third dilatable hoses with the same arrangements are placed at desired points in the hole with separate active pressure measuring hoses coming right up to the collar of the drill hole. There is, however, only one dilation hole which passes outside the extendable pipe and feeds into the space between the stainless steel tube and the inflatable hose. When there is more than one measuring point, the inflatable hoses are interconnected in series through suitable nipples.

On inflation, the inflation hose expands in diameter but contracts in length. As a result sufficient loose length is made available in the inflating nylon hose to avoid any tension of the inflating hose.

The results of some gas pressure measurements are given in Table 4.1. While measuring gas pressures, it is essential to ensure that the sealed chamber in which the gas pressure is measured is free of water. Frequently, it may be essential to incorporate a water separator into the gas pressure measurement system where the gas is pumped back into the chamber. Build-up of gas pressure with time gives an indication of the presence of water (Fig. 4.8). If the pressure build-up curve has a concave shape towards the pressure axis, then it is an indication of the presence of water in the sealed chamber and will give possibly the hydrostatic head of water in the seam. A convex-shaped curve (b) when obtained is an indication of a reliable test result.

Fig. 4.7. Triple packer system to monitor gas pressure in bore holes (after Lama, 1980).

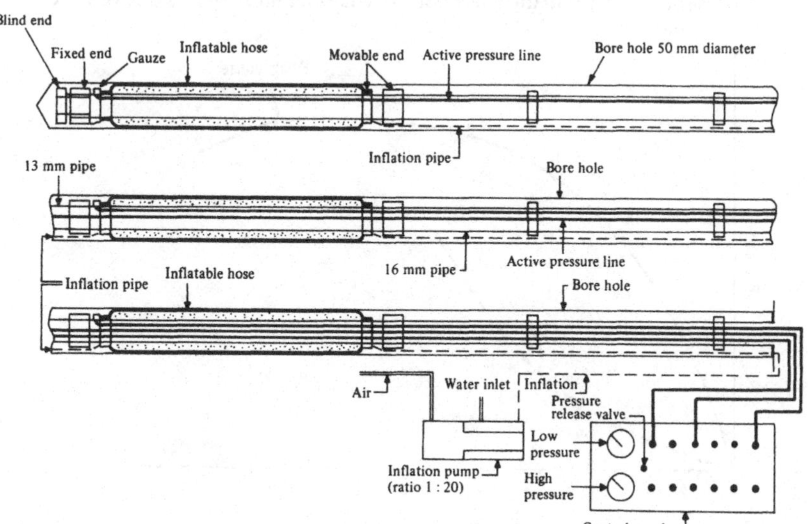

Table 4.1. *Gas pressures measured in some Australian and American collieries (after Lama and Bartosiewicz, 1983)*

Colliery (coal seam)	Gas	Depth (m)	Gas pressure (kPa)	Gas pressure / Hydrostatic head
Australia				
West Cliff (Bulli seam)	CH_4	480	3000	0.60
Appin (Bulli seam)	CH_4	520	4200	0.77
Leichhardt (Gemini seam)	CH_4	400	3640	0.91
Collinsville (Bowen seam)	CO_2	280	1500	0.54
Metropolitan (Bulli seam)	CO_2	610	3000	0.49
USA				
(Pocahontas 1403 coal bed)	CH_4	600	4330	0.72
(Pittsburgh coal seam)	CH_4	252	1360	0.54
Beatrix colliery	CH_4	470	4000	0.85

Note: The hydrostatic head is taken as equal to the depth without accounting for the depth of the water table.

Fig. 4.8. Pressure build-up in an isolated chamber in a coal seam. a – presence of water in the chamber; b – chamber more or less free of water.

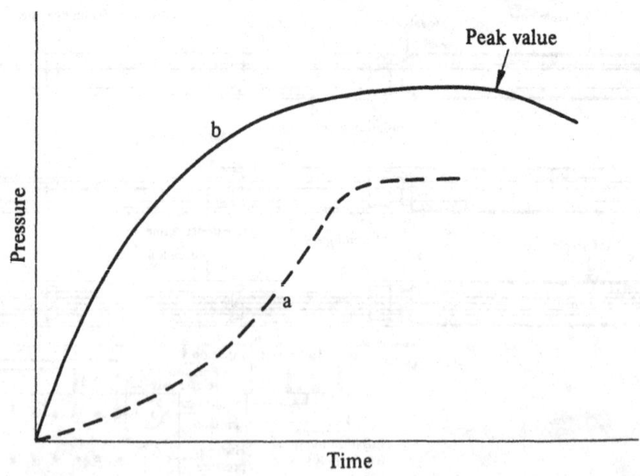

4.5 Emission of gas into mine workings

Emission of gas into mine workings depends upon a number of factors. The most important of these are:

(1) gas content of coal;
(2) permeability of coal;
(3) presence of a joint system;
(4) depth of workings;
(5) presence of other coal seams in the vicinity of the seam worked;
(6) thickness of the seam worked;
(7) method of mining; and
(8) method of ventilation.

The higher the gas content of coal, the greater the emission of gas into the mine openings. The gas content increases with depth but the permeability of the coal seams decreases. A decrease in permeability results in a lower rate of gas emission, but this effect is suppressed by the increased gas content. The presence of cleat in coal results in increased permeability. Permeability across the cleat is much higher. As a result, holes drilled face-on, i.e. intersecting at right angles to the main cleat, will deliver much more gas than those drilled butt-on. Also, headings driven face-on will experience high gas emissions from the ribs. A thicker seam not only has a higher gas content per unit surface area exposed, but also disturbs a wider area of rock surrounding it. As such, gas emissions into the working of thicker seams are comparatively higher. Similarly, caving methods result in a larger disturbance zone around excavations and give rise to higher gas emissions into mine workings. Gas enters the excavations not only from the coal seams worked but also from overlying and underlying seams, and other beds such as carbonaceous shales are sources of gas emission into the workings. Higher ventilation velocities, particularly when exhaust ventilation is practised, result in higher gas emissions because of the higher pressure differences required.

Higher rates of face advancement result in higher gas emission per unit time, although the total gas emission per unit tonne may be slightly lower.

Gas is not only emitted from the face but also from the broken coal moving on a conveyor. Under normal mining conditions, it is estimated that almost half of the gas contained in coal is emitted at the face and a quarter in the transport roadway before it reaches the surface. Gas emission in return roadways placed next to virgin areas experience higher gas emissions than those roadways placed between two flanking roadways. A roadway placed on the virgin side may have almost twice the gas emission of a roadway placed on the excavated side or central roadway.

The effect of increasing the fines in coal does not materially affect gas emission at the face because of the fact that coal contains a fairly dense fracture system (0.05 to 0.15 mm) (Kissell and Deul, 1974).

4.5.1 *Estimation of gas emission into development excavations*

For the purpose of determination of gas emission into development excavations, several methods have been developed and used in various countries. Basically two approaches have been adopted, namely,

(1) theoretical considerations in which gas emission into the excavations is calculated assuming laws of flow of gas through a porous medium; and
(2) empirical methods based upon measurements in mines.

Theoretical methods require field permeabilities, gas pressures and gas content estimations. Besides, there is no certainty that gas filtration laws (diffusion or Darcy's flow) are truly applicable to the emission of gas into mine excavations. Application of one or the other method must be tested based upon field observations. Errors introduced by transforming the results of one method to another may be of the order of 100 % or even more.

A short description of some of the most commonly used methods is given below.

Bożko method

According to this method (Pawiński, Roszkowski and Strzeminski, 1979), the methane emission can be calculated using the relationship

$$V_{CH_4} = \frac{\alpha w d l W_0 \gamma}{43200} \tag{4.15}$$

where V_{CH_4} = methane emission, m^3/minute;
α = factor depending upon the excavation development; for a single excavation, $\alpha = 2$, for two headings, $\alpha = 2 + z/l$ (<4) where z = length of cut-through;
l = depth of fracture zone (6 m);
d = thickness of seam, m;
w = rate of advance, m/month;
W_0 = methane content of seam, m^3/t; and
γ = density of coal (1.4 t/m^3).

Thus, for a rate of advance of 40 m/month, seam thickness of 2.2 m, and methane content of 8.5 m^3/t, the gas emitted will be

$$V_{CH_4} = \frac{2 \times 40 \times 2.2 \times 6 \times 8.5 \times 1.4}{43200} = 0.29 \text{ m}^3\text{/minute.}$$

For a rate of advance of 160 m/month

$$V_{CH_4} = \frac{2 \times 160 \times 2.2 \times 6 \times 8.5 \times 1.4}{43200} = 1.16 \text{ m}^3/\text{minute}.$$

This method assumes that gas emission is governed by the zone of fracturing alone. It gives acceptable results in stone headings or in carbonaceous shale at large depths where the effect of high stresses and low permeability of rock is limited to the fracture zone only.

Oriechowa method

According to this method (Pawiński *et al.*, 1979), methane emission into a heading can be calculated using the relationship

$$V_{CH_4} = 25\alpha d w V_p \bar{c} k_w \tag{4.16}$$

where V_{CH_4} = methane emission, m^3/minute;
α = 2 for a single roadway,
= $2 + 0.05l_1$ (<4) for two roadways where l_1 = distance between parallel roadways, m;
d = thickness of seam, m;
w = rate of advance, m/day;
V_p = quantity of gas liberated from a unit surface area of the exposed roadway, m^3/minute/m^2 (determined experimentally);
$\bar{c}$ = coefficient defining the increase in gas emission determined by the relationship $\bar{c} = 0.724 \ln(1.52t + 1) - 0.1t$ where t = time, months;
k_w = constant given by the equation $k_w = \dfrac{V_{CH_4} + V_{CH_{4\,max}}}{V_{CH_4}}$
where V_{CH_4} = gas liberated in the excavation, m^3/minute; and
$V_{CH_{4\,max}}$ = maximum gas liberated when blasting immediately after the blast, m^3/minute.

The use of the above method requires experimental determination of V_p of a freshly exposed surface. This value depends upon local conditions and may vary from 3×10^{-2} m^3/minute/m^2 to 1×10^{-4} m^3/minute/m^2, depending upon the gas content of the seam. For example, for a rate of advance of 1.5 m/day (w) in a 2 m thick seam (d), with gas emission rate (V_p) of 0.002 m^3/minute/m^2, the amount of gas liberated into a single excavation after a period of six months ($\bar{c} = 1.07$) will be

$$V_{CH_4} = 25 \times 2 \times 2 \times 1.5 \times 0.002 \times 1.07 \times 1.1$$
$$= 0.35 \text{ m}^3/\text{minute}.$$

Barbara experimental mine method

This method (Pawiński *et al.*, 1979) takes into account the gas liberated at the face from the coal (V_1), gas liberated from the face (V_2) and gas liberated from the ribs of the excavation (V_3) using the following empirical relationships:

$$V_{1CH_4} = \frac{wb(\Sigma d_w)\gamma(W_0 - W_k)}{1440}\ \text{m}^3/\text{minute} \tag{4.17}$$

$$V_{2CH_4} = b(\Sigma d_w)V_p\ \text{m}^3/\text{minute} \tag{4.18}$$

$$V_{3CH_4} = n(\Sigma d_w)wV_p(30k + n'0.5^{t-1})\ \text{m}^3/\text{minute} \tag{4.19}$$

and the total gas emitted is

$$V_{tCH_4} = V_{1CH_4} + V_{2CH_4} + V_{3CH_4}. \tag{4.20}$$

When $t > 6$ months

$$\begin{aligned} V_{tCH_4} &= \text{constant} \\ &= (\Sigma d_w)w \\ &\quad \times \left[\frac{\gamma b(W_0 - W_k)}{1440} + V_p\left(\frac{b}{w} + 59.0625n\right)\right]\ \text{m}^3/\text{minute} \end{aligned} \tag{4.21}$$

where d_w = total thickness of the coal bed exposed (i.e. neglecting any sandstone or shale bands), m;
b = width of the excavation, m;
γ = density of coal, t/m^3;
V_p = gas liberated from a unit area of exposed surface, m^3/minute/m^2;
n = number of surfaces in a roadway (for full thickness excavation $n = 2$);
W_0 = gas content of the coal, m^3/t, on an ash-free dry basis;
W_k = rest gas, m^3/t, on ash-free dry basis;
k = permeability, m^2;
n' = sequential number of the day after the start of drivage of the excavation, days; and
t = time of drivage of excavation, months ($t = n'/30$).

If the value of V_p is not known, then its value can be taken as follows:

For low gassy seams (gas content < 1 m^3/t), $V_p = 0.00009$ m^3/minute/m^2.

For medium gassy seams (gas content 1 to 4 m^3/t), $V_p = 0.00027$ m^3/minute/m^2.

For highly gassy seams (gas content 4 to 8 m^3/t), $V_p = 0.002$ m^3/minute/m^2.

For extremely gassy seams (gas content > 8 m^3/t), $V_p = 0.035$ m^3/minute/m^2.

For example, for a rate of advance (w) of 1.5 m/day in a seam with thickness (d_w) of 2 m having an excavation width (b) of 4 m, with gas content of coal (W_0) of 10 m^3/t, density of coal (γ) of 1.4, gas emission rate (V_p) of 0.004 m^3/minute/m^2 and rest gas (W_k) of 2 m^3/t, the gas liberated after the excavation has been driven for six months shall be

$$V_{1CH_4} = \frac{1.5 \times 4 \times 2 \times 1.4 \times (10 - 2)}{1440}$$
$$= 0.093 \text{ m}^3/\text{minute}$$
$$V_{2CH_4} = 4 \times 2 \times 0.004$$
$$= 0.032 \text{ m}^3/\text{minute}$$
$$V_{3CH_4} = 2 \times 2 \times 1.5 \times 0.004 \times 59.0625$$
$$= 1.4175 \text{ m}^3/\text{minute.}$$

Therefore

$$V_{tCH_4} = 0.093 + 0.032 + 1.4175$$
$$= 1.5425 \text{ m}^3/\text{minute.}$$

Pawiński method

A simplified solution of the flow of gas through a porous system assuming desorption of gas is given by Pawiński (1971). According to him, the amount of gas liberated into an excavation is given by the following equation:

$$V_{CH_4} = 2dw\left(\sqrt{\frac{mk}{\mu\pi p_a}}\right)\left(\frac{p_a^2 - p_0^2}{p_0}\right)t^{0.5} \tag{4.22}$$

where d = thickness of the seam or the height of the excavation when the seam is thicker than the height of the excavation, m;
w = rate of advance, m/s;
p_a = in-situ gas pressure in the seam, 10^5 Pa;
p_0 = air pressure in the excavation, 10^5 Pa;
m = porosity;
μ = viscosity of methane, Ns/m^2;
k = permeability, m^2; and
t = time elapsed after start of driving the heading, s.

For example, for a seam thickness of 2 m and a rate of advance of 1.5 m/day in a coal seam with a porosity of 3 %, an in-situ gas pressure of 3.5 $\times 10^5$ Pa, an air pressure of 1 $\times 10^5$ Pa, a viscosity of 1.1 $\times 10^{-5}$ Ns/m^2 and a permeability of 7 $\times 10^{-13}$ m^2, the gas liberated after 6 months will be

$$V_{CH_4} = \frac{2 \times 2 \times 1.5}{24 \times 60 \times 60} \times \left(\sqrt{\frac{0.03 \times 7 \times 10^{-13}}{1.1 \times 10^{-5} \times \pi \times 3.5 \times 10^5}}\right)$$
$$\times \left(\frac{(3.5 \times 10^5)^2 - (10^5)^2}{10^5}\right) \times (1.5552 \times 10^7)^{0.5}$$
$$= 0.0128 \text{ m}^3/\text{s} = 0.77 \text{ m}^3/\text{minute.}$$

Experimental techniques
Flow of gas into an excavation may be radial or linear depending upon the shape of the excavation and the thickness of the seam in relation to the diameter (height) of the excavation. Field observations have shown that gas emission from a heading at a constant rate of advance, w, can be approximated by an exponential function or a power function.
Power law. Pawiński *et al.* (1979) have used a power function of the type

$$V_{CH_4} = At^{\alpha} \tag{4.23}$$

and if a heading has been stopped, gas emission is given by

$$V_{CH_4} = At^{\alpha} - A\left(t - \frac{L}{w}\right)^{\alpha} \tag{4.24}$$

where V_{CH_4} = volume of gas emitted, m^3/minute;
t = time elapsed, days;
A and α = constants for a given area and depend upon local conditions;
w = rate of advance, m/day; and
L = length of the heading.

If observations are made over a period of time for gas emission from a heading advancing at a fixed rate, then the value of α can be calculated from the relationship

$$\alpha = \frac{\ln \dfrac{V_2}{V_1}}{\ln \dfrac{t_2}{t_1}}. \tag{4.25}$$

The value of A can be calculated from the relationship

$$A = \frac{V}{t^{\alpha}}. \tag{4.26}$$

While making observations, a ventilation station is set up at the return end of the heading and gas emission measurements are made over a period of time (daily, weekly or fortnightly) depending upon the rate of advance (Table 4.2). Since there is always a certain initial value of gas emission (V_0), unless observations are available right at the start of the heading, equations (4.25) and (4.26) can be modified as follows:

$$\alpha = \frac{\ln\left(\dfrac{V_i - V_0}{V_n - V_0}\right)}{\ln\left(\dfrac{t_i}{t_n}\right)} \tag{4.27}$$

$$A' = \frac{V_n - V_0}{t_n^\alpha}. \tag{4.28}$$

In calculating the value of α, it is advisable to take V_n as the final value of gas emission and substitute V_i by every successive value starting from the first observation. The corresponding values of t will be substituted for t_i.

Table 4.2. *Measurement of gas emission from a heading (after Pawiński et al., 1979)*

Observation No.	Date	No. of days elapsed	Length of heading (m)	Methane emitted (m^3/min)	Remarks
1	15.05	0	550	0.60	V_0
2	31.05	16 (t_i)	558	0.94	V_i
3	15.06	31	587	0.88	
4	1.07	47	627	1.22	
5	15.07	61	650	1.60	Heading stopped
6	1.08	78	650	1.52	Heading stopped
7	15.08	92	650	1.80	Heading stopped
8	3.09	111	663	1.90	
9	16.09	124	703	1.98	
10	3.10	141	733	1.80	
11	15.10	153	763	2.00	
12	29.10	167	771	1.96	
13 (1′)	15.11	184 (t_n)	777	2.10	(V_n) Heading stopped (final value)
14 (2′)	2.12	17	777	2.00	
15 (3′)	9.12	24	777	1.40	
16 (4′)	17.12	32	777	1.58	
17 (5′)	28.12	43	777	0.85	
18 (6′)	5.01	51	777	0.80	
19 (7′)	20.01	66	777	0.90	
20 (8′)	4.02	81	777	0.60	
21 (9′)	17.02	94	777	0.90	
22 (10′)	4.03	109	777	0.50	
23 (11′)	24.03	129	777	0.40	
24 (12′)	4.04	140	777	0.30	

V_0 = initial value of gas emission;
V_i = successive values of gas emission starting from the first observation;
V_n = final value of gas emission when heading stopped;
t_i = successive values of number of days elapsed starting from the first observation;
t_n = number of days elapsed when heading stopped.

The following example based upon the data in Table 4.2 makes it clear.

$$\alpha_2 = \frac{\ln\left(\frac{0.94 - 0.6}{2.1 - 0.6}\right)}{\ln\left(\frac{16}{184}\right)} = 0.61$$

$$\alpha_3 = \frac{\ln\left(\frac{0.88 - 0.6}{2.1 - 0.6}\right)}{\ln\left(\frac{31}{184}\right)} = 0.94$$

$$\alpha_4 = \frac{\ln\left(\frac{1.22 - 0.6}{2.1 - 0.6}\right)}{\ln\left(\frac{47}{184}\right)} = 0.65$$

$$\alpha_5 = \frac{\ln\left(\frac{1.6 - 0.6}{2.1 - 0.6}\right)}{\ln\left(\frac{61}{184}\right)} = 0.37$$

$$\vdots$$

$$\alpha_{12} = \frac{\ln\left(\frac{1.96 - 0.6}{2.1 - 0.6}\right)}{\ln\left(\frac{167}{184}\right)} = 1.01$$

The mean value of

$$\alpha = \frac{0.61 + 0.945 + 0.65 + 0.37 + \cdots + 1.01}{11}$$

$$= 0.56$$

$$A = \frac{(2.1 - 0.6)}{184^{0.56}} = 0.08.$$

While α is a constant for the conditions, the value of A changes with rate of advance. The new value of A' can be calculated by the relationship

$$A' = \frac{A \times w_1}{w}$$

where w_1 is the new rate of advance. Fig. 4.9 shows the experimental data with the power function fit as obtained above.

Exponential law. A simplified model assumes that the rate of gas emission is proportional to the gas present in an excavation at a given time, such that

$$\frac{\mathrm{d}V}{\mathrm{d}t} = -\alpha V_t$$

or

$$V = Ae^{-\alpha t} \tag{4.29}$$

where V = methane emission, litres/metre length;
t = time, days; and
A and α = constants.

Lama (1984*b*) has used this relationship to determine gas emission into development headings. The excavation is divided into a number of sections (50 m in length) and measurements are made on gas emission as a function of time. The amount of gas emitted from each section is plotted against the time elapsed after the drivage of the excavation (Fig. 4.10). Using the curve fitting technique, the values of A and α are calculated. Correlation coefficients of 0.6 to 0.8 have been found for elapsed periods of up to 240 days using equation (4.29).

Once the values of A and α are known, the gas emission for different rates of advance can be calculated by adding on the gas emitted in various sections of the developing roadway with time using the graphical technique. When the face is stopped, a decay curve is obtained since no new ribs are being created. When the rate of development changes, the

Fig. 4.9. Emission of gas into excavation. 1, measurements; 2, power function approximated to fit the data; 3, length of heading (after Pawiński *et al.*, 1979). 1–13 and 1′–12′ = observation numbers (see Table 4.2).

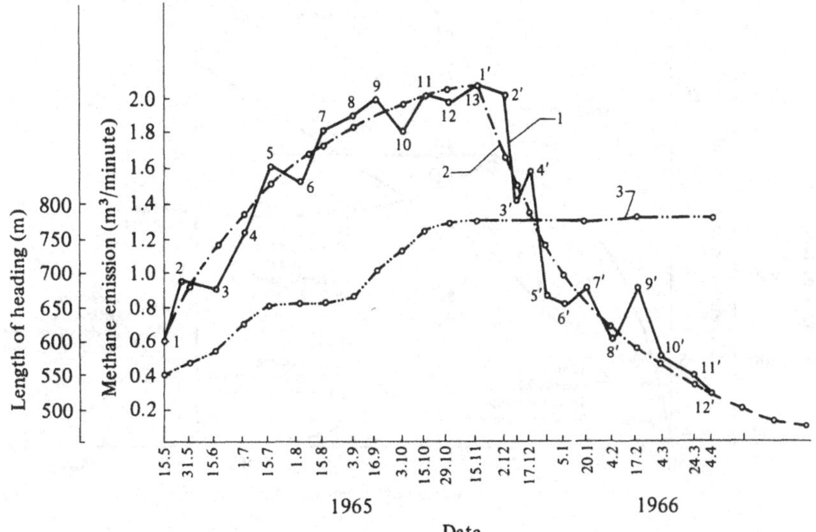

roadway is assumed to have stopped at this position and the new build-up curve for gas is calculated with the new development rate. The curves are then summed graphically. Fig. 4.10 shows an example of the results obtained with two different rates of development of a heading.

Irregularity of gas emission at the face of development headings

The total amount of gas liberated at the face from coal over a period is dependent upon the gas content of the coal and the rate of mining (equation 4.17). However, gas emission at the face is not regular but fluctuates over very wide limits depending upon the mean rate of mining and method of mining. With extraction using continuous miners, maximum gas emission occurs when cutting the face. When blasting is adopted, maximum gas concentration occurs 100 to 200 s after the blast. Peaks are higher in mining using blasting than using a continuous miner (Fig. 4.11). Machines using a thinner web give smaller peaks than machines using a thicker web.

In blasting, immediately at the time of initiation of the blast (0.3 to 0.9 s), the gas liberated is only about 1 to 2 % of the total gas present in the coal, which corresponds to the free gas present in the coal at the face. In the second stage, which includes completion of the blast, dislodgement of coal and emission of blast fumes, the gas emitted includes gas from the

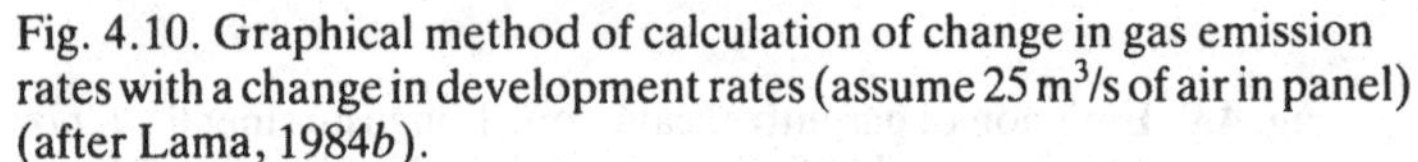

Fig. 4.10. Graphical method of calculation of change in gas emission rates with a change in development rates (assume 25 m^3/s of air in panel) (after Lama, 1984*b*).

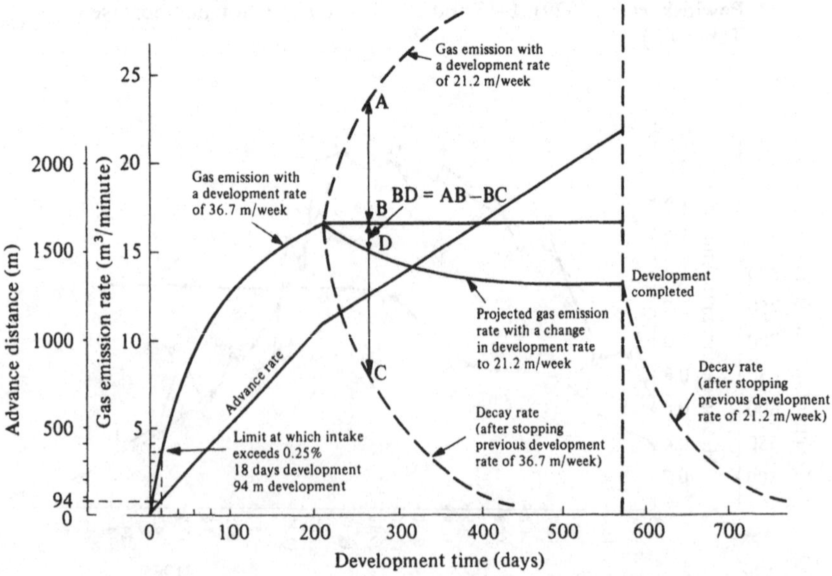

coal face which is governed by the pressure gradient, the depth of the cut (pull of the blast) and initial stage of desorption of gas from the coal. Most of the gas is emitted in this phase, which may extend up to 200 s from the initiation of the blast. Almost 50 to 75 % of the total gas is emitted in this stage.

The third stage of the process is associated with the loading of coal, which contributes to about 15 to 25 % of the total gas and extends from about 10 to 30 minutes. The fourth stage is the preparation stage for the next blast and this contributes to about 3 to 13 % of the total gas emitted.

Blasting produces higher gas emission at the face than machine cutting. Wetting of the face greatly reduces gas emission particularly when blasting is adopted using water stemming (Fig. 4.12).

When machine cutting, peak gas values occur at the point of cutting. These peaks are higher when cutting the intake edge of the face than the return edge because of problems of mixing of ventilation air and irregularity of air distribution at the face. Use of venturis, blowers or Coanda ducts can help greatly to reduce these peaks. While planning ventilation, it is essential to keep the peak values below statutory requirements (1.25 or 1 %). Measurements have shown that with properly designed mixing of air and gas at the face, the probability of such peaks occurring can be reduced to 1 in 30 000 (Kissell *et al.*, 1974).

Fig. 4.11. Methane emission when driving preparatory workings using: (*a*) drilling–blasting; (*b*) cutter–loader (after Myasnikov and Patrushev, 1981).

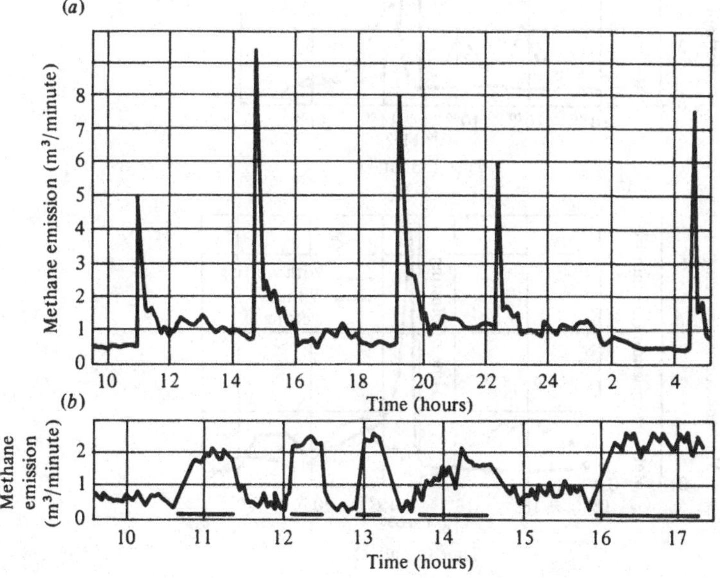

4.5.2 *Estimation of gas emission on longwall faces and pillar panels*

In longwall mining, gas is emitted from the seam mined as well as from the surrounding medium, which may contain very large amounts of gas particularly in the surrounding seams. Fissures develop around longwall panels. Fig. 4.13 shows a general schematic drawing of the fracture zones around a longwall.

The depth of the fracture zones in the roof may extend up to 100 m and in the floor up to 50 m. In cases where the width of the excavation is small, the height of these zones is limited but in cases of super-critical width the disturbance reaches the surface. The zone of primary caving (zone 1) is highly disturbed and coal seams and rock strata undergo large-scale

Fig. 4.12. Effect of wetting on methane emission after blasting operations in the Chertinskaya-Zapadnaya mine of Kuzbassugol' coal field: (*a*) in haulage drive along seam 5; (*b*) manway along seam 3 (after Myasnikov and Patrushev, 1981).

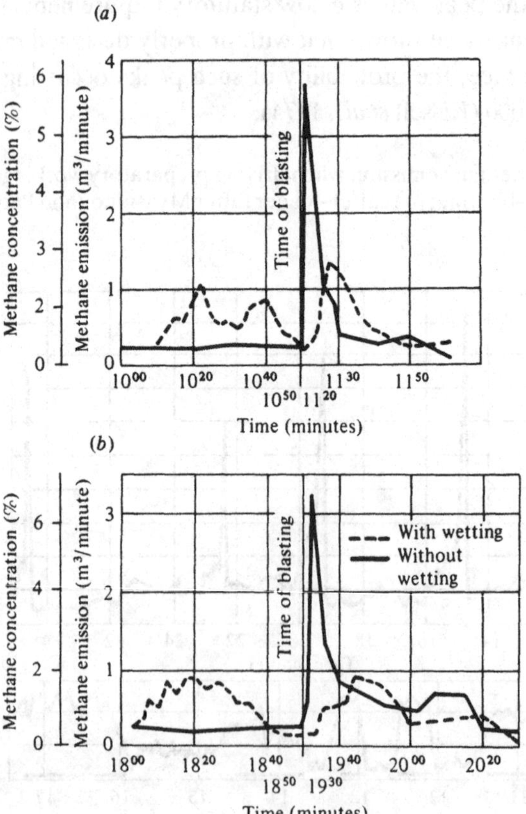

fracturing. Gas present in this zone almost completely enters the seam under extraction. The height of this zone extends to about five times the thickness of the seam when caving is adopted and to about two times when stowing is practised. The thickness of the zone of large disturbance in the floor (primary heave) is limited to about one to two times the thickness of the seam. Zone 2 in the roof and floor undergoes a comparatively smaller disturbance than zone 1. Beds open up and cavities are formed in these zones. Coal seams containing gas in these zones are highly emanative and gas from these zones enters slowly into the excavation in the seam under extraction. The permeability of the beds in these zones is highly anisotropic, being greater parallel to the bedding than normal to it. The limit of zone 3 in the floor starts at about 50 m or so and in the roof up to about 100 m. The permeability of the strata in these zones is not greatly disturbed and hence gas emission from these zones into the seam under extraction is negligible.

Fig. 4.13. Ground movement around a longwall face. t = thickness of seam.

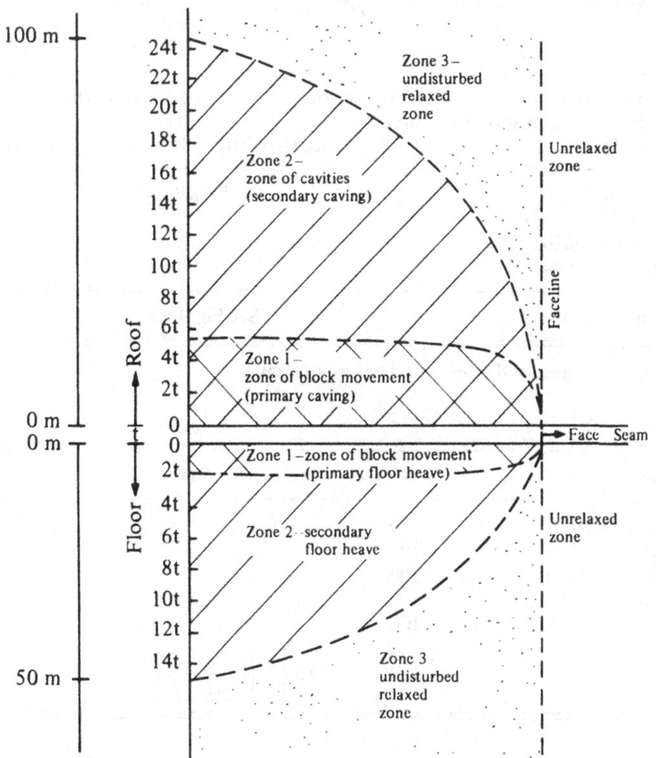

Table 4.3. *(a) Emission from adjacent seams; (b) emission from worked seam (after Boxho et al., 1980)*

(a)

	Emission zone		Degree of gas emission (%)	
Method	Roof	Floor	Roof	Floor
Gunther	Rectangular block $h = 100$ m	Rectangular block $h = 100$ m	100	$100 - h$
Jeger	Rectangular block $h = 165$ m	Rectangular block $h = 55$ m	90 up to ≈30 m[a] $\frac{C_{initial} - 2}{C_{initial}} 100$ from 30 to 120 m $\frac{C_{initial} - 2}{C_{initial}} 2(170 - h)$ from 120 to 170 m	90 down to 10 m[a] then $2(55 - h)$ from 10 to 55 m
	N.B. Modifications are used for different strata			
Flügge	Prism width l $h = 0.5l \tan \alpha_F$		$\frac{100l - 200h \cot \alpha_F}{l}$	
Schulz	Half cylinder diameter l	Rectangular block $h = 20$ m	$\frac{100}{l} \sqrt{(l^2 - 4h^2)}$	100 down to 5 m then $133 - 6.7h$
Winter	Unlimited block; practical limit taken at 10 % emission level		100 up to 20 m then $100 \exp[-\mu_W(h - 20)]$	100 down to 8 m then $100 \exp[-\mu_W(h - 8)]$
Lidin	Rectangular block $h = 60\,M(1.2 \pm \cos \alpha_f)$ gives maximum extent: + above; − below		$100(1 - h/h_{max})$	
Airey	Unlimited block		See Fig. 4.15	

[a] Depending on the thickness of the seam and nature of strata.

(b)

Method	Degree of gas emission (%)
Gunther/Jeger	50
Flügge	100
Schulz	100
Winter	100
Lidin	100
Airey	$100\left[1 - \exp\left\{-\left(\frac{x_0 + l_1}{wt_0}\right)^n\right\}\right]$

As the face advances, recompression of the strata behind the face takes place. This recompression closes the cracks about 150 to 250 m behind the face and as a result gas emission into the workings from the surrounding beds drops with advance of face. It is therefore clear that for successful drainage, holes must be placed in zone 2 in the roof and floor. When holes are placed in zone 1, then high leakage occurs, thereby diluting the captured gas.

The limit of the zones of disturbance laterally is usually 20 m. In the neighbourhood of the face line, the relaxed zone is narrower and it widens as the face moves away.

Several empirical methods have been developed to estimate gas emission into mine working when longwall (or wide front extraction) systems are adopted. Table 4.3 summarises the basic relationships governing the degree of gas emission as postulated by different authors (see Boxho *et al.*, 1980). Figs. 4.14, 4.15 and 4.16 show the degree of gas emission by different methods. For emission of gas from the worked seam all methods except Airey's method are independent of the rate of advance (Fig. 4.16). Jeger assumes 50 % gas emission from the worked seam which corresponds to about 18 m/week of advance of the face.

Emission of gas from the intervening strata between the coal seams is not allowed for except by Jeger and Gunther. Gunther assumes the gas content of stone to be 1 % of that of coal. Jeger assumes that the gas content of sandstone is about 10 % of that of coal and that of shales and fine-grained sandstones about 1 % of that of coal.

An example of the procedures for the calculation of gas emission into a longwall operation is given in Table 4.4. This table is arranged so that the gas make can be calculated for each gas source from the product of the gas content, the relative thickness and degree of gas emission divided by 100. It is assumed here that the strata have been previously undisturbed.

Notes to Table 4.3 on symbols, values and units:

All — h = distance normal to the bedding planes except in Lidin's method where h is the vertical distance from worked seam (m);
l = face length (m);

Flügge — α_F = angle of prism (45°, 54° or 58½° roof; 22½° floor);

Winter — μ_W = constant (0.016, 0.014 or 0.012 roof; 0.07, 0.05 or 0.03 floor);

Lidin — M = seam thickness (m);
α_f = slope of seam (°);

Airey — x_0 = constant (say, 4 m);
l_1 = face line to front abutment distance (say, 2 m);
W = rate of advance (m/s);
t_0 = 63 % emission time constant of minimum size lumps (at maximum stress line) in seam (say 5×10^5 s for bituminous coal);
n = constant (0.333 for bituminous coal); $C_i = 10$.

Fig. 4.14. Degree of gas emission from adjacent seams (results of Gunther, Jeger and Lidin for 1.5 m thick level seams, after Boxho *et al.*, 1980). For Jeger: $C_i = C_{initial}$; $h_1, h_1' = f$ (thickness of the seam and nature of the strata).

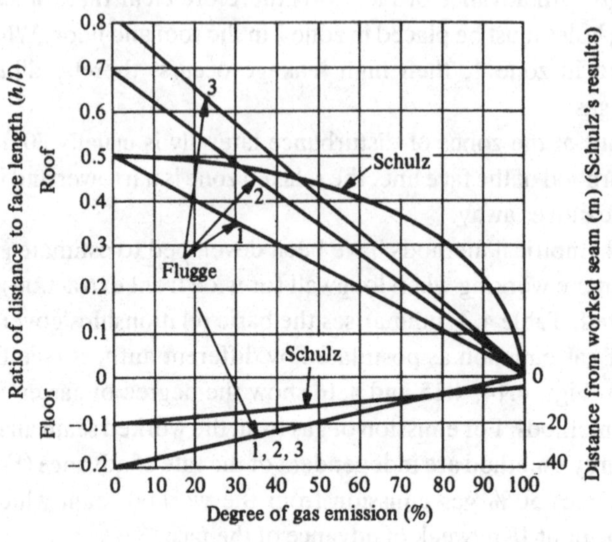

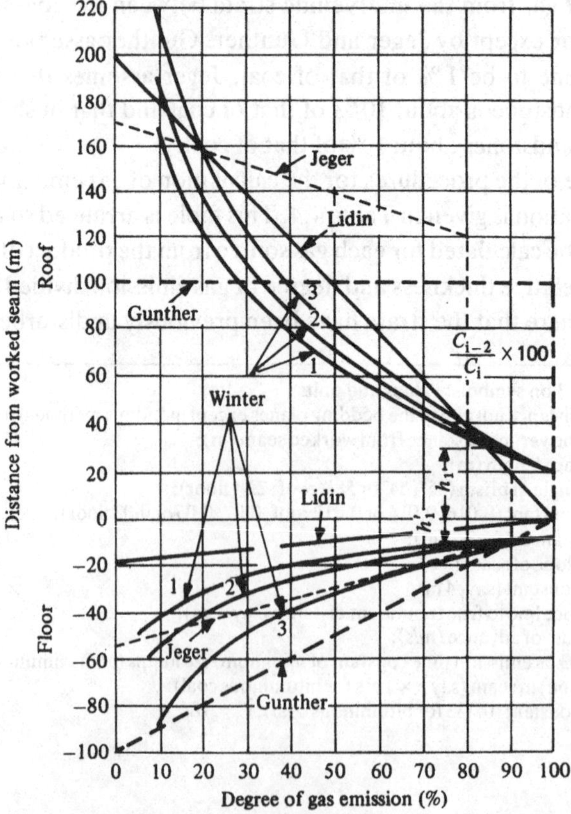

Fig. 4.15. Degree of gas emission from adjacent seams (after Airey, 1968). (*a*) Height correction in m before using 900 m curve; (*b*) degree of gas emission at 900 m.

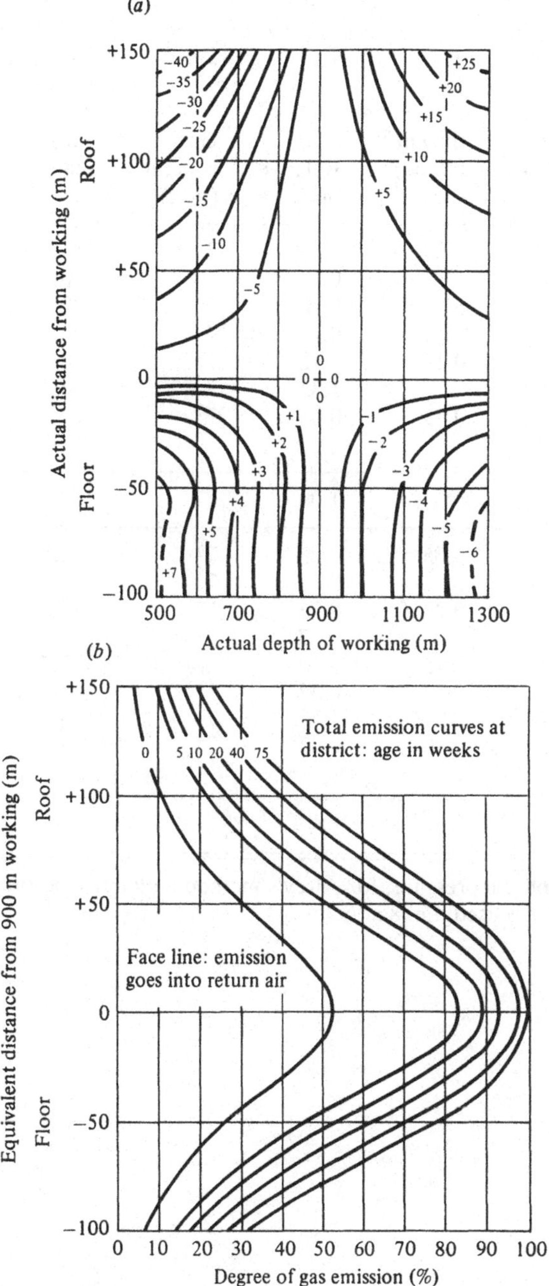

Table 4.4. *Prediction of gas emission by different methods (after Boxho et al., 1980)*

Gas source data						
1		2	3	4	5	6
Seam		Actual thickness (m)	Equivalent thickness (stone) (m)	Relative thickness	Gas content[a] (m^3/t)	Distance from worked seam (m)
Gustav		2.00		1.33	7	180
Hugo		0.20		0.13	7	110
	Stone	39	0.39	0.26	7[b]	80
Karl		1.35		0.90	7	60
	Stone	18	0.18	0.12	7[b]	50
Blücher		2.00		1.33	7	40
	Stone	39	0.39	0.26	7[b]	20
Roof total						
Wilhelm (worked)		1.50		1.00	7	0
	Stone	5	0.05	0.03	7[b]	2.5
Johann		0.50		0.33	7	5
	Stone	34	0.34	0.23	7[b]	17.5
Präsident		1.00		0.67	7	40
	Stone	{15 J 48	0.15 J 0.48	0.10 J 0.32}	7[b]	{47.5 J 65}
Dickebank		2.00		1.33	7	90
	Stone	10	0.10	0.07	7[b]	95
Floor total						
Overall total		—	—	—	—	—

[a] Desorbable to atmospheric pressure; for Airey's method a quantity desorbable to zero methane partial pressure is used, i.e. 8 m^3/t.
[b] Fictitious value.
J, Jeger's results.

Degree of gas emission[c] from each curve in %										
7	8	9	10	11	12	13	14	15	16	17
Flügge 1	Flügge 2	Flügge 3	Gunther	Jeger	Lidin	Schulz	Winter 1	Winter 2	Winter 3	Airey (75 weeks)
0	0	0	0	0	9	0	8	11	15	17
0	20	32.5	0	72	44.5	0	24	28	34	40
—	—	—	100	72	—	—	—	—	—	—
40	56.5	63	100	72	70	80	53	57	62	75
—	—	—	100	72	—	—	—	—	—	—
60	71	75.5	100	72	80	92	73	76	79	89
—	—	—	100	90	—	—	—	—	—	—
100	100	100	50	50	100	100	100	100	100	48
—	—	—	97.5	90	—	—	—	—	—	—
87.5	87.5	87.5	95	90	73	100	100	100	100	99
—	—	—	82.5	75	—	—	—	—	—	—
0	0	0	60	30	0	0	11	20	38	86
—	—	—	35	15	—	—	—	—	—	—
0	0	0	10	0	0	0	0	1.5	9	39
—	—	—	5	0	—	—	—	—	—	—
—	—	—	—	—	—	—	—	—	—	—

[c] Face length 200 m.
Worked seam thickness 1.5 m.
Dip of worked seam 0°.
Advance rate 20 m/week.

(*Table continued*)

Table 4.4 cont

Gas make[d] in m³/t										
18	19	20	21	22	23	24	25	26	27	28
Flügge 1	Flügge 2	Flügge 3	Gunther	Jeger	Lidin	Schulz	Winter 1	Winter 2	Winter 3	Airey
0	0	0	0	0	0.84	0	0.74	1.02	1.40	1.81
0	0.18	0.30	0	0.65	0.40	0	0.22	0.25	0.31	0.42
—	—	—	1.82	1.31	—	—	—	—	—	—
2.52	3.56	3.97	6.30	4.54	4.41	5.04	3.34	3.59	3.91	5.40
—	—	—	0.84	0.68	—	—	—	—	—	—
5.59	6.61	7.03	9.31	6.70	7.45	8.57	6.80	7.08	7.35	9.47
—	—	—	1.82	1.64	—	—	—	—	—	—
8.11	10.35	11.30	20.09	15.52	13.10	13.61	11.10	11.94	12.97	17.10
7.00	7.00	7.00	3.50	3.50	7.00	7.00	7.00	7.00	7.00	3.84
—	—	—	0.20	0.19	—	—	—	—	—	—
2.02	2.02	2.02	2.19	2.08	1.69	2.31	2.31	2.31	2.31	2.61
—	—	—	1.33	1.21	—	—	—	—	—	—
0	0	0	2.81	1.41	0	0	0.52	0.94	1.78	4.61
—	—	—	0.78	0.11	—	—	—	—	—	—
0	0	0	0.93	0	0	0	0	0.14	0.84	4.15
—	—	—	0.02	0	—	—	—	—	—	—
2.02	2.02	2.02	8.26	5.00	1.69	2.31	2.83	3.39	4.93	11.37
17.13	19.37	20.32	31.85	24.02	21.79	22.92	20.93	22.33	24.90	32.31

[d] Column 4 × column 5 × $\dfrac{\text{degree of gas emission}}{100}$.

First, the seams contributing to emission are listed in sequence in column 1 in Table 4.4. For Gunther's method, strata between seams are also included. Column 5 gives the gas content of each seam; where this is unknown for adjacent coal seams (as in this example), the value for the worked seam is used. The value used for Airey's method is taken as 1 m³/t greater than for the other methods. The 'relative thickness' is recorded in column 4; for each seam this is the ratio of the actual thickness (column 2) to the thickness extracted from the worked seam (for intervening stone the 'equivalent thickness' in column 3 is first calculated as 1 % of the actual thickness of stone). The distance of all seams and intervening strata from the worked seam are written in column 6 and these are used in

Fig. 4.16. Degree of gas emission of worked seam (after Airey, 1968).

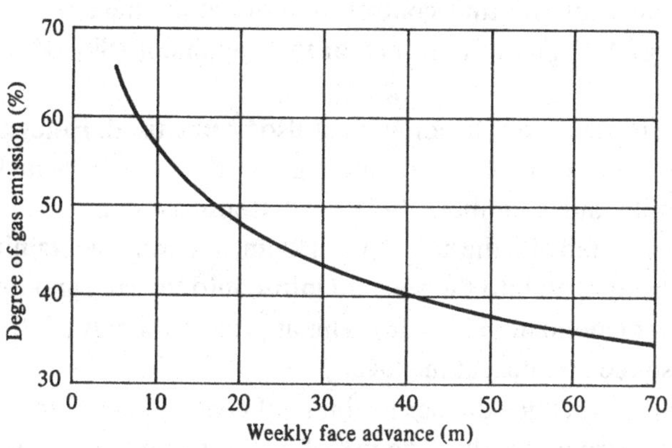

conjunction with Figs. 4.14 and 4.15 to obtain the percentage values of the Degree of Emission Function, which are recorded in columns 7–17, each column being for a different method or variation of a method.

Finally the gas make or specific gas emission (m^3/t) calculated for each source from the product of the gas content (column 5), the relative thickness (column 4) and the degree of emission (columns 7–17) divided by 100 are recorded in columns 18–28. Totals are obtained for roof and floor strata, and an overall total is finally worked out for each method.

4.6 Control of methane emission in mines

Commonly used methane control techniques are:

(1) dilution by ventilation;
(2) sealing; and
(3) methane drainage.

Ventilation is the most common method of dealing with methane gas hazard. The ventilation system must be flexible, allowing sufficient good quality air to be coursed into any working place where methane has been encountered, so that the gas can be diluted to safe limits and thus rendered relatively harmless. The method is applicable only in mines where the quantity of gas emitted is small to medium. The system breaks down when high gas emission is encountered because of high costs and unacceptable air velocities at the working places.

Cementation of small fissures or localised blowers is possible. This is only a temporary solution and can be effected, however, only when gas pressures are low. When a fissure is sealed at one point, it may allow advance of the working face but ground movement may result in cracking of the seal or in the formation of new cracks which will allow gas to escape.

Silica acid gel has been used to provide a very reliable seal of the fractures in a coal seam and thus control the flow of methane into the working area. Fig. 4.17 gives the results of this treatment (Slobod and Burcik, 1974).

Water infusion in advance of headings may also be practised. Holes are drilled into the coal seam ahead of mining, as the mining actually progresses (Cervik and Centibas, 1973, and Centibas *et al.*, 1974). Packers are then inserted into the holes to seal them, and water containing a wetting agent is pumped into the seam to infuse into the coal and thus block out the flow of methane (Fig. 4.18). This approach not only controls methane, but also reduces dust at the face.

The above methods reduce the permeability of coal and hence reduce the emission of gas from the ribs. The use of water is only a temporary solution since water will slowly escape from the coal back into the roadway, dry out and gas emission into the heading will occur again.

Drainage of gas is the only positive and reliable method of controlling high gas emission in mines. It is almost accepted that when the gas content of a seam exceeds 8 m^3/t, the pre-drainage or advance drainage of gas is essential for intensive and efficient mining techniques to be effective. When the gas emission into the mine from a seam exceeds 16 m^3/t of coal mined, then a general gas drainage system incorporating drainage of the goaf area (post-drainage) becomes essential to control gas emission in the mine.

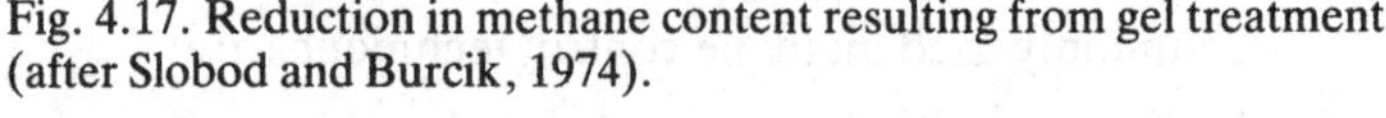

Fig. 4.17. Reduction in methane content resulting from gel treatment (after Slobod and Burcik, 1974).

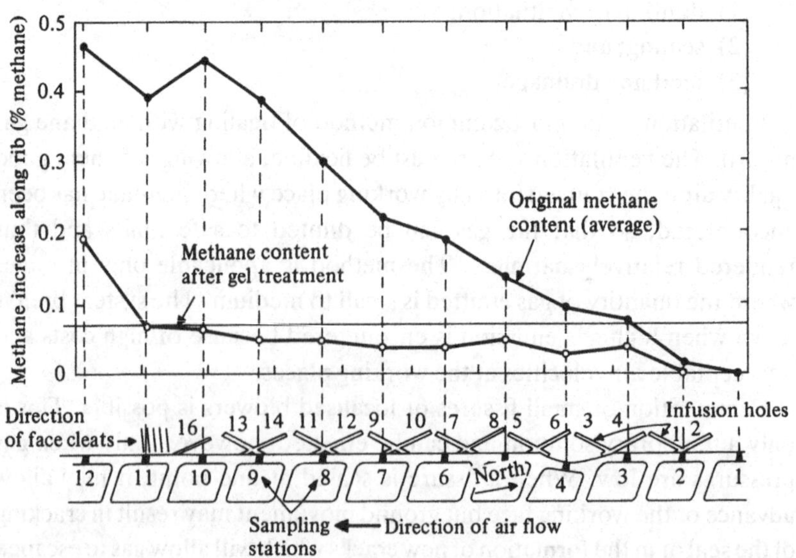

Pre- and post-drainage of gas has the following advantages:

(1) Increased safety and reduced risks of ignition and explosions.
(2) Seams which are liable to outbursts can be pre-drained, thereby reducing the risk of outbursts.
(3) Increased production as a result of fewer stoppages due to the gas percentage exceeding statutory requirements.
(4) Lower air volume flow rate requirements resulting in lower ventilation costs.
(5) Decreased air velocities at the faces resulting in lower dust concentrations at the faces.
(6) The methane gas drained can be utilised as a source of power, which pays back many times the cost of drainage.
(7) Pre-drainage has been shown to improve the roof conditions in certain mines as a result of the destressing of the coal seam due to contraction of the coal after desorption of gas.
(8) The long advance bore holes used in pre-drainage help in the exploration of areas and in the location of geological disturbances.

The only disadvantage associated with drainage is the cost involved in setting up a drainage system and the on-going costs of drilling, piping, etc.

Fig. 4.18. The water-infusion technique employs horizontal holes to pump water into the coal seam. The infused water then prevents gas emission.

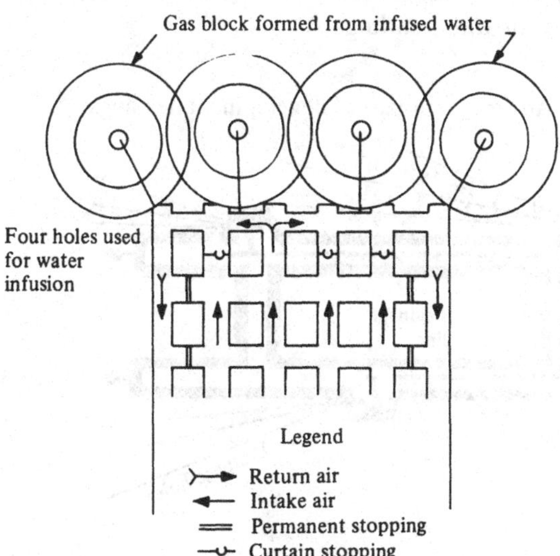

4.6.1 *Pre-drainage of gas from coal seams*

The pre-drainage of gas from coal seams is adopted in cases where high gas emission during the development of headings and at the longwall result in stoppages at the face, or in seams where outbursts of gas and coal endanger workers and materials.

Several methods have been developed for the pre-drainage of coal seams. These include:

(1) advance drainage;
(2) lateral drainage;
(3) drainage from the upper and lower lying seams;
(4) drainage from surface bore holes.

1. Advance drainage

Advance drainage is practised in cases where high gas emissions are experienced in the drivage of the first excavation. Holes are drilled to protect the heading under development and are placed at a low angle and maintained ahead of the face (Fig. 4.19). This helps to divert the movement of gas away from the face into the bore hole. The length of the holes can vary from 25 to 100 m and the diameter of the holes is usually limited to 100 mm.

The distance between the rib line of the heading and the hole should not be less than 5 m otherwise excessive leakage into the hole will occur. However, if this distance is increased beyond 15 m very little drainage from the face occurs and these holes are then not effective in protecting the heading. Continued suction applied to holes even after the heading has been driven past the end of the hole can greatly reduce or completely eliminate rib emission into headings.

Fig. 4.19. Advance drainage of development headings.

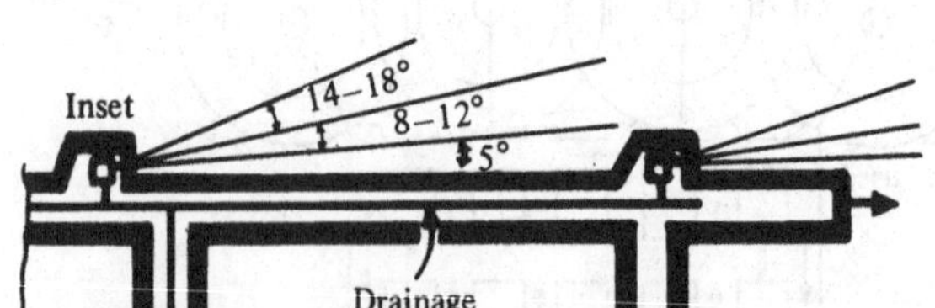

The drop in gas emission levels and reduced gas pressures ahead of the face can also help to reduce the danger of outbursts provided sufficient time is allowed for drainage.

2. Lateral drainage

Lateral drainage is applicable when an excavation is already available. Holes of up to 200 m in length can be drilled at right angles to the axis of an existing excavation (Fig. 4.20). The hole diameter is usually 50, 65 or 80 mm. For shorter holes (up to 100 m) a smaller diameter can be used. Longer holes are difficult to drill using small diameter bits and rods and the diameter usually used is 65 or 80 mm.

Holes longer than 200 m require special care during drilling in order to keep them within the seam and designed azimuth. The use of monitoring techniques such as the Eastman single shot and multishot surveying techniques or wireless bore hole survey tools (Geoscience Electronics Corp., USA) is recommended for this purpose.

To keep a long hole within the desired horizon requires experience in drilling. The most effective controls concern the thrust and the number of revolutions (Cervik, Fields and Aul, 1975, and Thakur and Poundstone, 1980). High thrust and a low number of revolutions tend to lift the hole upwards while lower thrust and a high number of revolutions tend to move the hole downwards. The maintenance of the azimuth is much more

Fig. 4.20. Lateral drainage.

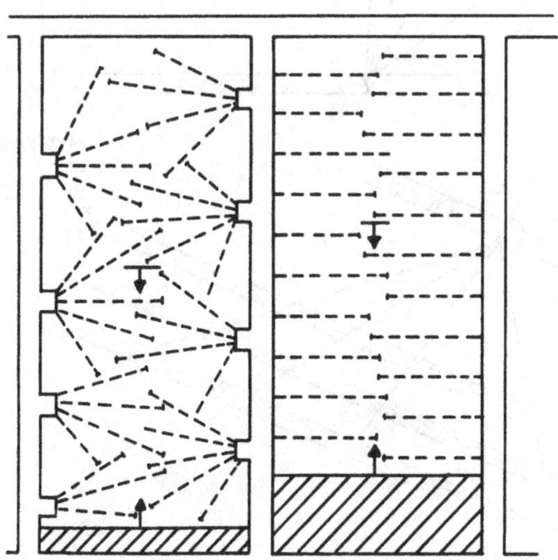

difficult. Holes tend to move to the right of the desired direction or follow the main cleat directions. The only reliable method is the use of down hole motors. However, the use of these in holes of up to 80 mm diameters greatly reduces drilling rates.

The spacing of the holes depends upon the permeability of the seam, the lead time available and the diameter of the holes. To optimise hole spacing, it is essential to conduct field measurements (Lama, 1980). In general a distance of 18 to 25 m is applicable in coal seams. The lead time under such conditions varies between three to six months with a vacuum of −15 to −20 kPa. Flow rates of the order of 3 to 10 litres/m/minute can be expected.

Fig. 4.21. Drainage from the upper and lower lying seams.

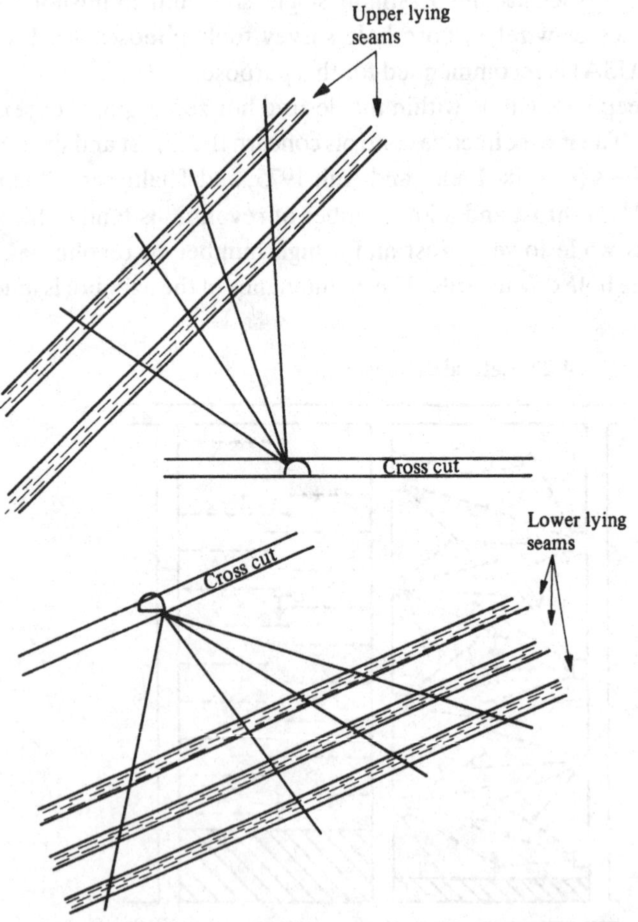

Flow rates are higher when the holes are drilled at right angles to the main cleat. When both the cleats in the coal are equally developed, holes should be drilled intersecting the two cleats.

3. Drainage from upper and lower lying seams

The drainage of upper and lower lying seams from cut-throughs placed in rock above or below the seam or from excavations placed in the upper or lower lying seams is common practice in the effort to reduce the gas content of highly gassy seams or seams liable to outbursts of gas and coal. Fig. 4.21 illustrates how the drainage holes are placed. Pre-drainage holes can be drilled from insets specially driven into the surrounding excavations. These holes are drilled in a fan pattern to cover the area and intersect the seam to form a triangular or square pattern, with the length of the sides varying from 15 to 25 m. All the holes are connected to a common manifold and monitored for gas quality and flow rates.

Drilling from the excavation placed below the seam or seams to be drained is much more effective because of the ease of drainage of water. If water is present in the holes, flow rates are greatly reduced. In such cases, larger diameter holes filled with an annular pipe can be used to drain water under gas pressure (Fig. 4.22). When valve V_1 is closed, the gas pressure builds up and pushes the water through the central pipe out. By opening valve V_2, water can be removed. Valve V_2 is closed when no more water comes out. Valve V_1 is then opened for continuous drainage.

Fig. 4.22. Draining of water from down holes. V_1 and V_2 are valves.

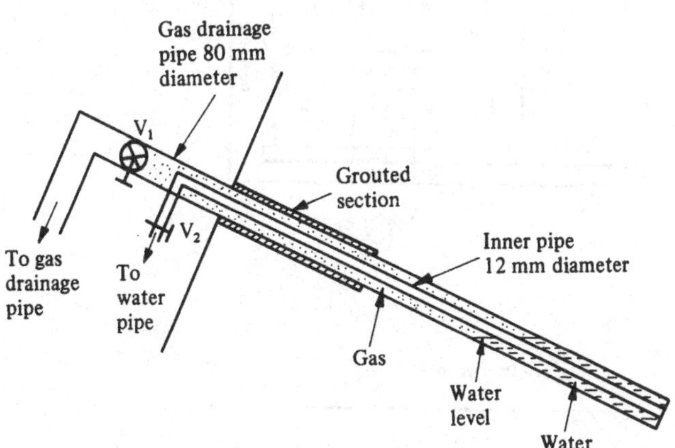

4. Drainage from surface bore holes

With developments in directional drilling and the maintenance of horizon control of bore holes, it is now becoming possible to drill a number of flanking (fan shaped) holes in a seam from a single bore hole drilled on the surface (Cervik and Elder, 1971, and Fields *et al.*, 1973). The use of hydraulic fracturing with propping sand where the fracture can be kept open and maintained within a coal seam has been tried in a few places with appreciably increased flow rates (Maksimovic, Elder and Kissell, 1977, and Steidl, 1978). These techniques are now developing at a fast rate although at this stage this method is not economical due to the high cost of drilling.

4.6.2 Post-drainage

Post-drainage is a common method of combating high gas emission in the goaf areas behind longwalls. Holes are drilled to meet the sources of gas emission above and below the coal seam in the roof and floor in the goaf. The arrangement of the bore holes depends upon the method of mining – advancing or retreating. With an advancing longwall, the roof drainage holes are drilled from the gate roadways as close to the face as possible and at right angles to the axis of the roadway (Fig. 4.23). These holes are drilled so that the bore hole just crosses the zone of

Fig. 4.23. Post-drainage on an advancing longwall face.

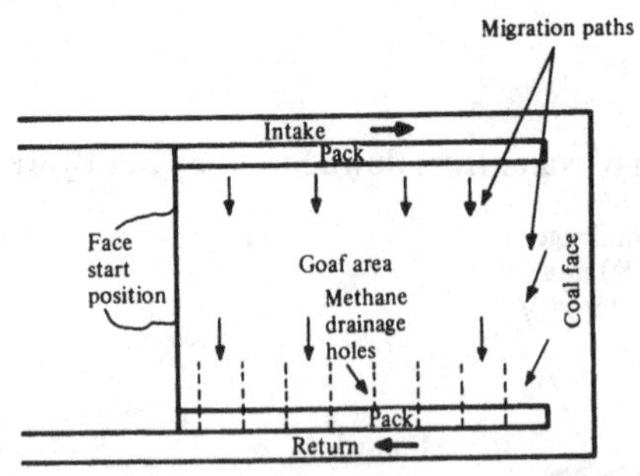

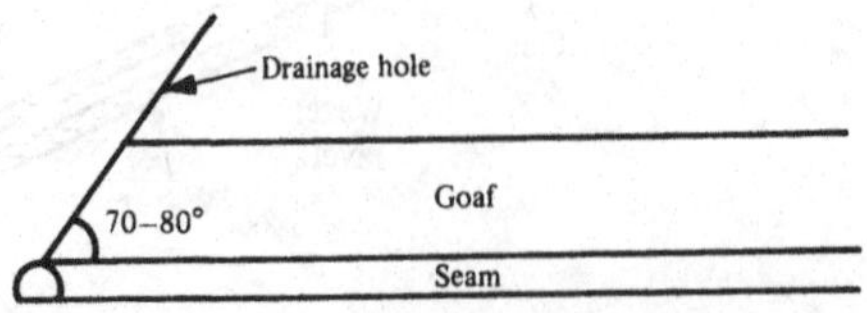

primary caving and lies in the zone of secondary caving (Fig. 4.13) but intersects the source of gas emission.

When it is not possible to drill the holes at right angles to the axis of the roadway, holes may be placed at some angle in front of the face inclined towards it or behind the face.

In retreating longwalls with single gate roadways, holes are always inclined towards the goaf because of the inaccessibility of the roadway (Fig. 4.24). In such cases bore holes need to be disconnected a few metres before the face reaches the collar of the hole unless it is possible to maintain them in the goaf. The holes are usually connected to plastic pipes laid in the floor. In such cases, holes can be drained when the face has gone past the collar of the hole. These pipes can then be turned off with valves provided at intervals.

When more than one heading is available to service the longwall gate roadways, drainage holes can be drilled from the roadway next to the gate roadway placed on the solid side. In such cases, holes can be maintained on suction over a much longer period of time (Fig. 4.25).

The efficiency of gas drainage in longwall operations depends upon local conditions and may vary from 30 to 60 %. In retreating faces with holes drilled from the secondary gate roadways placed next to gate roadway, efficiency is high (~70 %).

Fig. 4.24. Methane drainage with individual roof holes over a retreating face. (This bore hole will be disconnected before the face reaches it.)

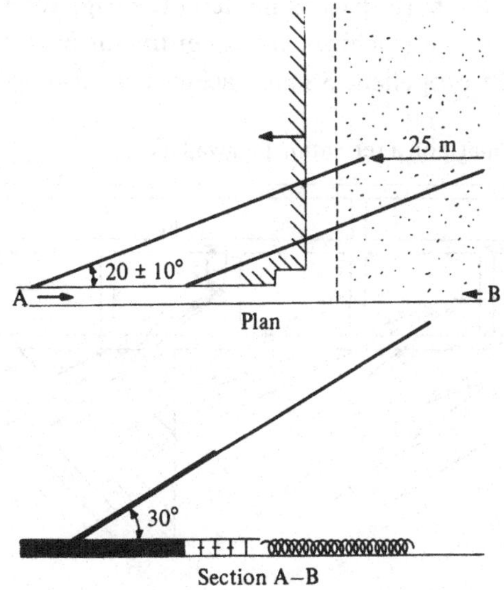

Floor holes are in general fewer in number than roof holes because less of the floor is disturbed. Floor holes are most effective when directed into the goaf and away from the face at intervals of 20 to 45 m with angles of 50 to 70° with the horizontal and penetrating 30 to 40 m into the floor. Floor holes drilled parallel to or towards the face are less effective. While angling holes away from the face, their direction should be chosen so that they intersect at right angles to the relaxation joints formed in the floor. Shallow-angled bore holes are effective when the lateral permeability of the beds in the floor is very high. The presence of water in the floor holes greatly reduces their effectiveness.

Sealed goaves or large expanding goaves can be drained by placing pipes in the goaves or drilling holes from roadways placed next to the goaves. In all such cases, it is essential to place the pipe close to the top end of the goaf (top end of the heading and rise side of the goaf) to capture the richest gas.

Goaves can also be drained by drilling holes from the surface into the goaf area. This method is economical only when the depth of the seam is small. Old abandoned mines can be a source of high gas accumulations and this gas has been trapped in some cases.

4.6.3 *Drilling of gas drainage holes*

The drilling of long holes for pre-drainage requires large drill machines (Acker Big John, USA, Diamec, Sweden, and Wirth, West Germany) while the drilling of post-drainage holes in longwall operations requires smaller, more compact machines, particularly when holes are drilled from gate roadways where space is limited (Turmag, West Germany, and Edeco, UK). Large machines are either tractor-mounted or tyre-mounted and are self-propelled. Small machines are usually skid-

Fig. 4.25. Post-drainage on a retreating longwall face.

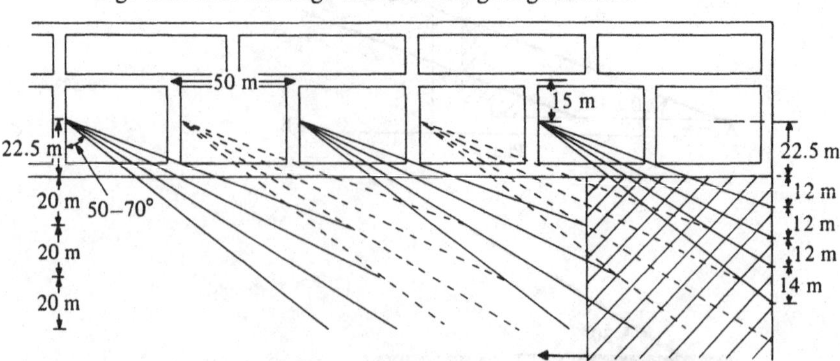

mounted. The length of the drill rod is usually about 3 m in large machines and 0.6 to 1.5 m in small machines. The diameter of the holes varies and depends upon the requirements, but is usually not greater than 120 mm, and is generally between 50 to 80 mm.

An arrangement of drilling drainage holes is shown in Fig. 4.26. In cases where high gas pressure is expected, the diameter of the hole from the collar to a depth of 10 m is enlarged and a stand pipe is grouted in and drilling is carried out through the grouted pipe. Cuttings and gas are separated at the stuffing box with gas fed to the drainage pipe and water recirculated.

Where high gas emission is not expected while drilling, the full length of the hole can be drilled. A stand pipe can then be inserted in the reamed section of the hole and sealed using densi tape, cement grout, resin, or any other sealing material. The pipe is then connected to the main drainage line. A general arrangement of a drainage hole installation is given in Fig. 4.27.

4.6.4 *Layout of surface gas drainage plant*

When the total quantity of gas drained is not large, it can be disposed of underground by feeding the gas directly into the main return of the mine so that it can be quickly diluted to within safe limits using a large amount of the return air. When very large gas emissions are expected, dilution of drained gas underground is not possible and the gas is transported to the surface through a main pipe installed in the return shaft, or through a specially drilled bore hole.

The diameter of the underground main and branch pipes feeding the various districts should be so selected that the velocity of gas and air plus gas mixtures flowing through these pipes at any particular stage does not exceed 16 m/s.

Fig. 4.26. Degasification hook-up, during and after drilling.

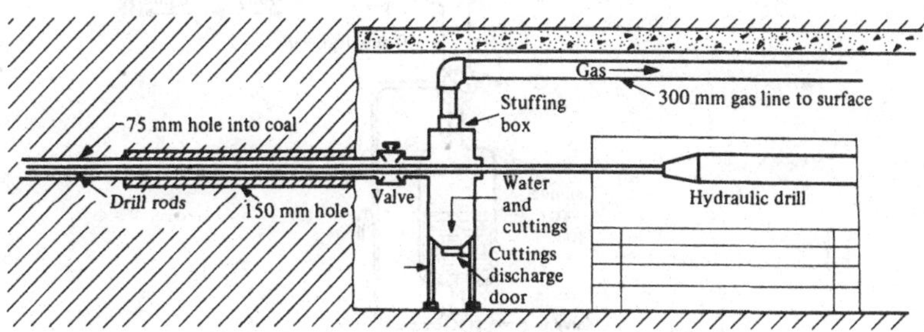

The general arrangement of the surface plant is shown in Fig. 4.28. In this case gas is delivered to the surface through a bore hole and passes through a water plus muck separator and a set of flame traps before entering the suction pumps or gas extractors. Gas is then passed on to a water separator and a discharge pipe. Gas is discharged above the fan evasee for quick dilution on a dilution stack. Gas quality monitoring is done on a continuous basis to ensure that no explosive mixture passes through the gas extractors.

The gas extractors used are water-sealed, positive displacement pumps which are designed in such a way that there is no metal to metal contact within the pump. The capacity of the surface drainage plant is such that it can accommodate not only the expected rate of gas emission, but cater for any sudden inrush of gas and for maintenance of the system. This is usually calculated by adding 30 % or one additional pump, whichever is less.

Regulation of the pump is determined by gas quality and vacuum. Pumps are run between −20 and −50 kPa and gas quality is not allowed to drop below 30 % (in rare cases 25 %). If the vacuum drops below 20 kPa, then additional pumps are switched on and if the vacuum increases beyond 50 kPa, then one of the pumps is switched off. The pumps are also stopped if the vacuum drops below 10 kPa, which is an indication that the gas line underground is damaged.

Fig. 4.27. Drainage hole installation. Valves C and D – butterfly valves; valves A and B – standard full open valves (after Lama, 1980).

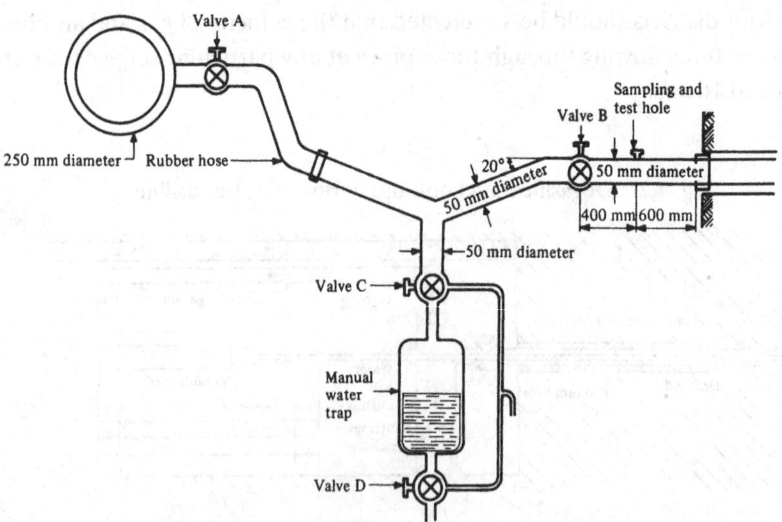

Pump station environments must be monitored for methane concentration, temperature of the various mechanical and electrical equipment, and safety features such as fire extinguishing equipment and alarm signals must be provided. Flame traps are provided on the inlet and discharge ends of the pipe ranges and safety valves are provided which discharge gas automatically into the exhaust stack if the pressure in the inlet line exceeds 5 kPa.

4.6.5 *Design of mine layout to combat high gas emission*

The layout of the district in mines with high gas emission requires special considerations. The essential feature in the design is that once gas is emitted into the mine working it should not be allowed to reach the working face and should be removed as quickly as possible. Recirculation of gases should be avoided as far as possible and pressure drops in the headings particularly between the intake and return headings should be kept to a minimum. This will reduce air losses and improve ventilation at the face.

In development headings, forcing ventilation using ducting is desirable in highly gassy mines. The return headings should be placed on the solid side. The amount of gas emission in the intakes can be greatly reduced by placing intakes in the middle and making outside headings the returns. Usually a minimum of three heading developments is advisable when

Fig. 4.28. Layout of surface plant for gas drainage system (after Marshall, Lama and Tomlinson, 1982). A, automatic pressure relief valve.

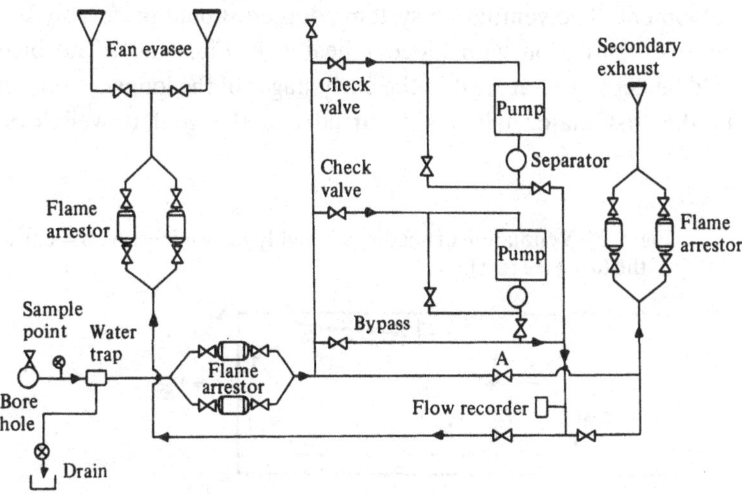

forcing ventilation using ducts is not practised. In such a case either a central intake with flange returns or two intakes with one return placed on the solid side should be used. The sequence of development should be such that the return heading is driven first followed by an intake heading with the cut-throughs puncturing the return side.

The distance between the end of the brattice (when using brattice ventilation), forcing duct or the exhaust duct (when using exhaust ventilation) should be kept to a minimum. A general rule is to keep the distance from the end of the duct to the face not more than $2\sqrt{A}$, where A is the cross-sectional area of the roadway. Venturis can be used to direct air to ventilate the far corner (A) of the face (Fig. 4.29). In extreme conditions, the return end (B) should be cut first followed by the far end with advancing of the brattice immediately after cutting end B. It is essential, however, that the brattice is not advanced too much to cause restriction to the flow of air. The area behind the brattice and between the rib should be kept clean. Rib dowelling greatly reduces rib spall and thereby reduces pressure losses.

Places of localised high gas emission such as blowouts, large fractures and certain faulted areas can be ventilated by using spiral ducts (Coanda duct) or blind ventilation ducts with small outlets placed along the length of the duct.

Sudden high gas emission has been experienced on longwall faces in many countries (UK, Australia, India, USA). This high gas emission invariably comes from the goaf areas at the return end of the face. The flooding of the return end can be controlled by extending the brattice line into the goaf behind the last chock.

Retreating longwall panels should be developed with a three-heading development. The ventilation system adopted should preferably be a Z system of ventilation with bleeder headings (Fig. 4.30). The bleeder should be kept open at least in the early stages of the longwall extraction until the first major fall has occurred and the goaf is well formed.

Fig. 4.29. Ventilation of headings in highly gassy mines. A, B = corners of the face (see text).

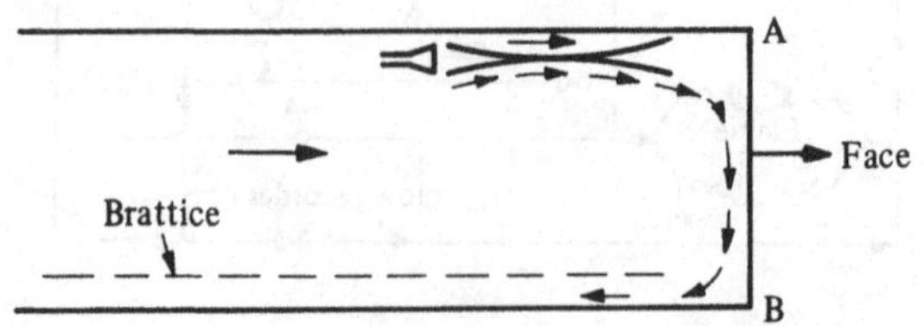

Regulation of the bleeder can greatly reduce the likelihood of sudden large amounts of gas released in the goaf invading the return gate end of the face.

Major falls in the goaf, particularly the first fall, can usually be predicted and is generally the place where high gas emission can be expected. Drainage holes should be placed closer together in these regions to cater for high gas emission. Similarly, longwall panels intersecting shear zones (strike-slip faults) are subject to high gas emission. When mining close to such structures, pre-drainage holes placed close together should be used. Some roof holes can be placed so that they open into the top end of the immediate roof so that in case of need they can be quickly opened. A 65 mm diameter hole at −20 kPa can suck out almost 10 m^3 of gas per minute when opened into a goaf.

4.6.6 *Combating layering of gas in roadways*

Methane layering can occur at the roof level for some tens of metres and ignition of these methane layers has caused a number of gas explosions. The velocity of the airflow is lower closer to the perimeter of a roadway than towards the centre. The combined effect of this low

Fig. 4.30. *Z* system of ventilation with bleeder return on retreating longwalls.

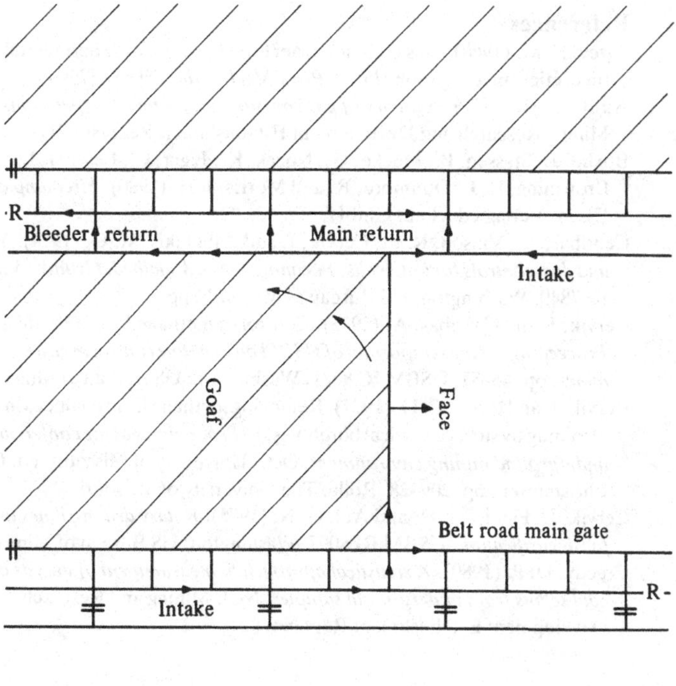

velocity of airflow close to the perimeter with the low density of methane (in comparison to air) emitted from the roof may result in the formation of methane layers at roof level. The main factors in the formation of methane layers are the gas emission rate, the velocity of airflow and the dimensions of the excavation. Methane emitted at roof levels mixes by turbulence and by drag against the underlying airflow. The degree of turbulence of an airstream increases with its velocity, and the degree of entrainment between a layer and the main airstream depends upon their relative velocity. If the air velocity is low, a layer may form.

The following equation may be used to indicate beforehand whether in any given roadway methane layering will occur:

$$v = \sqrt[3]{\frac{4.3Q}{W}} \qquad (4.30)$$

where v = velocity of airflow, m/s;
Q = rate of emission of methane into the roadway, m^3/s; and
W = width of the roadway, m.

If $v > 2$, layering will not occur. In steeply inclined, uphill ventilated roadways, a higher velocity is required.

To eliminate the possibility of layering, the basic requirement is to increase the airflow. The air velocity should be increased by constricting the airflow, for example, by placing a hurdle across the roadway at the floor.

4.7 References

Airey, E. M. (1968). Gas emission from broken coal. An experimental and theoretical investigation. *Int. J. Rock Mech. Min. Sci.*, **5**, 475–94.

Airey, E. M. (1971). *A theory of gas emission in coal mining operations*. NCB Mining Research and Development Establishment Rept No. 21.

Boxho, J., Stassen, P., Mucke, G., Noack, K., Jeger, C., Lescher, L., Browining, E. J., Dunmore, R. and Morris, I. H. (1980). *Firedamp drainage*. Essen: Verlag Gluckauf GmbH.

Centibas, A., Vinson, R. P., Cervik, J. and Zabetakis, M. G. (1974). *Methane and dust controls for longwalls: Pocahontas No. 3 coalbed, Grundy, Va*, USBM RI 7849. Washington: US Bureau of Mines. 16 pp.

Cervik, J. and Centibas, A. (1973). Control of methane by water infusion. In *Proceedings of the symposium of USBM on methane control in eastern US coal mines*, pp. 46–57, USBM IC 8621. Washington: US Bureau of Mines.

Cervik, J. and Elder, C. H. (1971). Removing methane from coalbeds in advance of mining by surface vertical boreholes. In *Proceedings of the conference on the underground mining environment*, Oct., University of Missouri, ed. C. R. Christiansen, pp. 209–28. Rolla: The University of Missouri.

Cervik, J., Fields, H. H. and Aul, G. N. (1975). *Rotary drilling holes in coalbed for degasification*, USBM RI 8097. Washington: US Bureau of Mines. 21 pp.

Creedy, D. P. (1980). *A statistical approach to measurement of the gas content of coal seams from underground samples*. NCB Mining and Research Development Establishment Rept No. 92.

Fields, H. H., Krickovic, S., Sainato, A. and Zabetakis, M. G. (1973). *Degasification of virgin Pittsburgh coalbed through a large borehole*, USBM RI 7800. Washington: US Bureau of Mines. 27 pp.

Flugge, G. (1971). Die Anwendung der Trogtheorie auf den Raum der Zusatzaus-gasung. *Gluckauf-Forschungshefte*, **32**, 122–9.

Gunther, J. (1965). Mécanisme et Prévision du Dégagement grisouteux. *Cerchar, Documents Techniques*, **11**, 697–703.

Kim, A. G. (1977). *Estimating methane content of bituminous coalbeds from adsorption data*, USBM RI 8245. Washington: US Bureau of Mines. 22 pp.

Kissell, F. N., Banfield, J. L., Jr, Dalzell, R. W. and Zabetakis, M. G. (1974). *Peak methane concentrations during coal mining*, USBM RI 7885. Washington: US Bureau of Mines.

Kissell, F. N. and Deul, M. (1974). Effect of coal breakage on methane emission. *Trans. Soc. Min. Engrs AIME*, **256**, No. 2, 182–4.

Kissell, F. N., McCulloch, C. M. and Elder, C. H. (1973). *The direct method of determining methane content of coalbeds for ventilation design*, USBM RI 7767. Washington: US Bureau of Mines. 17 pp.

Lama, R. D. (1980). *Drainage of methane from the solid at West Cliff colliery – optimisation of drainage hole design parameters.* Geomechanics of Coal Mining Report No. 18. Syndal: CSIRO Div. of Applied Geomechanics.

Lama, R. D. (1984*a*). *Investigations into drainage of gas on longwalls at West Cliff colliery.* Wollongong, Australia: Kembla Coal & Coke Pty Limited.

Lama, R. D. (1984*b*). *Gas emission and gas drainage requirements during early phases of development of longwall panels – Darkes Forest Mine (1984–1986).* Wollongong, Australia: Kembla Coal & Coke Pty Limited.

Lama, R. D. and Bartosiewicz, H. (1983). *An improved method of measurement of gas content of coal seams: studies on the Collinsville coal mine, Collinsville Coal Company, Collinsville, Queensland.* CSIRO Div. of Geomechanics Rept No. 54. Mount Waverley: CSIRO Div. of Geomechanics.

Liden, G. D. *et al.* (1965). Determination of the gassiness of coal seams and methods for the prediction of methane emission in coal mines of the USSR. *Restricted international conference of directors of safety in mines research, Sheffield.* English (SMRE) trans. 5158. Safety in Mines Research Establishment.

Maksimovic, S. D., Elder, C. H. and Kissell, F. N. (1977). *Hydraulic stimulation of surface borehole for gob degasification*, USBM RI 8228. Washington: US Bureau of Mines. 17 pp.

Marshall, P., Lama, R. D. and Tomlinson, E. (1982). Experiences on pre-drainage of gas at West Cliff colliery. In *The Australasian IMM. Illawarra branch symposium on seam gas drainage with particular reference to the working seam*, ed. A. J. Hargraves, pp. 141–56. Wollongong: Illawarra branch of the Australasian Institute of Mining and Metallurgy.

Myasnikov, A. A. and Patrushev, M. A. (1981). *Principles of coal mine ventilation planning.* New Delhi: Amerind Publishing Co. Pvt. Ltd.

Pawiński, J. (1971). Przepływy z wymianą masy i pędu w niektórych zagadnieniach przewietrzania Kopalń. *Zesz. nauk AGH*, **34**.

Pawiński, J., Roszkowski, J. and Strzeminski, J. (1979). *Przewietrzanie Kopalń.* Katowice: Wydawnictwo 'Slask'.

Schulz, P. (1959). Le Dégagement de Grisou du Charbon causé par l'exploitation. Une étude sur les Possibilités quantitatives de Captage et sa Prédétermination. *Rev. Univ. des Mines*, **102**, 41–58.

Slobod, R. L. and Burcik, E. J. (1974). Control of methane in coal mining with silica acid gel. *Min. Cong. J.*, **60**, No. 8, 48–50.

Steidl, P. F. (1978). *Foam stimulation to enhance production from degasification wells in the Pittsburgh coalbed*, USBM RI 8286. Washington: US Bureau of Mines. 10 pp.

Thakur, P. C. and Poundstone, W. N. (1980). Horizontal drilling technology for advance degasification. *Min. Eng.*, **32**, No. 6, 676–80.

Winter, K. (1976). Reichweite der Ausgasung im Einwirkungsbereich des Abbaus. *Gluckauf-Forschungshefte*, **37**, 22–7.

5

Dust in mine air and its control

Dust can be defined as finely divided solid matter and can be considered from two aspects: (1) its explosive properties and (2) its harmful physiological effects. Dust becomes a nuisance when it is in the form of clouds, reducing visibility, creating an uncomfortable environment (irritation of the eyes, ears, nose, throat and skin) and increasing equipment maintenance costs due to excessive wear and premature failure of components. Dust also causes higher mining costs by increasing the accident frequency and undue delays on days that it is too dusty to work.

Explosion hazard. The possibility of a dust explosion is always present in coal mines. In order to reduce this risk it is necessary to take action against the deposition of coal dust on roadways and to provide stone dust barriers, etc. This aspect is covered separately in Chapter 8.

Health risk. Due to the inhalation of fine dust particles and their retention in the alveoli of the lungs, there is a health risk, depending upon exposure time and the nature of the dust, particularly its concentration and physicochemical properties. The term 'pneumoconiosis' is used to describe all lung diseases caused by accumulation of insoluble dust in the lungs. The number of pneumoconiosis cases have gradually decreased with the application of improved dust control techniques but this condition still represents the greatest single health hazard at the present time.

5.1 Physiological effects of mineral dust

Inhaled air passes through the nose and throat into the trachea, which divides into two main branches, one going to each lung. The left lung is divided into two lobes whereas the right is divided into three (Fig. 5.1). These lobes are made up of a vast number of small alveoli (air sacs), each of which is the termination of the branches of the trachea known as the bronchi and bronchioles. Each alveolus is supplied with arterial and venous capillaries and a lymphatic duct.

The respiratory tract has a number of defence mechanisms to catch any dust which may be inhaled. The larger particles (>10 μm) are likely to be caught in the nasal passages or throat. Some of the particles passing through the trachea and bronchi are caught in their mucous secretions. This mucus is continually pushed upwards by numerous minute hairs (cilia), which work with a whip-like action, at about 1.25 cm a minute until it is discharged from the trachea. The remaining smaller particles (mostly <5 μm) may pass into the small alveoli. Here the mobile scavenger cells known as macrophages can engulf the particles and move into the bronchi to be ejected by the cilia. Other particles may, however, pass through the alveolar wall and remain in the lung tissue, or enter the lymphatic drainage system via the macrophages and reach the lymph glands which function as filters. Depending on the nature of the dust, a fibrotic reaction may take place within the lung tissue surrounding the dust particle. (Silica, asbestos, coal, diatomaceous earth, and silicates such as talc, kaolin and mica are known to produce fibrotic pneumoconiosis.) Lastly, many particles (<1 μm) remain suspended in the inhaled air and are exhaled.

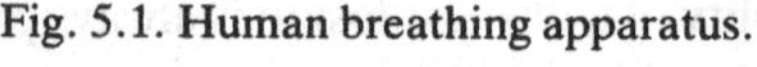

Fig. 5.1. Human breathing apparatus.

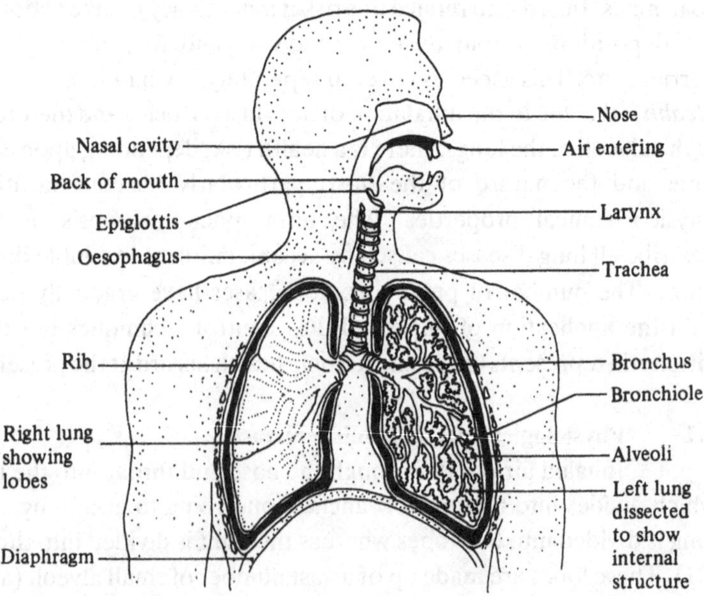

All the defence mechanisms are neither completely nor immediately effective so that the retention period of inhaled dust may range from hours to months. Where excessive concentrations of dust are encountered for long periods, the risk of a pathological response is greatly increased as defence mechanisms become overloaded. These pathological changes may alter the normal physiology, leading to a further deterioration in respiratory defence mechanisms and creating a condition of less effective lung clearance as the exposure continues.

With the deterioration of the lung tissue, pulmonary function is damaged. The symptoms may range from a shortness of breath to problems of oxygen and carbon dioxide diffusion or blood circulation, i.e. progressive diminution of working capacity, and eventually complete incapacity. There is also an increased susceptibility to tuberculosis.

Pulmonary function tests and radiological examinations have been coupled with industrial hygiene studies throughout the world in the last 50 or so years. (Radiological examinations are usually relied upon for detection and classification of the lung disease into various stages.) Threshold limit values (TLVs) of dust concentrations have been determined from the correlation of dust concentrations and medical findings.

5.1.1 Factors affecting the degree of health risk

(i) Physiological properties of dust

The properties of dust affecting the development and severity of lung diseases are: (a) composition, (b) size, and (c) concentration.

Composition. Free silica (also called crystalline silica comprising quartz, tridymite and cristobalite) is the most dangerous component of dust affecting the behaviour of alveolar macrophages. When these cells die they release a substance which is foreign to the body, producing an allergic type of reaction resulting in fibrosis. It must be emphasised that it is the composition of the dust reaching the lung and not that of the airborne dust which is important, since the composition of these two dusts can vary considerably.

Size. By conducting post mortem examinations of the lungs of numerous workers, it has been determined that pneumoconiosis is caused by dust particles below 5 μm. There is some evidence that the greatest danger is from particles ranging from 1 to 2 μm in size.

Only particles within the size* range of about 0.25 to 10 μm (called

* Size refers to aerodynamic size which is defined as the diameter of a unit density sphere having the same settling velocity as the particle in question.

respirable dust) enter the lung and only a percentage (depending upon the size of the particle and the person's retention characteristics) of those entering the lung are retained. In general, irregular particles fall more slowly than spheres of the same mass and for this reason irregular particles having a mass greater than that of a 10 μm diameter unit density sphere can be found in the lung. In some cases this effect can be serious. Asbestos can break readily into bundles of fibres having a diameter of only 0.02 to 0.12 μm with a length of 1 to 150 μm. Because of their very low sedimentation rate (depending upon the diameter), these fibres are not caught in the respiratory tract and are able to reach the lung where they are retained. Chrysotile has curved fibres and these are less likely to pass through the respiratory tract; it is probably for this reason that chrysotile is less likely to cause lung damage than amosite or crocidolite.

Concentration. Concentration of dust can be expressed as (i) mass of dust per unit volume of air, (ii) number of particles per unit volume, and (iii) surface area of particles per unit volume.

The mass concentration of respirable dust is the best single parameter to measure for assessment of the risk of pneumoconiosis from coal dust. For quartz dust, the surface area of the respirable particles is probably the best parameter to measure although it is usually measured with a gravimetric dust sampling instrument.

(ii) Time of exposure

The human respiratory system has a certain capacity for disposing of inhaled dust. Under overloaded conditions, larger lung dosages produce faster development of pneumoconiosis. Thus, the time of exposure to a certain dust concentration is an important factor in the development of pneumoconiosis.

The incidence of pneumoconiosis increases with increase in both the concentration and time of exposure and is correlated with the cumulative dust exposure calculated from the length of employment of the worker and the weighted average concentration of dust to which the worker is subjected on each shift.

5.1.2 *Threshold limit values*

Safe limits of dust concentration fixed in various countries have proved satisfactory in minimising the incidence of pneumoconiosis in mines. The TLVs (time-weighted average) are given for various countries in Table 5.1.

5.2 Airborne dust sampling, measurement and analysis

From the health point of view, the mass concentration of respirable dust is the main parameter to measure. It is important to use respirable dust samplers incorporating size selectors so that coarse non-respirable particles can be removed. The following figures recommended by the International Pneumoconiosis Conference held in Johannesburg in 1959 have been widely accepted as the basis for the design of elutriators aimed at the elimination of all but respirable dust in dust sampling instruments:

100 % of particles of size 1 μm equivalent diameter and less;
50 % of particles of size 5 μm equivalent diameter;
0 % of particles of size 7 μm equivalent diameter and greater.

The equivalent diameter of a particle is the diameter of a sphere of unit density having the same settling velocity in air as the particle studied.

All sampling instruments have some elements in common, such as an inlet (for entry of dust-laden air into the instrument), a dust separator (for separating the dust from the air, e.g. filter, impinger, thermal precipitator, electrostatic precipitator, elutriator, cyclone), a pump, flow control mechanism, flow rate measurement device and power supply.

Table 5.1. *Threshold limit (time-weighted average) values for various dusts in different countries (after Major, 1978, and Walton, 1978)*

Coal dust	Quartz-containing dust	Asbestos
Australia		
5 mg/m³ respirable dust containing ⩽5 % free silica	$\frac{25}{\%\text{ respirable free silica} + 5}\text{ mg/m}^3$	4 fibres/ml for chrysotile and amosite; 2 fibres/ml long-term average for other than crocidolite
UK		
8 mg/m³ – longwall faces	3 mg/m³ – drivages if quartz > 0.45 mg/m³ 6 mg/m³ – if quartz < 0.45 mg/m³ 6 mg/m³ – other places	0.2 fibres/ml for crocidolite when measured over a 10 minute period; 2 fibres/ml for other types of asbestos averaged over a 4 hour period
USA		
2 mg/m³ respirable dust containing ⩽5 % quartz	$\frac{10}{\%\text{ respirable quartz} + 2}\text{ mg/m}^3$	2 fibres/ml > 5 μm long (8 hour time-weighted average)

Sampling instruments can be classified according to the parameter measured and the principles used in its design. Many instruments had been developed for assessing dust concentration by the number of particles per unit volume. These instruments use three methods of dust collection, namely, thermal precipitation, settlement and impaction. Two instruments, the thermal precipitator and the konimeters, collect dust directly onto glass slides which are later examined under the microscope. The impingers collect dust as a liquid suspension and the settled dust is assessed by microscopic examination.

Instruments for assessing the total concentration of dust by mass of dust per unit volume of air are available using two methods of dust collection, namely, electrostatic precipitation and filtration. For assessing respirable dust, several instruments have been developed incorporating various types of size selectors.

5.2.1 *MRE gravimetric dust sampler type 113A*

This instrument (Fig. 5.2) was developed at the Mining Research Establishment of the National Coal Board (UK). An electric motor drives a diaphragm-type pump at a constant speed, drawing dust-laden air through the horizontal elutriator (here large particles fall out) and the filter disc, the clean air then being expelled through a smoothing device and a flowmeter. A counter geared to the pump gives the total airflow in litres, and it must be set to zero before each sampling period. The rate of

Fig. 5.2. MRE gravimetric dust sampler type 113A.

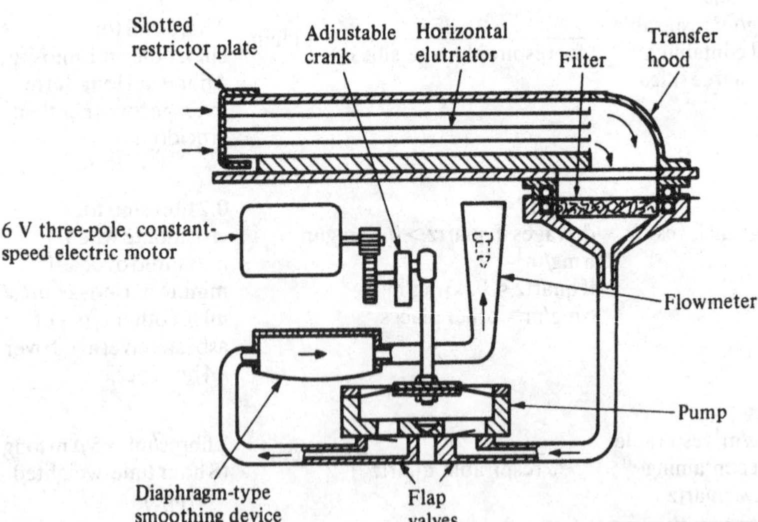

airflow is 2.5 litres/minute. The instrument can be used continuously for up to 12 hours.

The dust is collected on the filter, which may be of glass fibre paper or a sponge-like membrane material made of polyvinyl chloride. For simple measurement of mass concentration, filters are weighed before and after use. To avoid errors due to moisture changes, filters are conditioned after use and before weighing for about 20 minutes. The composition of the sample can be analysed using various methods, such as ashing, X-ray diffraction, or infra-red spectrophotometry for measurement of quartz or other mineral content.

A multi-shift gravimetric dust sampler has been developed at the Mining Research and Development Establishment (UK). This sampler has the capability of sampling continuously for five full working shifts (40 hours), or it can be programmed to sample any five shifts over a specified time interval. This has been achieved by reducing the number of elutriator plates in the type 113A instrument from 4 to 1 and correspondingly reducing the sampling rate from 2.5 to 0.625 litres/minute. Sequential sampling is achieved with the use of a built-in timer.

5.2.2 *The TBF 50 gravimetric dust sampler*

This instrument (Fig. 5.3) was developed at the Mining Research Establishment of the Steinkohlenbergbauverein (West Germany) (Breuer, 1975). It basically consists of two cyclones connected in series which separate the dust from the air into three fractions as follows.

Fig. 5.3. The TBF 50 gravimetric dust sampler.

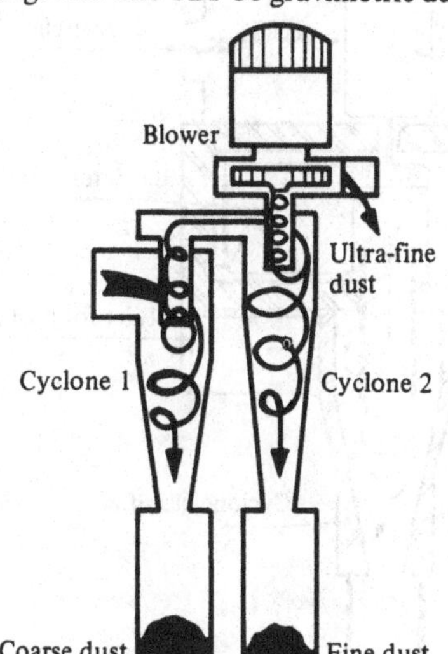

(1) The coarse dust separated in the first cyclone. This represents the proportion of dust deposited in the upper respiratory tract.
(2) The fine dust separated in the second cyclone.
(3) The fine dust not separated in the second cyclone. This corresponds to the small portion of respirable dust which is rejected with the exhaled air.

The first two fractions of dust separated in the cyclones is transferred for assessment into containers for coarse and fine dust by a stream of water. The third fraction can be separated out by a membrane filter, attached after the two cyclones, for special investigations.

The sampling rate is 50 litres/minute. The instrument can be operated by a compressed air ejector, a battery-operated fan or by a pump.

5.2.3 *The CPM3 gravimetric dust sampler*

This instrument (Fig. 5.4) was developed at the Centre d'Etudes

Fig. 5.4. The CPM3 mine dust sampler.

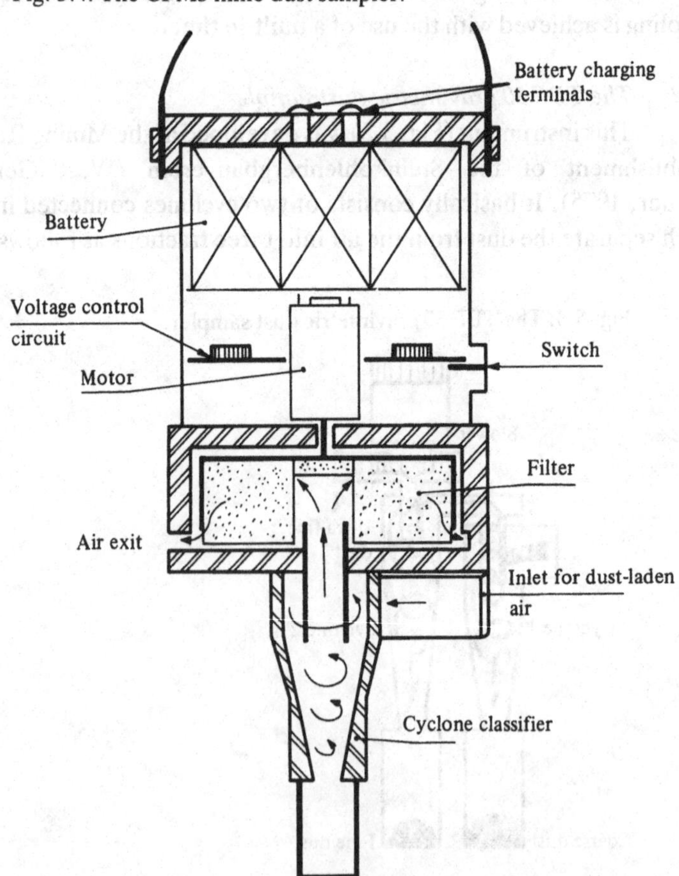

et Récherches des Charbonnages de France in collaboration with Mine Safety Appliances (Courbon, 1972). The instrument consists of a battery, a motor, a holder containing a polyurethane foam filter fixed to the motor shaft by magnetic system, and a cyclone. The filter–container assembly rotates inside a cylindrical casing. The cyclone separates out the largest particles normally deposited in the upper respiratory tract. The rotation of the filter–container assembly inside the casing produces the suction necessary for the cyclone to operate. The sampling rate is 50 litres/minute.

5.2.4 *Instruments assessing dust by light scatter*

Two instruments based on the measurement of scattered light have been developed in Germany and Britain. The German instrument is the Tyndallometer TM Digital for respirable dust developed by the Bergbau Forschung and manufactured by Leitz. The readout is digital and takes about 15 s to be displayed. The instrument can be linked to a recording and integrating device, and can be operated either with a battery pack or from a constant voltage supply. No air-moving device is used to induce air into the instrument. A sensing chamber is opened to allow passage of the dust-laden air into the illuminated area.

The 'Simslin' instrument was developed at the Safety in Mines Research Establishment in the UK and is manufactured by Rotheroe & Mitchell (Leck and Harris, 1978). Some details of the self-contained Mk II instrument are given in Fig. 5.5. A small pump draws the dust-laden air at the rate of 0.625 litres/minute through a parallel plate elutriator to remove the non-respirable dust, through a light beam and then a filter so that the dust can be used for further analysis. The use of a filter also enables instrument calibration.

The air containing the respirable dust is drawn through the photometer via two tubes. The space between the tubes forms the light-scattering region. Here the dust-laden air is constrained to flow in a well-defined

Fig. 5.5. Simslin II instrument for measuring dust content of air by light scatter.

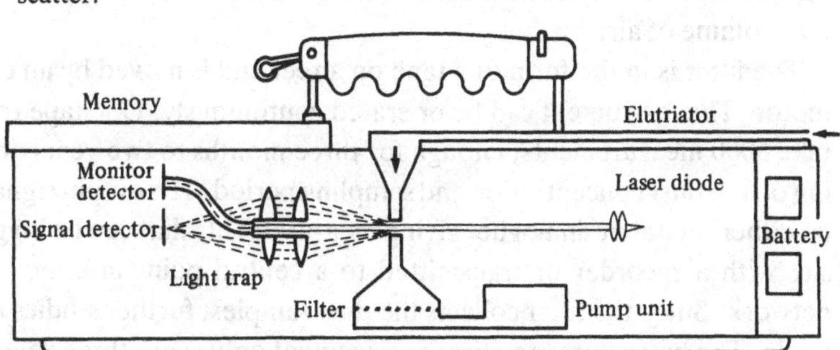

column by a sheath of recirculated clean air. A small laser diode emits infra-red radiation which is collected and focussed into a primary beam by the condenser lens. This beam is directed along the axis of the photometer towards the column of dust-laden air in the scattering region. A small fraction of the incident radiation is scattered by the dust in this region. Some of the scattered radiation is collected by the lens system and focussed onto the central detector. A light trap is used to prevent the primary beam from reaching the central detector.

The unscattered primary beam enters the trap via a light pipe to a second identical detector which is used to generate a reference signal.

A digital display of instantaneous concentration is given. It also calculates automatically and displays the running average concentration every 15 minutes. Other features include programmable automatic operation, and an interchangeable solid-state memory unit (Simstor) which allows a reading to be stored every 15 s for later playback to a chart recorder or data processing system where a detailed analysis of results can be conducted.

5.2.5 *Instruments assessing dust by beta-absorption*

Two instruments based on the beta-absorption method of dust measurement have been developed in the UK and the USA. The British instrument is the Philips PW 9790 developed by Pye Unicam. In operation, radiation from a low-level beta radiation source is passed through a glass-fibre filter and measured with a Geiger–Muller detector. An electronic updown counter counts the output pulses over a pre-set period. After passing a known volume of dust-laden air through the same section of the filter, depositing dust, the filter is exposed once again to beta radiation, and the output pulses from the detector are used to count down from the number attained with a clean filter towards zero. The dust collected on the filter absorbs more radiation and the count takes a longer time than before. The difference between up-and-down-counting periods is directly related to the mass of dust on the filter. Since the volume of air sampled is known, the dust concentration can be expressed as mass per unit volume of air.

The filter is in the form of a tape on a reel and is moved by an electric motor. The instrument can be operated continuously. One tape reel can take 5000 measurements, enough for three months to two years depending on the dust concentration and sampling period. The output signals can be either digital or analogue, giving measurements that can be logged on site with a recorder or transmitted to a central point in a monitoring network. Since the filter collects the dust samples, further studies can be made. The beta source requires replacement only every three years.

The American instrument is the GCA Model RDM-201 developed by the GCA Corporation. This incorporates a cyclone pre-collector where the respirable dust is collected on a thin, grease-coated plastic by impaction.

The beta-absorption method is reported to be unaffected by the physical, chemical or optical properties of the dust.

5.2.6 *Personal samplers*

Several personal samplers have been developed in recent years to assess the risk dust exposure represents to the health of a worker. The sampler consists of two parts: a motorised pump unit and a cyclone elutriator/filter unit, the two being connected by a length of tube. In the case of Simpeds respirable dust sampler, the pump unit is mounted on the cap lamp battery whereas the cyclone unit is mounted on the helmet by the side of the lamp (Gwatkin and Odgen, 1979). A cross-section of the cyclone is given in Fig. 5.6. The dust-laden air is drawn up the vertical entry tube and enters the cyclone chamber tangentially. The respirable fraction of the dust is collected on the filter near the top and the

Fig. 5.6. Cross-section of the Simpeds cyclone (after Gwatkin and Odgen, 1979).

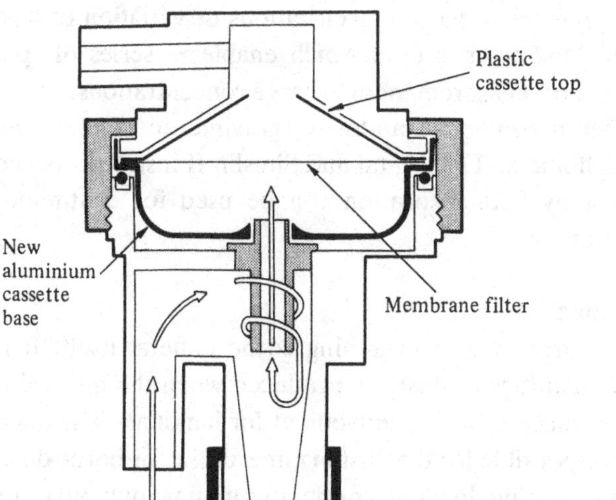

non-respirable fraction in the grit pot at the bottom. The filter is weighed to determine the respirable mass of dust per unit volume of air and analysed by X-ray spectroscopy for silica content of the dust. The main advantage of a personal sampler is that the miner wears the sampler, permitting sampling of the dust cloud to which he is exposed. It also provides an estimate of the respirable dust in the atmosphere for periods of up to 8 hours.

Sampling of respirable dust at strategic points throughout the mine should be done by positional samplers and the relationship between the sampler's exposure and the workers' exposure established and checked periodically using both positional and personal samplers. Personal samplers can be distributed amongst a representative number of workers.

5.2.7 *Dust measurements for control purposes*

For control purposes, relative assessments of the dust are required. The instrument selected depends on the dust concentration, type of mining operation, available sampling time, and the type of concentration, i.e. peak or average, to be measured.

For high dust concentrations, thermal precipitators or gravimetric samplers are required. For intermittent or limited-duration measurements, the konimeter is ideal. For continuous operations, any of the three, i.e. konimeter, thermal precipitator and gravimetric sampler, can be used. For measurement of peak concentrations or variation of dust concentrations, the konimeter is used which enables a series of spot samples to be taken. For measurement of average concentrations, any of the three – konimeter, thermal precipitator and gravimetric sampler – can be used. The Tyndallometer TM Digital and Simslin II instruments and those assessing dust by beta-absorption can be used for continuous monitoring with advantage.

5.3 Sources of dust

The main source of dust in mining is the mineral itself. It is obvious that large quantities of dust are produced when the mineral is broken down and reduced to a size convenient for handling. The main mining operations responsible for the dust in mine air, i.e. airborne dust, are: drilling, cutting, blasting, loading, continuous mining, dumping cars and drawing chutes.

5.4 Control of dust

There are four rules regarding dust control:

(1) produce a minimum amount of dust and prevent it from becoming airborne at its source;

(2) dilute it as soon as possible by ventilation;
(3) separate it;
(4) avoid it.

Every effort should be made to prevent the formation of dust as well as to prevent it from becoming airborne at its source. Water is mainly used for this purpose. In order to suppress airborne dust by spraying, it is necessary to spray the material before it is broken up, or to confine any dust by a curtain of water sprays. An adequate quantity of good quality water supplied through a well-designed spray system is essential. The pump selected should be powerful enough to give the quantity and pressure required. An appropriate pipe size is also important. A non-clogging filter system is also required to maintain the quality of water.

Smaller particles of water are more efficient in suppressing dust (Fig. 5.7). If the droplet diameter is much greater than the dust particle, the dust particle simply follows the airstream lines around the droplet, and little or no contact occurs. If, on the other hand, the water droplet is about the same size as the dust particle, contact occurs as the dust particle tries to follow the stream lines. Thus the probability of impaction increases as the size of the water droplet decreases. For a given spray nozzle, the collection of small particles increases as the pressure increases. At a given

Fig. 5.7. Airflow around large and small water droplets (after Schowengerdt and Brown, 1976).

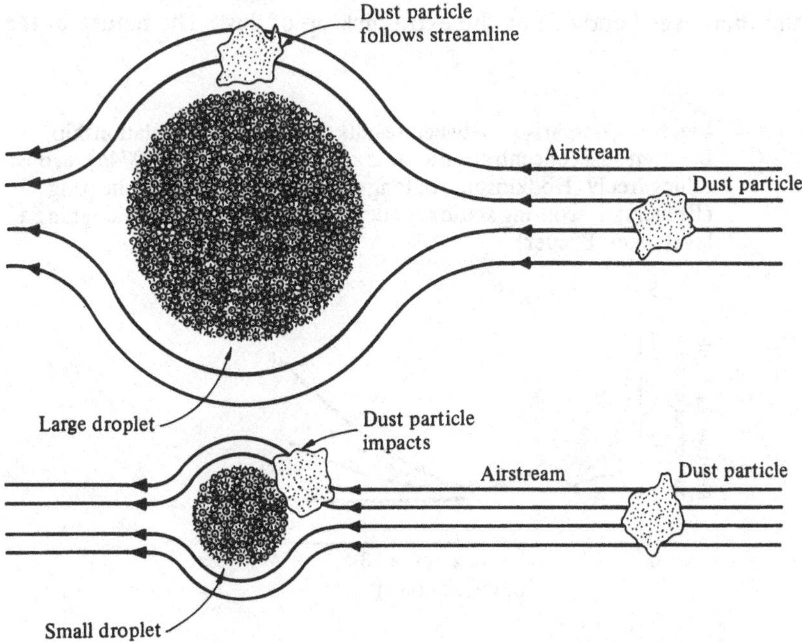

pressure, the collection increases as the nozzle design is changed so as to produce smaller droplets.

Spray nozzles are of two types: hydraulic atomising and pneumatic fogging. The spray pattern may be a flat fan, an elliptical jet, or solid or hollow cone configurations. The hollow cone pattern reduces dust better than the solid cone sprays. At the same pressure, pneumatic sprays are more efficient.

Excessive use of water should be avoided as it can result in the deterioration of climatic conditions as well as in problems in the preparation of the coal, for example it can cause sticking to the screens and bunkers.

The use of a wetting agent when properly selected and in appropriate concentrations can reduce the dust concentration by as much as 25 %. Certain agents can enhance dust production.

Dilution by ventilation is required when dust suppression measures at its source fail. By supplying an adequate airflow, the dust concentration can be reduced to safe levels. However, it may not be possible or economical to supply the necessary airflow. Also, increasing air velocities beyond a certain limit (>4 m/s) can cause normally stagnant dust to become airborne.

Figs. 5.8 and 5.9 show the relationship between dust concentration and air velocity. As the velocity is increased from very low levels, the dust concentration decreases steadily by dilution. However, as the velocity increases further and further, the dilution effect is gradually counteracted and then overshadowed by the extra pick-up of dust. The nature of the

Fig. 5.8. Comparison between results of studies of the relationship between dust concentration and air velocity (after Anon, 1974*b*). a, dust falling freely (Hodkinson); b, longwall coal face (Hall); c, shearing (Breuer); d, scouring stationary deposits (Hodkinson); e, ploughing a level seam (Breuer).

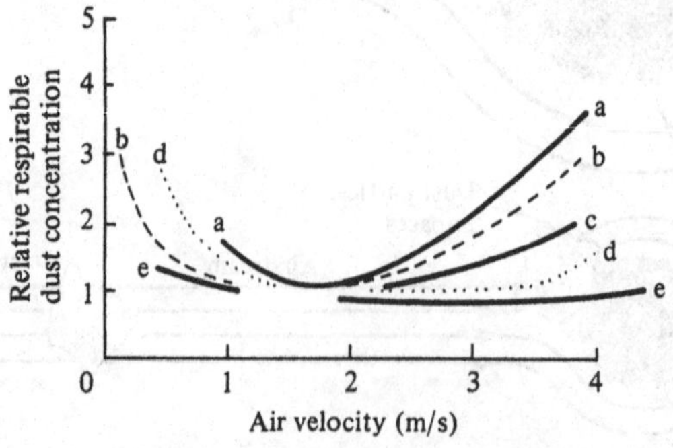

dust source, the moisture content and size of the particles have some effect on the results. If the air velocity is less than 1.5 m/s, dust concentration can be reduced by increasing the air velocity up to about 4 m/s.

Dust separators can be used to remove airborne dust from the air. The methods available for dust separation are:

(1) gravity methods;
(2) centrifugal methods;
(3) filtration;
(4) wet scrubbers;
(5) electrostatic precipitators.

Gravity collectors remove dust by allowing a horizontal current of dust-laden air to enter a relatively stagnant zone of airflow where the dust particles are separated and deposited into a bunker. The main disadvantages are the relatively large size of the apparatus and its ability to cope only with large particles, say, >100 μm.

Centrifugal collectors force the dust-laden air to undergo a number of revolutions in a circular chamber. The dust particles are carried to the wall of the chamber and fall to the base, where they are withdrawn through a valve or gate. About 75 to 90 % of the particles down to 10 μm in size may be separated out in these collectors. Centrifugal collectors with a conical shape are called cyclones. A typical cyclone is shown in Fig. 5.10. Normally, they are unable to remove a sufficiently large proportion of the particles of less than 10 μm to provide a satisfactory respirable atmosphere.

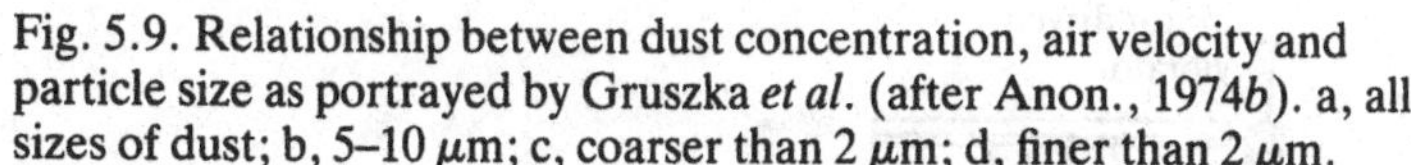

Fig. 5.9. Relationship between dust concentration, air velocity and particle size as portrayed by Gruszka *et al.* (after Anon., 1974*b*). a, all sizes of dust; b, 5–10 μm; c, coarser than 2 μm; d, finer than 2 μm.

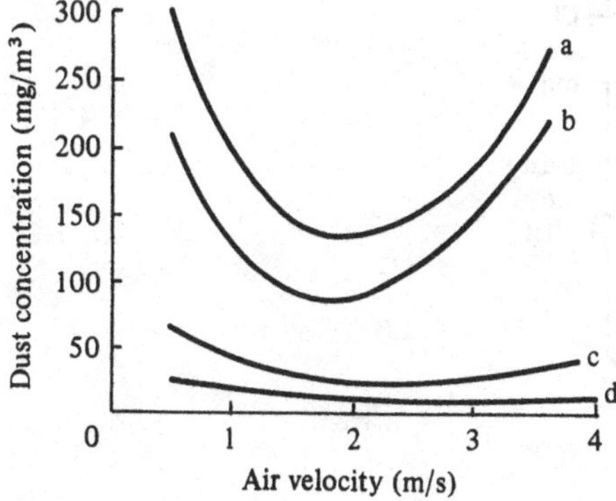

Fig. 5.10. Low-pressure cyclone (after Soderberg, 1970).

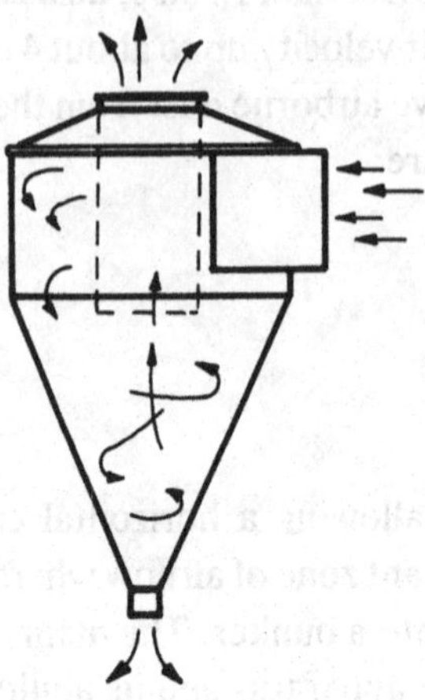

Fig. 5.11. Principle of operation of shaker collector (after Soderberg, 1970).

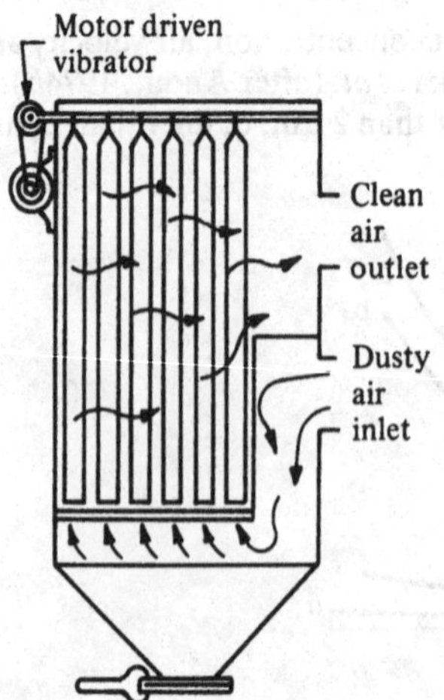

Airborne dust can be removed from the air when it is passed through a permeable, porous material which traps particles. A typical collector using a filter of fabric is shown in Fig. 5.11. These can have very high efficiencies. The material used in the filters may be woven or felted fabric, in the form of flat envelopes or sleeves. Friction during filtration may produce high electrostatic spark discharges, which can be reduced by incorporating conducting elements in the filter material. The dust deposit from the fabric can be removed by changing the filter element, mechanical rapping/shaking, reverse air jet, or pulse jet.

The simplest form of wet scrubber is the spray tower (Fig. 5.12), in which water spraying downwards meets the dust-laden air flowing upwards. The other types are self-induced spray scrubbers, venturi scrubbers and rotary scrubbers. Wet scrubbers of various types are favoured in coal mines.

A fibrous bed scrubber consists of a fan, a short duct at the fan inlet, and a rectangular housing that holds a thick bed or mat of knitted wire mesh. Dust-laden air is drawn through the bed of wire mesh. Spray nozzles are installed in the duct upstream of the mesh to wet it.

A cyclone scrubber consists of a bank of cylindrical cyclone tubes. Stationary vanes in each tube induce the air to spin and dust particles are thrown to the side of the tube where either they can be bled off in a separate airstream or they impact on the wetted vanes or side wall of tube. Water sprays are placed upstream of the panel to insure complete wetting.

A wetted fan scrubber consists of a fan with water spraying into the fan inlet. In the case of a wetted brush scrubber, a bed of wire or plastic brushes is added to the inlet side of the basic wetted fan scrubber. Water sprays are located upstream of the brushes to insure complete wetting.

Fig. 5.12. A spray tower (after Soderberg, 1970).

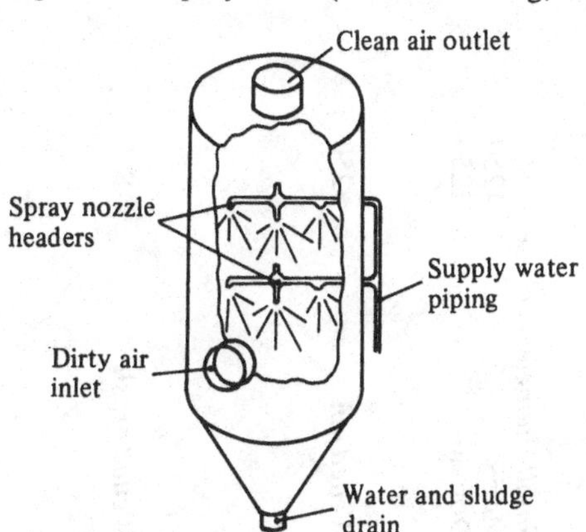

Table 5.2. *Summary comparison of maximum dust collection efficiency (after Divers and Janosik, 1978)*

Scrubber type	Differential pressure across fan (Pa)	Water rate (litres/minute)	Water pressure (kPa gauge)	Average maximum efficiency ($<7\ \mu m$)	Advantages	Disadvantages
Flooded bed	1100	10	830	95	Highest dust collection efficiency	May clog where large ($>25\ \mu m$) particulate is encountered
Small-diameter cyclone						
Merix	1250	21	485	91	Can be used as both scrubber and eliminator; thus gives minimum size if eliminator must be used	Dust collection efficiency is limited
Donaldson	1250	21	485	89		
Wetted fan	250	10	965	87	Minimum size and cost	Spray water and particulate can cause fan blade erosion. (Component layout of other scrubbers can prevent this problem)
Wetted brush	850	15	310	SS = 88 PP = 82	Small size and low cost. Higher efficiency than wetted fan, or lower water use than wetted fan	Fan blade erosion

SS = Stainless steel; PP = polypropylene.

Special requirements	Recommended scrubber type	
	First choice	Second choice
High efficiency	Flooded fibrous bed	Small-diameter cyclone
Minimum size	Wetted fan	Wetted wire brush
Coarse or sticky particulate	Small-diameter cyclone with maximum water	Wetted wire brush with maximum water
Low water consumption (less than 3.8 litres/minute per 0.5 m^3/s)	Small-diameter cyclone	Flooded fibrous bed

Table 5.2 gives a summary of efficiency, advantages, disadvantages and recommended use for each type of scrubber. The fibrous bed scrubber achieves respirable dust collection efficiencies of 95 %. The cyclone scrubbers have about 92 % efficiency and are best where excess water is a problem. The wetted brush scrubbers have 88 % efficiency at low cost and the wetted fan scrubbers are slightly less efficient at 87 % but offer the smallest size and cost.

The other scrubbers include the Joy microdyne scrubber, and the Mining Research and Development Establishment's irrigated fabric filter and rotovent.

In an electrostatic separator (Fig. 5.13), a stream of electrons at right angles to the direction of the airflow passes from a number of negative electrodes (may be plain wire, star section or pointed), maintained at a high voltage (25 to 50 kV), to earthed positive electrode plates. Dust particles acquire a negative charge, are attracted to the positive electrodes and are removed from these by intermittent mechanical vibration or by

Fig. 5.13. Electrostatic precipitator: principle of operation (after Moyes and Reeves, 1974).

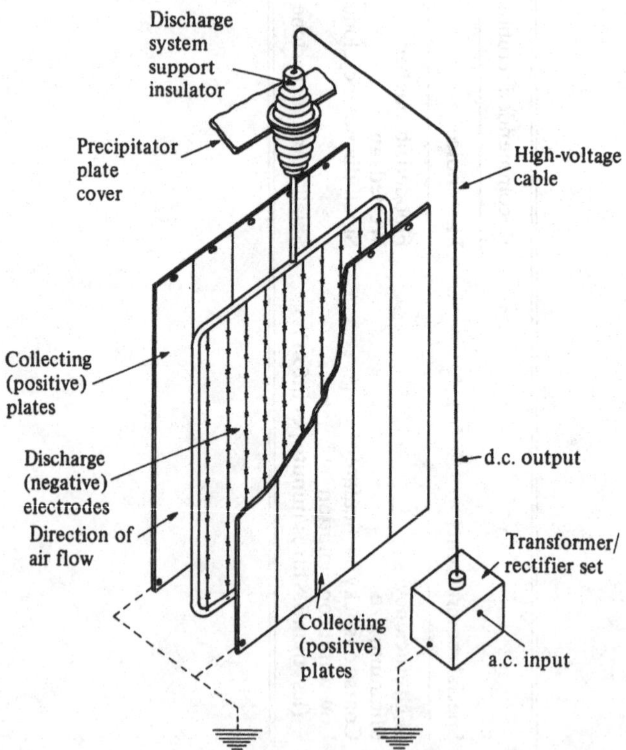

water flushing. The system can collect submicron sized particles with high efficiency and with relatively low energy consumption, but is not suitable for the moist conditions usually prevailing underground.

When workers are exposed to high concentrations of dust for only short periods of time, such as blowing-over operations in sinking shafts, respirators should be used as a protection measure. Modern respirators are very efficient (approximately 96 %), reasonably comfortable to wear and do not cause breathing difficulty. Recently, a special 'Airstream Anti-Dust' helmet incorporating a small axial flow fan powered by a battery has been developed by the Safety in Mines Research Establishment of the UK. It is a three in one headset consisting of helmet, face visor and personal dust extractor. Fresh air is blown across the face by the fan, after first being filtered.

Removal of workers is the most effective way of preventing exposure to dust. It is applied mainly after blasting by prescribing a certain minimum re-entry period.

5.4.1 Control measures for individual mining operations

Drilling

Drilling is done wet when possible. If it has to be done dry, exhaust ventilation and dust collectors are used to capture the dust. An air–water mist is to be preferred for blowing out holes.

Blasting

Prior to blasting, the area is sprayed with water. Water stemming can be used, and spray mist can be used during blasting. Ventilation removes the dust from the area. Muck piles are sprayed after blasting. A certain minimum re-entry period is prescribed. In the case of coal, compressed air breaking gives a larger sized product and less fines.

Mucking

Muck piles should be wetted down frequently.

Dumping

The dump entry should be maintained under exhaust ventilation. Water sprays should be used.

In the case of ore and waste passes, proper planning, extraction fan installation and a tipping-point door design are important to control the dust created at rock passes and tips. Broken rock falling through vertical or steeply inclined passes acts like a piston and pushes dust-laden air through openings below it such as dumping points and access doors.

Surges in vertical passes are very powerful and develop pressures that can bounce open heavy dump doors. Siting ore and waste passes close together, connecting them at each level and sufficiently above any control chutes so that surges can recirculate and pressure can be reduced at dumping points, and isolating such connections from haulage ways by airtight access doors (Fig. 5.14) can minimise and/or relieve surges. All dump openings should be covered with sturdy, mechanically operated doors to eliminate all unnecessary air leakages. It should also be planned to exhaust sufficient air from the system so that air flows into the system (velocity about 1 m/s) through all the remaining leakages and also dump openings while dumping. The exhaust should be discharged either directly into a return air circuit or filtered to remove the dust to prevent contamination of airways and travel ways. The alternative is to position the passes in headings driven off main cross cuts and apply local exhaust dust collection.

Hauling

Cars and haulage ways should be wet down adequately. Rock with a moisture content of only 1 % by mass produces significantly less dust during transportation than rock transported under dry conditions. A moisture content of 5 % is usually aimed for minimum dust production during transportation. Surface-active agents can be used for allaying dust in mine haulage ways.

Crushing

At the crusher mouth, spray nozzles should be used. Also the crusher mouth should be enclosed and maintained under exhaust ventilation. At

Fig. 5.14. Bypass and dump layout at ore and waste passes (after Anon., 1974*a*).

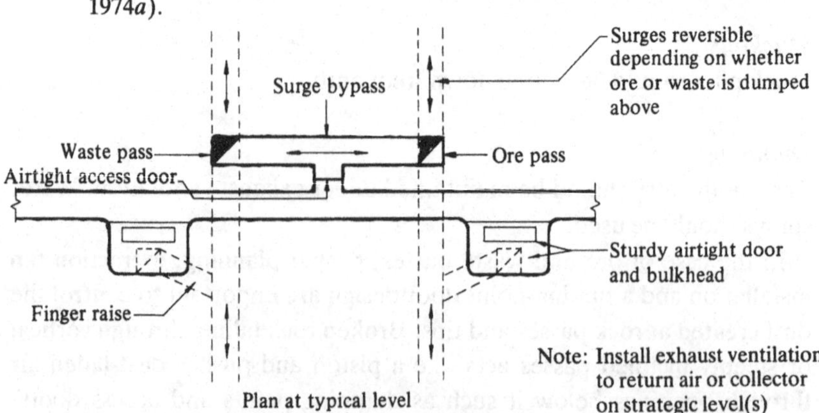

floor level, an airtight seal should be made and exhaust ventilation should be used under the crusher to confine airborne dust. The exhaust take-off should be positioned far away from the crusher discharge and the dust-laden air cleaned by means of a filter or discharged into a return airway.

Loading pocket

Dust is dispersed when measuring pockets and skips are being filled and the problem is serious if the shaft is a fresh airway. Ventilated exhaust hoods mounted on the pockets, chutes and skips can be used to confine the dust. The dust-laden air should be cleaned with a filter or discharged directly into a return airway. Water sprays may also be used but only sparingly. Impingement curtains can also be used for dust collection.

Skip dump into mine ore bin

Surges of dust-laden air escape from a mine bin every time a skip is dumped. The bin should be covered to confine dust except for an opening no larger than is necessary to dump the ore. Sufficient exhaust ventilation should be maintained from the top of the bin near the periphery so that air flows into the system through leakages and openings and the dust-laden air passes to a collector. The dust-laden air should be cleaned with the collector before being discharged into the atmosphere.

Coal cutting

There are several factors influencing the amount of dust produced, its size distribution pattern and its dispersion properties in the airstream. A high-rank coal containing a high percentage of vitrinite produces more dust in comparison to a low-rank coal. In certain cases, such as pre-drainage of methane and undermining of seams and standing pillars, the coal dries up resulting in high dust concentrations during mining. The material being cut also influences the amount of dust produced: cutting in sandstone or shale can produce up to 20 times as much fine dust as the same operation with coal.

Coals can be divided into different classes depending upon the dust produced. The coal 'dustiness index' can be determined in a cabinet (Fig. 5.15) specified in the American Society for Testing and Materials Standard D 547. The bulk coal is mixed and divided into 20 kg subsamples. A sample is transferred to the top of the cabinet. After replacing the lid of the cabinet, the slide holding the coal in the top is removed allowing the sample to fall to the base. After 5 s, the top slide is removed and two other slides are inserted (one on top of the other) to collect all the dust that settles within 2 minutes and within 10 minutes of the coal falling to the base of the cabinet. The mass in grams of the dust collected in 2 minutes (i.e. on the top slide) is determined and multiplied by a factor F, to give the coarse dustiness index, where

$$F = (1\ \text{t})/(\text{mass of coal sample}) = 1000\ \text{kg}/20\ \text{kg} = 50.$$

The same is done for the dust which is collected between 2 minutes and 8 minutes. This dustiness index is called the fine or float dustiness index.

Shearers. Most of the dust comes from the top and bottom of the cut and it is very important to choose the correct drum size and to control the horizon effectively. The cutter's picks should be sharp and in good condition. Depth of cut, pick line spacing and drum rotational speed, i.e. pick speed, influence the amount of dust raised. Fig. 5.16 shows the relationship between dust raised, depth of cut and pick speed. If a shearer can be used at the lower drum rotational speed while the haulage rate is kept at the same level, the depth of cut increased and pick speed reduced, a substantial reduction in dust concentration can be achieved. The drum should have the minimum number of picks to be consistent with good

Fig. 5.15. Schematic of the dust deposition cabinet (after Membrey, 1984).

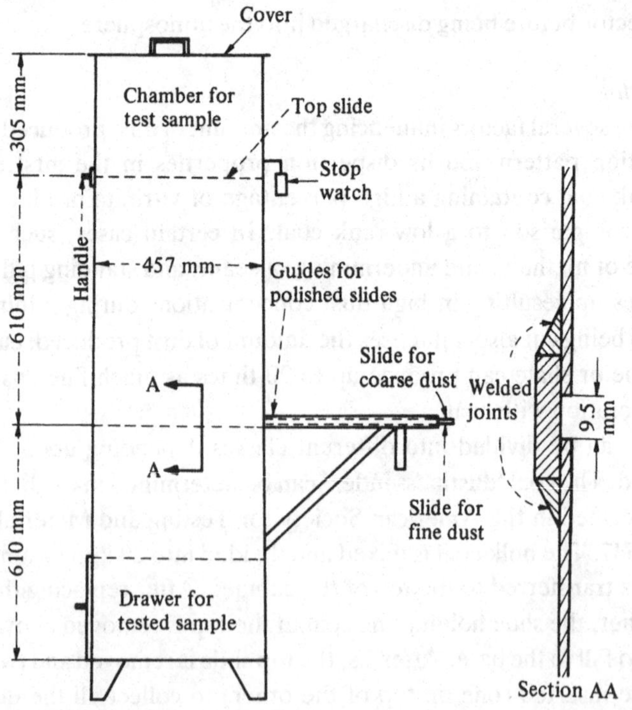

cutting. A pre-cutter in some cases can also help to reduce dust. The use of cowls to help screen the dust source from high velocity ventilation air can aid reductions in airborne dust.

Water sprays are also used on shearers. The water should be directed as a jet or as a coarse spray as close to the pick as is possible. Pick-face flushing (Fig. 5.17) on shearers reduces dust to about one-half of that dispersed from a machine fitted with external sprays. Externally mounted sprays can also be used effectively to deal with dust arising from flushing ahead of the drum or from falling tops.

On very dusty faces, dust produced on the longwall can be deflected back onto the coal face by mounting suitable venturis on the shearer. Attempts have also been made to incorporate dust collector systems within the shearer drum, using a number of axial tubes in the body of the drum (Fig. 5.18). Dust is sucked through the tubes and deposited at the bottom. A reduction in dust produced at the cutting of the longwall of almost 90 % has been claimed.

Impingement curtains can also be used on the coal face.

Fig. 5.16. Gross airborne dust versus depth of cut at different pick speeds (after Anon., 1974*b*). Key to pick speeds: circles, 3 m/s; crosses, 2 m/s; squares, 1 m/s; triangles, 0.25 m/s.

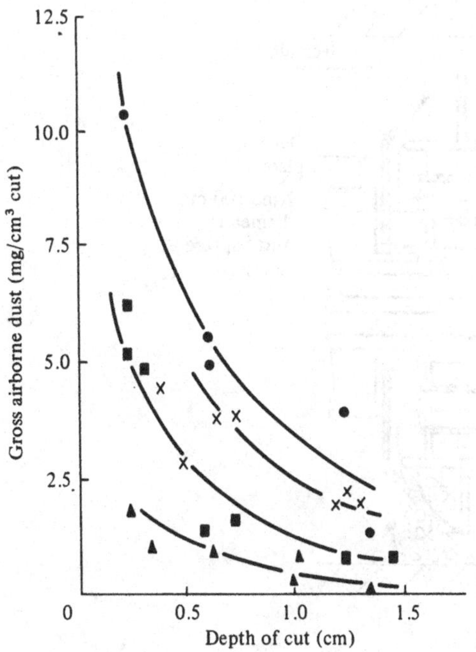

Fig. 5.17. Pick-face flushing.

Fig. 5.18. Cross-section of a dust extraction drum (after Hamilton, French and Browning, 1983).

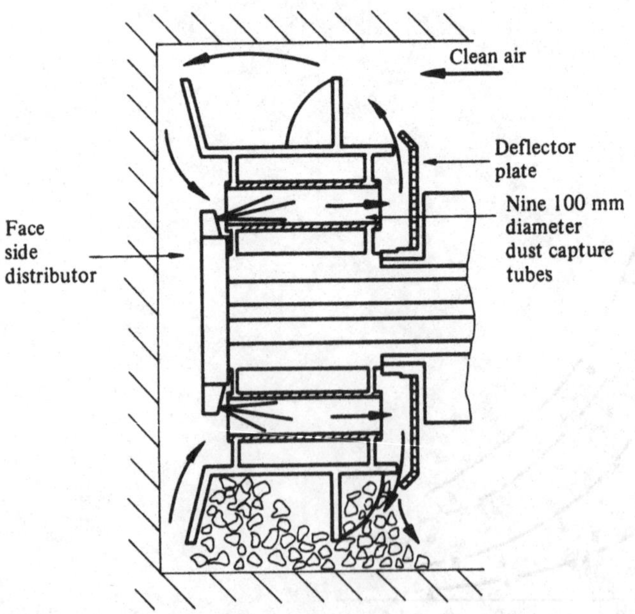

Continuous miners. About three-fifths of the respirable dust produced at the face is formed during the sumping operation, and the rest during shearing and loading. Lowering the rotational speed of the cutting bits and increasing the depth of cut can significantly reduce the amount of respirable dust produced. Although the lower rotational cutting speed would decrease coal production, the greater depth of cut should increase it to maintain normal production rate.

Water sprays are commonly used on continuous miners. An arrangement of spray nozzles for the US Bureau of Mines investigation is shown in Fig. 5.19. During sumping, top and bottom sprays have been found to be equally effective in dust suppression and dust has been reduced by about 13 %. However, in shearing and loading, bottom sprays have been

Fig. 5.19. Location of spray nozzles on a continuous miner (after Courtney, Jayaraman and Behum, 1978).

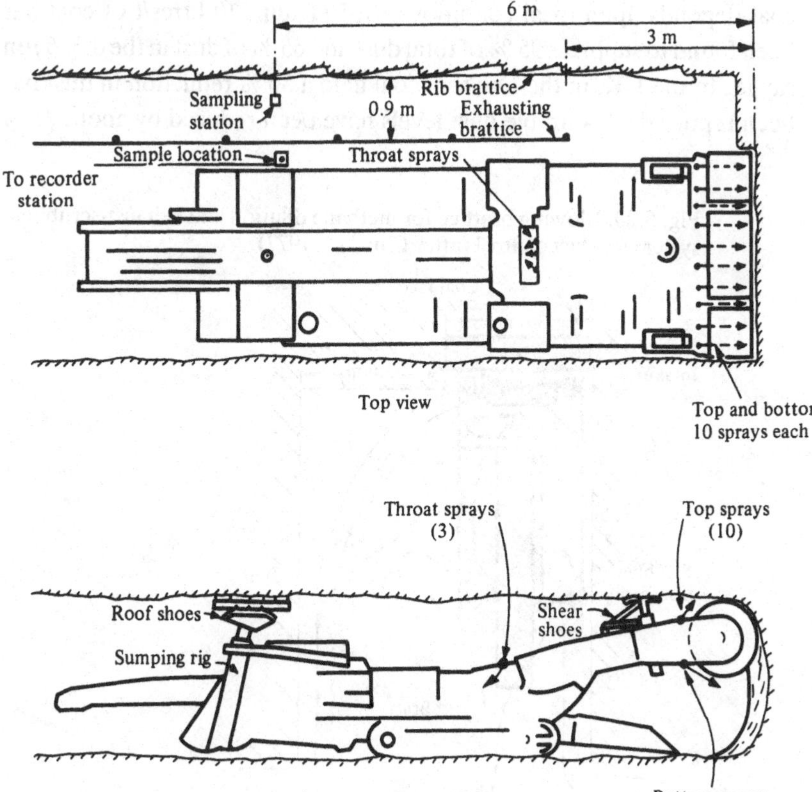

found to reduce dust by 61 % and are about 50 % more effective than top sprays. The combination of top and bottom sprays has been found to be more effective than the separate systems, but water consumption is increased by 60 %. Inclusion of throat sprays has decreased dust reduction by removing some of the water from the top and bottom sprays where it has been found to be more effective.

A dust collector can also be fitted to a continuous miner. A blowing brattice, combined with an exhaust–scrubber system, assures a good airflow near the face for methane dilution while filtering dust (Fig. 5.20).

Water infusion. Water infusion of coal seams to suppress respirable dust is widely practised in West Germany, France, Belgium and to some extent in the UK, USSR and USA. In West Germany, almost 50 % of the longwall faces are infused and in the northern coal fields of France almost 89 % of the coal produced comes from infused faces. The success of infusion has varied depending upon the mining conditions, type of coal seam and the technique practised. The amount of water pumped into the coal depends upon local conditions. In Belgium, 10 litres/t of coal has been found to suppress 95 % of total dust and 65 % of dust in the 0.5–5 μm range. In the UK, in the Durham coal field a 50 % reduction in dust has been reported. Respirable dust levels have been reduced by about 75 %

Fig. 5.20. Blowing brattice for methane dilution and exhaust-scrubber system for dust control (after Chironis, 1977).

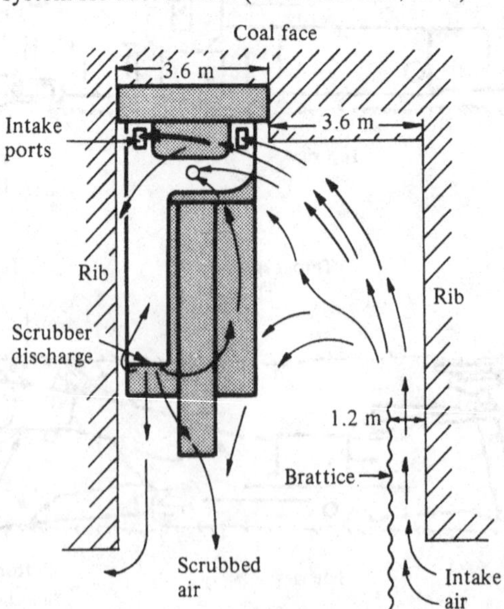

on a longwall section in the Pocahontas No. 3 coal seam and by about 50 % on a development section in the Upper Kittanning coal seam, but no statistically reliable reduction has been found in the respirable dust levels in the Pittsburgh coal seam in the USA.

Water infusion in a coal seam may be done at the development faces (Fig. 5.21) or in the longwall block from advance roadways (Fig. 5.22). The length of the holes may be short (about 2–5 m) or long (up to 100 m). The purpose of water infusion is to increase the humidity of the coal just before excavation. Water circulates in the coal through a network of fissures; it moves under the influence of capillary attraction and gravity towards free surfaces of the coal mass, where significant evaporation takes place in coal seams even after the water feeding has stopped. Therefore the feeding of water should not be prematurely halted. It is preferable to infuse a given quantity of water at a slow rate rather than at a rapid rate in a short time. The second method only causes circulation of the water in the largest fissures or in the fractures caused by the high water pressures used, whereas in the first method the water is better distributed, having time to penetrate a network of the finest fissures by capillary action. High rates of infusion under high pressures also lead to dynamic loading of the roof and floor, causing severe ground control problems. In low-pressure infusion, a water pressure of 800–1000 kPa is usually enough. In long holes, pressures up to 2000 kPa can be used. Flow rate is

Fig. 5.21. Water-infused section (after Cervik, Sainato and Deul, 1977).

regulated from about 0.2 to 3.0 litres/minute. Increasing the moisture content to about 3 % is usually enough to reduce the dust to acceptable levels. Water-assisted cutting reduces dust considerably; more than a 90 % reduction has been claimed.

Conveying

The main sources of dust are at: (1) transfer points, (2) the drive unit and loop take-up, (3) the conveyor belt and idlers, and (4) the tension end. Good mining practice requires a smooth flow of ore or coal, a minimum drop, enclosing the transfer points and maintaining them under exhaust ventilation and the wetting of the material at least 15 m before the transfer point. (It is best to apply water at the start of the conveyor system.) The

Fig. 5.22. Infusion of a retreating longwall from the rib sides of the panel. (*a*) – one hole; (*b*) – two holes (after Cervik *et al.*, 1977).

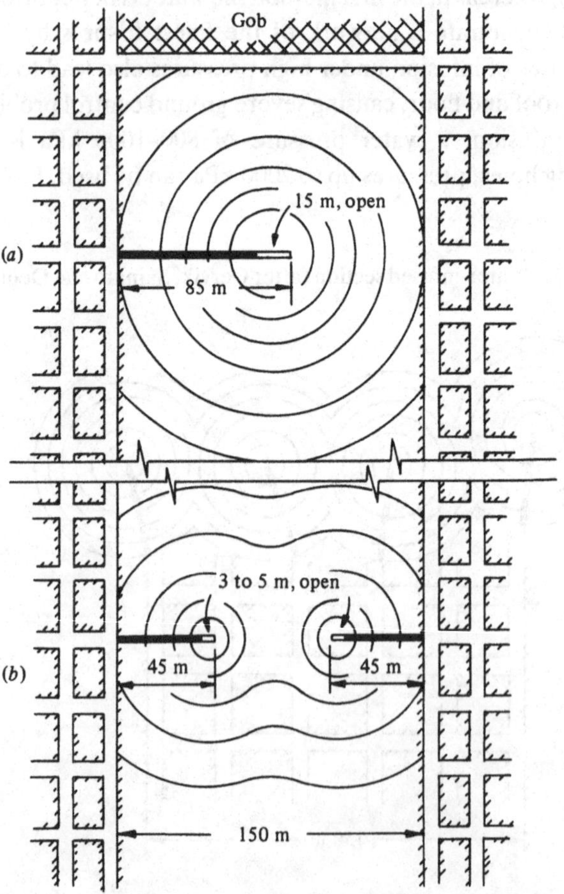

dust-laden air should be cleaned with a filter or discharged into a return airway.

Dust arising at the drive units and the tension end can be reduced by using belt scrapers and ploughs and by sealing the belt joints. Bottom-belt sprays directed at the upper surface of the bottom belt can also be used. They should provide a spray whenever the belt is running, using autospray units. Water should be sparingly used so that there are no problems with belt slip or the build-up of the sludge.

Spillage along the roadway and at transfer points should be wetted down before loading on to the belt. Otherwise, it should be loaded into strong paper bags and brought to the surface in empty mine cars. Vacuum cleaners can also be used although the track required for transporting the cleaner is bulky. The other problems are the disposal of the collected dust, the danger of static build-up and the fire and explosion hazard due to the mass of fine dust collected.

If possible, a homotropal ventilation system may be used either completely or just in the main airways which contain conveyors. Since the air travels in the same direction as the mineral, homotropal ventilation reduces the relative velocity of the air and mineral and thus dust pick-up. Any dust raised and produced at transfer and loading points will flow out of the mine instead of through a working face.

A free-standing dust collector, consisting of a fan and filter, can also be installed in the roadway so that the dust entering the working face is reduced.

Impingement curtains of various materials such as hessian and spun yarn, fibre, nylon or felt and muslin impregnated with a sticky substance can also be used in intake gates as dust collectors, to reduce intake contamination. Regular maintenance of the curtains is essential. Where they cannot be cleaned on site by washing with water, they should be placed in a sealed container and sent to the surface for disposal. Some curtains can cause excessive pressure losses by reducing air volume flow rates through the roadways.

Advancing powered supports

Considerable quantities of dust can get into the air stream when powered supports are advanced. Two methods can be used for dust suppression – (1) water spraying on the supports and (2) using strips of plastic sheeting bratticing over the tops of supports.

Roof-bolt drilling

Dust suppression techniques for roof-bolt drilling include (1) water flushing, (2) dry internal dust collection and (3) a water mist system.

Stone dusting

It is better to apply stone dust during non-production shifts especially when compressed air or mechanical equipment is used for its application. Burst stone dust bags should not be left on the floor in the roadways. Stone dust should be stored on site out of the main air stream and travelling routes. (Stone dust barriers also should be advanced during non-production shifts. If possible, monorails may be used for moving the barriers forward without unloading the stone dust.)

Driving headings

Exhaust ventilation should always be regarded as the main means of controlling dust in a heading drivage. The duct should be kept as near as is practicable to the face of the heading. The forward air velocity round the machine should be not less than 0.5 m/s so that the dust-laden air keeps away from the machine operator. A forcing overlap system can also be used. The short exhaust duct with a dust filter and fan is moved continuously with the machine.

Dust in roadways

There may be roadways where the floors are covered by excessive dust through which workers have to travel in large numbers, causing comparatively dense dust clouds to be raised and made airborne. To prevent dispersion of the floor dust, the moisture content of the material needs to be kept at about 10 % so that the dust remains bound together. Water can be used for this purpose but dries out rapidly depending on the air velocity and humidity, necessitating repeated wetting which could have an adverse effect on the floor strata.

Hygroscopic salts such as calcium chloride or sodium chloride can be used to maintain the moisture content of the dust at the required level of 10 %. The quantity required decreases as the air humidity increases. Table 5.3 shows the calcium chloride content required to obtain adequate consolidation for floor dust of 25 to 50 mm average thickness.

The roadway is first cleaned of loose dust, and stone dust is applied to the floor. The floor is made level and smooth so that the dust layer is limited to 25 mm thickness. The floor is then wetted to a depth of at least 25 mm. A wetting agent may also be used. Flaked calcium chloride is then spread over the floor. Treatment is required again after about 6 months. Gloves must be worn to handle calcium chloride.

The hygroscopic action of calcium chloride makes dust particles adhere by extracting moisture from the atmosphere. Calcium chloride is corrosive and rails, ropes, etc. should be kept out of contact with the salt.

In the case of sodium chloride, the humidity of the atmosphere plays a role. If it is greater than 75 %, the extraction of moisture from the atmosphere due to hygroscopic action is sufficient for the adhesion of dust. If the humidity is less than 75 %, periodic rewetting is necessary for consolidating the floor dust; the atmosphere extracts moisture from the salt solution, producing crystallisation. The action which causes adhesion of the dust is the crystallisation and subsequent recrystallisation of the salt following the periodic rewetting. The rewetting process drives some of the dust into the salt layer and other dust particles stick lightly to the viscous surface. As the wet salt dries, recrystallisation occurs which makes the surface dust particles adhere to the crystalline structure effectively sealing the dust in the salt below.

Sodium chloride is cheaper than calcium chloride but is effective for a shorter period than calcium chloride. Sodium chloride is also corrosive.

Dust from stock piles

Coal stored in surface stock piles produces a lot of dust in the vicinity, particularly in dry weather in areas of high wind velocities. The design of the shape of the stock pile can greatly help to reduce dust and erosion. The height of the stock pile should be equal to or less than the structures surrounding it. The use of fences equal to or slightly higher than the stock piles can protect stock piles from high wind velocities.

A system of spray sprinklers to keep the surface of the stock piles wet can reduce dust production greatly if the moisture content of the surface layer can be kept to about 7 %. Automatic sprinkler systems which are activated by the moisture content of the surface layer of the stock piles are highly recommended.

Table 5.3. *Calcium chloride content required to obtain adequate consolidation (after Anon., 1973)*

Relative humidity of air	Calcium chloride content of dust required to maintain 10 % moisture (%)	Weight of calcium chloride flakes (kg/m^2)
40	7	3.8
50	6	3.3
60	5	2.7
70	4	2.2
80	3	1.6
90	2	1.1

Where coal is to be stored in stock piles over extended periods of time, compaction of the surface and sealing it with some form of sealant (e.g. bitumen) not only preserves the quality of coal, but is very effective in controlling dust.

5.4.2 *Medical supervision*

Periodic medical examinations including X-rays should be conducted on all mine workers. They should also be conducted when a person enters and leaves employment, for his protection as well as for the protection of the company. The diseases caused by dust can be arrested if detected early enough and relieved by changing occupations.

5.5 References

Anon. (1973). *Control of harmful dust in mines*. London: NCB Mining Department.

Anon. (1974*a*). *Design guidelines for dust control at mine shafts and surface operations*, 2nd edn. Toronto: The Mining Association of Canada.

Anon. (1974*b*). *Dust control.* Stanhope Bretby: NCB Mining Research and Development Establishment. 20 pp.

Breuer, H. (1975). TBF 50 and Tyndallometer TM digital – two instruments supplementing each other for the occupational hygienic and technical assessment of dust conditions. In *International mine ventilation congress*, Johannesburg, ed. A. Hemp and F. H. Lancaster, pp. 445–52. Johannesburg: The Mine Ventilation Society of South Africa.

Cervik, J., Sainato, A. and Deul, M. (1977). *Water infusion of coalbeds for methane and dust control*, USBM RI 8241. Washington: US Bureau of Mines. 27 pp.

Chironis, N. P. (1977). Peabody's mines find better way to reduce dust. *Coal Age*, **82**, 106–10.

Courbon, P. (1972). Dust weight sampling. In *Proceedings of the conference on technical measures of dust prevention and suppression in mines*, Luxembourg, 1972, pp. 165–78. Luxembourg: Commission of the European Communities Centre for Information and Documentation.

Courtney, W. G., Jayaraman, N. I. and Behum, P. (1978). *Effect of water sprays for respirable dust suppression with a research continuous-mining machine*, USBM RI 8283. Washington: US Bureau of Mines. 10 pp.

Divers, E. F. and Janosik, J. J. (1978). *Comparison of five types of low-energy scrubbers for dust control*, USBM RI 8289. Washington: US Bureau of Mines. 38 pp.

Gwatkin, G. and Odgen, T. L. (1979). The Simpeds respirable dust sampler. *Coll. Guard.*, **227**, 326–31.

Hamilton, R. J., French, A. G. and Browning, E. J. (1983). The extraction drum: benefits in dust and firedamp control. *Proceedings of the 20th international conference of safety in mines research institutes*, Sheffield, UK, paper A5. Safety in Mines Research Establishment.

Leck, M. J. and Harris, G. W. (1978). Simslin II. *Coll. Guard.*, **226**, 676–7.

Major, G. (1978). Occupational hygiene standards and the interpretation of environmental measurements. *Proceedings of the 2nd Australian pneumoconiosis conference*, Sydney, Feb., pp. B.3.1–B.3.12.

Membrey, W. B. (1984). Dustiness versus total moisture. In *Proceedings of the symposium on dust suppression*, Univ. of Newcastle, pp. 5-1–5-50. Newcastle, Australia: Univ. of Newcastle.

Moyes, A. J. and Reeves, E. G. (1974). Economic comparison of dust-collection methods in the quarrying industry. In *Minerals and the environment*, pp. 579–88. London: Institution of Mining and Metallurgy.

Schowengerdt, F. D. and Brown, J. T. (1976). Colorado School of Mines tackles control of respirable coal dust. *Coal Age*, **81**, 129–31.

Soderberg, H. E. (1970). Dust collection equipment – theory and operation. In *Proceedings of the symposium on respirable coal mine dust*, Washington, Nov. 1969, pp. 183–206, USBM IC 8458. Washington: US Bureau of Mines.

Walton, W. H. (1978). International research programmes in the pneumoconioses – legislation, standards, medical aspects, etc. *Proceedings of the 2nd Australian pneumoconiosis conference*, Sydney, Feb., pp. A.3.1–A.3.21.

6

Mine radiation and its control

6.1 Radiation hazard

The radiation in mines results primarily from the presence of radon, a gaseous decay product of the uranium series found in all uranium mines and in varying degrees in certain other types of mines such as fluorspar mines in Newfoundland and iron ore mines in Sweden and Britain. (Natural decay is the process whereby an atom of a heavy radioactive element breaks up, produces a new element and releases energy in the form of an alpha- or beta-particle or a gamma ray.) Radon is odourless, colourless, tasteless and has a specific gravity relative to air of 7.665. The other gas of concern is thoron, a member of the radioactive decay chain headed by thorium. Since thoron has a short half-life (52 s), a small amount of it reaches the mine air before decay takes place. In

Fig. 6.1. Uranium decay series, U, uranium; Th, thorium; Pa, protactinium; Ra, radium; Rn, radon; Po, polonium; Pb, lead; Bi, bismuth; Tl, thallium; At, astatine.

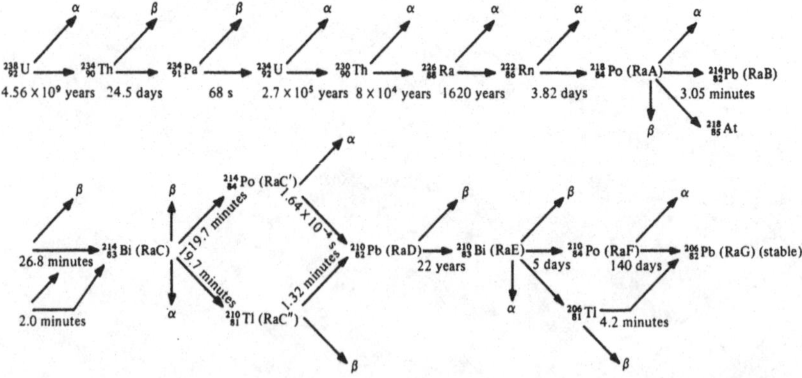

situations where thorium concentrations are high, such as in some Alaskan and Canadian uranium mines, it can be hazardous.

The uranium decay series is given in Fig. 6.1 and the details of the radon sub-series are given in Table 6.1. All the decay products of the uranium series are solids except radon, so all the solid products in the series before radon remain within the rock structure. When radon is produced, it enters the mine atmosphere either directly, along cracks and fissures, or by transportation in water.

On entering the mine atmosphere, radon continues to decay to form airborne radioactive 'daughters' which are usually positively charged, atomic-sized particles. Because of their size and nature, the radon daughters tend to attach to respirable dust and to other free surfaces in the mine atmosphere.

When a miner inhales the air, radon diffuses from the lungs into the blood and part of the dust is deposited in the lungs and breathing passages where the attached radon daughters continue to decay, emitting alpha radiation which damages the lung tissue. Radon is eliminated from the body mainly in exhaled air. Most radon daughters decay in the body before they can be excreted. Thus the radon daughters rather than the radon gas constitute the major health threat to miners in mines with radiation problems.

There is a relationship between cumulative exposure to radiation by miners in underground mines – particularly uranium mines – and excess risk of lung cancer. Smoking adds to the already high lung cancer risk for miners who work in mines with radiation problems.

Table 6.1. *Radon sub-series (after Dungey, Hore and Waller, 1979)*

Name	Symbol	Half-life	Type of emission	Principle particle energy (MeV)	Atoms per 100 pCi
Radium	– $^{226}_{88}Ra$	1620 years	α	4.78	2.7×10^{11}
Radon	– $^{222}_{86}Rn$	3.82 days	α	5.49	1.77×10^{6}
Radium A	RaA $^{218}_{84}Po$	3.05 minutes	α, β	6.00	977
Radium B	RaB $^{214}_{82}Pb$	26.8 minutes	β	0.65	8580
Radium C	RaC $^{214}_{83}Bi$	19.7 minutes	α, β	5.44, 1.65	6310
Radium C′	RaC′ $^{214}_{84}Po$	0.164 ms	α	7.68	8×10^{4}
Radium C″	RaC″ $^{210}_{81}Tl$	1.32 minutes	β	1.96	
Radium D	RaD $^{210}_{82}Pb$	22 years	β	0.02	3.7×10^{9}
Radium E	RaE $^{210}_{83}Bi$	5.0 days	α, β	4.7, 1.16	2.3×10^{6}

6.2 Radiation dosage

The radiation dose delivered to the lungs by inhaled radon daughters depends upon many factors, such as dustiness of the mine air, the length of time the air has been in circulation, the breathing rate, mucus in the bronchial passages, and the physiology of the bronchial passages as affected by smoking or infection.

The unit for the measurement of the radiation level is the working level (WL) which is defined as any combination of radon daughters contained in one litre of air that will release 1.3×10^5 MeV (million electron volts) of energy during decay to $^{210}_{82}$Pb. At equilibrium, one WL is equivalent to a radon concentration of 100×10^{-12} Ci/litre (1 Ci is equivalent to 3.7×10^{10} disintegrations per second). The WL is an exposure level and not a dose rate. For radiation dosage purposes, i.e. for calculating cumulative exposure, the WL is multiplied by the total time of exposure. In the USA the working level month (WLM) based on a working month of 170 hours has been adopted as the unit for dosage or cumulative exposure. The maximum cumulative dose of radiation permitted in the USA is now 4 WLM per year. (This standard was based on study data showing low lung cancer mortality rates among miners exposed to fewer than 120 WLM – the cumulative total for a miner who is limited to 4 WLM a year and works for 30 years.) For example, a miner may work a full 12 months at a WL of 0.33, or for 4 months at a WL of 1 and for the next 8 months in a radon-free area.

6.2.1 *Measurement of working level*

The WL is dependent upon the concentration of various radon daughters. Therefore the growth of the WL is a function of the growth rate of the radon daughters. The growth of RaA, RaB, RaC and WL is shown in Fig. 6.2. After about 120 minutes, the WL reaches equilibrium, i.e. the radon daughters are being produced by their parent at the same rate as they are decaying.

The atmospheric concentration of radon daughters may be measured by drawing a known volume of air through a suitable filter (membrane or polystyrene or glass-fibre filter) and counting the alpha activity on the filter at a measured time after sampling. From this information, the radon daughter concentration in the air sampled may be calculated in WLs.

Kusnetz method

This method consists of drawing a sample of known volume of mine air (between 100 and 250 litres) through an open-faced filter with a vacuum pump. After about 40 to 90 minutes, an alpha-particle count is taken. At the time of measurement, the ratio of the concentrations of the various radon daughters is assumed to be constant and the WL may be determined from the measurement of the total alpha activity. One major disadvantage of this method is the assumption of perfect equilibrium – errors of $< \pm 25\%$ may be expected. Various 'modified Kusnetz' methods exist which compensate for the assumption of equilibrium. They generally involve taking alpha activity counts at several points during the growth of the daughter activity and making corrections for non-equilibrium. Another drawback of the Kusnetz method is that for any measurement a minimum of 45 minutes is required to obtain a result.

Personal dosimeters

Integrating instruments have recently been developed for personal dosimetry (Stafford, 1979). There are three main types of personal dosimeters.

(i) Thermoluminescence (TL) dosimeter. In this, radon daughters are collected on filter paper, using a calibrated mechanical air pump. The radioactive decay of alpha-particles on the filter is registered on luminescent material.

(ii) Track etch film dosimeter. This measures alpha activity from radon daughters in ventilation air in its immediate vicinity and does not require an air pump. The principle used involves chemical changes in a sensitive plastic film due to radiation energy.

(iii) Electronic dosimeter. In this, radon daughters are collected on filter paper, using an air pump. Alpha disintegration signals are processed

Fig. 6.2. Growth of radon daughters and weeking level (WL) (after Evans, 1969).

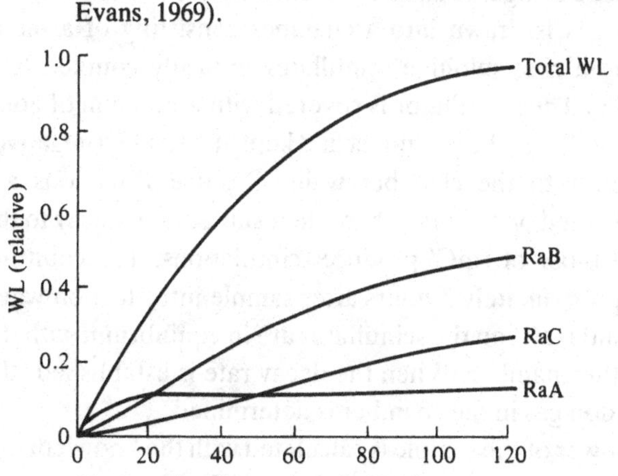

and stored electronically until the instrument is read out by an electronic reader.

The TL dosimeter is accurate and suitable for long sampling periods. Its disadvantages include a relatively high cost, problems of filter clogging and the need for a constant airflow. The advantages of the track etch film dosimeter are that it doesn't require an air pump and can be exposed for a fairly long time. Its drawbacks are that the readout technique is difficult and that it is less accurate than the other two instruments. The main advantage of the electronic dosimeter is its accuracy. Its disadvantages are its relatively high cost, the need for a constant airflow, and limits on the length of the sampling period.

Continuous working level measuring instruments for area monitoring

These instruments rely either on gross alpha or beta counting. Beta detection is the more suitable of the two methods.

6.3 Radon gas emission and sources

Radon gas is contained in the interconnecting interstitial spaces of the rock and diffuses to the surface of mine openings. The rate of diffusion varies from one rock to another. Grade, porosity and permeability are important factors influencing the radon emission rate. Porosity determines the storage capacity of the rock whereas permeability governs the rate of movement of the radon from its point of origin in the rock to the mine atmosphere. Porosity also affects the amount of radon released when rock is blasted.

The emanating power of representative samples of ore and waste can be measured in the laboratory. The samples are placed in radon-tight steel cells where the radon gas is allowed to build up for about 24 hours. The filtered radon gas is drawn into a chamber consisting of a metal chamber containing a zinc sulphide scintillator optically coupled to a photomultiplier tube. The scintillator is covered with a thin film of gold, silver or aluminium foil to make it conduct and kept at a 1000 V (negative) potential with respect to the chamber walls. Positive RaA ions are electrostatically collected on the film where their subsequent decay to the alpha-particles and those of RaC′ produce scintillations. The counting rate is measured approximately 3 hours after sample introduction when the atoms of RaA and RaC′ on the scintillator are in equilibrium with the atoms of radon in the chamber. When the decay rate is established, the concentration of radon gas in the chamber is determined.

The emanating power of the sample is calculated with the knowledge of the volume of the rock sample (cm^3), the volume of the air sample drawn from the steel cell (cm^3) and the radon concentration (pCi/litre). The emanation rate ($pCi/cm^2/s$) can be calculated if the surface area of the sample is measured.

The in-situ radon emanation rate of a rock surface can be determined by emplacing steel chambers tightly to rock faces, evacuating air samples from the chambers after equilibrium is reached and then establishing the radon gas concentration in a Lucas chamber. From the information gained, the emanation rate of the wall rock surface is calculated.

Table 6.2 gives the radon emanation rates obtained at four mines by the Mining Enforcement and Safety Administration (USA) (Rock and Beckman, 1977).

Water flowing into mine workings is another source of radon emanation. The flow of interstitial water into a mine may contribute more radon than the diffusion process of interstitial radon gas.

The determination of the radon concentration in water is normally carried out by liberating the gas from solution in a radon bubbler (Fig. 6.3). The water sample is collected in an evacuated bubbler tube and is left there for several days to reach equilibrium. The tube is then connected to an evacuated scintillation flask and clean air is drawn through the sample; a fritted glass disc is used to give an even distribution of fine bubbles. The activity of the resultant radon–air mixture is then determined with a photomultiplier/counter system.

Table 6.3 gives the results of some ground water samples.

Survey of an inactive mine gives measurements of the natural changes in radon emanation from the intake to the exhaust portals. These natural changes (up to 200 %) in radon emanation are due to barometric pressure variation. These measurements also serve as the base levels.

Table 6.2. *Radon emanation rate measurements obtained at four mines by the Mining Enforcement and Safety Administration (USA) (after Rock and Beckman, 1977)*

Mine	Airflow (m^3/s)	Number of readings	Estimated ore grade (percentage of U_3O_8)	Rock type	Radon emanation rate ($pCi/m^2/s$)
1	3.8	6	0.20	Medium to coarse-grained	142
1	2.8	4	0.20	sandstone	143
3	5.9	24	0.05	50% sandstone, 50% siltstone, highly fractured	263
4	4.0	7	0.30	Gneiss, with fractured zone	537

Table 6.3. *Ground water analyses (after Rock and Beckman, 1977)*

Mine	Water flow (litre/minute)	Radon-222 (pCi/litre)	Radium-226[a] (pCi/litre)	Potential[b] radon release (pCi/minute)
1	3.8	4413	—	16700
	1.9	2730	—	5170
	3.8	1256	—	4750
	5.7	650	—	3690
	([c])	212	—	—
	([c])	173	228	—
	([c])	864	—	—
2	None	—	—	—
3	None	—	—	—
4	95	2.3×10^6	5700	218×10^6
	379	1.8×10^6	28600	681×10^6
5	3.8	6200	440	23400
	1.9	3130	410	5900
	5.7	1600	1075	9000
	([c])	280	110	—
	([c])	2850	2975	—

[a] 1 pCi/litre Ra-226 will generate 0.5 pCi/litre Rn-222 in 3.8 days.
[b] From Rn-222 content only.
[c] Not measurable.

Fig. 6.3. A radon bubbler (after Dungey, Hore and Waller, 1979).

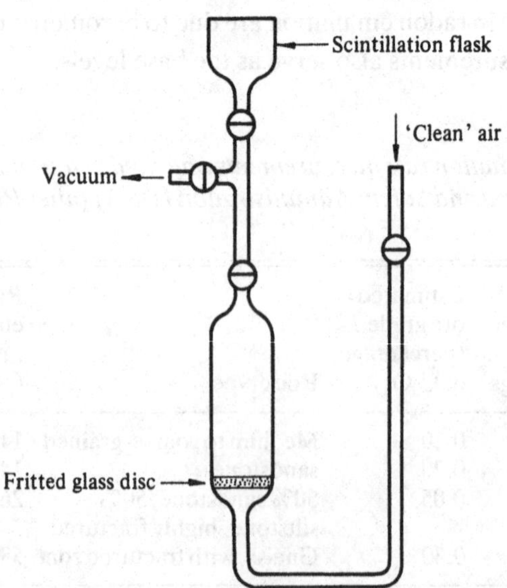

Blasting gives the largest momentary increase (about 400 %) in radon concentration, but the concentration reduces quickly in a well-ventilated stope to about a 40 % increase. This is due to the increased surface area of the loose ore.

Sealed, mined-out areas can be another source of radon when the barometric pressure decreases.

Drill holes provide further surfaces from which radon could emanate, as well as passage for it and its daughters from inactive to active areas.

6.3.1 *Radon concentrations and working levels in ventilated areas*

Air volume flow rate through a mine, the distribution of airflow within the mine, and the radon emanation rate are the primary factors affecting the radon concentrations and WLs in ventilated areas. The total air volume flow rate through the mine determines the average time air takes to travel from the inlet to the production area and then to the outside of the mine. This time is called the air-transit or residence time and during this period, radon daughters accumulate.

Tracer gas techniques can be used for studying air-transit times in mine ventilation systems.

6.4 Control of radiation

The main method of controlling radon and daughter concentrations in mines is ventilation. It is essential to maintain a low radon concentration through dilution with clean air and to allow the radon a short residence time (10–15 minutes) so that only 10–20 % of the daughters are produced in the mine atmosphere.

To calculate ventilation requirements for a working district in the mine, the radon levels likely to be encountered and the 'age' of the air supplied to the district have to be estimated with confidence. The air volume flow rate requirements for three stoping methods are given in Table 6.4.

Table 6.4. *Air volume flow rate requirements in different mining areas (after O'Brien, 1979)*

Mining areas	$m^3/s/t$	Characteristics
Jumbo mining	0.091	Six or seven separate working places per area
Longhole mining	0.104	Large stope area, decreased air velocities, more broken ore in each blast
Pillar mining	0.156	Large open stopes, slow movement of air, increase in 'residence' time of air requires flooding with large air volume flow rates

The following factors must be considered when designing a ventilation system for a mine with radiation problems:

(1) Primary fresh air distribution systems should be routed through waste rock where possible, or in some cases through low-grade ore.
(2) Main distribution airways should be kept as free from mining activity as is practicable, so that relatively high air velocities can be readily maintained.
(3) Mining areas should be separated into different ventilation blocks, affording better control capabilities. A split system of ventilation should be used so that the relatively uncontaminated intake air is proportioned between mining areas according to requirements. Air from each area is collected in an isolated, return airway and exhausted to the surface without contaminating other active mining areas.
(4) Maintaining the necessary pressure profile in the mine is important. Employing both downcast and upcast fans, a positive pressure can be maintained in the working area, preventing recirculation from worked-out areas which are maintained at a lower pressure.
(5) Blasted ore should be removed as soon as possible to eliminate radon daughter formation.
(6) Where possible the haulage ways should be confined to the district return airways to reduce the dust and diesel exhaust levels in the workings, and to optimise ventilation requirements.
(7) Ore passes, conveyor ways and crushing stations should be ventilated so that exhaust air can be directed to the return air system quickly.
(8) At major dust sources, such as grizzly dumping points, crusher and ore passes, which would contaminate, wholly or partly, the mine's intake air circuits, large, bag-type dust collectors, 5 to 20 m^3/s each, should be used.
(9) In areas near the margins of mining activity where exposure to dust and radiation cannot be controlled by increased air volume flow rate, miners should be given respiratory devices, e.g. crews operating haulage equipment, jumbo drills, crushers, conveyor belts, rock breakers, working in pillar extraction areas or tending skips at the loading pockets. Powered respirators, e.g. airstream helmets, are particularly successful in reducing the concentration of radon and its daughters breathed in by the wearer.

(10) The repair garages should be positively ventilated by controlled air volume flow rate.
(11) Lunchrooms should be supplied with ultra-filtered air from small, bag-type dust collectors (0.3 to 0.4 m^3/s each).
(12) Ventilation stoppings should be built as airtight as possible. Use of sealants on stoppings and air pressure control behind stoppings can reduce leakage.
(13) Water entering the mine should be either stopped or piped out as quickly as possible. In the case of diamond drill holes, either grout or a plug may be used.
(14) Open drill holes from inactive to active areas should be sealed as close to the input airflow as possible. In case the inactive area is at a lower pressure than the input air carrier, the hole should be left open.
(15) The use of sealants is envisaged primarily in mine openings carrying fresh air to the working areas so that uncontaminated air is delivered to the working areas. The sealants are also excellent for special cases such as lunchrooms, shops and haulage ways driven in ore.
(16) Where areas that do not provide ventilation to the mine are found to be a major source of contamination, it may be possible in the future to adopt a principle similar to that of methane drainage, as used in the coal mining industry, to tap and dispose of that contamination.

Some poor work practices that add to radiation problems are (Stafford, 1979):

(1) Parking a large piece of equipment, like a truck or shuttle car, in such a way as to block an intake air passage while the operator takes a break, resulting in accumulation of high radon daughter working levels in large sections of a mine in a few minutes.
(2) Entering working places to check blasted areas, extend ventilation ducting or wet down muck piles before switching on ventilation fans.
(3) Shutting down ventilation fans because miners are cold or are irritated by the fan noise.
(4) Putting in ventilation ducting after, rather than before, installing a work platform or drill staging in a raise when setting up a drilling area.
(5) Shortcircuiting a ventilation system by carelessly leaving an air control door blocked open or closed.
(6) Leaving ventilation ducting too far from a work area; also failure to repair torn or otherwise damaged ventilation ducting.

6.4.1 Work force management

Controlling cumulative radiation exposure of miners by limiting their occupancy time in high risk areas and rotating them to low risk areas can be a useful practice. For this, cumulative radiation exposure has to be known for each miner and alternative working places have to be available. Large mining operations could provide alternative working places without much difficulty. To estimate the cumulative exposure of the miners to radon daughters, the average concentration of radon daughters in each working area of the mine and the time spent by the miners in each working area have to be known from the mine records.

6.5 References

Dungey, C. J., Hore, J. and Waller, M. D. (1979). An investigation into control of radon and its daughter products in some Cornish mine atmospheres. *Trans. Instn Min. Metall.*, Sec. A, **88**, A35–A43.

Evans, H. D. (1969). Engineers' guide to the elementary behaviour of radon daughters. *Health Phys.*, **17**, 229–52.

O'Brien, N. (1979). Environmental protection in underground uranium mining. *Min. Mag.*, **140**, 14–25.

Rock, R. L. and Beckman, R. T. (1977). *Measurements of some of the factors which influence the radon daughter health hazard.* Mine Enforcement and Safety Administration Inf. Rept 1067, 12 pp.

Stafford, S. (1979). Mine radiation control efforts are waged on variety of fronts. *Mine Safety & Health*, **4**, No. 1, 2–9, 28.

7

Mine climate and its control

The term 'mine climate' refers to the prevailing conditions of temperature, humidity* and velocity of the air in a mine. In deep mines, intensively mechanised mines and working areas farther from shafts, climatic conditions deteriorate due to increases in temperature and humidity, influencing the health, safety and efficiency of the miners. In general, in comparable situations, as temperature and humidity increase, accident rate increases and working efficiency and hence production decreases.

7.1 Physiological effects of mine climate

Food taken in by a person breaks down by chemical and bacterial action and combines with oxygen to provide mechanical power in the muscles and body (i.e. metabolic) heat to maintain the body's temperature at about 36.9 °C for proper physiological function. If the rate of heat loss to the surrounding atmosphere is equal to the rate of metabolic heat generation, a heat balance exists, maintaining the body at 36.9 °C. If the rate of heat loss to the surrounding atmosphere exceeds the rate of metabolic heat generation, the heart rate slows down, blood vessels contract and shivering occurs. If, despite these reactions, the rate of heat loss continues to exceed the rate of metabolic heat generation, the body core temperature begins to fall, leading to hypothermia and possible collapse.

If the rate of metabolic heat generation exceeds the rate of heat loss to the surrounding atmosphere, there are three possible outcomes depending upon the degree of imbalance and the time involved.

* Humidity is determined from the dry- and wet-bulb temperatures.

(1) The heart rate rises, increasing the blood flow to the skin and also the amount of sweating. The higher skin temperature and the cooling effect of the evaporation of perspiration may then restore the heat balance without too much discomfort.
(2) If the regulatory mechanism experiences considerable strain in attempting to increase body cooling, considerable discomfort and inefficient work result.
(3) If the imbalance continues, the body gains heat resulting in hyperpyrexia and possible heat stroke.

The initial symptoms of heat imbalance are psychological and are as follows:

(1) loss of interest in the job and frequent rests;
(2) a wish to complete the work quickly;
(3) irritability;
(4) reduced degree of concentration;
(5) lack of care and attention.

The physical symptoms start with lethargy and a feeling of tiredness. Continued imbalance results in headaches, nausea, sickness and then coma, terminating in death.

7.1.1 *Mechanism of heat loss*

The human body loses heat to the surrounding atmosphere by radiation, convection and evaporation. Cooling by radiation to the surrounding air and solid surfaces occurs if their dry-bulb temperature is less than the skin temperature. Cooling by convection depends upon the dry-bulb temperature of the air and its velocity. Cooling by evaporation depends upon wet-bulb temperature, air velocity and dry-bulb temperature. (In hot and humid environments, evaporative cooling of the body is the most important one.) All three forms of cooling are also affected slightly by air pressure.

7.2 Measurement of air cooling power

Wet-bulb temperature can be used as an indicator of air cooling power. A whirling hygrometer consisting of dry- and wet-bulb thermometers is used to measure the dry- and wet-bulb temperatures. The dry-bulb temperature is that of the atmosphere. At the wet bulb the temperature is reduced by evaporation depending upon the dryness of the air.

The cotton sleeve (cover) round the wet bulb keeps moist by capillary action from a reservoir of distilled water. The hygrometer is rotated at a steady rate of about 180 rev/minute for a minute, and this repeated until a steady reading is obtained. Each time, the wet-bulb thermometer is read first, then the dry-bulb thermometer.

There is some analogy for potential cooling between a wet-bulb thermometer and a sweating person. The upper limit of the wet-bulb temperature which could prevail without affecting body temperature was found to be 31 °C for a clothed person at rest and in still air; 34 °C for the same person if the air velocity was increased to 50 m/minute; and 25 °C if the person was performing moderate work in still air (Belding, 1970). The objection to the use of this instrument is that the wet-bulb temperatures are obtained by whirling thermometers, whereas a person is not subject to whirling. Also, convection and evaporative properties for a person are quite different from those of a thermometer bulb.

A kata thermometer can also be used for assessing air cooling power. This spirit thermometer graduated between 35 and 37.8 °C has a large bulb. After raising the temperature of the thermometer with hot water and wiping it dry, it is suspended and the time it takes to cool from 37.8 to 35 °C is measured with a stop watch. The operation is repeated with a wetted sleeve. The time in seconds thus obtained is multiplied by a factor (given by the manufacturer) to determine the combined effect of radiation and convection on the cooling of the dry bulb, and the combined effect of radiation, convection and evaporation on cooling of the wet bulb. Although the kata thermometer gives a better indication of the cooling power of air in comparison to the wet-bulb thermometer, the analogy between the kata thermometer and a person is not very good because kata thermometers have a smaller volume to surface area ratio compared to the human body. The instrument is not popular because of the fragility of the large bulb and the necessity of carrying hot water in a thermos flask for use underground.

Because of the limitations of the wet-bulb thermometer and the kata thermometer to measure air cooling power, American workers have introduced an index called the 'effective temperature' index based on observed effects on the human body. The effective temperature is that temperature of still saturated air which gives the same sensation under the condition of the said dry- and wet-bulb temperatures and air velocity. In related experiments, three environmental variables – air temperature, humidity and air velocity – were changed in a room over a range of values and subjects were asked how comfortable they felt during the experiment in comparison to the environment in a room with no air movement, 100 % humidity and various air temperatures. From the information obtained, a standard chart was constructed which could predict the relative feeling of sensation for a person exposed to various combinations of humidity, air velocity and air temperature.

Two scales are available, the basic scale for persons stripped to the waist and the normal scale for persons wearing normal indoor clothing. The former scale is more applicable to mining conditions and is given in Fig. 7.1. To determine the effective temperature from the chart, a line is drawn connecting the wet- and dry-bulb temperatures. The intersection of this line with the air velocity contour gives the effective temperature of the atmosphere.

In the USA the wet-bulb globe temperature index has been adopted as a standard for air cooling power. Three temperatures, namely, dry-bulb temperature, wet-bulb temperature and black-globe temperature are used to calculate the index. The standard set of instruments used in the field for wet-bulb globe temperature measurements is shown in Fig. 7.2. Wet-bulb temperature is measured by a thermometer with a wetted sleeve hung in the environment (whirling hygrometer readings are not acceptable) and the black-globe temperature is measured by a thermometer placed in the centre of a hollow black globe. The thermometer bulbs are held at about 1.2 m from the floor.

The following formulas are used to calculate the wet-bulb globe temperature index (t_{wg}). For indoor evaluation, i.e. when solar radiation is absent,

$$t_{wg} = 0.7t_w + 0.3t_g. \tag{7.1}$$

Fig. 7.1. Chart of effective temperature (basic) (after NCB, 1979).

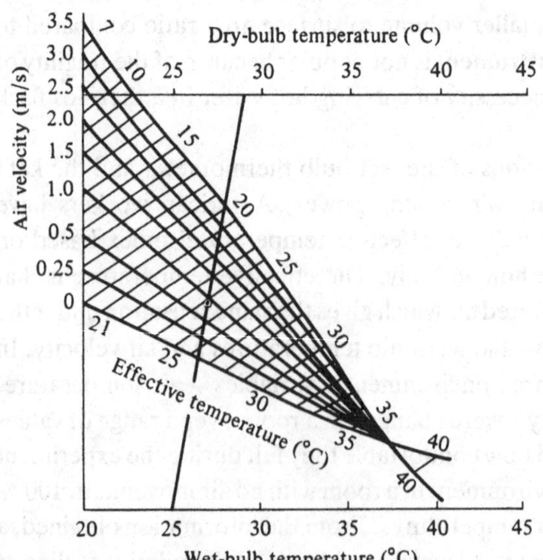

and for outdoor evaluation

$$t_{wg} = 0.7t_w + 0.2t_g + 0.1t_d \quad (7.2)$$

where t_w = wet-bulb temperature;
t_g = black-globe temperature; and
t_d = dry-bulb temperature of air.

The American Conference of Governmental Industrial Hygienists has adopted the following threshold limit values for t_{wg} for a moderate work load in hot environments (ACGIH, 1978):

Continuous work	26.7 °C (80.0 °F)
75 % work–25 % rest each hour	28.0 °C (82.4 °F)
50 % work–50 % rest each hour	29.4 °C (84.9 °F)
25 % work–75 % rest each hour	31.1 °C (88.0 °F)

Higher exposures are permissible if certain work practices and medical surveillance are in effect to insure that the worker's body temperature does not exceed 38.0 °C (100.4 °F).

Fig. 7.2. Instruments for wet-bulb globe temperature measurements (after Misaqi *et al.*, 1976).

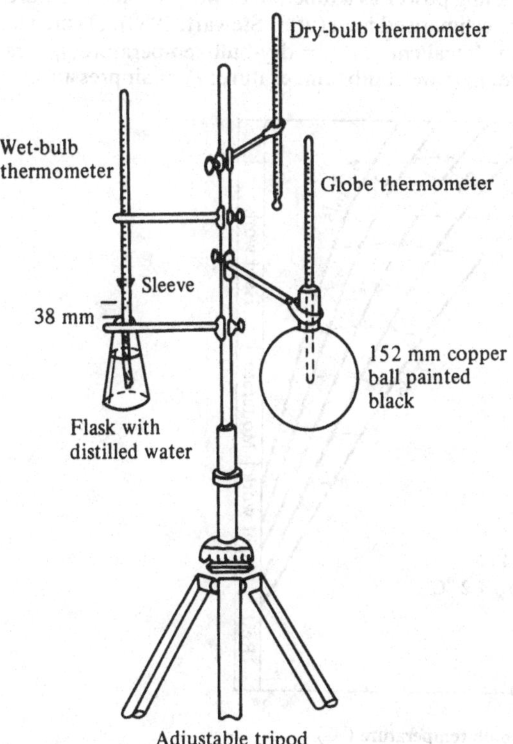

Using the physical principles of heat transfer from humans, the South African Chamber of Mines has developed an index called the 'cooling power' index (Stewart, 1980). The rate of cooling experienced by the person is called the cooling power of the environment and is dependent upon

(1) wet-bulb temperature, t_w;
(2) dry-bulb temperature, t_d;
(3) radiant temperature, t_r;
(4) air velocity, V;
(5) air pressure, P_a; and
(6) skin temperature, t_s.

The first five are the climatic parameters. By assuming values for the less significant parameters ($t_d = t_r = t_w + 2\,°C$, $P_a = 100$ kPa and $t_s = 35\,°C$) nomograms (Figs. 7.3 and 7.4) have been drawn giving the relationships between cooling power and wet-bulb temperature and air velocity for acclimatised* and unacclimatised men. (For other mining conditions,

* See section 7.4 for details.

Fig. 7.3. Cooling power as a function of wet-bulb temperature and air velocity for acclimatised men (after Stewart, 1980). Dashed lines, wet kata readings (mcal/cm^2·s). t_d = dry-bulb temperature; t_r = radiant temperature; t_w = wet-bulb temperature; P_a = air pressure.

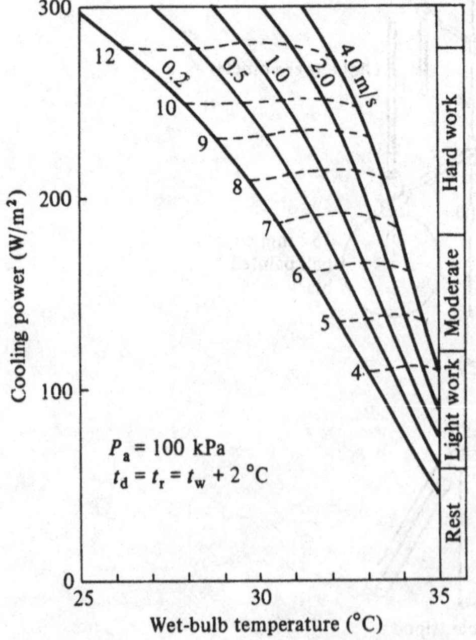

appropriate values may be assumed and corresponding nomograms drawn.)

Metabolic heat generation rates of 115, 180 and 280 W/m^2 are typical of light work (e.g. winch operation, sweeping, fitting), moderate work (e.g. building ventilation walls, operating box holes, building roof supports, drilling, team leaders' activities) and hard work (e.g. shovelling rock, pushing ore cars), respectively. Men can reach thermal equilibrium with a skin temperature of 35 °C if the point defining the environment in which they are working lies above the line relevant to their work rate.

Relative importance of air velocity and wet-bulb temperature: In underground working places, air velocities are generally low (0.5–1 m/s). In the region of low air velocity, the cooling power of the environment increases rapidly with increasing air velocity, but only slowly with decreasing wet-bulb temperature. At high air velocities, the beneficial effect of an increase in air velocity diminishes. The benefits of increasing air velocities above 2 m/s are small.

Fig. 7.4. Cooling power as a function of wet-bulb temperature and air velocity for unacclimatised men (after Stewart, 1980). Dashed lines, wet kata readings (mcal/cm$^2 \cdot$s). t_d = dry-bulb temperature; t_r = radiant temperature; t_w = wet-bulb temperature; P_a = air pressure.

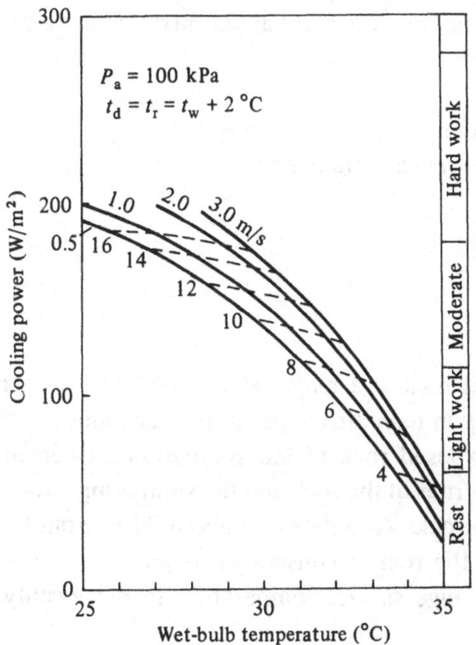

Comparison with wet kata readings: Lines of equal kata readings are also given in Figs. 7.3 and 7.4. In Fig. 7.3, there is a good correlation between wet kata readings and cooling power values, i.e. wet kata readings can be used to measure the cooling power of the environment for acclimatised men. Minimum wet kata readings of 4.5, 7 and 12 are necessary to perform light, moderate and hard work, respectively, for acclimatised men.

In Fig. 7.4, there is not such a good correlation. However, minimum kata readings of 9 and 16 are adequate for the safety of unacclimatised men performing light and moderate work respectively.

Comparison with effective temperature index: When lines of effective temperatures were plotted on figures giving the relationships between cooling power and wet-bulb temperature and air velocity, they were found to be not at all horizontal and it can be concluded that there is no correlation between cooling power and effective temperature.

7.3 Factors affecting the mine climate

The most important parameters of climate influencing the cooling power of the environment are wet-bulb temperature, dry-bulb temperature, air pressure and air velocity. The first three parameters are measured and used to determine the psychrometric state of air–water vapour mixtures from tables or charts, or by using a computer program. (Some details about the psychrometrics of air–water vapour mixtures are given in Appendix III.)

7.3.1 Sources of heat in mines

The principal sources of heat in mines are

(1) exposed rock;
(2) adiabatic compression; and
(3) machinery.

1. Exposed rock

The heat flow from exposed rock is a major source of heat in deep underground mines, resulting in high air temperatures. The amount of heat transferred from the exposed rock to the ventilation air current depends on the thermal properties of the rock and the ventilating air.

Thermal properties of the rock: At a depth of about 20 m from the surface, the temperature of the rock is constant throughout the year, equivalent to the annual average surface temperature in the locality.

Above this level, the rock temperature varies seasonally with the air temperature variations at the earth's surface and below this level, the rock temperature rises at a rate called the geothermal gradient. The geothermal gradient varies according to the type of rock formations and is inversely proportional to the thermal conductivity of the rock. (A review of the methods of determining the thermal conductivity of rock has been given by Patnaik and Singh, 1978.) The geothermal flow of heat from the hot core of the earth is about 0.05 W/m^2 and is practically constant over most of the earth's surface. The geothermal gradient, gg, in °C/m, is calculated as follows:

$$gg = \frac{0.05\ (\mathrm{W/m^2})}{\text{thermal conductivity}\ (\mathrm{W/m \cdot {}^\circ C})}. \tag{7.3}$$

It can also be calculated from the measured temperatures at various depths.

The rock temperature, t_r, in °C, at a depth D_1 m is calculated from the following equation:

$$t_r = t_c + (D_1 - D)gg \tag{7.4}$$

where t_c = constant temperature of the rock near the surface, °C;
D = depth of the constant temperature zone, m; and
gg = geothermal gradient, °C/m.

The typical values of geothermal gradients (°C/100 m) are as follows:

Broken Hill, NSW, Australia	1.97
Ontario, Canada	1.22
Hungary	5.00
Kolar Gold Fields, India	1.10
Witwatersrand, South Africa	0.80
UK and Europe	1.00–3.00
USA	0.91–3.65

2. Adiabatic compression

As the air descends in the downcast shaft, it gets compressed by the increasing column of air and consequently gains heat. If the vapour content of the air remains constant, the air flow is considered frictionless and there is no heat transfer between the shaft walls and the air, the process is known as adiabatic compression.

The increase in dry-bulb temperatures due to adiabatic compression per 100 m depth is calculated as follows:

$$\frac{\text{increase in heat content of 1 kg of air}}{\text{specific heat of air}}$$

$$= \frac{\text{mass} \times \text{gravitational acceleration} \times \text{distance}}{\text{specific heat of air}}$$

$$= \frac{(1 \times 9.81 \times 100/1000)(\text{kJ/kg})}{0.996(\text{kJ/kg}\cdot{}^{\circ}\text{C})}$$

$$= 0.99\ {}^{\circ}\text{C}.$$

In the upcast shaft the air temperature decreases due to adiabatic expansion.

In wet shafts, there may be considerable cooling due to moisture evaporation (i.e. the rise in dry-bulb temperature may be less than that calculated due to adiabatic compression) but the wet-bulb temperature may rise. Heat from the shaft walls, compressed air pipes and electric cables also contribute to the rise in air temperature although it will be eventually small because of the long-term cooling effect of air on the sources and the relatively short time of travel in the shaft. The combined effect of these sources is very difficult to calculate. The seasonal and even daily variation of air temperatures at the top of the downcast shaft further complicates the issue.

Figs. 7.5 and 7.6 show increments in dry- and wet-bulb temperatures in 1000 m shafts for two different British coal mines of similar depth.

Fig. 7.5. Increments in dry-bulb temperature in 1000 m shafts in British coal mines (after Verma, 1979). Crosses, colliery A; circles, colliery B.

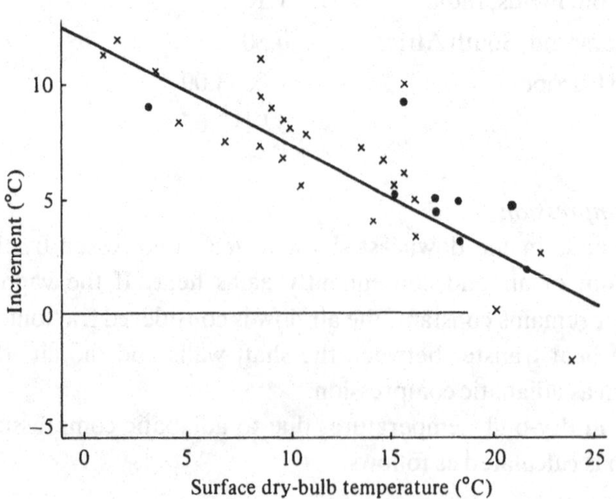

3. Machinery

All machines contribute to the heat load of the mine. Whether the machines are run by diesel engine, electric motor or compressed air, a non-mechanical portion of the work produced by them is converted to heat through friction and electrical losses.

In the case of diesel locomotives, these consume about 0.24 kg of fuel per kWhour. Since the calorific value of the fuel is about 44 MJ/kg, the total energy in the quantity of fuel used becomes:

$$\begin{aligned} \text{total energy} &= 0.24 \times 44 \times 10^3 \text{ kJ/h per kW} \\ &= 10560 \text{ kJ/h per kW} \\ &= 176 \text{ kJ/minute per kW} \\ &= \text{approximately } 3 \text{ kJ/s per kW} \\ &= 3 \text{ kW/kW.} \end{aligned}$$

Of this energy, 1 kW is converted into mechanical energy. The remaining 2 kW appears as heat in the exhaust gas and in the engine itself. On average, the rate of heat production is about 1 kW per rated kW because of the intermittent operations of the locomotive.

In the case of conveyors, about 12 % of the input energy is lost at the motor and gearbox and the rest is lost at the idlers, while with auxiliary fans, all the power input is dissipated as heat into the air in which the motor is situated.

The average machine power in UK coal mines is 1.51 kWhours/t/day (range 1.10–2.32).

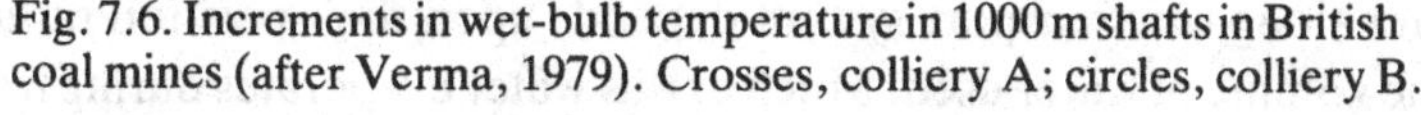

Fig. 7.6. Increments in wet-bulb temperature in 1000 m shafts in British coal mines (after Verma, 1979). Crosses, colliery A; circles, colliery B.

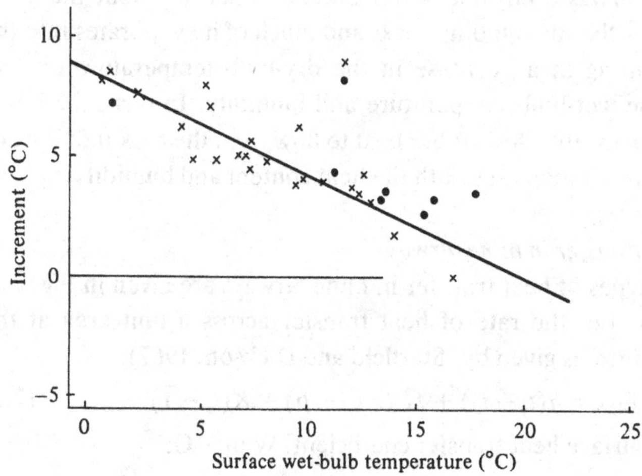

Other sources of heat

Exothermic reactions such as the oxidation of minerals (pyrite, sulphide ores and coal) and cement curing, explosives, movement of rock masses (work against friction), compressed air lines, electric cables, lighting equipment, friction between the air and walls of the airways and turbulence resulting from the irregularities of the airway shape are some of the other sources of heat. Heat from cement curing could be significant if large amounts are used for stope filling or shotcreting; it would be liberated for a relatively short time. Most of the heat released by explosives is removed from the mine by the ventilation air current during the re-entry period; so heat from explosives is not an important factor. Heat from the movement of rock masses is absorbed by the rock mass itself and little of this is transferred to the ventilating air current. In the case of compressed air, this is usually pre-cooled before it enters the mine so that some of the heat is removed. (Cooling also removes moisture from the compressed air.) The temperature of the compressed air going down a mine shaft also increases due to adiabatic compression. Because of the greater density of the compressed air, the pressure rise and so temperature rise is much more than for ventilation air (5 to 7 times) and there could be a transfer of heat from the compressed air line to the ventilation air. Heat from air friction and turbulence is insignificant.

7.3.2 *Sources of moisture*

The main sources of moisture are dust suppression water and ground water. The effectiveness of sprays on mineral-breaking machines and along the transport system depends upon their position rather than the volume of water used. Much of this water evaporates during transport along intake airways. Ground water enters a mine at about the same temperature as the surrounding rock, and much of it evaporates into the mine air resulting in a decrease in the dry-bulb temperature and an increase in the wet-bulb temperature and humidity. In turn, the lower dry-bulb temperature allows more heat to flow from the rock into the air. Thus ground water increases both the heat content and humidity.

7.3.3 *Heat transfer in mine airways*

The types of heat transfer in mine airways are given in Fig. 7.7. The heat flux, i.e. the rate of heat transfer across a unit area at the air–rock interface, is given by (Starfield and Dickson, 1967):

$$\text{Heat flux} = h(t_s - t_d) + fLE(p_s - p) + K(t_s - \overline{t}) \qquad (7.5)$$

where h = surface heat transfer coefficient, W/m$^2\cdot$°C;

t_s = temperature of rock surface, °C;
t_d = dry-bulb temperature of air, °C;
f = wetness factor;
L = latent heat of evaporation of water, J/kg;
E = coefficient of mass transfer, $kg/m^2 \cdot s \cdot kPa$;
p_s = saturated vapour pressure at temperature of the wet portion of rock in °C, kPa;
p = partial pressure of water vapour in the air, kPa;
K = overall transfer coefficient for radiation from dry surfaces to wet surfaces, $W/m^2 \cdot °C$;
$\bar{t}$ = average temperature of surface to which t_s is radiating, °C.

A review of the theoretical methods available for the prediction of heat pick-up and air temperatures in intake haulages in mines has been given by Vost (1982).

7.4 Control of mine climate

The various possible ways of reducing or avoiding the ill effects of poor climatic conditions are given in this section.

Acclimatisation

Repeated exposure to a hot and humid environment can increase the body's tolerance to it. Acclimatisation is a long-term response or adjustment of an individual to an identified stress. An acclimatised person can perform many tasks in a hot and humid work site where a non-acclimatised person cannot work.

Fig. 7.7. Types of heat transfer in a mine airway. Cond = conduction; Conv = convection; E = evaporation; R = radiation (after Vost, 1982).

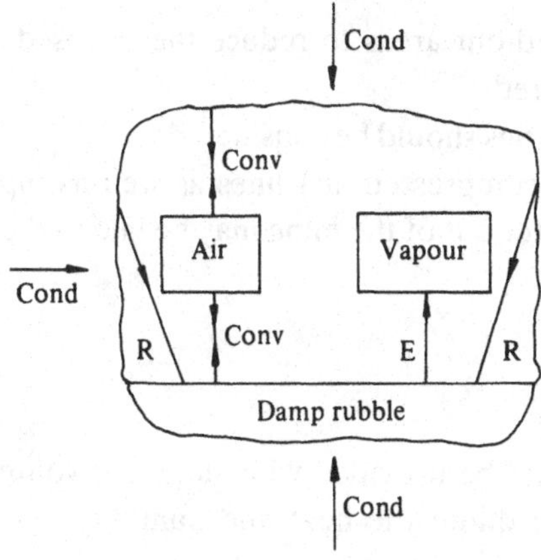

A person should be given enough time to adjust to a hot and humid work site (for example, where the wet-bulb globe temperature exceeds 26.7 °C or 80 °F). The following 6-day acclimatisation procedure has been recommended by Misaqi *et al.* (1976):

First day: 50 % exposure
Second day: 60 %
Third day: 70 %
Fourth day: 80 %
Fifth day: 90 %
Sixth day: 100 %

Acclimatised persons returning after 9 or more consecutive calendar days of leave should undergo a 4 day acclimatisation schedule as follows:

First day: 50 % exposure
Second day: 60 %
Third day: 90 %
Fourth day: 100 %

Medical surveillance during acclimatisation is also recommended. During the first few days, salt loss could easily exceed the intake, resulting in dehydration, and salt and water must be made readily available.

Mining practice

(1) A mining method that allows better use of ventilation should preferably be selected. A retreat system of mining, concentration of working areas, faster rates of mining, short-gate length panels and an advance system of mining without stable hole are preferred.
(2) Filling in the worked-out areas to reduce the exposed surface area may be considered.
(3) Sealing off old workings should be considered.
(4) Service (water and compressed air) lines as well as piping for pumping ground water out of the mine may be laid in the upcast shafts.

Ventilation

(1) Working areas should be provided with sufficient volume flow rates of fresh air to dilute the heat and humidity and give a

required air velocity for necessary effective temperature. It is not very profitable to plan air velocity above about 3.5 m/s.

(2) Exhaust fans on the surface or discharging directly into the upcast system are preferred to avoid the addition of heat (due to fans) to the air entering the mine.

(3) A split system of ventilation should be considered.

(4) Usually, the conveying system is in the intake airway. Homotropal ventilation, i.e. conveying the mineral in the same direction as the ventilated air by siting the conveyor in the return airway, may be considered. Two intake airways may also be considered.

(5) Fresh, cool air may be ducted to localised parts if necessary.

(6) Small, compressed air or electric airmovers may be used for local increases in air velocity.

(7) Partial recirculation of the air may be considered. It is important to see that the fresh air is sufficient to dilute the pollutants, including heat.

(8) Y, W and double Z systems of ventilation may be considered (Figs. 7.8 and 7.9). By employing these systems, it has been possible to delay the installation of air cooling equipment in some West German coal mines.

(9) In the case of headings, a forcing system of auxiliary ventilation is preferred to deal with climatic problems at the face. If dust is also a problem, a forcing overlap system (either conventional or recirculation) is selected so that climate may be controlled by a forcing main duct and dust by an exhausting overlap duct leading to a dust filter.

Machinery

(1) When selecting machines, properties such as heat generation, the possibility of heat shielding and isolation, automation and remote controls should be considered.

(2) Machines should be maintained adequately so that minimum amount of power is consumed.

(3) The minimum number of machines required should be used.

(4) The work of machines should be spread over a maximum time period and peak loadings avoided whenever possible.

(5) Conveyors should not be allowed to run empty for long periods of time.

Fig. 7.8. Y ventilation (*a*) one intake airway (after Mucke, 1973). Length of face – 200 m; length of working – 1500 m; thickness – 1.5 m; rate of advance – 4 m/day; electrical capacity – 700 kW. (*b*) Two intake airways (after Voss, 1975).

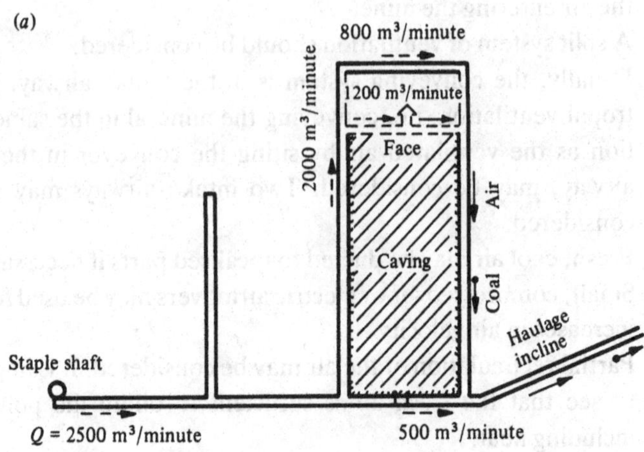

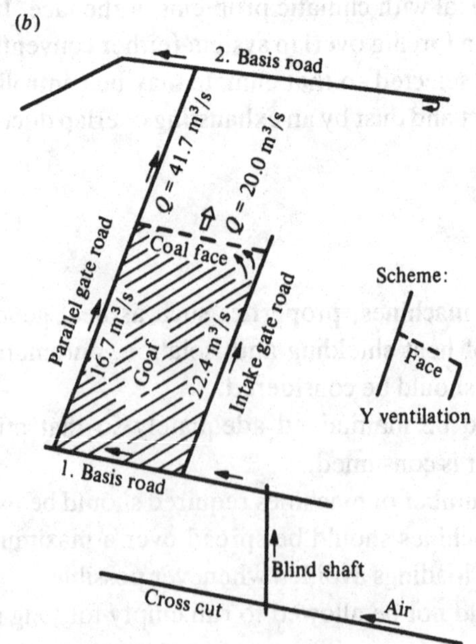

Fig. 7.9. (*a*) W ventilation, (*b*) double Z ventilation (after Voss, 1975).

(*a*)

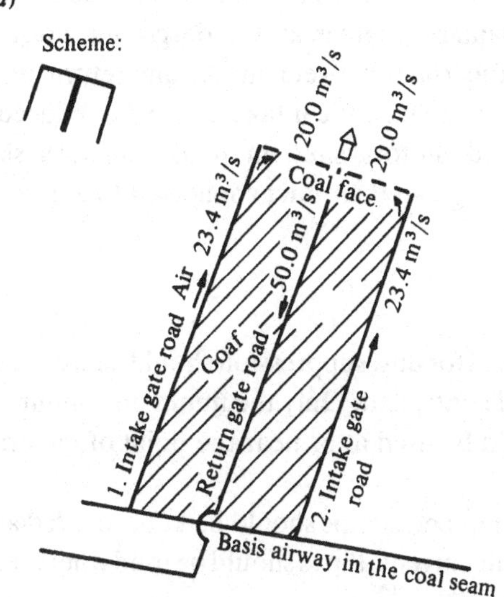

(*b*)

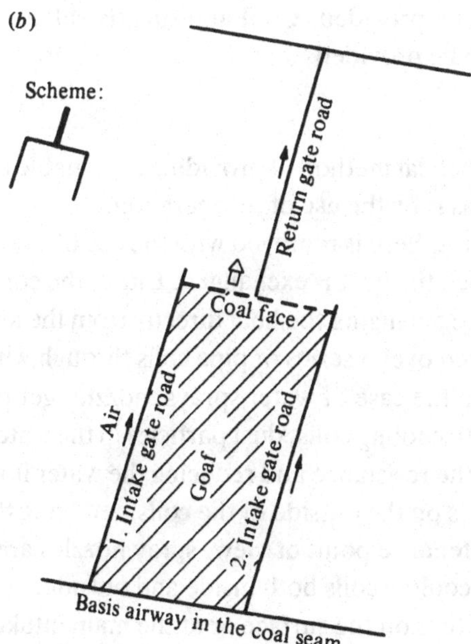

(6) The motors selected should not be unduly overrated and should be run only when necessary.
(7) Electric motors, transformers and machinery such as compressors, pumps, haulage engines and underground hoists may be sited in connecting roads between intake and return airways and thus the pick-up of heat by the intake air can be reduced.
(8) If possible, fixed motors sited in intake airways should be shrouded and a regulated air duct connected to a nearby return airway.

Control of moisture

(1) Water sprays used for dust suppression should be well positioned, maintained and controlled. Only the minimum amount of water necessary should be used at or near the point of mineral breakage.
(2) Drainage channels and sumps should be kept covered and out of the ventilating air current. Pipes should be used wherever practicable.

Work practice

If necessary, reasonably short work periods and frequent breaks, i.e. a work–rest programme, may have to be instituted. Cool rest areas at strategic locations should be provided. Cool and slightly salted (0.1 %) drinking water should also be provided.

Refrigeration

In some cases, the only practical method of providing reasonable climatic conditions in working areas is by the use of refrigeration.

Heat from the mine atmosphere is removed with the use of cold water. There are two ways in which the heat is exchanged. Either the cold water is sprayed into the hot air, exchanging the heat directly from the air to the water or the hot air is passed over a series of pipe coils through which the cold water is circulated. In the case of water sprays, nozzles get plugged by dust particles, while with cooling coils, dust particles in the water build up in the coils, increasing the resistance and reducing the water flow rate. Also, dust in the air collects on the outside of the coils, reducing the heat exchange. From the maintenance point of view, spray nozzles are easier and quicker to clean than cooling coils both inside and outside.

Cooling of air is done either on the surface or at the main intake levels or near or in the stopes. The primary bulk air coolers on the intake levels normally remain in one position for many years. The layout of a typical spray chamber is shown in Fig. 7.10. The cold water coming from the refrigeration plant is used directly in the primary sprays. The rate of flow to these sprays (and thus to the spray chamber as a whole) is controlled by a temperature-sensitive valve that regulates the temperature of the air leaving the spray chamber to a pre-set value.

The water collected in the sump of the primary sprays is pumped through a second set of (secondary) sprays. A submerged low dam wall separates the sumps of the primary and secondary sprays. It stops the free mixing of the water in the two sumps but allows an overflow between them.

The water collected in the sump of the secondary sprays is pumped back to the refrigeration plant. The rate of cold water coming from the refrigeration plant to the primary sprays must be equal to the rate of water being pumped from the sump of the secondary sprays to the refrigeration plant. To achieve this, some water from the delivery side of the pump is sent to the third set of nozzles. This flow is controlled by a modulating float valve that maintains a constant level of water in the dam.

Fig. 7.11 illustrates a typical positioning of nozzles and Fig. 7.12 shows a design for (droplet) eliminator plates. The plates can be made of aluminium, fibreglass or other rigid plastic, or asbestos cement.

The secondary semi-bulk air coolers essentially consist of sets of parallel finned tubes through which the cold water passes. Fig. 7.13 shows a very large cooler for the bulk cooling of intake air entering a stope. Such coolers may consist of up to 16 sets of coils often arranged in layered pairs, sited horizontally to save height and to help self-cleaning of the coils by water condensing upon them. Fig. 7.14 shows a schematic layout of another underground air-cooling plant.

Fig. 7.10. Layout of typical 2½ stage spray chamber (after Bluhm and Whillier, 1978). 1, Return pipeline; 2, pump; 3, level control valve (modulating); 4, dam wall; 5, final stage sprays (1/2 stage); 6, secondary sprays; 7, submersible pump; 8, low dam wall; 9, primary sprays; 10, float valve (high level shut-off); 11, temperature control valve; 12, supply pipe (cold water); 13, dam wall; 14, eliminator plates (to drain into dam).

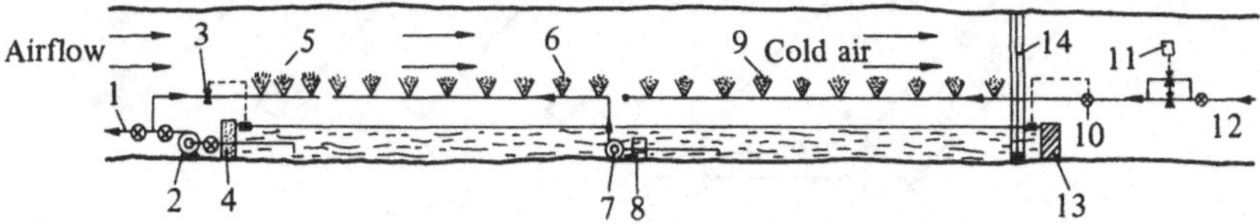

Fig. 7.11. Typical spray pattern in a spray chamber (after Bluhm and Whillier, 1978).

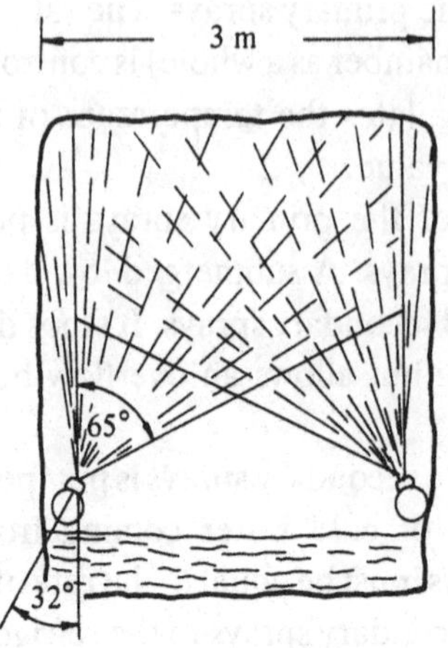

Fig. 7.12. Design for eliminator plates (plan view) in a spray chamber (after Bluhm and Whillier, 1978).

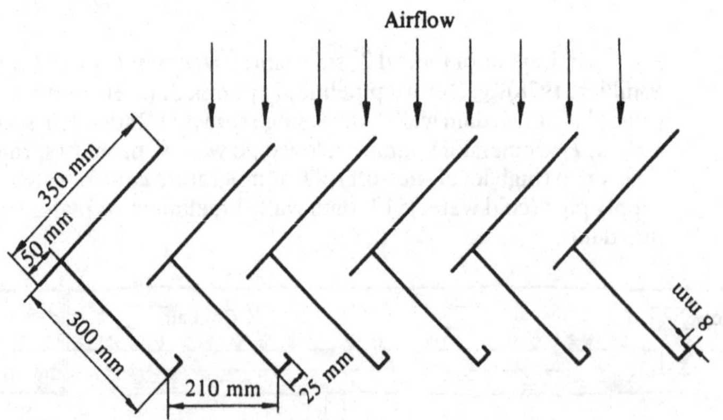

Fig. 7.13. Type of unit used to cool air entering a stope (after Stroh, 1974). Air in: 31.0/32.5 °C; air out: 26 °C saturated; volume: 30 m^3/s; H_2O in: 7 °C; H_2O out: 18 °C; flow: 16.3 litres/s.

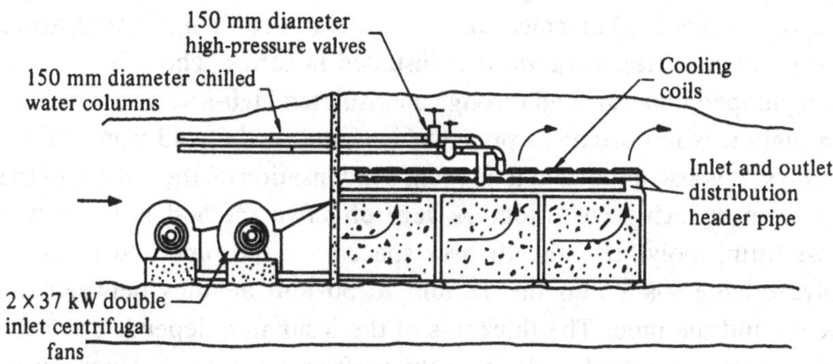

Fig. 7.14. (*a*) Schematic layout of underground air-cooling plant; (*b*) location in system (after Laird and Harris, 1973).

(*a*)

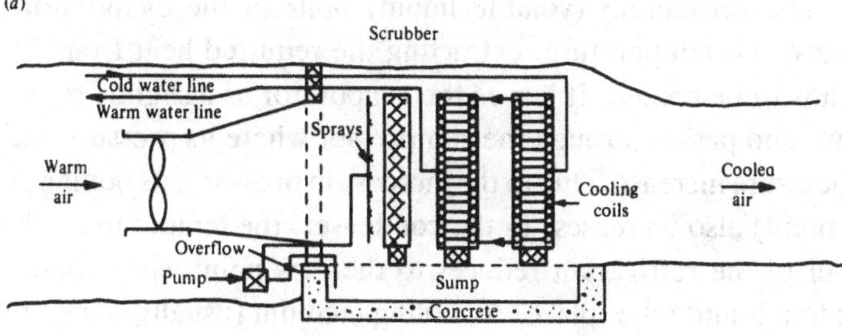

(*b*)

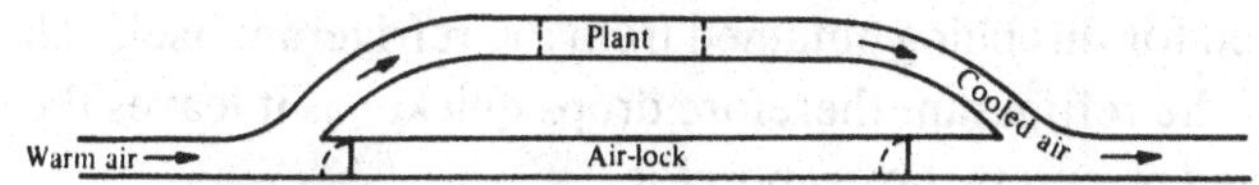

The tertiary (stope) coolers are much smaller units and consist typically of three 100 W coils, trolley-mounted to facilitate their moving up to the face. A 20 kW fan draws air through the unit and forces conditioned air through ventilation ducting to the working face.

A common way of cooling water is on the surface in a cooling tower (by evaporative cooling) or a mechanical refrigeration unit or a cooling tower and mechanical refrigeration unit installed in series. The cold water is then pumped underground through an insulated high-pressure pipe line. (Insulation is important to prevent heat gain of the cold water in the airways. It is also important to prevent condensation on the outside of the pipes, so reducing corrosion. A very effective method is to draw a loose-fitting polythene sheath over spacers on the pipe and to inject polyurethane foam into the 12 mm to 30 mm annulus between the sheath and the pipe. The thickness of the insulation depends upon the size of pipe. This work is done on the surface, leaving the flanges to be insulated after installation underground.) The water then goes either directly to large, more permanent air-cooling plants (which cool air for large sections of the mine) or to heat exchangers (that use the high-pressure cold water from the surface) to cool low-pressure water which is then pumped to smaller air-cooling plants located near the working areas.

Mechanical refrigeration units can also be installed underground to provide cool water to low-pressure air-cooling plants located near the working areas. The hot condenser water is either pumped to the surface or to an exhaust airway where the heat is no problem.

Mechanical refrigeration. The basic refrigeration cycle is given in Fig. 7.15. The refrigerant (volatile liquid) boils in the evaporator at low pressure and temperature, extracting the required heat from the water which is to be cooled. It leaves the evaporator as a slightly superheated vapour and passes through the compressor where its pressure and hence temperature increase. Due to the increase in pressure, its boiling point (or dew point) also increases. In the condenser, the temperature of the hot vapour of the refrigerant reduces to the dew point when condensation back to a liquid takes place. A cooling medium (usually water) removes the heat of condensation from the condenser. The refrigerant now in liquid form is at a high pressure and it then passes through the expansion valve where the pressure drops causing the liquid to boil, the heat required for this being obtained from the refrigerant itself. The temperature of the refrigerant therefore drops quickly as it leaves the expansion valve and re-enters the evaporator.

The evaporator and condenser are very similar and are of shell and tube construction. The finned tubes are normally made of cupro-nickel and carry water inside. The water side of the tubes gets fouled owing to scaling and corrosion, reducing the heat transfer particularly in the condenser. Where there is a large difference between evaporator and condenser temperatures, multi-stage compressors, often with interstage cooling, are used.

Ammonia is the most commonly used refrigerant in surface plants. As it is toxic, the fluorinated hydrocarbon refrigerants (R11 and R12) are used in underground plants (McPherson, 1976).

For surface installations, cooling towers usually supply the condenser water. For underground installations, mine drainage water (if it has sufficient cooling capacity) or cool water from either surface or underground cooling towers is used.

Underground cooling towers (spray chambers). Cooling towers supply cool water to the condenser as well as being used to remove heat from the water returning from the condenser and release it into the mine return air. These are situated very close to upcast shafts and in some cases they form

Fig. 7.15. Basic refrigeration cycle.

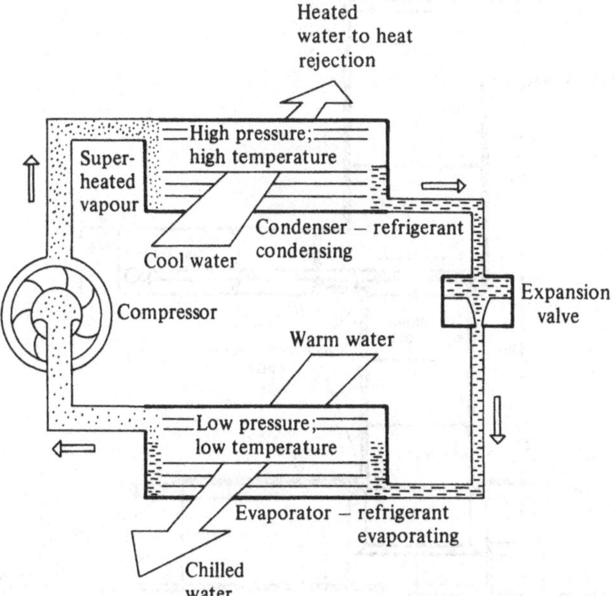

the sump of the shaft. A vertical counterflow cooling tower is given in Fig. 7.16. (Horizontal and inclined spray chambers are also used for this purpose.) Some 80 to 90 % of the heat transfer (from the water to the ascending airflow) takes place by evaporation of the falling droplets of water, the remaining being by convective cooling.

The height of the tower usually lies between 12 and 25 m from sprays to sump. The air rising through the tower normally has a velocity of 2.5 to 7.5 m/s. If the air velocity exceeds about 9 m/s then undue carry-over of water droplets takes place, resulting in an excessive loss of water from the condenser circuit. Carry-over is also reduced by incorporating an enlarged section above the sprays with a vertical distance of two tower diameters and a cross-sectional area of about twice that of the tower.

The water is normally cooled by some 7 to 9 °C during its fall through the tower. The water flow rate is usually limited to 12 litres/s/m^2 to prevent channelling of the water. Non-clogging-type water sprays pointing upward are used so that air–water contact time is increased by 0.5 to 1 s.

The water lost from the condenser circuit by carry-over and evaporation is replaced by a make-up of fresh water. A further bleed-off is

Fig. 7.16. Underground vertical counterflow cooling tower (after Stroh, 1974).

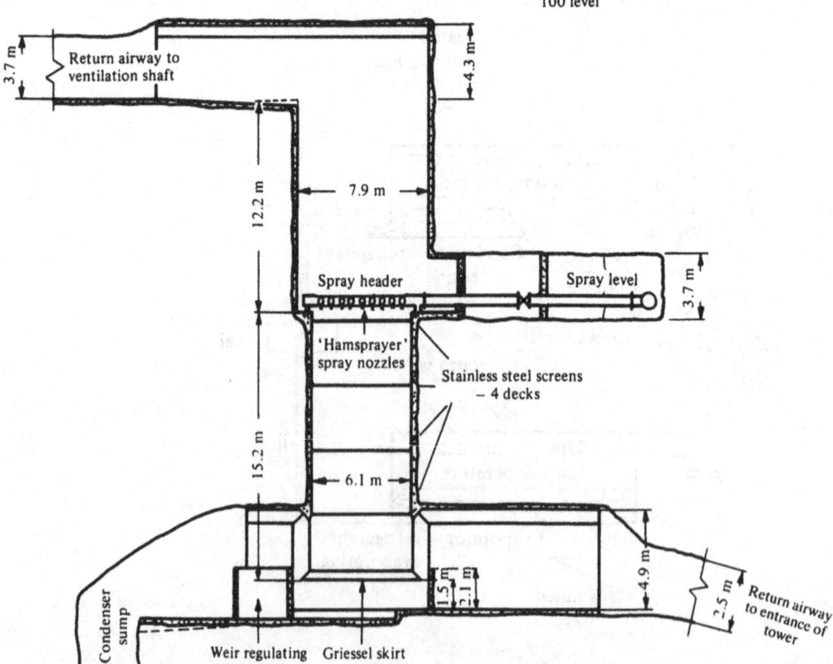

arranged and replaced by fresh water to prevent the build-up of solids and dissolved salts in the condenser tubes. Chemicals are added to soften the water and to control the growth of algae.

Two or more screens are usually suspended in the cooling tower. These help to maintain a uniform airflow as well as breaking up any channelling of the water.

Cooling of the service water. Water is commonly used for dust suppression. If cold water is used for this purpose, it would remove most of the heat from the rock directly, so that the temperature of the ventilation air need not increase in passing through the workings.

Service water is usually cooled on the surface and the main pipes are insulated to reduce the effects of the heat pick-up by cold service water to as little as 0.3 °C per km of pipe length. Due to adiabatic compression, the water temperature increases at the rate of 2.33 °C per km of depth and this increase in temperature can be obviated by using a water turbine. A turbine with an efficiency of only 70 % would reduce an increase in temperature of 0.7 °C per km of depth and would provide a large fraction of the energy needed to pump the service water back to the surface after it has been used in the mine.

Microclimate suits

Microclimate suits have been developed by the Chamber of Mines of South Africa and they can decrease the physiological effects of heat and humidity. The suit consists of an inner poncho and an outer insulating jacket. The poncho has two walls between which are 28 sealed water bags, and is made of a plastic-coated nylon material. The water bags are made from polythene tubing.

The outer jacket is made by stitching together two layers of woven nylon material coated with plastic, enclosing a 10 mm thick sheet of plastic foam.

The suits are hung in a freezer room before use. A thick woollen or cotton vest should be worn underneath the pre-frozen suit.

The rate of work determines the time the suit can be used and at a high work rate it must be changed after 2 hours.

7.5 Overlap systems of auxiliary ventilation with recirculation

The comparative advantages of forcing and exhausting systems of auxiliary ventilation are as follows:

(1) The high velocity of the discharged air at the face from a forcing system produces a high scouring effect, breaking up any tendency of methane to layer back from the face. It also results in a cooling effect by lowering the effective temperature.

(2) With a forcing system, the methane emitted outbye in the heading is carried away from the face. With an exhausting system, it is taken into the face area where there is the greatest danger of ignition.
(3) With a forcing system, unsupported flexible ducting (lower resistance, lower fan duty, lower cost and easier transport) may be used. With an exhausting system, either rigid or supported flexible ducting has to be used to withstand the high suction pressure developed by the auxiliary fan.
(4) With a forcing system, the air reaching the face usually takes up less heat and moisture. Leaks in the ducting are more easily detected.
(5) With an exhausting system, airborne dust and blasting fumes are taken away from all workers in the heading. It maintains fresh air within the heading and it is possible to filter the airborne dust. The main disadvantage is that there is little control of the airflow over the face of the heading.

In summary, a forcing system is selected when methane is the main problem and an exhausting system is selected when the airborne dust is the main concern. If the main problem is due to heat and humidity, then a forcing system is selected.

Conventional overlap systems

The conventional overlap systems have the advantages of both the forcing and exhausting systems. With an exhausting overlap system, blasting fumes may be removed by an exhausting main duct, with methane being swept from the face by a forcing overlap duct. When a heading machine is used, a forcing overlap system is selected so that methane may be controlled by a forcing main duct and airborne dust by an exhausting overlap duct leading to a dust filter.

The main disadvantage of the conventional overlap systems is that there is a reduction in the air volume flow rate (hence velocity) which may give methane layering problems in the heading over the overlap length. Because of this, the overlap fan should not take more than 50 % of the available air. This limitation also prevents uncontrolled recirculation.

In the forcing overlap system, up to half only of the dust-laden air can be filtered. The overlap fan must be reasonably silenced and interlocked with the main auxiliary fan.

Recirculation

The term ‘recirculation’ means that the same air flows more than once past a given point.

When air is recirculated, the air re-entering a place carries back part or all of the contaminants that have already been produced in that place. Because of this, unacceptably high concentrations of contaminants may develop. To avoid this problem, recirculation is prohibited by government regulation.

The concentration of the contaminant could build up indefinitely if the recirculation is in a self-contained closed circuit. If the recirculation is not in a closed circuit and at the same time the re-entering air is continuously mixed with an adequate fresh airflow, the concentration of the contaminant at any point in the recirculating circuit does not build up indefinitely and stabilises at a limited value.

The concentration of the contaminant depends only on the rate of contaminant generation into the area and the volume flow rate of fresh airflow through it.

In Fig. 7.17, a forcing overlap system of auxiliary ventilation for a long heading with controlled recirculation of air is shown. $Q_R = (1.5 \text{ to } 2) \times Q_F$, where Q_R = volume flow rate of air handled by the exhaust fan and Q_F = volume flow rate of air at the end of the forcing duct.

Under steady-state conditions, the concentration of the contaminant in the air leaving the heading must be C/Q_F where C = rate of make of contaminant in the heading and Q_F = volume flow rate of air at the end of the forcing duct.

The concentration of the contaminant in the recirculating air must also be C/Q_F since it forms part of the same air handled by the exhaust fan. The rate of contaminant recirculation is:

$$\frac{(Q_R - Q_F)C}{Q_F} = \frac{Q_R C}{Q_F} - C. \qquad (7.6)$$

The rate of addition of contaminant at the face is C so that the entrance rate of contaminant into the exhaust duct is $Q_R C/Q_F$ and the resulting contaminant concentration in the exhaust duct is C/Q_F.

Fig. 7.17. Forcing overlap system with controlled recirculation. Q_R = volume flow rate of air handled by the exhaust fan; Q_F = volume flow rate of air at the end of the forcing duct. $Q_R > Q_F$. (After Allan, 1983.)

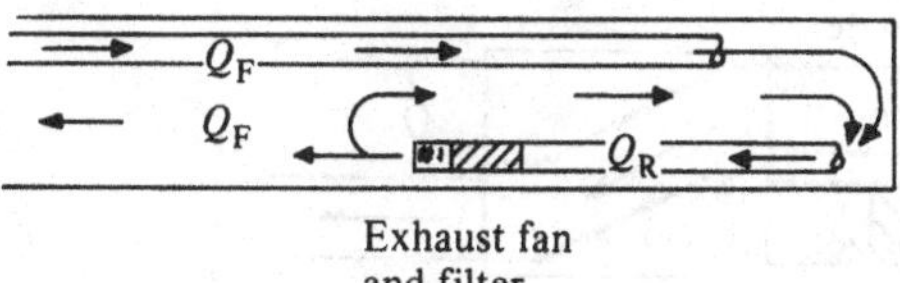

In the case of dust filter usage, the following equations apply, where D = rate of make of dust in the heading and R = rate of dust collection.

The dust concentration in the air leaving the heading must be equal to $(D - R)/Q_F$. The dust concentration in the recirculating air must also be $(D - R)/Q_F$. The rate of dust recirculation is:

$$\frac{(Q_R - Q_F)(D - R)}{Q_F} = \frac{Q_R(D - R)}{Q_F} - (D - R). \tag{7.7}$$

The rate of addition of dust at the face is D so that the entrance rate of dust into the exhaust duct is

$$\frac{Q_R(D - R)}{Q_F} - (D - R) + D = \frac{Q_R(D - R)}{Q_F} + R. \tag{7.8}$$

Since the rate of dust collection by the dust collector is R, the dust leaves the exhaust duct at the rate of

$$\frac{Q_R(D - R)}{Q_F}$$

and the dust concentration is $(D - R)/Q_F$.

The highest dust concentration in the system is in the air entering the exhaust duct end and is given by

$$\frac{(D - R)}{Q_F} + \frac{R}{Q_R}. \tag{7.9}$$

In comparison with a conventional overlap system, this recirculation system has the following advantages:

(1) A higher air velocity in the inbye end of the heading (i.e. the length of heading from the end of the forcing duct to the face) to prevent methane layering.
(2) Lower effective temperatures at the inbye end of the heading.
(3) Improved dust conditions in the heading.

Fig. 7.18. Lingard box (after Allan, 1983).

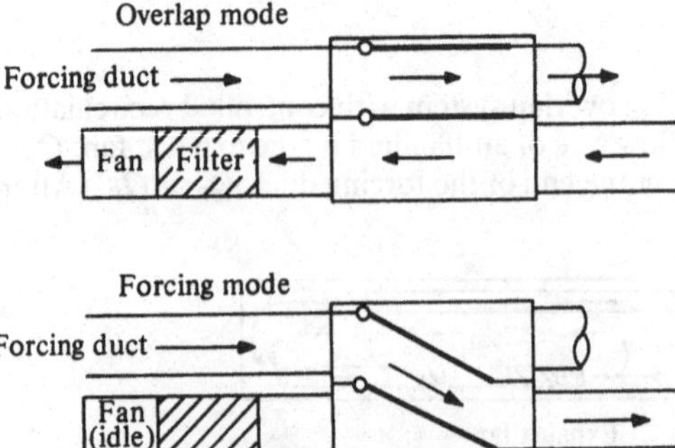

Lingard box (Fig. 7.18)
This equipment allows either a forcing or a forcing overlap system (with or without recirculation) to be used in the same installation. During non-production periods (i.e. when the machine is not cutting) or in the event of failure of the inbye system, a normal forcing system is used. During production periods, a forcing overlap system is used. In the box, there are two doors operated electrohydraulically. The position of the doors determines the two modes of operation of the system.

7.6 References

ACGIH. (1978). *Threshold limit values for chemical substances and physical agents in the workroom environment with intended changes for 1978.* Cincinnati, Ohio: American Conference of Governmental Industrial Hygienists. 94 pp.

Allan, J. A. (1983). A review of controlled recirculation ventilation systems in U.K. collieries. *CIM Bull.*, **76**, No. 849, 83–7.

Belding, H. S. (1970). The search for a universal heat stress index. In *Physical and behavioral temperature regulation*, ed. J. D. Hardy, A. P. Gagge and J. A. J. Stolwijik, pp. 193–202. Springfield, Illinois: C. C. Thomas Publ.

Bluhm, S. J. and Whillier, A. (1978). The design of spray chambers for bulk cooling of the air in mines. *J. S. Afr. Inst. Min. Metall.*, **79**, 1–9.

Laird, F. J. Jr and Harris, E. J. (1973). Mine air conditioning. In *SME mining engineering handbook*, vol. 1, ed. I. A. Given, pp. 16.71–16.84. New York: SME of the American Institute of Mining, Metallurgical and Petroleum Engineers.

McPherson, M. J. (1976). The heat problem underground with particular reference to South African gold mines. *Min. Engr.*, **135**, 391–405.

Misaqi, F. L., Inderberg, J. G., Blumenstein, P. D. and Naiman, T. (1976). *Heat stress in hot U.S. mines and criteria for standards for mining in hot environments.* Mining Enforcement and Safety Administration Inf. Rept 1048. 67 pp.

Mucke, G. (1973). Efficacy of measures for the air conditioning of workings. *15th international conference on safety in mines research*, Karlovy Vary, 18–21 Sept. (Safety in Mines Research Establishment transl. No. 6275.)

NCB (1979). *Ventilation in coal mines: a handbook for colliery ventilation officers.* London: NCB Mining Department.

Patnaik, N. K. and Singh, R. D. (1978). A critical review of the methods of determining the thermal conductivity of rocks. *J. Mines, Metals and Fuels*, **26**, 239–46, 74–82.

Starfield, A. M. and Dickson, A. J. (1967). A study of heat transfer and moisture pick-up in mine airways. *J. S. Afr. Inst. Min. Metall.*, **68**, 211–34.

Stewart, J. M. (1980). The use of heat transfer and limiting physiological criteria as a basis for setting heat stress limits. In *2nd international mine ventilation congress*, Reno, NV, Nov. 1979, ed. P. M. Jones, pp. 556–71. New York: SME of the American Institute of Mining, Metallurgical and Petroleum Engineers.

Stroh, R. M. (1974). The refrigeration systems of Western Deep Levels Limited. *J. Mine Vent. Soc. S. Afr.*, **27**, No. 1, 7–18.

Verma, Y. K. (1979). Mining and mine climate in the year 2000 AD. *Min. Engr.*, **139**, 225–33.

Voss, J. (1975). Control of the mine climate in deep coal mines. In *International mine ventilation congress*, Johannesburg, ed. R. Hemp and F. H. Lancaster, pp. 331–38. Johannesburg: The Mine Ventilation Society of South Africa.
Vost, K. R. (1982). The prediction of air temperatures in intake haulages in mines. *J. S. Afr. Inst. Min. Metall.*, **82**, 316–28.

8

Mine explosions and their control

An explosion is an intensely rapid combustion of a substance or a mixture of substances (gas, liquid or solid) forming largely or entirely gaseous substances with the development of high pressure and heat. The main explosive substances are methane and coal dust in coal mines and sulphide dust in metal mines.

8.1 Methane and coal dust

Almost all coals release a certain amount of methane although the quantity released varies from a few cubic centimetres to almost 100 m^3 per tonne of coal mined. Methane when mixed with air is explosive within the range of approximately 5 to 15 % and is most explosive when its percentage in air is 10. Although when measured against other explosive gases, methane may be regarded as difficult to ignite, it is merely a comparative assessment and in mining situations no comfort should be derived from this.

Coal when mined leads to the production of coal dust. When compared with the dusts of many metals and synthetic plastic materials coal dusts are not highly inflammable. The concentration of particles of coal dust in air required to make it explosive must be exceedingly high and this level is unlikely to be reached except by some violent disturbance (e.g. methane explosion or blasting). Nevertheless, the nature of coal mining is such that two potentially explosive substances, methane and coal dust, are either released or formed.

Coal mine explosions are invariably initiated by ignition of methane through a variety of causes, usually at or near the coal face. Most of the ignitions spend themselves harmlessly in a local accumulation of gas and do not propagate. Some, however, lead to large-scale damages and fatalities.

The main danger of methane ignition underground is that it may produce a large enough blast to raise coal dust into the air, reaching concentrations high enough to be ignited by the burning of methane and thereby adding to the force of the blast so that more and more coal dust is raised as fuel for the flame, leading to a propagating coal dust explosion (Fig. 8.1).

The explosion of a methane–air mixture is in the nature of a rapid inflammation rather than a true explosion. The flame speed is limited to a few tens of metres per second in the majority of cases and its movement is undulatory or vibratory in character. The presence of obstructions increases the speed of flame and should the speed be sufficient to ignite the unburnt methane in front by adiabatic compression a single intense oscillation traverses the whole mass of the mixture, resulting in its detonation. Though rare, such detonations do occur and the rapid inflammation of methane is a general phenomenon. However, when a true detonation occurs tremendous damage may be caused.

Generally, the ignition of methane raises the coal dust cloud and the burning of the methane ignites it (Fig. 8.1*b*). The flame then propagates away from the point of ignition and may soon become a propagating coal dust explosion. The flame front is preceded by a compression blast wave followed by another pressure wave. A coal dust explosion will continue to extend indefinitely as long as sufficient fuel is present and no barrier of inert material is placed effectively in its path. Flame speeds observed experimentally cover a wide range, from about 20 to 2000 m/s. In most cases, however, the speeds are of the order of 200 m/s.

Fig. 8.1. The development of an explosion.

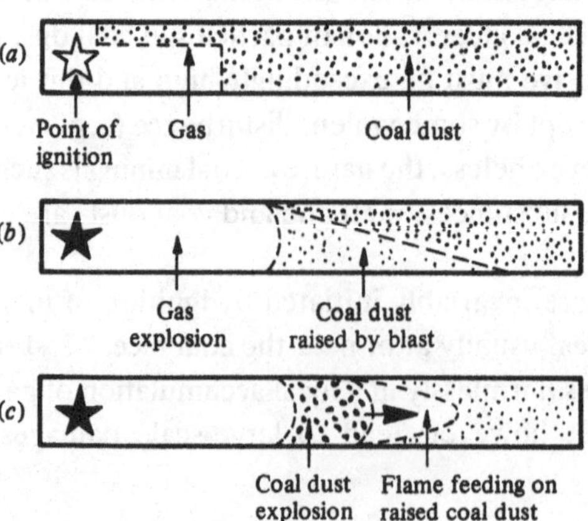

8.2 Methane explosions

Methane explosions are characterised by two distinct phases, the direct blast and the indirect blast or backlash. In the direct blast, a pressure wave travels ahead of the explosion flame. The backlash is caused by the vacuum arising out of cooling of gases and condensation of water vapour behind the exploding flame and is of less intensity than the direct blast and traverses the same path backwards. When the concentration of methane is greater than 9.5 %, two types of explosion flames appear, the primary and secondary flames. The primary flame propagates and consumes the entire available oxygen. The secondary flame is produced by the later burning of the unburnt gas with the help of the oxygen supplied by the backlash. It propagates slowly in the opposite direction to that of the primary flame at a lower velocity than the primary flame.

8.2.1 Causes of ignition of methane

The three principal sources of ignition in modern mining are working machinery (friction, sparking and impact), blasting and electricity in descending order of importance. The other sources are naked lights (in most countries, all coal mines are now designated 'safety-lamp' mines), the use of forbidden articles such as safety matches, defective safety lamps, open fires (possibly from the spontaneous combustion of coal), hand blows and falling materials.

Working machinery

Certain materials and processes associated with modern mining may give rise to hot surfaces or incendive sparks which can ignite methane. For example, modern coal-mining machines have fast-moving picks which can create a risk of ignition if they strike some rock other than coal. The hazardous rocks are those containing appreciable amounts of granular quartz or pyrite. Mudstones, shales, clays, slates, limestones and ironstones are unlikely to produce sparking when hit with machine picks.

Many ignitions have been caused by the machine losing its proper horizon and allowing the picks to cut into the roof or floor. They can, therefore, be classed as avoidable by proper horizon control. Sparking due to the picks striking quartzitic or pyritic material associated with intrusions, faults, rolls and washouts or such material present within the seam itself (silica bands) is impossible to avoid when the whole of the seam is extracted. The various methods for protection are the reduction of methane concentration through improved face ventilation, advance drainage and gas emission control, improved ventilation of cutting area, e.g. hollow shaft drum, reduction of the incendivity of tool bits (cutting tools) through improved metallurgy of the bit or through reduced bit speed, better pick design, water spray quenching systems and an ignition-quenching device incorporated into the design of the coal-mining machine.

Adequate ventilation as close as possible to the pick point is vitally important. This has to be achieved by a combination of design of the machine to make the best use of the face ventilation and by ancillary ventilation directed at the vicinity of the pick points. Face ends (advancing longwalls) should be kept down to a minimum length and adequately ventilated using ancillary ventilation devices such as venturis and fans. The accumulation of methane layering and freshly cut small coal should be avoided in areas where hot spots are likely to occur. Water-powered venturis have also proved to be effective in ventilating the vicinity of the cutting head of the heading machine.

The selection of the cutter pick material is important. Several materials have been considered as alternatives for the steel shank and the tungsten carbide insert of a coal cutter pick. These include nickel alloys (e.g. Rene 41), heat-treated stainless steels (e.g. 17-4 PH), nickel-bonded tungsten carbide and nickel-copper-bonded titanium diboride which are less incendive than the usual 4130 steel.

The possibility of ignition increases with speed of the cutter pick. The risk of ignition can be minimised by more powerful cutting machines taking deeper cuts at lower pick speeds. The essential feature is to have a pick that is cutting and not rubbing the rock. Low-wear material at the tip of the pick is a prime requirement in pick design. Point attack picks are better than radial picks. At any given cutting speeds worn picks increase the probability of sparking.

The probability of ignition occurring when picks cut into an incendive material can be reduced by the application of water behind the pick. Such an application of water is the reverse of what has been found best for dust suppression, i.e. pick face flushing. Therefore, if water is to be relied on to minimise frictional ignition both methods of applying the water are required.

Ignition-quenching devices are being developed for incorporation into the cutting machines. At present these devices are feasible for mine applications where the spatial or geometrical limitations are not a problem. The three main components of the ignition-quenching device are a flame-detection unit, an extinguishing system and an extinguishant dispersal system. Research indicates that an ultraviolet-type flame detector provides an adequate response. Hybrid agents of Halon 1301 and dry chemical (e.g. $KHCO_3$) or water are highly effective. An explosive-activated dispersal system is necessary to insure the rapid extinguishment required. Also, the toxicity of the extinguishants does not seem to be a serious problem if the quenching device is quick and effective and the flame diameter is small.

Blasting

Most of the ignitions here are caused by faulty blasting cables, defective exploders and malpractices, i.e. non-compliance with the blasting regulations. Regular checking and maintenance of the equipment and strict adherence to the regulations will overcome this problem. The development of new, less incendive explosives has helped the safety record.

Electricity

In this category most of the ignitions are caused by damage to trailing cables, faulty equipment and cable couplers. Electrical equipment can be made safe for use in flammable atmospheres principally by two methods: low-power equipment can be made intrinsically safe and high-power equipment can be made flameproof.

A very small amount of electrical energy (equivalent to one-thousandth of that required to light a small torch bulb for one second) will ignite methane. Despite this, a large variety of electrical equipment can be made ‘intrinsically safe’ – that is to say, can be designed and constructed so that a spark produced anywhere in the circuit, say, by a breaking wire connection, would not be capable of igniting gas. Relays, signalling systems, delay firing and instruments come into this category.

Protection of switchgear operating large motors is not possible by intrinsically safe devices and the protection becomes more mechanical than electrical. Boxes which house switches must be strong enough to withstand an internal explosion and have flange gaps narrow and long enough to prevent gas that has seeped inside and become ignited from igniting a flammable atmosphere surrounding the boxes.

Light alloys

Light alloys of magnesium and/or aluminium when struck against rusty steel create sparks. Materials which can cause sparks themselves generate further heat by a combustion process. For this reason such alloys should not be used in hazardous environments. In many countries these alloys are not allowed in underground mines.

8.2.2 *Prevention of methane explosions*

No practical measures exist for arresting methane explosions in coal mines. Only preventative measures against this hazard can be adopted. The various measures are: (1) against accumulation of dangerous methane mixtures in mine workings from the beginning and (2) against ignition of methane mixtures.

8.3 Coal dust explosions

Although coal dust itself can be ignited given a high enough energy source, coal dust explosions usually follow accidental ignition of a methane pocket or roof layer, which may develop sufficient violence to pick up, disperse and ignite coal dust lying on the surfaces of mine openings. A self-generating dust explosion can then develop and propagate over large distances in the mine. The violence of the explosion increases with the increase in the distance it travels and with the amount of fuel it picks up on the way. The velocity of propagation and the dynamic pressure also increase with the distance traversed and the quantity of fuel available. Even the weakest explosions have a speed of up to 30 m/s and a dynamic pressure of the order of 1 kPa. Particularly violent explosions reach a flame velocity of the order of 1000 m/s, and a dynamic pressure of the order of 1700 kPa. The maximum velocity of the flame which corresponds to the detonation of coal dust is of the order of 2000 m/s.

Certain conditions must exist before a coal dust explosion can occur and propagate. The coal dust must be:

(1) of a composition which includes sufficient volatile matter;
(2) of the necessary fineness;
(3) in a dispersible condition;
(4) present in sufficient quantity so that if raised as a cloud in the mine air it will be present above a certain minimum concentration;
(5) raised into a cloud by some agency; and
(6) ignited by an initiating flame of sufficient intensity.

In a coal dust explosion, two types of pressure waves are observed, one which precedes the flame with a steeply rising front of short duration forming a shock wave, followed by one or more pulses with peaks lasting for a second or more, after the passage of the flame. The magnitudes of the peaks may vary from a few to about 700 kPa. Between the shock wave and the flame front there is an advancing air blast which, in raising dust from the roadway surfaces, provides fuel to propagate the explosion further. Flame speeds may range from between 20 and 2000 m/s. In the majority of cases the flame speed has been found to be of the order of 180 m/s.

8.3.1 Factors affecting the explosibility of coal dust

The explosibility of a dust is generally defined as the quantity of inert material required to be mixed with a unit weight of coal dust to prevent the propagation of flame through the mixture. It depends mainly on the ratio between the inert and the combustible content. This ratio is known as the explosibility index, Z, and may be defined by the equation

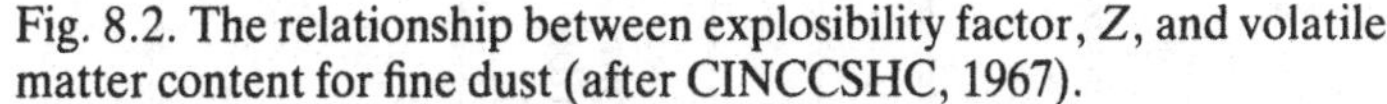

$$Z = \frac{S}{100 - S} \qquad (8.1)$$

where S = the percentage of inert matter that must be present to ensure that the dust will just not explode, and is determined by tests.

Volatile matter

The explosibility index of a coal increases as the volatile matter content of the coal increases (Fig. 8.2). Coals of 25 % volatile matter content and above should all be regarded as highly explosive and therefore require the same standard of protection. For coals below this volatile matter content, a reduction of 5 % incombustible content for each 5 % range of volatile matter content below 25 % down to 10 % volatile matter can be made. Coal dust explosions have never been observed in coal dust with a volatile matter content of $\leqslant$10 %. Distillation of volatiles by the flame is probably responsible for the propagation of coal dust explosions.

Presence of methane even in quantities less than can be detected by the flame safety lamp increases the explosibility of road dust appreciably. For each 1 % of methane in air, the coal content of the road dust must be reduced by one-fifth, i.e. the incombustible content S % must be raised by $(100 - S\ \%)/5$. Table 8.1 shows the increased incombustible requirements in a mine atmosphere containing methane.

Fig. 8.2. The relationship between explosibility factor, Z, and volatile matter content for fine dust (after CINCCSHC, 1967).

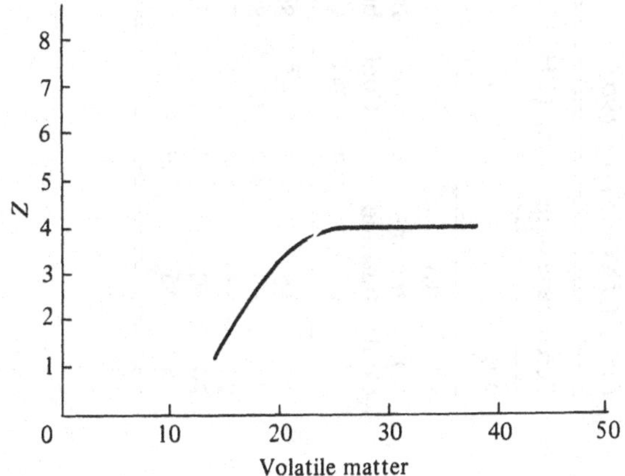

Table 8.1. *Percentage of coal, incombustible content and ratio of incombustible content to coal dust in corresponding limit mixtures*
(after CINCCSHC, 1967)

Percentage of firedamp (CH_4 + air)																	
0 %			½ %			1 %			1½ %			2 %			2½ %		
(a) Coal (%)	(b) Incombustible (%)	b/a	Coal (%)	Incombustible (%)	b/a	Coal (%)	Incombustible (%)	b/a	Coal (%)	Incombustible (%)	b/a	Coal (%)	Incombustible (%)	b/a	Coal (%)	Incombustible (%)	b/a
15	85	5.7	13.5	86.5	6.4	12	88	7.3	10.5	89.5	8.5	9	91	10.1	7.5	92.5	12.3
20	80	4.0	18	82	4.6	16	84	5.2	14	86	6.1	12	88	7.3	10	90	9.0
25	75	3.0	22.5	77.5	3.4	20	80	4.0	17.5	82.5	4.7	15	85	5.7	12.5	87.5	7.0
30	70	2.3	27	73	2.7	24	76	3.2	21	79	3.8	18	82	4.6	15	85	5.7
35	65	1.9	31.5	68.5	2.2	28	72	2.6	24.5	75.5	3.1	21	79	3.8	17.5	82.5	4.7

Fineness of the dust

For practical purposes coal particles coarser than 240 μm do not take part in the coal dust explosion although the temperature to which the particles are subjected influences the critical size. A considerable quantity of dust (between 3 and 5 % of the coal mined) less than 240 μm in size is generated by cutting, drilling and blasting operations.

Fineness substantially increases the explosibility of coal dust, the explosibility increasing linearly with the specific surface area. The dust deposited on the roof and sides is not only finer but more dispersible than that deposited on the floor. Much of the dust on the roof and sides is less than 75 μm in size.

Dispersibility of the dust

Dispersibility is the ease with which dust deposited is raised into the air by a given blast to produce a cloud of sufficient concentration for a flame to propagate, and it is influenced by particle size, the interparticle adhesion and adhesion between the particles and the underlying surface. The presence of moisture has a considerable effect. Adhesion is reduced by the vibration caused by the shock wave preceding the flame front of an explosion. The strength of the emanating blast and the turbulence created by supports and constrictions in the roadway are important factors affecting dispersion.

The agencies in a mine capable of creating a blast sufficient to raise a dust cloud are: (1) a methane explosion which not only creates a sufficient blast but also provides the initiating flame; (2) extensive caving in of the roof, giving rise to an airblast capable of blowing out stoppings and resulting in changes in the ventilation circuit and flooding excavations with unexpected methane concentrations; (3) the accidental detonation of explosives in an underground store; and (4) the bursting of compressed air mains, producing large turbulent velocities.

Dust concentration

The minimum concentration of dust in a cloud capable of propagating an explosion varies between 20 and 50 g/m^3 of air. A layer of dust no thicker than a thin sheet of paper on the roadway surfaces when dispersed is sufficient to give the required concentration. Coal mining operations inevitably give rise to dust settling on roadway surfaces in excess of the minimum required.

8.3.2 Lines of defence against coal dust explosions

Because mining operations unavoidably and continually generate more coal dust than the minimum required to give rise to a coal dust explosion there are no absolutely certain means of fighting coal dust explosions. Besides, in gassy mines, it is not possible to prevent emission of methane from the exposed coal faces so that a source of ignition and initiation is always present and small percentages of methane add considerably to the danger. This statement is borne out by the fact that despite measures taken to prevent coal dust explosions they still occur, although over the last 50 years the number of explosions even with increase in mechanisation and concentration of output has been greatly reduced by proper planning of the defence measures.

The following lines of defence can be distinguished in fighting coal dust explosions:

(1) fighting the coal dust at the point of its origin so that its generation is minimised or so that it is not allowed to become airborne;
(2) fighting the hazard of coal dust ignition;
(3) fighting the development of a coal dust explosion;
(4) stopping the propagation of a coal dust explosion.

1. Fighting the coal dust at the point of its origin

The most obvious and most practicable method lies in the use of water in cutting, drilling, blasting and loading operations so as to wet the dust as it is formed and to send it out of the mine with the broken coal before it dries. This is no more than good housekeeping and relies on supervision and discipline for effectiveness. The cleaning up of spillage and accumulations in haulages and in particular belt conveyor roadways must be properly organised. The thoroughness with which this is done greatly assists in the neutralisation of the remaining coal dust.

2. Fighting the hazard of coal dust ignition

Almost every one of the many coal dust explosions that have occurred in the history of coal mining has been initiated by a methane explosion. The explosion of as little as 0.4 m^3 of methane is sufficient to initiate a coal dust explosion. The precautions against methane and methane explosions are therefore of paramount importance. The primary precaution is good ventilation. Ventilation aspects and precautions against the ignition of methane have already been dealt with (section 4.6.5, Design of mine layout to combat high gas emission, section 4.6.6, Combating layering of gas in roadways, and section 8.2.1, Causes of ignition of methane).

The siting of explosives stores, and supervision over the transport, storage and use of explosives are important in reducing the risk of explosions.

Maintenance of electrical equipment, following safe procedures during the repair of equipment underground and strict adherence to the regulations prescribed in the use of equipment are absolutely essential. Many explosions have been initiated by faulty cables or damage to the cables while flitting machines. It is essential that appropriate steps are taken to avoid damage to cables and that replacement of damaged cables is carried out.

3. Fighting the propagation of a coal dust explosion

(a) Stone dusting of workings. Stone dusting (or rock dusting) of workings is the most common method used for protection against the propagation of an explosion. Usually about 0.7 to 0.8 kg of dust is required to be dispersed per tonne of coal mined. The stone dust used in coal mines is mostly finely ground limestone. Unfortunately this method has the following drawbacks:

(i) During the working shift, coal dust is continuously being deposited on haulage ways and other roadways whereas invariably they are only stone dusted from time to time. During a coal dust explosion it is the dust cloud raised in the air that decides its development. The composition of this cloud is influenced only by the topmost layer of dust. In effect, the overall dust in the workings can have a very high incombustible content and still the workings are not protected against the propagation of an explosion.

(ii) The difficulties in stone dusting efficiently mostly occur in workings where the coal dust production is intense (e.g. belt conveyor roadways and in particular the conveyor structure cannot be stone dusted).

(iii) The density of stone dust is about twice as high as that of coal dust; therefore the latter is more difficult to disperse.

(iv) For there to be sufficient protection against the propagation of explosion when very dangerous coal dust is present, a very high incombustible solids content is necessary in the whole cross-section of the roadways.

(v) Supervision of efficient stone dusting of the workings is difficult.

(vi) The regulations determining the requirements for stone dusting (Table 8.2) are, for the most part, insufficient in the case of very dangerous coal dust and a strong initiation of explosion.

(vii) Large-scale use of stone dust contaminates the coal, which is particularly undesirable from the economic point of view in the case of coking coal.

In spite of these drawbacks, stone dusting has proved to be effective as is shown by world statistics. Since the introduction of this method the number of disastrous explosions has been greatly reduced.

Table 8.2. *Stone dusting requirements in various countries*

Country		S (%)	Ratio of incombustible content to coal dust $\left(\frac{S}{100-S}\right)$	Remarks
		Not less than	Not less than	
Austria	1959	0	0	All roads are wetted
Belgium	1965	60–78	1.5–3.5	Volatile content 14 % to more than 26 %
Canada	1959	65	1.9	Add 1 % for each 0.1 % methane
Czechoslovakia	1957	80	4.0	
France	1965	50–70	1.0–2.3	Volatile content less than 22 % to more than 29 %
Germany	1965	80	4.0	Unless treated with paste
Holland	1965	65–80	1.9–4.0	For methane from 0–1.5 %
Italy	1965	50–75	1.0–3.0	Volatile content less than 20 % to more than 35 %
Poland	1959	70–80	2.3–4.0	Non-gassy–gassy
South Africa	1959	50	1.0	
United Kingdom	1961	50–75	1.0–3.0	Volatile content up to 20.35 % and above
USA (Federal)	1960	65	1.9	Add 1 % for each 0.1 % of methane
USSR	1953	60–75	1.5–3.0	Non-gassy–gassy

S = the percentage of inert matter that must be present to ensure that the coal dust will not explode.

Transport and handling of stone dust: Stone dust is available in 50 kg bags although bulk deliveries to the coal mines can be made if facilities are available at the mine site. The price per tonne of stone dust supplied in bags is much higher than that received in bulk. Handling of stone dust in bags is costly because of the large amount of labour required and the loss of stone dust caused by breakage of bags when handling, damage caused by rain and other factors. This loss may range from 10 to 20 %. If silo facilities are available at the mine site, stone dust can be transported down the mine by piping the stone dust into the mine down the shaft or bore hole or by providing special bulk trailer tankers or tip trailers for transport of the dust from the silo to the face.

Dispersion of dust underground: In many old mines, dust is dispersed manually or by small portable venturis. More recently mechanised dusting units have become very common. These consist of a tank to hold the dust, a mechanism to mobilise the stone dust and a discharge unit in the form of a hose or spreader. The steel tank keeps the stone dust 'fluidised' by introducing pressurised air into the tank to prevent it from clogging. Stone dust from the fluidised storage tank is introduced by means of a screwfeed or a venturi into an airstream, supplied by a blower. The airstream carries the stone dust through a flexible hose to a discharge nozzle for manual application or to a spreader which discharges the stone dust automatically over a fixed area.

Discharge of stone dust for in-section stone dusting is at the rate of around 7.5 kg of stone dust per minute; this amount will cover approximately 7 m^2 of a roadway. The distribution of stone dust becomes inefficient once the discharge rate exceeds 30 kg/minute because of difficulties in operation, in spreading and excessive spilling.

Several machines are available for underground use. These include large-capacity machines (25 t) which are powered mechanically through a prime mover and can cover large distances. These are used mostly for dusting the main intake and return roadways. Medium-capacity machines (about 1 t) are used for dusting within a section, and stationary machines (trickle dusters) are used for dusting return airways. These units have an output of 1 kg/minute and can dust a roadway length of up to 500 m but need to have an air velocity of about 3 m/s.

(b) Spraying of coal dust with water. The spraying of coal dust with water is the oldest method of keeping it down. The drying up of this water as a consequence of the normal ventilation of the workings, making it ineffective, is considered to be its main drawback. In effect, the spraying of water was abandoned long ago in nearly all coal-producing countries as a principal measure for protection against the propagation of explosion.

In some countries, for example Poland, the regulations allow the use of water. For water to be effective, it must wet the dust well. This can happen only if it has a sufficient amount of inherent moisture. The wettability of the dust can be increased by adding surface-active agents. The use of water is justified if the roadway is damp or wet to such a degree that the relative air humidity exceeds 95 %. If the workings contain moist or wet coal dust, they are not suitable for stone dusting since the stone dust rapidly loses its dispersibility.

(c) Paste method. To combat superficial depositions of airborne coal dust requiring frequent stone dusting, studies have been conducted where the coal dust is wetted and bound using hygroscopic solutions with the addition of wetting agents in the form of a paste with which the roadways are covered. The main components of the pastes are:

(1) a solution of hygroscopic salt containing a predetermined amount of magnesium chloride;
(2) a non-ionic surface-active agent present in a concentration of approximately 1 %;
(3) a thickener, mainly composed of hydrated magnesium oxide of gel structure, which is formed by the conversion of calcium hydroxide in a reaction with the dissolved magnesium salt.

Such pastes have the form of a pulp. The paste sets the deposited coal dust and makes it non-dispersible.

Since there are certain difficulties in managing the fluid paste, hygroscopic agents in the form of dry powders can be used in the same manner as stone dust. The powder contains $CaCl_2 . H_2O$ and about 3 % of the non-ionic surface-active agent. The powder is dispersed on the surface of the workings and absorbs moisture from air to form the paste.

The drawbacks of this method are corrosion, a slippery working surface and objects and equipment covered with the paste. The use of pastes for fighting coal dust explosions is statutory in the coal mines of the German Federal Republic. For the use of stone dust the special permission of the mining authorities is necessary.

4. Arresting the propagation of coal dust explosions

When there is a very great length of workings to be protected against the propagation of a coal dust explosion, it is practically impossible to carry out perfectly in all workings the requirements of regulations for prevention of this hazard. Since the regulations themselves are not foolproof, one must accept the possibility of initiation and propagation of an explosion. Arresting the propagation of an explosion is possible using barriers. The barriers work by dispersing a large amount of non-flammable material into the path of the flame thereby depriving it of fuel while at the same time extracting heat from it and cooling it down. As a result the explosion flame is spent up.

There are basically two types of barriers used in coal mines which are differentiated by the type of the extinguishant used. These are stone dust barriers and water barriers. Overturning or tipping of the barrier is essential. The degree of dispersion of stone dust or water depends on the conditions of the development of the coal dust explosion, which vary widely. To stop an explosion, the barrier must be overturned at an optimum time, just before the arrival of the flame and the extinguishant dispersed appropriately. If the barrier is overturned too late the flame may simply pass through it. On the other hand, accidental, low shock not associated with explosions may overturn the barrier too early, making it ineffective. Also, the amount of extinguishant required to arrest a propagating flame depends upon the size of the flame. Violent explosions require more extinguishant and a heavier barrier, but heavier barriers are more difficult to overturn and hence are easily skipped by weaker explosions. The amount of stone dust or water on the barrier sufficient to stop the explosion varies within a wide range from some tens to about 400 kg/m^2.

Generally speaking the most difficult explosions to stop are those at the extremes, i.e. the weakest and the strongest.

Stone dust barriers. A stone dust barrier consists of a series of shelves on which a predetermined quantity of stone dust is placed. These shelves form the basic elements of the barrier. Different types of shelves have been investigated. The criterion of effectiveness of these shelves is that they should be tipped at the low dynamic pressures so important for arresting very weak explosions. Fig. 8.3 shows a design of shelves which has proved very effective. Barriers made with this type of shelf are

Fig. 8.3. Stone dust barrier at the Barbara experimental mine, Poland (after Cybulski, 1972).

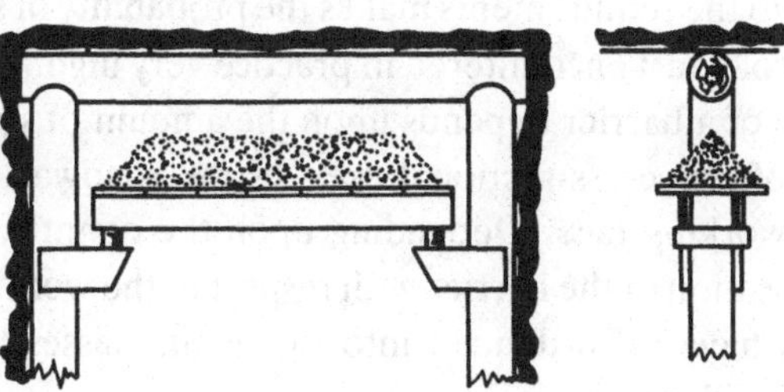

activated even by the weakest coal dust explosions. It must be stressed that the very weak explosions are initiated by accumulations of methane at the roof level. A knowledge of the minimum dynamic pressure activating a given type of barrier determines whether it may be efficient in the case of very weak explosions. With very violent coal dust explosions the time intervals between the moment of approach of the blast wave and the moment of approach of the flame front to the barrier are very short, which, in addition to the other properties of violent explosions, creates difficulties in stopping them by barrier. As mentioned earlier, an explosion must travel some distance to pick up enough energy to tip a stone dust barrier. It is necessary to design a stone dust barrier that is not activated by shocks produced by normal mining operations. From operational considerations, stone dust barriers can stop the explosion at a distance not shorter than 60 m from the spot of initiation of explosion. The optimal distance of the barrier from the spot of initiation also depends, among other things, on the amount of dust on the shelves, on the characteristics of the initiation and on the development of the explosion. However, the optimal distance of the first barrier, called the primary barrier, from the spot of initiation of the explosion is about 100 m. This is difficult to maintain under operational conditions. The spacing of the barriers is such that the coal dust explosion should not cover a distance longer than 200 m without encountering a barrier. In this way the range of the disaster is limited.

The operation of barriers around bends is important. The most advantageous location of a stone dust barrier is at the shortest distance from the bend. It is found that explosions are difficult to stop if they develop in a roadway with a smaller cross-section and pass around the bend to a roadway with a greater cross-section. Then, with the increase in the distance of the barrier from the bend, the difficulty of stopping the explosion increases very sensibly.

The effectiveness of stone dust barriers is largely increased by stone dusting of the zone around the barrier up to the incombustible solid content of at least 85 %. Stone dusting of the zone where the barrier is installed according to the requirements makes the probability of stopping every explosion of coal dust encountered in practice very high.

The effectiveness of a barrier depends upon the amount of stone dust per square metre of the cross-sectional area of the roadway and the distance from the working face. Depending upon the quantity of dust required, and the location of the barrier with respect to the working face, stone dust barriers have been divided into two main classes, primary barriers and secondary barriers.

The purpose of the primary (sometimes called light or auxiliary) barriers is to arrest a weak explosion as close as possible to the face and hence these barriers are comparatively lighter than the secondary barriers, needing lower stone dust quantities, and are placed 60–100 m from the face.

The secondary (sometimes called main or heavy) barriers are placed at a larger distance (<400 m) from the face and at the panel entries and are designed to arrest strong explosions which have travelled a fair distance from their initiation.

Table 8.3 gives a schedule of the quantities of stone dust required under regulations in various countries.

It is not possible to load a single shelf with the required quantity of dust. Several shelves have to be placed close to each other so as to meet the stipulated stone dust requirements for each barrier.

In low roadways there are difficulties in installing barriers along the full width of the roadways because they can cause an obstruction to the movement of traffic and also the possibility of their being tipped over by moving traffic is high. In such cases lateral barriers are installed which occupy only a part of the width of the roadway. They are sufficiently efficient if the normal shelves, of type given in Fig. 8.3, occupy 50 % of the

Table 8.3. *A schedule of the quantities of stone dust required on barriers under regulations in various countries*

Country	Name of the barrier	Quantity of stone dust (kg/m^2)
Great Britain	Primary	100
	Secondary	100
	Light	100–200
	Heavy	400
Czechoslovakia	Auxiliary	200
	Main	400
France		400
German Democratic Republic	Auxiliary	100
	Main	400
Federal Republic of Germany	—	400
Poland	Non-methane seams:	
	Auxiliary	200
	Main	200
	Methane seams:	
	Auxiliary	400
	Main	400
USSR	—	400

width of the excavated roadway. In cases of exceptional difficulties this width can be reduced still further. Figs. 8.4 and 8.5 show these narrow lateral barriers, which occupy only about 30 % of the roadway width and have proved to be very effective in experimental galleries both with very weak and violent explosions.

Meeting requirements relative to stone dusting is practically impossible in some cases. This frequently happens, for instance, in belt conveyor roadways. For such conditions so-called spaced barriers have been introduced. These consist of particular shelves installed at distances so as to meet the condition that the volumetric stone dust concentration of the barrier is not lower than 1 kg/m^3. The distance between the shelves depends, of course, on the size of the roadway. For common conditions it is about 20 m. The individual shelves should contain not less than 5 kg of stone dust for each square metre of the roadway cross-sectional area. Owing to the variable speed of the flame, only some shelves will be triggered at the appropriate time. Also, spaced barriers stop the coal dust explosion over a longer distance than do the ordinary barriers. This is very disadvantageous because the longer the flame travels, the greater are the effects of the disaster. Therefore, spaced barriers cannot be replaced by ordinary barriers, but should only be used instead of stone dusting when this is difficult to do.

Water barriers. Water barriers were invented by Taffanel and owe their rapid development and frequent use not only to comprehensive testing, but also, to a large extent, to the development of plastics with properties suitable for the operation of barriers. They consist of plastic containers filled with water. Figs. 8.6 and 8.7 illustrate the arrangement of the containers held by framelike holders in the mine roadway. The containers can be suspended from a support or from the roof. The possibility of suspending water containers in various ways is the great advantage of water barriers, as it permits the distribution of the containers in the free space in the upper part of mine roadways which are frequently occupied with the ducts, pipe lines, etc. which make it difficult to place stone dust shelves there. Whether the water containers are held in framelike holders,

Fig. 8.4. A lateral stepped stone dust barrier (after Cybulski, 1972).

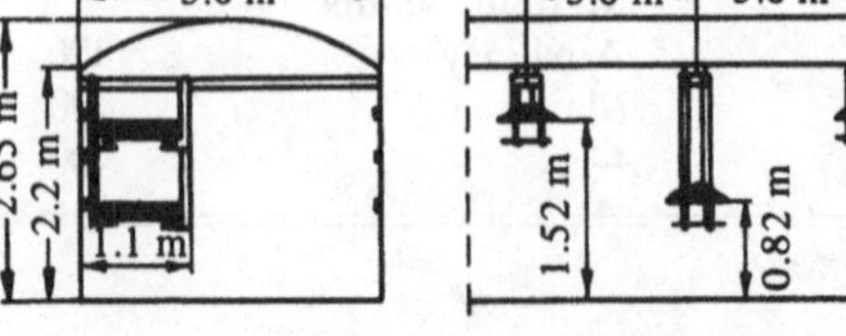

Fig. 8.5. A lateral stepped water barrier (after Cybulski, 1972).

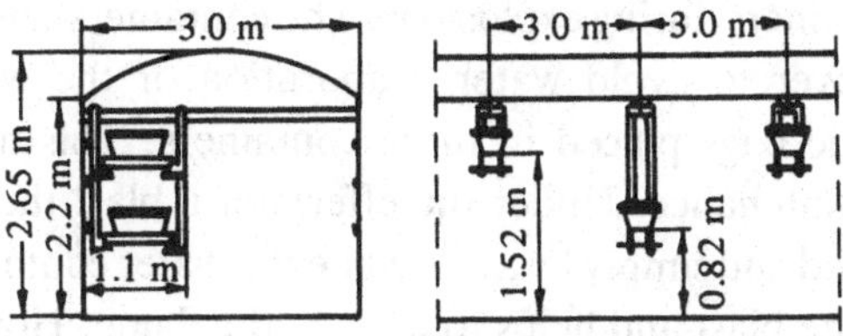

Fig. 8.6. The ways in which water containers can be arranged: (*a*) suspended on roof bars; (*b*) on framelike holders; (*c*) on brackets; (*d*) on frames (after Cybulski, 1975).

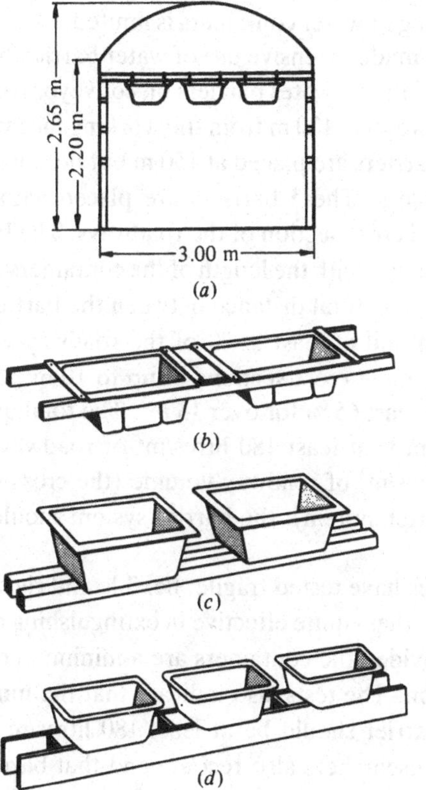

Fig. 8.7. The ways of suspending water containers: (*a*), (*b*), (*c*) under the roof; (*d*) under the roof bars (after Cybulski, 1975).

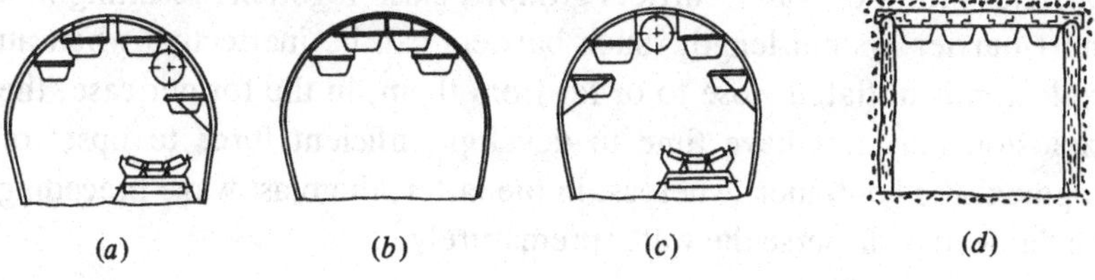

or suspended from a support depends upon the principle of the functioning of these water containers during explosions. The containers can either be covered with a cover to avoid water evaporation or the water is enclosed in thin plastic bags placed in these containers. This method requires minimum maintenance. Under the effects of a blast the water containers are destroyed and simply fly away while the water contained in them is dispersed by the blast and helps to quench the flame. However, with very weak explosions the rigidly mounted containers are not destroyed and consequently the explosion passes along the barrier without being stopped. Accordingly, installation of water containers on supports similar to those in stone dust barriers is more reliable. Thus, the advantage of a rigid mounting of water containers is limited.

The West Germans have made extensive use of water barriers in their coal mines. In longwall operations, water barriers on conveyor roads are installed at 50 m but not more than 120 m from the working face while in bord and pillar workings, barriers are placed at 150 m but not more than 400 m from the working faces. These barriers are placed with water containers in the upper third cross-section of the roadways. The barriers are rigidly held in support frames with the length of the containers at right angles to the roadway axis. The total distance between the barriers and ribs does not exceed 1.5 m and at least 35 % of the roadway width is covered by the barriers for mine cross-sections of up to 10 m^2, at least 50 % for up to 15 m^2 and at least 65 % for over 15 m^2. The total quantity of water in a barrier system is at least 180 litres/m^2 of roadway cross-section and at least 400 litres/m^3 of roadway volume (the cross-section times the length of the barrier system); the barrier system should be at least 20 m long.

Other European countries have tested fragile, flexible and rigid water troughs and have also found them quite effective in extinguishing moderate-strength explosions provided the containers are a minimum of 45 m from the explosion initiation. The tests also indicate that the minimum quantity of water in the barrier should be at least 180 litres/m^2. Both Polish and West German researchers also recommend that barriers be placed a minimum of 25 m beyond a 45° bend and 50 m beyond a 90° or larger change in roadway direction.

The West German water barrier regulations refer to concentrated barriers – rows of water barriers grouped close together, resulting in a short barrier system length. Such barriers can be ineffective when an explosion is initiated close to or far from them; in the former case, the explosion may not have time to develop sufficient force to upset or fragment the container, whereas, in the latter, the blast wave preceding the flame may disperse the water prematurely.

The Europeans have also studied distributed water barriers. In this system, the distance between the rows of barriers is increased and the total barrier system is lengthened, thus increasing the possibility that a portion of the system will be at some optimum distance from the explosion source. Good results have been obtained with distributed barrier systems; water densities as low as 10 litres/m^2 of cross-section per row of barriers with rows greater than 18 m apart were successful in arresting explosions propagating at speeds of up to 365 m/s.

The US Bureau of Mines has developed modified water barriers for suppressing weak explosions. One type is a modification of a conventional container barrier and depends on the dynamic pressure generated ahead of an explosion to tilt the container to release its water and suppress the explosion. The principal feature of the modified barrier is a hinged faceplate that is attached to the water container. The hinged faceplate functions to make the barrier operate rapidly and at low wind speeds. Fig. 8.8 illustrates the modified passive water barrier. The container is made of PVC, is about 280 mm high, 480 mm wide, 760 mm long, and holds approximately 80 litres of water. The hinged faceplate is made of 3 mm thick corrugated PVC. A thin strip of vinyl hinges the base of the faceplate to the base of the container. The container is supported near the roof by two bars (25 mm square) mounted transversely to the roadway. The blast generated ahead of the propagating flame exerts a force on the faceplate, which moves back against the container in an arc, turning on its hinge. As the faceplate lip leaves the support bar, the container drops, swinging about the rear container lip, which is forced against its support bar by the blast, and the water is dispersed into the atmosphere.

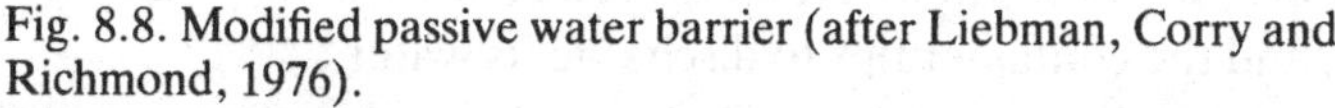
Fig. 8.8. Modified passive water barrier (after Liebman, Corry and Richmond, 1976).

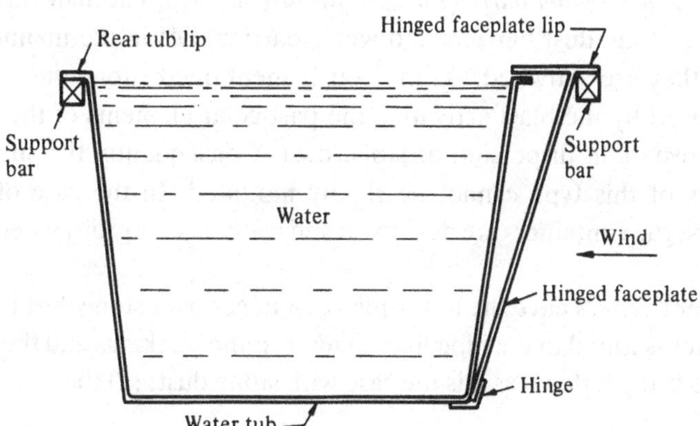

When the container is filled with water, a horizontal force of about 70 N applied at the midpoint of the faceplate is just sufficient to start the motion of the plate. When the coefficient of friction between the faceplate and the support bar is increased by fastening a thin strip of neoprene to the underside of the lip, the horizontal force required to start the motion is about 270 N. The former supporting mode, in which the faceplate lip is in direct contact with the support bar, is recommended only in those regions of a coal mine where the chances of the container being bumped are remote. The latter support method, using a material such as neoprene to increase the frictional resistance between the faceplate lip and the support bar, is preferred in all other locations.

The other type of modified water barrier operates in response to an increase in the static pressure developed ahead of an explosion. Tests indicate that the first barrier begins to release its water at air speeds as low as 15 m/s and the second barrier will operate at a rise in the static pressure of as little as 3.5 kPa. Both barriers were found to be effective in suppressing propagation of coal dust explosions at speeds as low as 30 m/s.

These static-pressure-sensitive barriers operate by responding to static pressure instead of the usual dynamic pressure. This results in a large force advantage for the operation of the barrier, especially for slow-moving dust explosions.

Fig. 8.9 shows a pressure-sensitive device for a static-pressure-sensitive water barrier. A pressure-activated device is fixed to the face of an 85 litre German PVC container. In practice the barrier is suspended near the roof of the mine by three short sections of wire rope: two wires attached to the rear top corners of the container and one to the top of the pressure-sensitive device. In operation, a small rise in the static pressure pushes the small diaphragm and rod inwards to push back the spring-loaded retaining pin, which trips the swivel clamp to release the wire rope attached to the hook, and the container tips to discharge its water.

Comparison of water barriers with stone dust barriers. The main difference between stone dust barriers and water barriers lies in the manner with which they are activated. As previously mentioned, stone dust barriers are started by the blast activating the particular elements of the barrier which results in dispersion of stone dust. Consequently the shelves of barriers of this type cannot be rigidly mounted. In the case of water barriers, the containers are destroyed and water is simply dispersed by the blast.

Water barriers have the following advantages over stone dust barriers: (1) water is found in the pipe lines in many mine workings and there is no need to bring it down as it is the case with stone dust; (2) they are easy to install as the containers can be easily filled or replaced; (3) water does not lose its properties with the passage of time as does ordinary stone dust, which loses its dispersibility especially in wet or damp workings.

In general, both types of barriers are equally efficient in stopping explosions. Under certain conditions, however, the water barriers are more efficient than the stone dust barriers, while in other cases stone dust barriers are more efficient. In order to avoid misunderstandings, it is to be stressed that by the higher or lower efficiency of the barrier is meant the amount of stone dust or water per 1 m^2 of the roadway cross-sectional area required to stop an explosion.

When the time interval between the activation of the water barrier and the arrival of flame to it is relatively long (exceeds 3 s), the water barrier is not effective. This is the major disadvantage of water barriers in comparison to stone dust barriers. In addition, the water barriers are sensibly less efficient if they are installed close to the bends of roadways. They operate correctly if they are installed at a distance of some tens of metres from the bend. As spaced barriers or lateral barriers the water barriers show a similar efficiency to stone dust barriers.

Triggered barriers. In the last decade considerable research has been carried out in stopping the explosion at the place of its initiation using modern flame-sensing devices and triggered or automatic barriers. Passive stone dust and water barriers, requiring explosion-induced wind forces for operation, are limited in use to situations where there is adequate room to install them.

Fig. 8.9. A pressure-sensitive device for a static-pressure-sensitive water barrier (after Liebman *et al.*, 1976).

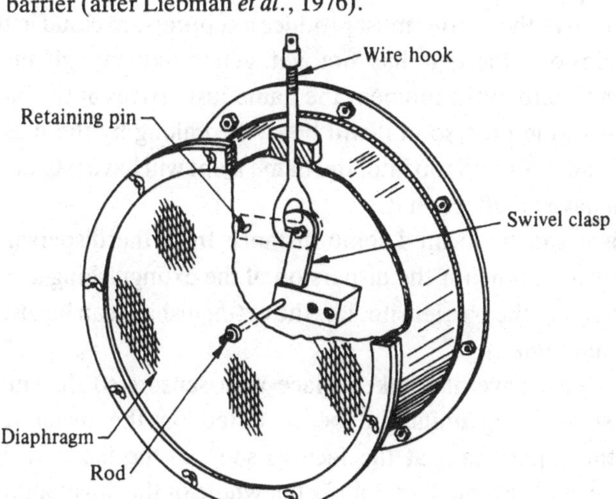

The other limitations are:

(1) they are ineffective against gas explosions;
(2) they may fail when the explosion is weak;
(3) considerable labour is involved in moving them forward as the workings move forward.

To overcome these limitations, triggered barriers have been developed.

A triggered barrier differs from a passive barrier in that it separates the functions of explosion detection and suppression, and the dispersion of the extinguishing agent is made independent of the methane or coal dust explosion that it is designed to suppress.

The general scheme of a triggered barrier consists of three main elements:

(1) a sensing device reacting to a selected explosion parameter (pressure, temperature, optical radiation frequency, etc.);
(2) an executing system which under the effect of the detector impulse activates the barrier and disperses the extinguishing material in the direction of the flame source;
(3) an extinguishing agent which may be limestone dust, sodium bicarbonate, water or an inert gas such as nitrogen, carbon dioxide, etc.

The explosion is detected by a sensor and this triggers the executing system, dispersing the extinguishing agent into the path of the explosion. The executing system comprises a container for the suppressant and a source of energy to eject it. The sources of energy may be a Cardox cylinder, a detonating explosive (Cordtex), compressed nitrogen, Halon 1301 and propellant explosives.

To be effective the barrier must produce a suppressant cloud in the path of the explosion where it has not yet generated enough energy to constitute a hazard at the moment the flame just arrives at the barrier. If the cloud is formed too soon it will be carried along by the blast of the explosion, and if formed too late, the flame front will have passed and the barrier will have no effect on it.

The sensor can be located some distance from the dispersal unit to provide sufficient time for the dispersion of the extinguishing agent prior to flame arrival at the barrier site, i.e. the extinguishant can be discharged at an optimum time.

Developments have also taken place with sensors so that miniature explosion-suppressing units can be mounted on the machine which disperses the suppressant at the face as soon as the flame is detected (within a few milliseconds), engulfing the whole of the developing flame.

Fig. 8.10. Low-pressure-powered disperser developed by the US Bureau of Mines (after Liebman and Richmond, 1978).

Fig. 8.10 shows the US Bureau of Mines low-pressure dispersing unit. The disperser has a spherical vessel, approximately 33 cm in diameter, and has been found to operate satisfactorily when 70 % filled with water with the void space pressurised with nitrogen. The material is released at the base of the vessel through a 7.6 cm wide burst diaphragm and then exited through a multiple nozzle or a 7.6 cm wide pipe.

Water appears to be the superior extinguishant, keeping the toxic effects and problems of cleaning up after such a barrier has been activated to a minimum.

Spectacular results have been obtained in stopping methane explosions with these barriers. It is to be expected that in the near future automatic barriers will contribute remarkably to the considerable progress made in combating not only coal dust explosions, but also methane explosions.

Automatic barriers are complicated devices and their long-term reliability has yet to be established; the orthodox barriers which in many cases have proved their qualities in mines will continue to be used for a long time in addition to the automatic barriers.

8.4 Sulphide dust explosions

Blasting in copper pyrite mines generates sulphide dust, which can explode and create hazardous conditions. The explosibility of sulphide (pyrite) dust depends upon its fineness. The most explosive dust is that with a size of about 0.1 mm. The finer dust is less explosive and particles of less than 0.1 mm are almost inert. Sulphide dust, like coal dust, explodes only when it is found in a suspended state in the mine air. The sulphide dust–air mixtures become an explosion danger if the dust concentration is between 0.25 and 1.5 g/m^3.

The explosion of sulphide dust takes place only during blasting when a large amount of dust is produced at the face. If an explosion occurs at this stage, it spreads to distances which usually do not exceed 35 to 40 m.

The most effective measures for preventing sulphide dust explosions are: drilling blast holes with the aid of water (sometimes with dust-wetting additions); sprinkling faces and the places of blasting with water before blasting; and providing good ventilation of faces.

8.5 References

CINCCSHC (Coal Industry National Consultative Council Safety and Health Committee) (1967). *Final report of the working party on coal-dust explosions.* London: NCB. 100 pp.

Cybulski, W. (1972). Present-day state of fighting coal dust explosions in mines. *Proc. Int. Symp. Dust Explosion Risks in Mines and Industry*, Karlovy Vary,

Czechoslovakia, 11–13 Oct., 1972. Geneva: International Security Association of the Prevention of Occupational Risks in the Chemical Industry. 26 pp.

Cybulski, W. (1975). *Coal dust explosions and their suppression*. Warsaw: Foreign Scientific Publications Department of the National Centre for Scientific and Economic Information.

Liebman, I., Corry, J. and Richmond, J. K. (1976). *Water barriers for suppressing coal dust explosions*, USBM RI 8170. Washington: US Bureau of Mines. 26 pp.

Liebman, I. and Richmond, J. K. (1978). Coal dust explosion barriers. In *Proceedings of the Bureau of Mines technology transfer seminars on coal mine fire and explosion prevention*, Pittsburgh and Denver, pp. 15–26, USBM IC 8768. Washington: US Bureau of Mines.

9

Mine fires and their control

An outbreak of fire in the underground workings of a mine poses not only a direct threat from the fire itself but also an invisible and immediate threat from carbon monoxide poisoning and an explosion, particularly in gassy coal mines, both to persons working underground at the time of the outbreak and to those involved in the subsequent rescue and fire fighting. On many occasions an outbreak of fire has required the sealing off of substantial areas of mine and sometimes the whole of the mine.

Fires in mines may be divided into two types:

(1) open fires caused as a result of ignition of a combustible material; and

(2) fires resulting from spontaneous combustion of coal and sulphide ores.

9.1 Open fires

For an outbreak of fire there must be: a combustible material, sufficient oxygen and a source of heat sufficient to cause combustion. These are generally referred to as the necessary fire triangle. Since sufficient oxygen is always available in the mine air, the problem is tackled by taking 'prevention is better than cure' precautions to reduce the availability and flammability of combustible materials which could take part in a fire and against potential sources of heat.

9.1.1 Combustible materials

The combustible materials encountered underground are as follows:

(1) methane;

(2) coal and coal dust;

(3) timber (wooden supports);

(4) PVC and other similar materials (belts, brattice cloth, insulation on cables, tyres);
(5) flammable fluids and greases (hydraulic oils, lubricants);
(6) others (paper, rags).

Numbers 1 and 2 are naturally present in coal mines and numbers 3 to 6 are taken underground for use.

Precautions

Good ventilation, supplemented with good control of methane emission, is required to reduce the danger of methane inflammation. Coal dust should be regularly cleaned up and not allowed to accumulate in the vicinity of any potential source of heat. This applies particularly to underneath belt conveyor systems and transfer points. All unused, old or discarded combustible materials such as wooden prop stubs, oils, greases, cleaning rags and old belts should not be discarded underground but transported back to the surface for safe disposal. Whenever practicable, fire-resistant materials exhibiting self-extinguishing properties should be used in underground mines.

To retard fires, timber is impregnated with inorganic as well as organic chemicals. Ammonium salts and sodium tetraborate are very effective fire retardants but they are liable to leaching out under underground conditions. Organic chemicals are leach-resistant, offering less of a corrosion risk, but are very expensive.

Fire-resistant materials such as neoprene, styrene butadiene (SBR) and PVCs are used in the manufacture of conveyor belts, hoses, electric cables and V-belts for underground use. Neoprene is the most difficult to ignite. Flexible ventilation ducting and brattice cloth impregnated with fire-resistant materials are also available.

Four classes of fire-resistant hydraulic fluids are available:

HS-A–oil-in-water emulsions containing a maximum of 20 % combustible material. These usually contain 95 % water.

HS-B–water-in-oil emulsions containing a maximum of 60 % combustible material. These usually contain 40 to 45 % water.

HS-C–water-glycol solutions. These usually contain at least 35 % water.

HS-D–water-free fluids. These usually refer to phosphate-ester-containing fluids.

For hydraulic supports, both HS-A and HS-B class fluids are used. For continuous miners, shuttle cars and roof bolters, HS-B class fluid is used. Scoop couplings use HS-D class fluid. Traction-type fluid couplings mainly use water.

Classes of fires

There are four classes of fires depending upon the kind of combustible material involved. Class A fires involve combustible materials such as paper, wood, cloth, rubber and plastics. These materials leave ashes or residue. Class B fires involve flammable liquids, gases and greases such as petrol, oils, paints and solvents. Class C fires involve live electrical equipment such as motors, switches and appliances. Class D fires involve combustible metals such as magnesium, titanium and sodium. In reality, a fire rarely involves only one class of combustible materials. An underground fire that starts in hydraulic fluids or lubricants (class B) can spread to timber and other class A combustible materials. A fire of class B diesel fuel may spread to class A material in hoselines. A class C fire invariably involves class A insulating materials.

9.1.2 *Potential sources of heat for fires*

Heat can be generated by:

(1) conversion of mechanical energy, e.g. friction between the jammed belt and the driving pulley, cutter picks striking an incendive rock;
(2) conversion of electrical energy, e.g. shortcircuiting and overloading, electrostatic discharge;
(3) chemical reaction, e.g. explosives, aluminium/magnesium alloys, i.e. thermite reaction.

All mechanical and electrical equipment should be properly installed, regularly inspected and maintained.

Belt conveyors

Fires are caused by improper operation of the belt conveyor itself or by some external sources. Friction between the belt and the driving drum (overloading or improper tension), a stalled idler (bearing failure), and other materials such as coal, timber and rock (misalignment) and a thruster brake stuck partly in the 'on' position due to either a mechanical failure, electrical fault or accumulation of dust can cause a fire.

The obvious answers lie in good maintenance (e.g. lubrication of drives, idlers and proper tensioning of the belt) and installing a control system that incorporates a means of detecting dangerous conditions (e.g. belt slip, torn belts, misalignment on the gearhead, brake off before restarting, blocked chutes) and stopping the belt if necessary.

Cutting machines

Frictional ignition due to the cutting picks striking roof rocks, particularly when they contain high silica, are the single largest cause of methane fires and explosions. The main precaution is the provision and maintenance of good standards of ventilation, backed up with positive ventilation of the machine where necessary. Developments in automatic monitoring of cutting horizons using nuclear sensors have resulted in greatly reducing the hazard.

Electricity

The presence of electricity is the greatest fire hazard in modern mines. Intrinsically safe or flameproof electrical equipment backed up with sensitive earth-leakage and overloading protection should be used to overcome the problem of shortcircuits or overloading. Non-metallic materials used underground (e.g. insulation covering on electric cables, conveyor belting, flexible ventilation ducting, brattice cloth, hoses – particularly compressed air hoses) should be anti-static to overcome electrostatic discharge. Overheating of electrical equipment can also cause a fire, so an adequate flow of air to cool all electrical equipment is essential.

Cables, especially trailing cables, can be easily damaged (e.g. by falls of rock, blasting, coming into contact with machine picks) causing fusing and arcing. All cables should be properly 'hung-up' to prevent damage and tested regularly.

Explosives and blasting

The chance of a fire from the ignition of methane is small if blasting regulations are strictly enforced.

Aluminium/magnesium alloys

The use of these alloys, particularly with high percentages of these metals, is forbidden in many countries and should be avoided.

Cutting torches

Cutting torches (oxyacetylene as well as electrical) can cause fires. Special cutting bays are normally set aside for this type of work, but if it is necessary to use a cutting torch out of these bays, the following precautions should be taken:

(1) have some means of extinguishing a fire immediately available;
(2) always have another person to observe where the molten slag is falling;
(3) wet the area thoroughly after cutting;
(4) examine the area thoroughly to ensure that no smouldering is taking place.

9.1.3 *Detection of a fire*

Small localised fires will never be entirely eliminated. The problem is to prevent a small containable fire from developing into a major one. Early detection of a fire is the essential first step. In many mines, fire patrols are used for this purpose, although they have serious limitations. They are present only part of the time in limited areas near the working places; they are absent most of the time in remote areas and older workings; sealed areas and goaves remain unpatrolled. Therefore, fire patrols must be assisted by automatic fire detection instruments.

The detector types that are available may be classified as follows:

(1) thermal detectors;
(2) flame detectors;
(3) products of combustion detectors
 (i) gas detectors;
 (ii) smoke detectors.

Thermal detectors

Thermal detectors respond to the temperature or the rate of temperature increase at a point, or along a continuous line. Examples of these types of detectors are the fusible alloy plug of a sprinkler head, thermocouples, bimetallic strips, twisted wire with insulation that melts at a given temperature, and a variety of such devices. The main attributes of the thermal detectors are their ruggedness, freedom from false alarms and their relatively low cost. However, they are not very sensitive and have to be close to the fire to activate the alarm. They are also susceptible to the velocity of the airflow and great care should be taken in the placement of detectors.

Flame detectors

Flame detectors make use of the light spectrum emitted by a flame. The detectors are designed to react to the invisible part of the spectrum which is usually absent in the illuminating equipment (e.g. lamps) used in underground mines. There are two types of flame detectors: ultraviolet radiation and infra-red radiation. Since they respond to the light emitted by a fire or flame, the detectors must be in the path of the emitted radiant energy from the fire or from the surfaces heated by the fire. They can also be activated by welding arcs and other bright flashes of light, so that siting should be carefully selected.

Products of combustion detectors

Any fire generates products of combustion and these are carried to regions far removed from the fire by the mine ventilation or by the fire's own convection currents. Examples of combustion products that can be detected by routine methods are carbon monoxide, carbon dioxide, visible smoke, invisible smoke, i.e. submicron-sized particles, and a variety of products of pyrolysis.

Gas detectors. The detection of carbon monoxide is of prime importance in fire detection. The most successful instruments have a spectral gas analyser (infra-red analyser) and are very sensitive.

A timber fire gives rise to low carbon monoxide production in the early, oxygen-rich stages of burning but a high carbon dioxide content. Carbon dioxide measurement by an infra-red gas analyser has been used successfully for detecting fires in South African gold mines.

Smoke detectors. These are of two types. The optical type uses light extinction or scattering due to smoke to detect its presence. This type of detector consists of a light source, the beam of which traverses a smoke chamber to a detector. The sensitivity of these detectors is generally low and they are useful in areas where dense smoke clouds are generated (e.g. overloaded electric cables).

The ionisation type of detector senses the effect of smoke on the air-ionisation current of a radioactive diode. This type consists of a radioactive source, a set of electrodes and an electrometer to sense the reduction in ion current that results from the presence of smoke.

Alarm limits

Setting up of alarm limits using these detectors requires the analysis of mine return air to determine the base levels. These base levels may be quite different from mine to mine, and in a large mine they may differ widely between its various districts. The alarm limits must also take into consideration the effect of various other factors such as blasting, which gives out enormous amounts of carbon dioxide, carbon monoxide and smoke. Electronic delays can help suppress the activation of alarms in these circumstances since the increased output signals are of a shorter duration and can be easily identified.

9.1.4 *Fire extinguishing agents*

Once an open fire has been detected, the first step to control it is to use an extinguishing agent. The five common fire extinguishing agents are dry chemical, water, foam, halogenated hydrocarbons (halon) and carbon dioxide. Each extinguishing agent has specific capabilities and limitations that affect its usefulness in different situations, such as class of fire, location, presence of different types of materials and direction of movement of products of combustion. An extinguishing agent should be chosen according to the worst fire possible in a given situation.

Dry chemical

There are three types of dry chemicals:

(1) sodium bicarbonate based;
(2) potassium bicarbonate based, known as Purple-K dry chemical; and
(3) monoammonium phosphate, known as Multi-purpose dry chemical.

Both sodium bicarbonate and potassium bicarbonate based chemicals are only effective on B and C class fires. Potassium bicarbonate chemicals are better than sodium bicarbonate based chemicals. The monoammonium phosphate based chemical is effective on A, B and C class fires, and for underground applications this chemical is specified where A, B and C classes of fire are encountered.

The dry chemical extinguisher offers a rapid discharge of relatively short duration. It generally provides the quickest knockdown of a fire, but it has little cooling capacity and hence there is a possibility of re-ignition.

Water

Water is the most commonly used fire extinguishant for underground mines but cannot be used in class C fires because it conducts electricity and can injure or kill personnel operating the extinguishers and can also cause serious damage to the electrical equipment. In the case of class B fires, water can be used only with specialised equipment by highly trained fire fighters.

Foam

Foam consists of an aggregate of bubbles which is generated by blowing air through a screen or net which is suitably wetted with water containing a surface-active agent such as carboxymethyl cellulose. The bubbles serve to carry water to the fire as foam. When the foam hits the fire the bubbles break, releasing the water. As the water evaporates into steam, it results in a drop in the oxygen percentage and thereby helps to extinguish the fire. Furthermore, the conversion of water to steam absorbs a lot of heat, resulting in cooling. When enough foam accumulates, it forms a barrier to prevent the spread of fire.

If the ratio of air to water in the foam is higher than 100 to 1 (on a volume/volume basis), the foam is considered to be of 'high expansion'. A ratio lower than 100 to 1 indicates that the foam is low or medium expansion.

Foam is extremely effective for seeping into inaccessible places. It has the advantage of effectively blanketing everything within range and consequently provides slow but relatively certain extinguishment. Although foam has low water requirements it is relatively expensive.

Halogenated hydrocarbons (halon)

Halon is a colourless, odourless, non-toxic, non-corrosive and electrically non-conductive gas. It is generally used in enclosed areas to produce a 5 to 7 % concentration of the gas. Because halon is a gas, and is non-corrosive and leaves no residue to clean up, it is used frequently in the case of complex electrical installations where other agents might do as much or more damage than the fire itself.

Carbon dioxide

As carbon dioxide is heavier than air, it can be used to blanket the fire, preventing oxygen from reaching it.

9.1.5 Fire extinguishing equipment

Fire extinguishing equipment can be divided into two main categories: (1) portable extinguishers and (2) fixed fire extinguishing systems.

Portable fire extinguishers

An extinguisher is an appliance containing an extinguishing medium which can be expelled by the action of internal pressure which may either be stored or obtained by a chemical reaction. Portable extinguishers are designed to be carried and operated by hand and weigh not more than 25 kg. They are of different shapes, metals and methods of construction.

Dry chemical extinguishers. These consist of cylindrical containers filled to a certain level with dry chemical. Depending upon the size, the amount of dry chemical varies from 3 to 12 kg. The nozzle is usually attached to the end of a length of hose and a device is fitted to prevent accidental discharge. The expelling agent is either a carbon dioxide gas cartridge activated by depressing a plunger in the cap, or compressed air at 1100 kPa which is activated by depressing a valve mechanism located in the neck of the extinguisher body. A valve mechanism can be released, if necessary, to stop the discharge of chemical.

Water extinguishers. There are three different types of water extinguishers: soda acid, gas pressure and stored pressure.

A soda acid water extinguisher is a chemical reaction type extinguisher in which two chemicals are allowed to react to produce a propellant gas when the operating mechanism is activated. A cylindrical or conical container with a capacity of 4.5 to 9.1 litres is filled to a certain level with water containing sodium bicarbonate. The extinguisher also contains a glass phial of sulphuric acid. When the acid is released (either by striking the plunger to break the phial or by inverting the cylinder so that a lead stopper falls away from the neck of the acid phial), it mixes with the solution and reacts to produce gas. The pressure of the gas expels the water solution. The resulting stream is directed on to the fire either from a nozzle or from the end of a length of hose attached to the extinguisher.

In gas pressure extinguishers the pressure is produced by means of a compressed gas (carbon dioxide) which is released from a gas cartridge fitted into the extinguisher. The cartridge, which is fitted with a sealing device, is pierced when the extinguisher is operated. The nozzle is attached to the body of the extinguisher or to the end of a length of hose attached to it. The pressure of the gas expels the water.

The stored-pressure type extinguisher is relatively new. The body of the extinguisher is permanently pressurised and the expellant gas is stored with the extinguishing agent in the body of the extinguisher. It consists of a cylindrical container partly filled with 9.1 litres of water with a brass screw cap which contains the operating valve. Air is pumped into the container until a pressure of approximately 600 to 750 kPa is attained. A dial gauge on some models indicates when recharging is necessary.

Foam extinguishers. There are two different types of foam extinguishers: chemical foam and air foam.

The chemical foam extinguisher is a chemical reaction type of extinguisher. It consists of a cylindrical container filled with a solution of aluminium sulphate in water. A brass screw cap on the top of the cylinder sometimes contains a sealing valve and when this is opened and the extinguisher is inverted (or if there is no sealing valve, when the extinguisher is inverted) the chemicals mix to produce foam and a gas which acts as an expellant.

There are several versions of the air foam type of extinguishers in use, but all produce foam from a solution of foam compound and water. The solution is stored either in the body of the extinguisher or in a hermetically sealed inner container. The capacity of the extinguisher is usually 9.1 litres. The pressure to force it out comes either from carbon dioxide gas contained in a small metal cartridge, or from air introduced into the extinguisher and compressed at 1050 kPa. This extinguisher is more rapid in action than the chemical foam type and is operated in the upright position. A small, foam-making branch is attached to the end of the length of hose and is pointed towards the fire and the plunger in the cap depressed (a safety cap on the plunger may have to be removed first). As the liquid comes out of the tube it creates a venturi effect which sucks in air, creating mechanical foam. Almost 8.5 times as much foam as the capacity of the extinguisher is produced.

Carbon dioxide extinguishers. Portable extinguishers are available in various sizes up to 9 kg capacity. The extinguisher consists of a cylindrical container filled with liquefied carbon dioxide to approximately two-thirds of its capacity and fitted with a control valve. A distinctive feature of the extinguisher is the discharge horn (made of insulating material) which is fitted so that it can be discharged at right angles to the cylinder. It is designed with the orifice end wider than the end attached to the cylinder to allow for the rapid expansion of the gas as it enters the atmosphere and its temperature rises. If the horn was not this shape, the gas might freeze in the tube and block the extinguisher. The release valve and discharge horn are made to operate efficiently at temperatures of up to 75 °C dry bulb and 40 °C wet bulb. Some horns are fitted with a shut-off control. A device is fitted to prevent accidental discharge.

Halon extinguishers. This extinguisher consists of a cylindrical container filled with liquid halon. It is expelled by a pressurising medium contained in the extinguisher directly above the liquid either as a jet or as a spray. The extinguisher capacity is usually measured by weight.

There are several kinds of halon. The lower volatility of bromochlorodifluoromethane (BCF) (Halon 1211) is favoured over bromotrifluoromethane (BTM) (Halon 1301) in portable extinguishers. Since a large quantity of BCF is discharged as a liquid it can be directed more effectively at the fire and the discharge stream is affected less by wind and other air currents. Halon extinguishers are not allowed underground.

Fixed fire extinguishing systems

A fixed fire extinguishing system is located in such a position as to extinguish a fire caused by a specific hazard or site. Nozzles are fixed and the flow of the extinguishing agent (generally dry chemical, water or foam) is directed onto critical areas of the hazard.

The kinds of hazards requiring fire extinguishing systems usually have several common characteristics. The hazard is fixed and its parameters are generally defined. There is a relatively high fire risk involved and/or the piece of equipment is of relatively high value, either in terms of its cost or its importance to the continued operation of the mine.

Systems can also be designed to shut off power and sound alarms at the same time as they are extinguishing the fire. The parameters of the system, its requirements and type of extinguishing agent used depend upon the type of equipment, its location and working conditions.

Face machinery. There are two main fire extinguishants in use for face equipment: dry chemical and water.

All dry chemical systems have a storage tank for the chemical and a storage tank for the propellant gas used to move the chemical. The system is activated automatically or manually. The driving gas and chemical mixture move through piping to nozzles where the chemical is directed onto the hazardous locations. The systems are completely self-contained, i.e. they require no external source of agent or power. However, these systems have several disadvantages: they lack follow up capability. If the fire re-ignites, the system cannot be used to put it out a second time. When dry chemicals are used, the equipment must be cleaned of the chemical after use of the system. The operator must move toward inlet air after triggering the system to avoid the cloud of chemical used. If the fire has developed beyond the machine, the system does not have the capacity to extinguish it. Also it is difficult to check whether or not these systems are functioning correctly.

Water systems are mainly used on face machines equipped with a water line. These systems are easy to check functionally. They are easy to design for use in case the fire develops beyond the machine. Cleaning of the machine is not required after use of the system. However, the system is not self-contained as it requires connection to a water line.

Dry chemical systems are normally used on cutters, drills, loaders, roof bolters and shuttle cars. Water systems are used mainly on continuous miners. Figs. 9.1 and 9.2 show a water fire extinguishing system for a continuous miner and a dry chemical fire extinguishing system for a shuttle car, respectively.

Mobile mining equipment. Many pieces of mining equipment currently have fixed, manually activated fire extinguishing systems installed on them. Automatic systems have been developed where a thermal point or wire sensor detects the fire and activates the system. Fig. 9.3 illustrates a system using thermal point source sensors. The system can be used on coal cutters and load-haul-dump machines. This system can be fitted to an existing fixed, manually activated fire extinguishing system but is self-contained, i.e. it has its own power source and will work even if there is an electrical fire.

Fig. 9.1. A continuous miner water extinguishing system (after Greene, 1973).

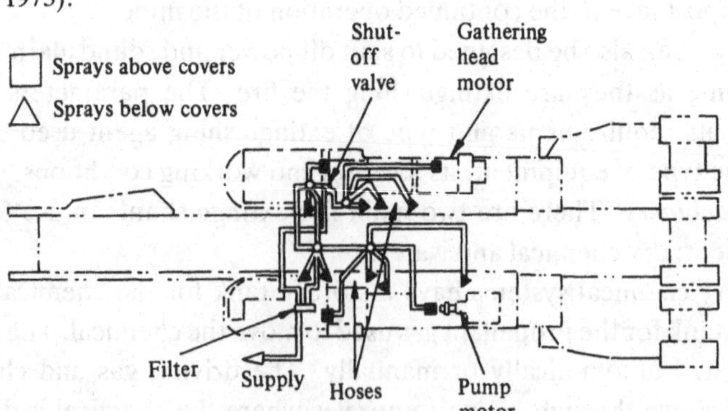

Fig. 9.2. A shuttle car dry chemical fire extinguishing system (after Greene, 1973).

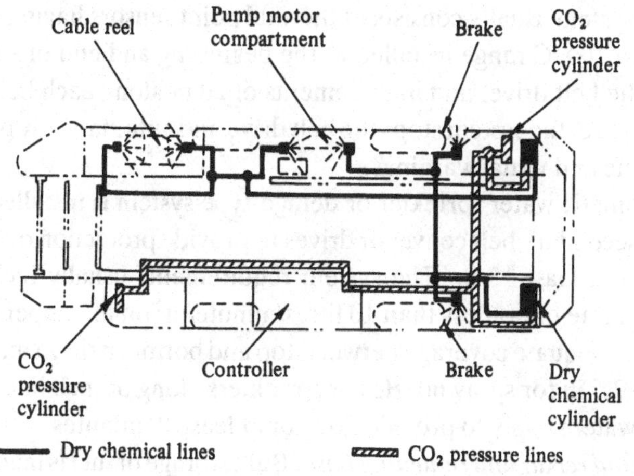

Fig. 9.3. Hardware for the automation of an existing fire extinguishing system using thermal point source sensors (after Bickel, 1977).

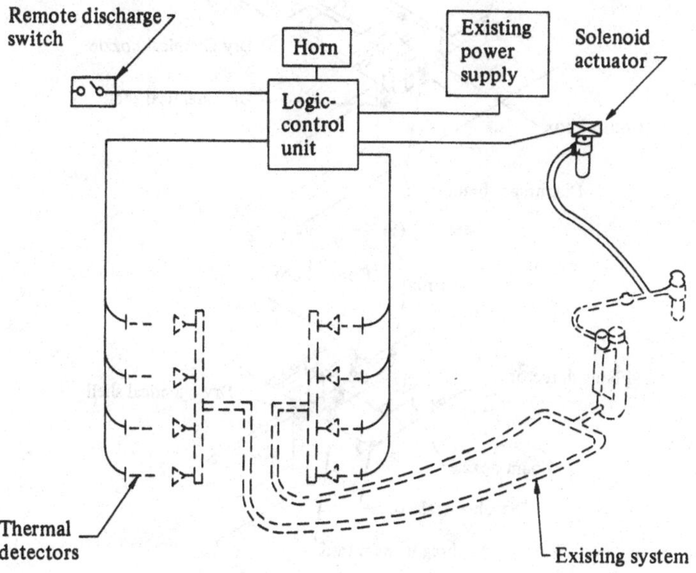

Belt conveyors. Many countries have promulgated regulations requiring the installation of a fire protection system for belt conveyors. The fire detection system usually consists of thermal point sensors having a rating in the 60 to 100 °C range installed at the beginning and end of each belt flight, at the belt drive, and in increments of 20 m along each belt flight. In case of a fire, the system stops the belt drive and automatically provides both audible and visual warnings.

An automatic water sprinkler or deluge-type system is installed at the main and secondary belt conveyor drives to provide protection over a belt distance of at least 15 m. The system requirements usually include an application rate of not less than 10 litres/minute/m^2 on the upper surface of the belt, adequate coverage between top and bottom belts, a maximum spacing of 2.5 m for spray nozzles or sprinklers along branch lines, and a sufficient water supply to provide flow for at least 10 minutes.

Underground repair and refuelling base. Bulk storage of fuel is maintained on the surface and only a limited quantity (some 750 to 2000 litres) is piped to storage tanks at the repair base at any one time. Adequate protection of this area is essential using a dry chemical or high-expansion foam. Fig.

Fig. 9.4. (*a*) Underground fuel storage area fire protection; (*b*) underground fuel transfer area fire protection (after Pomroy, 1977).

Flame detector
Foam nozzle
Warning
Dry chemical nozzle
Dry chemical shell
Control box
Diaphragm foam tank

(*a*)

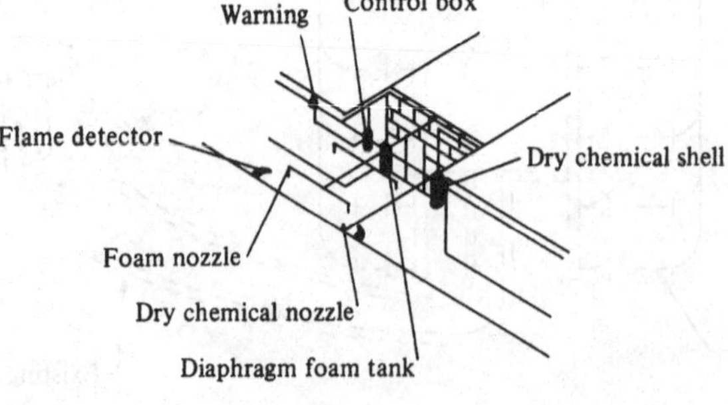

(*b*)

9.4 shows a system developed by the US Bureau of Mines for the fuel storage and transfer areas. The fire is detected with ultraviolet optical flame detectors which have a millisecond response to fire conditions, and a twin-agent system is used which rapidly extinguishes the fire with dry chemical and secures the fuel against re-ignition with foam.

Shaft and shaft station areas. A fire protection system developed by the US Bureau of Mines for shaft and shaft station areas is shown in Figs. 9.5 and 9.6. It utilises thermal, carbon monoxide and ionised-particle smoke detectors, and remotely controlled smoke doors and sprinklers. The surface control unit of the system receives the fire warning signal via multiplex wiring through two separate routings. The underground control units of the system then are activated from the surface to warn the miners at each shaft station. A stench system would then be activated by surface personnel to warn other miners.

The sprinklers and smoke doors can be opened or closed from either their local control unit or the master control unit on the surface. Fire warning alarms and lights and mine evacuation signals at the underground units can be controlled from the surface control unit.

Fig. 9.5. Elevation view of the fire protection system for the shaft and shaft station areas (after Johnson, 1976). Shaft station system: 3 CO sensors (CO); 1 ionisation sensor (I); 2 vent doors; water/foam extinguisher (ext); thermal wire (90 m) (T); control alarm circuit (C). Shaft system: 1 CO sensor; 1 ionisation sensor; 1 thermal wire (90 m); 1 control alarm circuit; 1 water/foam extinguisher.

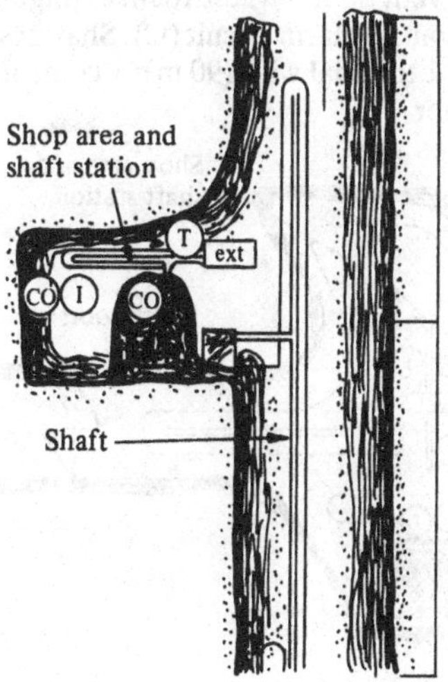

9.1.6 *Combating fires*

To fight fires, one or more types of equipment should be provided at suitable locations in mines. Besides portable extinguishers, foam-generating machines, rock dusters and bags of rock dust are stored at vital points. Many mines also provide fire fighting trains that contain large fire extinguishers of the dry chemical type, water cars, rock dust, brattice cloth, water hoses and water connections to be used for no other purpose than fire fighting. In addition to this mobile equipment, many mines have laid down several kilometres of pipe line with valves available for connection to water hoses at frequent intervals.

Effects of a fire

When fighting a fire it is necessary to understand its characteristics, the way it can develop and the speed with which it can spread, the effect of it on the ventilation system and conversely, the effect of the ventilation system on it.

Fire is always accompanied by smoke. The amount and density of the smoke varies according to the type of materials burning and increases progressively with the intensity of the fire. Smoke obscures vision; can choke persons working in it; and is unpleasant.

A number of toxic gases may be released as products of combustion depending upon the combustible materials. Carbon monoxide is the

Fig. 9.6. Plan view of the fire protection system for the shaft and shaft station areas (after Johnson, 1976). Shaft station system: 3 CO sensors (CO); 1 ionisation sensor (I); 2 vent doors; water/foam extinguisher (ext); thermal wire (90 m) (T); control alarm circuit (C). Shaft system: 1 CO sensor; 1 ionisation sensor; 1 thermal wire (90 m); 1 control alarm circuit; 1 water/foam extinguisher.

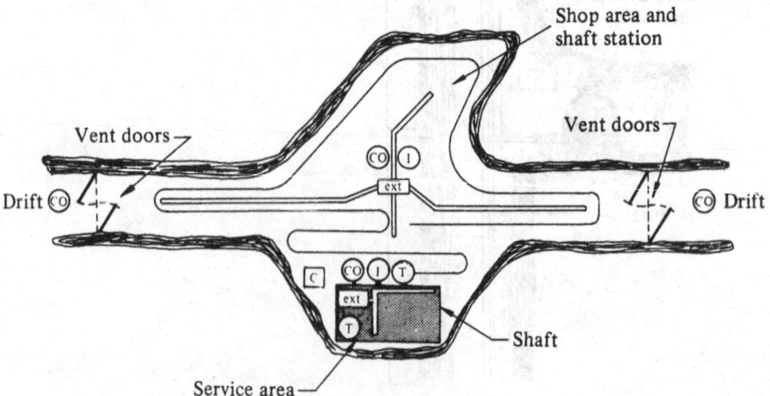

principal toxic gas encountered in mine fires. The concentration depends on the intensity of the fire. In the case of small fires, the toxicity level is not likely to endanger life. In the case of large fires, there is a possibility of oxygen deficiency and excessive carbon dioxide on the downwind side.

Damage to the support system and excessive heat can result in roof falls in the fire area. Roof falls can seriously impede fire fighting not only by blocking the roadway but also by breaking water pipes and other services. As a result of falls, the roadways may be choked, upsetting the ventilation system of the mine.

The heat generated by a mine fire can also affect the ventilation system. A small fire (e.g. resulting from frictional heat at a conveyor idler) is unlikely to have any significant effect on the stability of the main ventilation, but a severe fire can affect it by constriction and buoyancy.

The constriction or throttle effect is brought about in two ways. The rise in temperature causes expansion of the air volume. The release of products of combustion add to the volume. Additional vapours may be released due to heating of the mineral and the rocks. The effect of this is to oppose the ventilation but is generally limited to the vicinity of the fire.

The buoyancy effect is caused by the heat from a fire lowering the density of the atmosphere on the downwind side of it. In the case of a level roadway, the effect is negligible but in a dipping roadway, it opposes ventilation, and in a rising roadway, it assists ventilation. Fig. 9.7 compares the combined constriction and buoyancy effects of a large (severe) open fire in a level, a dipping and a rising roadway.

Consider a severe fire in a main intake dipping roadway (Fig. 9.8*a*) where the fire tends to oppose the ventilation. If the pressure across the doors (A) is high and the buoyancy pressure low, there will be a comparatively small reduction in the volume rate of air flowing over the fire and no possibility of reversal. Conversely, if the buoyancy pressure is high compared with pressure across the doors (A), then there will be the risk of a reversal of the ventilation. (A large and intense fire may produce an explosive mixture of fire gases irrespective of whether the mine is normally very gassy or not. Hence the importance of taking all necessary action to ensure that the fire gases are not allowed to reverse over the fire.)

Consider a severe fire in one of two parallel, dipping and interconnected intake roadways (Fig. 9.8*b*). The fire will decrease ventilation in the intake in which the fire is located and increase ventilation in the other one. The most important consideration in this type of situation is to see that there is no reversal of ventilation airflow through the fire. This can be achieved by regulating the ventilation in the adjacent intake roadway and increasing the airflow in the roadway where the fire is. However, at the same time, it is important not to oversupply the fire with air or otherwise its activity will increase.

If the main intake roadways are rising instead of dipping, then the ventilation through the fire will increase and decrease in the adjacent parallel intake. In this type of situation, the airflow through the fire should be regulated by the construction of a regulator as near as possible to the fire on the upwind side.

Fig. 9.9 considers a fire in one of two descensionally ventilated longwall faces. The fire will oppose and thereby reduce the airflow through the fire and can result in a reversal of the ventilation in district A (with the consequent risk of an explosion) and recirculation in district B. In this

Fig. 9.7. The effects of a large (severe) open mine fire in mine roadways (after Willett, 1973). Buoyancy effect: (*a*) Level road – negligible (assuming level downwind); (*b*) dipping road – opposes ventilation; (*c*) rising road – assists ventilation. The magnitude of buoyancy effect depends on the temperature of the fire and the roadway inclination. Constriction effect: due to expansion of the atmosphere in the fire zone, the effect by itself is to reduce the ventilation in all cases. The reduction is related to (1) original air volume flow rate; (2) fire temperature; (3) ventilation system resistance. Notes: With severe fires, the buoyancy effect is dominant in inclined roadways: (1) rising roadways – ventilation is increased through fire; danger of recirculation in inbye split. (2) Dipping roadways – ventilation is reduced through fire; danger of reversal through fire.

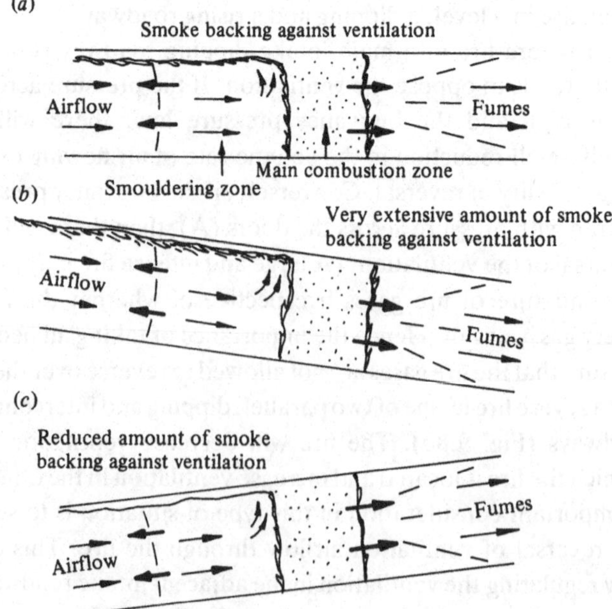

Fig. 9.8. Effect of a severe fire in a dipping roadway (after Willett, 1973). (*a*) In a main intake dipping roadway where the fire tends to oppose the ventilation; (*b*) in one of two parallel, dipping interconnected intake roadways. Note for (*a*): there is a risk of reversal if (1) doors 'A' are sealed; (2) main fan is stopped. Note for (*b*): there is a danger of reversal through fire area resulting in (1) increasing ventilation in adjacent intake; (2) some recirculation of the fire gases. Action: install regulator (R) at point 'B' to maintain stability of vent.

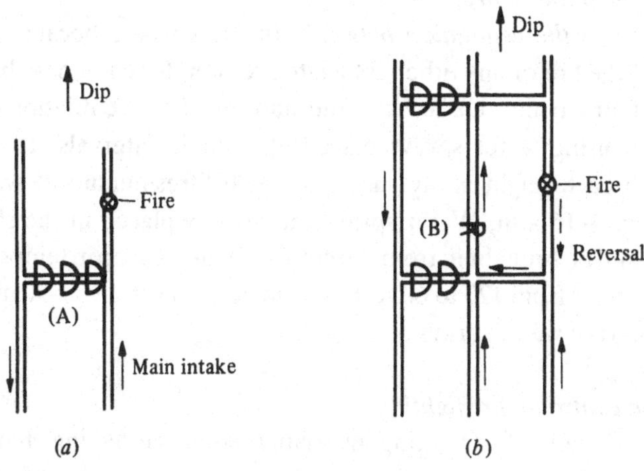

Fig. 9.9. Effect of a severe fire on a descensionally ventilated longwall face (after Willett, 1973). Note: The fire will reduce and possibly reverse ventilation in district A and increase ventilation in district B. Regulators (R) are necessary in (B) to stabilise ventilation.

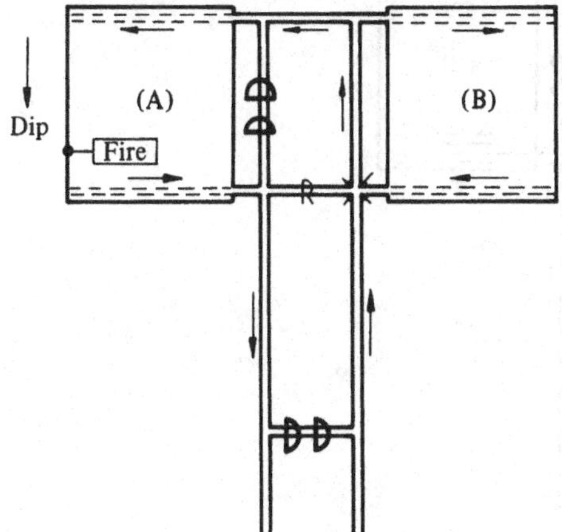

case, airflow in district B should be regulated so that the direction of the airflow in district A becomes stable.

If the longwalls are ascensionally ventilated (Fig. 9.10), the fire will increase the airflow in district A and decrease it in district B. In extreme cases it can lead to a reversal in district B. The situation can be controlled by regulating the airflow in district A. This is also essential to reduce airflow to the point of fire.

Stabilisation of the ventilation network. In cases where because of the location of the fire or any other associated reason there is a possibility of reversal of air current due to heat, the stability of the ventilation can be maintained using water sprays placed at suitable intervals. It usually requires three to eight sprays using 300–600 litres/minute of water at pressures of 4–16 atm. Water sprays can also be placed in the shaft or roadways and be brought in from the cut-throughs. A drop in temperature of the return air from 120 to 60 °C has been noticed within 60–90 minutes after the start of water sprays.

Ventilation control in fire fighting

As indicated above, fire fighting invariably requires effecting changes in the mine ventilation system so that the fire does not spread beyond control, the workers can be safely withdrawn and the danger of explosion

Fig. 9.10. Effect of a severe fire on an ascensionally ventilated longwall face (after Willett, 1973). Note: The fire will increase volume flow rate in district A with risk of possible reversal in district B. Action: regulate on district A. In all cases of severe fire in an ascensionally ventilated roadway, the procedure should be to decrease the flow through the fire to at least the original volume flow rate.

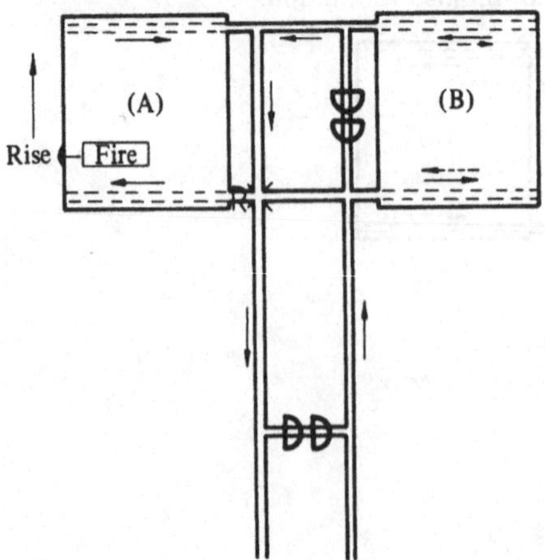

does not arise. Various methods of control of ventilation in fire fighting include:

(1) regulation of air volume flow rates;
(2) shortcircuiting;
(3) isolation of fire;
(4) reversal of mine ventilation.

The choice of any one of the above methods depends on the location of the fire with respect to the ventilation system, assessment of its effect on the ventilation system and the position of the workers at the time of the fire.

When a fire is in a main airway, one may slow down the mine fan, adjust the damper in the fan drift, or redistribute airflows so as to obtain a reduced airflow to the fire.

Shortcircuiting the fire gases by opening ventilation doors will lead the fire gases directly to the return airflow without having to go through other mine workings. One may also shortcircuit the fresh airflow going to a fire so that the fire is starved of oxygen. Simultaneously with shortcircuiting, any doors in the shortcircuited area should be closed to prevent airflows in the area. The disadvantages of shortcircuiting are that in gassy mines accumulation of methane takes place and that the load on the mine fan increases. This condition must be checked before effecting shortcircuiting.

Complete isolation of a fire is possible in mines which are divided into fire sections by means of fire doors and preparatory stoppings.

The reversal of mine ventilation following a mine fire is a controversial subject. Besides other factors, type of fan and type of doors installed underground must be considered.

As a general rule, the main mine fan should not be reversed unless the fire is at the junction of splits, in the downcast shaft or near the shaft bottom of the downcast shaft. When a fire is in the downcast shaft, reversal of ventilation due to the heating of air may be to the advantage of workers.

The type of fan is important in that on reversal it will operate at a lower efficiency, resulting in lower air volume flow rates and lower pressure drops which may result in the formation of flammable atmospheres inside the mine.

Reversal either requires triple doors or specially designed doors which are provided with latches so that they do not open on reversal.

It should be remembered that withdrawal of workers is the foremost concern of the mine management before any one of the above methods of controlling ventilation is adopted and also that the method of ventilation chosen can be modified during fire fighting if conditions so warrant.

Extinguishing a fire

To extinguish a fire in its incipient stage, almost any method is effective. The most common is a shovel full of rock dust and the use of multipurpose dry chemical from a fire extinguisher directed at the fire.

Once a fire develops, it can spread rapidly. Handheld fire extinguishers are of no use for extinguishing a well-developed fire. The only effective extinguishing agents are water and foam.

Use of water has certain limitations. For example, a solid stream of water must not be used on energised electrical equipment. Water can react with the burning coal which is at a high temperature, producing explosive hydrogen and toxic carbon monoxide and this possibility must be considered. The steam produced when fighting a fire with water can cause roof falls which can retard fire control activities. Fog, foam and spray are not as hazardous as a solid stream of water.

Besides the points discussed in the section entitled 'Effects of a fire' (p. 280), local conditions must be considered. Smoke backs up against the ventilating airflow, interfering with approach and control activities. Increasing air velocity to reduce smoke roll back can intensify other problems in fire control. Advancing behind a transverse brattice cloth with a continuous application of water applied as a fog is the best way to approach a live fire. If this fails, the fire has to be attacked from a distance using high-expansion foam.

The possibility of a gas explosion always exists. During a fire in a coal mine combustible gases accumulate from four sources: (1) normal methane liberated from coal; (2) volatile gases driven from the coal by heat; (3) incomplete combustion and (4) water–gas reaction (steam in contact with hot carbon). In the case of large fires, the best defence is to control the airflow over the fire so that the concentration of combustible gases is kept low. Stopping and starting the airflow (or stopping and starting the fan) creates a serious explosion hazard.

When a fire breaks out in an intake roadway, both the associated intakes and returns downwind are contaminated by the fumes. If there is no reversal of ventilation, there is no risk of methane from the working places being drawn over the fire. By comparison, when a fire breaks out in a return roadway, the intakes are not contaminated by the fumes (assuming no reversal of ventilation). This is very important as far as the safety and withdrawal of the workers is concerned. However, since the methane from any inbye working place must pass through the fire, it is essential to maintain the ventilation at a level so as to ensure that it is adequately diluted.

Fires in dead ends are difficult to tackle and can seldom be approached without respiratory equipment. Fortunately, ventilation of these areas can be easily controlled. Besides, they can be easily isolated and their effect on the rest of the mine is minimal. Most face machinery fires occur in a dead end. High-expansion foam is the best extinguishant to deal with such fires.

If the fire appears to be getting out of hand or there is evidence that an explosive mixture of gases might reach the fire, it is necessary to seal off the area.

9.2 Spontaneous combustion of coal

Coal undergoes slow oxidation on exposure to air at ambient temperatures, with the evolution of heat, gases and moisture. The heat generated, if not dissipated, gives rise to an increase in the temperature of the coal. As the temperature of the coal rises, the rate of oxidation increases. If this is allowed to proceed unchecked it can eventually result in the ignition of the coal. This oxidation process is known as spontaneous combustion or spontaneous heating or self-heating. Self-heating, therefore, occurs when the rate of heat generation exceeds the rate of heat dissipation, the former being determined by the rate and heat of oxidation.

During recent years there has been a renewed interest in the spontaneous combustion of coal in all coal mining countries particularly because of the use of caving methods and the thicker seams being mined. Large-scale bulk storage and bulk transport of coal have also become more important with the increase in coal trade.

9.2.1 Evaluation of the potential of coal for spontaneous combustion

Several methods have been used to evaluate the potential of coal for spontaneous combustion but none is clearly superior. The most common methods used are described below.

Oxygen absorption

In this method, a coal sample is placed in a container and oxygen or air is added to it. The amount of oxygen absorbed by the coal is estimated from the analysis of the gaseous reaction products. The temperature increase per unit of oxygen consumed indicates the potential of coal for spontaneous combustion.

Heating rate/crossing-point temperature
In this method, a coal sample is placed in a bath and heated at a constant rate. Initially, the temperature of the coal lags behind the temperature of the bath but as coal begins to self-heat, the temperature of the coal first coincides with and then exceeds the temperature of the bath. The crossing-point temperature is known as the 'relative ignition temperature'. Usually, the crossing-point temperature is used as a measure of the potential of coal for spontaneous combustion although the index based on the ratio of heating rate to crossing-point temperature is more suitable because the spontaneous combustion potential of coal not only depends on the ignition temperature but also on the rate of heat generation.

Adiabatic calorimetry
In this method, a coal sample is placed in an insulated bath, and the whole system is heated to a pre-selected temperature. Oxygen or air is then added to it and oxidation of the coal raises its temperature. Since no heat is lost to the surroundings, the change in the temperature of the coal in a given time, the time needed to reach a pre-selected temperature, or the amount of heat generated per unit time indicates the potential of coal for spontaneous combustion.

Isothermal calorimetry
In this method, a coal sample is placed in a large bath held at a constant temperature. Heat generated in the coal sample due to spontaneous combustion is measured by thermocouples and dissipated in the relatively large heat sink. The amount of heat generated per unit time gives an indication of the potential of coal for spontaneous combustion.

9.2.2 *Factors contributing to spontaneous combustion*

Coal characteristics

Some coals are more prone to spontaneous combustion than others. The rate of oxidation of coal depends upon many factors, including rank, presence of pyrite, particle size, moisture content, temperature, extent of previous oxidation of coal and the composition of the ambient air.

It is generally accepted that as the rank of coal decreases, the risk of spontaneous combustion increases.

The presence of pyrite increases the potential of coal for spontaneous combustion, particularly when the pyrite concentration exceeds 2 % and when it is very finely distributed. Pyrite accelerates spontaneous combustion by swelling and causing disintegration of the coal mass, thereby increasing the surface area available for oxidation.

The smaller the coal particle, the greater the exposed surface area and the greater the tendency toward spontaneous combustion. Friable coals which produce a considerable amount of fines when mined are more vulnerable to spontaneous combustion.

The changes in moisture content of the coal affect the potential of coal for spontaneous combustion. It has been found that the rate of oxidation increases with an increase in moisture content. Also, wetting is an exothermic process and drying is an endothermic process.

Airflow rate

For spontaneous combustion to develop, the rate of heat generation should be more than the rate of heat dissipation. At very high airflow rates almost unlimited oxygen for the oxidation of coal is available but dissipation of the heat generated by oxidation is very efficient. A low flow rate restricts the amount of oxygen available, but does not allow the heat generated to be dissipated. A critical flow rate is one that provides sufficient oxygen for widespread oxidation but does not dissipate the heat generated.

Geological factors

The presence of faults in coal seams often contributes to the development of heating in coal mines by allowing air and water to migrate into the coal seams. Zones of weakness which usually develop in the area around the faults also aid in the development of heating.

The temperatures of the strata increase with depth. Therefore, the oxidation rate will increase with depth, making deeper seams more vulnerable to spontaneous combustion. On the other hand, the higher rank of coal found in these seams decreases the chances of heating.

Thick coal seams are often considered to have more potential for spontaneous combustion because the working of these seams is invariably accompanied by high losses of coal in the goaf areas. The low thermal conductivity of coal compared with that of shale or sandstone is also a contributory factor.

When a coal seam under a shallow overburden is mined, the goaf areas become connected to the surface by cracks and fissures. Air and water from the surface can gain access to the coal and increase the potential for spontaneous combustion. Similarly, when multi-seams in close proximity are worked, the cracks and fissures developed in the intervening strata increase the potential for spontaneous combustion of the surrounding unmined seams, particularly the undermined seams.

Mining practice

Some of the most common places where spontaneous heatings occur are goaf areas and unconsolidated wastes, pack walls with a high proportion of coal, the edges of goaves where high strata pressure causes crushing, roof falls and floor heaves, crushed pillars, regulators, doors and air crossings and constrictions in the roadways.

Coal left in goaf areas is very liable to spontaneous combustion as the air movement there is very sluggish, and any heat generated as a result of oxidation will not be removed.

In coal mines, coal is left in the roof and/or floor to support the weak adjoining strata or bands of inferior quality coal which are left unmined. However, on long standing, roof falls and floor heaves occur causing large-scale crushing of the left coal and creating conditions susceptible for heating.

Pillars that have been standing for a long time are prone to heating, particularly when they are liable to crushing.

Regulators, doors and air crossings are points of high air leakage, the air moving through the fractures in the solid coal around them. The greater the pressure difference across them, the greater the leakage. Constrictions of mine roadways also cause leakage of air. Changes in ventilation, either intentional or accidental, may cause excessive air leakages or may suddenly bring moist air into contact with dry coal.

Goaf areas, where a large amount of coal is left and particularly where a bleeder ventilation system is used to clear gas from the goaf, present optimal conditions for spontaneous heating.

Incubation period

The term 'incubation period' generally implies the time required for the oxidation of coal, in suitable circumstances, to cause a rise in temperature to its ignition point. It depends on the characteristics of the coal, the air leakage and the heat accumulation in the environment. For low-rank coals, the time period generally varies between 3 and 6 months, but with high-rank coals the period varies between 9 and 18 months. The incubation period can be extended by reducing fissuration and/or air leakage. Under adverse conditions, the period can be less than 2 weeks, especially with low-rank coals.

9.2.3 *Prevention of spontaneous combustion*

Prevention of spontaneous combustion is based on two factors: (1) elimination of coal from the area and (2) control of ventilation so as to exclude oxygen entirely from the area, or to supply a sufficient flow of air to dissipate the heat efficiently as it is generated and before a critical temperature is reached. The methods adopted depend upon the local situation.

Mining layout

When designing mining layouts for seams liable to spontaneous heating it is essential that the general layout of the mine is simple and that each area can be quickly and effectively sealed off. The relative positions of the various districts in the seam and surrounding seams must also be taken into account. It is essential to follow descending order of extraction when mining multiple seams.

The panel system is an appropriate one for mining seams liable to spontaneous combustion. This system facilitates effective sealing with a few stoppings. The size and configuration of the panels depend upon the method of mining, the seam contours and other geological considerations. If necessary, the panels must be of a size which would permit complete extraction within the incubation period. The size of panel barriers needs to be sufficient for stability.

When working seams by the bord and pillar method, the size of the pillars must be sufficient to avoid excessive crushing. This also applies to coal pillars left at the start of longwall faces.

When working a seam by a longwall, the retreating method is preferable as it eliminates leakage currents through the goaf area.

On completion of production from a panel, reclamation of material should be completed without delay and the panel adequately sealed as quickly as possible.

Air leakage

As far as is practicable, the formation of leakage paths should be minimised by providing adequate support, e.g. adequately sized pillars and good gateside packs. If this is not sufficient to prevent air leakage, leakage paths should be sealed off by sealant coating or injection.

Fractures extending to the surface offer a source of air leakage into sealed areas. Artificial sealing from the surface, usually by sand, can prevent such leakage.

Doors, regulators and stoppings should be properly sited. Unnecessary stopping and starting of main and booster fans should be avoided. When a panel has ceased production and is to be stopped off, the ventilation pressure difference should be balanced across the old panel. Balancing the ventilation pressure is not a substitute but a complementary requirement for effective stoppings.

Inhibitors

In storage areas and surface stock piles, certain chemical agents can be applied to the coal surface which can hinder the penetration of oxygen into the coal by sealing the surface pores and thereby stopping initiation of auto-oxidation of coal at ambient temperatures. Surface stock piles can also be sealed off by consolidation and bitumen. Stock piles can be so

9.2.4 *Detection of spontaneous combustion*

The development of heating underground is accompanied by the progressive appearance of:

(1) haze formed when air heated by an incipient fire meets colder air;
(2) sweating or condensation on the roof and exposed surfaces due to the moisture formed by combustion;
(3) goaf stink or fire stink with a characteristic smell, variously described as musty, oily, petrolic, aromatic or tarry;
(4) smoke in airways; and
(5) fire.

In the past, reliance has been placed on these indications for the detection of spontaneous combustion, although it has never been satisfactory for the reason that the spontaneous combustion must have reached an advanced stage, thus seriously limiting the time available for control, reclamation of equipment and sealing off.

Modern methods of early detection of spontaneous combustion are based on changes in air composition. The oxidation leading to the spontaneous combustion of coal consumes oxygen from the air and produces carbon dioxide and carbon monoxide. Carbon dioxide is produced in much greater quantities than carbon monoxide but its presence cannot be used as an indication of the onset of spontaneous combustion because of the high base levels in fresh air (3000 ppm) which make small changes undetectable. On the other hand, there is no carbon monoxide in fresh air and virtually none in a panel intake so that a change in level of a few parts per million can mean a severalfold increase.

Exhausts from diesel engines and blasting fumes are two common sources of carbon monoxide underground but their effects can be distinguished from a gradual increase or trend due to spontaneous combustion because they are basically intermittent in nature.

In panels where ventilation conditions are steady, even a small change in the concentration of carbon monoxide in the return airway may be sufficient to detect a spontaneous heating condition. Fluctuations in ventilation affect the concentration of carbon monoxide by dilution but an allowance for this can be made by calculating either the carbon monoxide/oxygen deficiency ratio or the actual production of carbon monoxide.

Carbon monoxide/oxygen deficiency ratio (Graham's ratio)

The calculation of this ratio depends on the constant ratio of oxygen to nitrogen in fresh air. The formula for the calculation is:

$$\frac{CO}{O_2 \text{ def.}} \text{ ratio} = \frac{100CO}{\frac{20.93}{79.04} N_2 - O_2} = \frac{100CO}{0.265N_2 - O_2}$$

where CO, N_2 and O_2 are the percentages of the gases present at any given time in a sample of air coming from the suspected area in a mine.

Every mine and every panel has its own typical value or 'norm' for the make of carbon monoxide and for the carbon monoxide/oxygen deficiency ratio depending on the oxidation of the coal and the conditions in which it is mined. Any analysis showing a higher value than the norm determined should be followed by resampling. Confirmation of continuous increasc warrants immediate investigation underground.

Typical values of the carbon monoxide/oxygen deficiency ratio for underground coal mines are given below:

0.4 or less – normal value

0.5 – necessity for a thorough check-up

1.0 – heating is almost certain

2.0 – heating is serious, with or without the presence of active fire

3.0 – active fire surely exists

Continuous monitoring of carbon monoxide in mine air

Automatic monitoring for carbon monoxide is done in mines susceptible to heating. Automatic monitoring also permits the determination of carbon monoxide trends and absolute values using microprocessors without the need to relate them to oxygen deficiency.

Continuous monitoring of carbon monoxide at a number of strategic points in the mine can give timely warning of the onset of spontaneous combustion and permit delineation of areas in a mine. Computerised data collection systems with graphic displays and a continuous graphical record permit easy recognition of the changes in background levels and enable exhausts from diesel equipment or other sources to be distinguished.

Two types of analysers are available for continuous monitoring of carbon monoxide in the air: (1) the infra-red analyser and (2) the electrochemical analyser. Only the infra-red analyser is available in a form approved for use in underground coal mines.

There are two systems used in monitoring. In one system, the analysers are installed at various points underground and they either record the percentage of carbon monoxide on site or telemeter the results to some convenient point underground or on the surface. In the other system, lengths of tube are installed from the sampling points to the surface and the samples drawn through these tubes are analysed sequentially. This system is known as the tube bundle system.

The main advantage of installing on-site analysers underground lies in the immediate availability of results. But analysers are dedicated instruments and can monitor only carbon monoxide. The advantage of the tube bundle system is that it provides a sample for analysis on the surface which can be analysed for all gases. The limitation of this system is the delay between the air entering the tube at the sampling point and its subsequent analysis on the surface. For detecting spontaneous combustion, a delay of one or possibly two hours in getting the results of the samples is not a serious matter because spontaneous combustion has a relatively long incubation period.

Generally, for large installations involving many sampling points, the tube bundle system is much less expensive than a system in which each point has a separate analyser. The costs of pneumatic tubing are normally comparable with the wiring costs for analysers installed underground; however, the tube bundle system requires only one analyser, whereas the other system requires an analyser at each point underground. This reduces the cost of the tube bundle system substantially. Moreover, maintenance costs for a single analyser and pumping station are lower than for a system containing many individual analysers, each of which must be periodically checked, cleaned, or adjusted for sensitivity. (However, when the system is to be used for monitoring ventilation during a sealing-off operation, on-site analysers are far superior due to the instant availability of results.)

9.2.5 *Control of spontaneous combustion*

The method adopted for dealing with spontaneous combustion once it has occurred must depend upon the position and intensity of the heating, the likelihood of accumulation of inflammable gas and the accessibility of the heating from the point of view of ventilation and treatment. The three basic methods of control are:

(1) the extraction of the hot coal;
(2) the use of extinguishing agents; and
(3) the exclusion of oxygen from the affected area.

When the seat of heating is accessible to the existing transport system, the heated coal may be dug out and removed from the mine. Under such circumstances care is usually taken to prevent the coal from catching fire while in transport by covering it with stone dust liberally as it is loaded.

The disturbance of heated coal, which has been near its ignition temperature, often results in its inflammation. Steps must be taken to protect workers loading burning coal.

Water under pressure as a means of controlling underground heatings must be used with caution particularly when there is no through ventilation because this would generally only aggravate the fire and introduce the risk of ignition due to a semi-water gas/producer gas reaction. Bentonite slurry, if available, may be used instead of water.

The final expedient in dealing with the control of heatings underground is the sealing off of an area, thus isolating it from the rest of the mine. The object of sealing-off is to prevent further access of oxygen to the site and if done effectively there will be a gradual diminution of the amount of oxygen available until the stage is reached where the atmosphere within the sealed area will no longer support combustion.

In the case of some fires in the goaf area, when sealing-off has failed, the void may be filled with water, cement stone dust, bentonite, sand, rubble and grout, chemical grouts, clay slurry, and inert gases (carbon dioxide or nitrogen).

Injection of inert gases

Inert gases can be used to control fires. The purpose of introducing the inert gas is to reduce the oxygen in the air around the seat of the fire to such an extent that the fire is reduced to a harmless level or is put out altogether, and to convert the explosive atmosphere of gas and air into a non-explosive one. The inert gas can be introduced directly into the seat of the fire if, for instance, it is not necessary to interrupt the ventilation of the face and gates. The ventilation currents which flow towards the heating area must be, as far as is possible, replaced by inert gas completely.

Inert gases such as carbon dioxide, gases produced after complete combustion and nitrogen have all been used to fight mine fires. Carbon dioxide is delivered to the site in liquid form and is then vapourised. This method has not been used to any significant extent for the following reasons:

(1) carbon dioxide is not available in large quantities;
(2) the permissible maximum carbon dioxide content of 1 % by volume would be very rapidly reached and exceeded on the return side of the fire;
(3) carbon dioxide is absorbed by rock;
(4) carbon dioxide is reduced to carbon monoxide if the temperature at the seat of the fire is high.

Combustion generators, fuel oil or kerosene-fired jet units can be used to generate combustion gases. The disadvantages are as follows:

(1) there is a danger of fire and explosions from the use of the equipment, particularly when kerosene is used;
(2) the carbon dioxide content is between 15 and 18 % by volume;
(3) the residual CO, SO_2 and NO_x contents are large;
(4) the combustion gases have to be cooled.

Because of these disadvantages, combustion gases are also not popular.

Nitrogen has been the most commonly used inert gas and has many advantages. It is simple to deliver in liquid form. It can be gasified in vapourisers with a rated output of up to 300 m^3/minute quite easily and this quantity can be raised even further by operating units in parallel. Nitrogen is produced from air and is a pure inert gas. It is cold and therefore presents no problems with cooling, as is the case with other processes where the gas is produced catalytically or by chemical reaction.

However, liquid nitrogen as the base material is not available in unlimited quantities, and it is not suitable for storage over long periods. It is, therefore, not impossible that it might be difficult to obtain in sufficient quantities if large amounts (more than 500 m^3/minute for several days) are needed; this may be the case if there are several fires at the same time, or if there is a really large fire. Moreover, nitrogen can only be introduced into the mine by piping it from the surface, which means that a great deal of preparatory work on the surface and underground is required in each case. However, though nitrogen is slightly less expensive than carbon dioxide it is more expensive than combustion gases.

9.3 Spontaneous oxidation and combustion of sulphide ores in underground mines

Oxidising sulphides, particularly in combination with combustibles such as timber, pose a serious problem to the health and safety of workers and to the efficient production of ore in metal mines mining sulphide ores. Of all the sulphides, the iron sulphides are the most reactive, and the reactivity increases when more than one type of sulphide is present. The greater the percentage of sulphides in the ore, the greater the production of heat. Water is necessary for oxidation, although too much water acts to cool the sulphides and reduce the reaction rate.

Before starting to mine, a thorough assessment of a potentially reactive ore body should be carried out. Important considerations include the geological structures, composition of ground waters and the chemical behaviour of the sulphide minerals.

Premature drainage, dewatering and ventilation of excessively large regions of porous ground containing sulphides should be avoided to lessen the risk of heating. If reactive sulphides are associated with the ore body, initial problems are likely to arise in proximity to faulted or brecciated zones. Periodic analyses of gases from rock strata and mine water will provide forewarning of underground oxidation as indicated by low oxygen and high carbon dioxide content, the presence of increased quantities of soluble solids from carbonates, and higher iron and metal content.

Good housekeeping, by removing various types of rubbish and combustible waste from the mine workings, must always be practised, as should other standard underground fire prevention techniques. Blasted ore should be kept dry under conditions of low humidity and good ventilation, which will retard oxidation. Preventing air from penetrating the sulphide surface can be accomplished by raising the moisture content of the muck to about 10 %. However, wet, sticky ore is often difficult to handle and the additional water adds to the cost of transportation. Packaging of ore in the mine, unless the ore is extremely valuable, is uneconomical.

During the transport of reactive ore, changing ambient conditions should be avoided. Fluctuating moisture content accelerates oxidation. Unnecessary vibration and movement of ore acts to create fresh reactive surfaces, generate friction and help remove oxide products. Air-conditioning of ore storage and crushing areas may be effective.

Control of ventilation appears today to be the most economical and effective means of preventing and controlling oxidation of sulphide ores underground. The mine should be designed for large volume flow rates of air since air should not be stagnant in slowly oxidising areas. Large airflows are also needed when dealing with hot sulphides to dilute the large quantities of gases produced. Some preventive ventilation techniques directed towards minimising the increase in the temperature of the air in its passage to the work place are: (1) reducing the area of exposed rock (shorter circuits); (2) reducing the heat transfer from rock by having dry airways and (3) decreasing energy loss by smoothing the airways.

In the case of spontaneous oxidation of sulphide ores, spraying and flooding with water is an inexpensive method of effective treatment. The water acts to prevent oxygen access by filling pore spaces. As already mentioned, the presence of a small percentage of moisture accelerates oxidation. Thus, how and when water is applied is crucial. The use of water as a coolant for already hot sulphides should be avoided unless chemical combustion inhibitors are introduced or rapid inundation can be achieved. An insufficient amount of water will accelerate oxidation, mobilise soluble products and sulphur, and consequently increase steam and sulphur dioxide formation.

Although the application of water as a spray is commonly used to control rapid oxidation of sulphides in underground mines, its effect is relatively small and short-lived. This treatment is effective in dealing quickly with isolated hot spots in the mine. However, unless flooding is used, water is relatively ineffective in dealing with more extensive, long-term oxidising areas.

In applying water it is better to use a large volume for a short time rather than a small volume for a long time. In extreme cases of spontaneous combustion, the fire spreads so rapidly and so extensively that flooding of the area becomes necessary. As a result of flooding, excessive damage is done to the mine and it should be used as a last resort. Even after draining the flood water and re-establishing the ventilation, oxidation and combustion may break out again.

The addition of chemical additives to water is useful although these have been very restrictively used in underground mines. Lime is commonly added to water in treating reactive sulphides. At Mount Isa Mines (Australia) a calcium hydroxide suspension in water is applied to coat the reactive sulphides with limonite, thereby restricting rapid oxidation. However, the lime neutralises the acid solutions produced during oxidation and thus promotes oxidation and heating. For pyrrhotite ores, various types of metallic corrosion inhibitors, including compounds containing dichromate, benzoate and quinolate, have been effective in reducing reactivity.

Reactive ore should be mined out as soon as possible, but until the ore is removed, cooling by means of introducing large volume flow rates of air at low pressures or locally cooled air is advisable. The object is to cool the reacting area so that appreciable oxidation cannot occur. Air-conditioning is applied only when all other ventilation techniques of decreasing air temperature have been investigated and tried. Without air-conditioning, volume flow rates of air above the normal design levels must be circulated to keep the oxidising areas cool. At Mount Isa Mines, estimated volume flow rate requirements for each particular mining method are given below.

Mining method	Air circulation per tonne of ore per day (m^3/minute)
Sub-level open stoping	1.8
Narrow shrink stoping	2.4
Mechanised cut and fill stoping	2.3
Sub-level caving	2.8–3.4

Additional allowances must be made for service areas, crushers and ore passes as well as for deeper mining. Control and cost effectiveness make ventilation the most widely used strategy for controlling sulphide oxidation in large areas in deep underground sulphide mines.

9.4 Sealing off fires underground

When more direct methods of combating a fire or heating underground have proved unsuccessful or are considered impracticable or unsafe, it is necessary to seal off the area from the rest of the mine.

Wherever there is the possibility that methane may accumulate near, or migrate towards, the site of the fire, it is strongly desirable that the volume flow rate of air is maintained and monitored at a level to ensure safety during the building of the seals. Ventilation can be reduced only if it is safe. This can be ascertained from analysis of the changing composition of the atmosphere throughout the district. These considerations also apply to the danger of explosion from 'fire gases' which may accumulate near fires involving large amounts of heated material. It is very important that the fire gases should not be blown back over the fire as a result of even momentary reversal of ventilation.

The likelihood of an inflammable atmosphere building up after sealing-off depends mainly on the rates of accumulation of inflammable gas and of consumption of oxygen by the fire. Usually the explosion hazard builds up to a peak and then subsides. From the known rate of accumulation of methane in conjunction with analyses of the atmosphere within the sealed area, an estimate can be made of the danger period.

The sealing-off operation should be carried out quickly, such as by the closing of a door or its equivalent, and all workers should then be withdrawn from the danger area for a period sufficiently long to ensure that the danger period is past. Stoppings should be explosion-proof, and also appropriate precautions should be taken against the development of a coal dust explosion within the roadway being sealed off. Both the stopping itself and the ground in which it is sited must be made satisfactorily airtight, and air pressures on the stoppings must be carefully balanced to minimise leakage. Space for the provision of pressure-balancing chambers is often desirable. Consideration must also be given to possible need for re-entry into the area, efficacy of the seal and provisions for obtaining information on the conditions within the sealed area.

The following factors should be taken into account when deciding the location of stoppings in relation to the seat of fire:

(1) Normal make of methane in the district and the places of its emission in relation to the site of the fire.

(2) The ventilation layout, i.e. volume flow rates and distribution of airflow and the likely changes occurring as a result of restriction arising during construction of the stoppings.
(3) The dipping and rising roadways in the area and their locations with respect to the seat of the fire.
(4) The effect of the above factors on the accumulation of methane where it might ignite.
(5) The possible effect of the fire on the ventilation. This might slow, or even reverse, the general ventilation – and moreover must inevitably increase the likelihood of the accumulation of methane.
(6) The presence of coal dust. If there is the possibility of a coal dust explosion, the site for the stopping(s) will need to be moved still further away than otherwise might be the case. This assessment must be made for each roadway depending upon the provision of stone dust barriers for arresting such an explosion.

While fighting fires in a gassy mine, it is essential that the percentage of methane in the ventilation reaching the seat of fire must be continuously monitored. Where the increase in the percentage of methane exceeds 2 %, the source of the methane must be located and appropriate steps taken to stem its flow into the fire zone. The maximum percentage of methane reaching the fire zone should not exceed 3 %. If the methane percentage cannot be controlled, it is essential to force more air into the ventilation stream reaching the fire by regulating other circuits.

It is essential that air samples be taken at places as close as possible (both inbye and outbye) to the seat of fire at time intervals of not more than 15 minutes and analysed for methane. The explosibility of the fire gases (Appendix IV) must be calculated from samples taken on the outbye side of the fire at intervals of no more than 30 minutes. The development of the fire must be monitored by sampling the products of combustion at intervals of not less than an hour and Graham's ratio (see p. 292) for the fire should be calculated. (An increase in its value represents a higher fire activity and a drop indicates that the fire is being controlled. In case of an active fire, its value may vary between 25 and 60 and where there is an absence of a fire it drops to 0.6.) The possibility of methane layering cannot be excluded in gassy mines and these must be monitored in areas close to the fire. Existing water or air pipes can be used for collecting samples for this purpose.

For mine air analyses and monitoring in emergency situations such as mine fires, mine explosions and recovery operations, mobile gas analysis laboratories are increasingly used (Haven and Hattingh, 1981; Mac-

Kenzie-Wood and Ellis, 1981). A typical laboratory would consist of analysers for oxygen, carbon monoxide, carbon dioxide, methane and nitrogen. The actual analysis of the gas sample takes about 5 minutes.

When methane drainage is practised, it is essential to see that no changes are made to the methane drainage system of the mine. Methane drainage must be continued even if the fire area is in the active methane drainage zone except when the percentage of methane in the pipe lines in the area itself drops to the range of 25 to 30 %. This drop certainly represents damage to the drainage lines as a result of the fire. In such a case it is important to locate the damage and cut off the line from the rest of the drainage system. The following precautions should be taken when the drainage lines run through the fire zone:

(1) monitor the percentage of methane in the drainage line continuously;
(2) check the state of the pipe line if possible;
(3) establish alternative routes for the pipe line so as to bypass the fire zone;
(4) if required, cool the pipe lines using water sprays and protect these from direct heat.

Breakage of pipe lines resulting in the leakage of gas into the ventilation circuit inbye of the fire can present problems particularly when it dilutes the air to the explosive limits. If the pipe breaks at the place of fire or in a region of high temperature, then it can burn slowly without causing any danger until its percentage drops to the explosive limit. If both the ends of the pipe are open, the explosion will travel to the other end. It is therefore important to close tightly the far end of the pipe. If a length of pipe line suddenly breaks at some distance inbye of the fire area, the methane in the pipe line can be released depending upon the ventilation pressure difference at its two ends. The build-up of methane percentage in the air depends upon the cross-sectional area of the pipe, the percentage of gas in the pipe and the cross-sectional area of the roadway, and can be calculated using the following relationship:

$$\eta = 0.0078 d^2 c_{CH_4} (101 - p)/A \qquad (9.1)$$

where η = increase in the percentage of methane in the airway, %;
c_{CH_4} = percentage of methane in the pipe;
A = cross-sectional area of the roadway, m^2;
d = diameter of the pipe, m; and
p = vacuum in the pipe range, kPa.

The length of this increased methane column is equal to the length of the broken pipe.

9.4.1 *Minimum flow rate of air required at the seat of fire*

The minimum flow rate of air required over a fire zone must be calculated using the Coward triangle (Appendix IV). However, sometimes a quick calculation can be made using the following equation:

$$Q_{a_m} > \frac{Q_g(100 - L_L)}{L_L - c_{CH_4}} \tag{9.2}$$

$$Q_g = \frac{Q_a c_g}{100}$$

where Q_{a_m} = minimum air volume flow rate, m^3/minute;
L_L = lower limit of explosibility of the mixture flowing over the seat of the fire as determined by analysis, %;
c_{CH_4} = initial concentration of inflammable gases (mostly methane) in the air fed to the district, %;
Q_g = volume flow rate of inflammable gases (CO, H_2, C_xH_y, CH_4, etc.) as determined from analysis of samples taken at the outbye end of the fire, m^3/minute;
c_g = sum total of the percentage of the inflammable gases, %, as determined in the volume flow rate Q_a (m^3/minute) over the the fire zone;
Q_a = volume flow rate of air over the fire zone, m^3/minute.

In some cases, there may be a need to increase the volume flow rate and care must be taken that this change does not cause reversal.

Calculation of the period for which the atmosphere in a sealed area remains explosive and after the lapse of which there is no danger of explosion is decisive when fighting fires and sealing off an area.

When the volume flow rate of air is reduced below the minimum volume flow rate (Q_{a_m}) then the critical time, t_{cl} (minutes), at which the lower explosive limit of the atmosphere behind the seals is reached with airflow below the minimum flow, and the time t_{cu} (minutes) at which the upper explosive limit is reached after the area has been completely sealed off can be calculated from the following relationships:

$$t_{cl} = \frac{V}{Q_a + Q_g} \times \ln\left(\frac{Q_a c_g + Q_g - (Q_a + Q_g)c_{CH_4}}{Q_a c_g + Q_g - (Q_a + Q_g)L_L}\right) \text{ minutes} \tag{9.3}$$

$$t_{cu} = \frac{V_2}{Q_g} \times \ln\left(\frac{100 - c_g}{100 - L_U}\right) \text{ minutes} \tag{9.4}$$

where V = total volume of the sealed area both intake side of fire V_1 and return side of fire V_2, i.e. $V = V_1 + V_2$;

Q_a = volume flow rate of air through the fire area, m^3/minute;
c_g = sum total of the percentage of inflammable gases flowing through the fire zone ($CH_4 + CO + H_2 + C_xH_y$), %;
Q_g = total amount of all inflammable gases formed and present at the return side of the fire zone, m^3/minute;
$Q_g = Q_a c_g/100$, m^3/minute;
L_L = lower limit of explosibility of the mixture flowing over the fire zone as determined from the analysis, %;
L_U = upper limit of explosibility of the mixture flowing over the fire zone as determined from analysis, %;
c_{CH_4} = initial concentration of methane in the air, %.

Time elapsed (t_0) for the percentage of oxygen in the sealed area to drop to a certain critical value $c_{O_{2c}}$ is given by

$$t_0 = \frac{V_1}{Q_g} \times \ln\left(\frac{c_{O_2}}{c_{O_{2c}}}\right) \tag{9.5}$$

where c_{O_2} = percentage of oxygen in the area at the time of sealing as determined by analysis;
$c_{O_{2c}}$ = critical value of the O_2 percentage in the area such that the mixture is no more explosive as determined by Coward triangle.

9.4.2 *Design and construction of explosion-proof stopping*

It is desirable in designing explosion-proof stoppings to assume that pressures of 140 to 350 kPa may be developed.

The larger the cross-sectional area of the roadway, the longer and stronger must be the stopping. The cross-sectional area of the roadway determines the total force to be resisted while the length and the shear strength of the contact surface between the side of the roadway and the material used in stopping construction determines its shear resistance.

The traditional type of explosion-proof stopping has been one of sandbags incorporating reinforcing materials. In more recent years, however, excellent results have been obtained by building stoppings of gypsum, using pump packs. Stoppings built of this material which are in length equivalent to the average of the height and the width of the roadway have been found to provide explosion-proof stopping.

Prevention of an explosion during the sealing-off operations is achieved by incorporating in each stopping an opening (tunnel or tube) which allows sufficient ventilation. When steel tubes are used it is necessary for these to be of the order of 750 mm or more in diameter and where square tunnels are used the cross-section is usually of the order of 900 mm square.

It is important to make arrangements to ensure that the tube or tunnel can be closed as quickly as possible once the stopping is completed. This is achieved by providing a stout steel door and housing it at the inbye end of the tunnel or tube and by fitting a flange to the outbye end of the tube or, by filling the tunnel with sandbags.

It is necessary to seal all the stoppings simultaneously and to withdraw all the workers until the danger of an explosion is thought to be passed. The waiting period during which no workers are allowed to return to the stoppings should not be less than 24 hours unless indicated otherwise by the monitoring system. Before final sealing, power circuits, signalling, lighting and telephone circuits, etc. extending inside the area to be sealed should be disconnected. The area should be heavily stone dusted and precautions taken to ensure that in the event of a methane explosion during sealing-off this area is not extended by a coal dust explosion towards the stoppings under construction.

A check should be carried out to make sure old roads leading into the district are properly sealed off.

The question of whether doors within the district should be left open or closed should be considered from the point of view of the safety of the workers during sealing-off operations, the effect of the fire and the requirements for subsequent re-opening. A record of all conditions and procedures adopted should be kept and stored in a safe place.

A sealed district or sealed areas in gassy mines which may be opened at some future stage need to be sampled at intervals. Sampling of a sealed area is done by building into the stopping a sampling pipe placed close to the roof of the seam. The sampling tube is normally a 6 mm copper one. Where it passes through the body of the stopping it needs to be protected by being run through a wider bore steel pipe.

Sampling the atmosphere behind the stopping may well have to proceed for a period of months or longer. To minimise the contamination of these samples by air drawn in by the 'breathing' that must occur over such a period of time, the sample line should be extended as far inbye as possible. This sample pipe can only be laid in the early stages of the building of a stopping.

If there is an existing water or air pipe which passes close to the fire it can be broken near the fire and used as a sampling pipe. Sampling can be carried out with aspirator pumps, and samples can be assessed either at the site by detector tubes or in the laboratory. If the mine has a remote gas monitoring system, the sampling points at the stopping can be connected to the monitoring system for continuous readings.

9.4.3 *Pressure balancing across stoppings*

When an area is sealed by two or more stoppings, the pressures on the outer faces of each stopping should, as far as is practicable, be balanced to prevent circulation of air through the sealed area, particularly past the seat of the fire.

The pressures on the stoppings can be efficiently balanced by removing or adjusting doors in roadways connecting the outbye sides of the stoppings (Fig. 9.11*a*). If it is possible to erect further stoppings in the intake and return roadways outbye the open connecting roadway, further isolating the system from the mine ventilation, a perfect balancing of the pressures on the faces of the inner stoppings can be achieved (Fig. 9.11*b*).

Where the roadways lead to a sealed area which cannot be connected by a pressure balancing roadway, it is possible to balance the pressures on the stoppings by means of a ventilating duct (Fig. 9.11*c*). The pressure on stopping *C* has been equalised with that on stoppings A and B by means of a duct. The doors D must be sufficiently airtight.

Often there is no convenient connecting roadway which can be used as a pressure-balancing roadway and the stoppings are too apart for it to be practicable to connect them by pressure-balancing ducts. In each cases, it is necessary to use a pressure chamber for each individual stopping.

A pressure chamber is constructed by erecting a wall a short distance outbye of the stopping – say, 3 to 6 m away, and arranging a way to raise

Fig. 9.11. Methods of balancing pressures on stoppings by pressure equalising roadways and ducts (after Willett, 1973). (*a*) By removing doors in a roadway connecting the outbye sides of the stoppings; (*b*) by erecting further stoppings in the intake and return roadways outbye of the open connecting roadway; (*c*) by a ventilation duct. A, B, C = stoppings; D = doors.

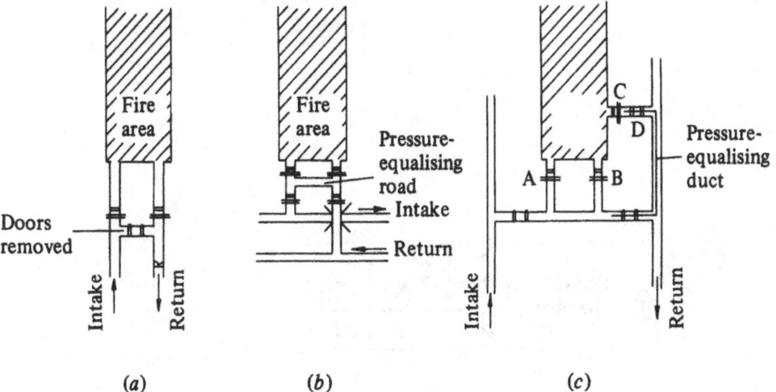

or lower the pressure of air in the chamber. The pressure in the chamber is controlled by means of a compressed air ejector or injector or by an auxiliary fan. Methods of doing this are shown in Fig. 9.12.

9.4.4 *Remote sealing systems*

To overcome the risks involved in building stoppings underground, remote sealing systems have been developed that are capable of working at a depth of almost 500 m. This technique consists of pumping fly ash underground from a surface bore hole (or from a remote area in the mine using a pipe line if already available). The area is surveyed through the bore hole (150 mm diameter), using closed-circuit television for visual observation of the downhole passage dimensions and a sonar ranging system for assisting in probing when dense smoke prevents visual observation. The probing system includes a rotator for downhole correlation with the mine map and an engineering survey at the surface. The information obtained during preliminary probing is used for establishing the suitability of bore hole locations, sealing material requirements, location of potential obstructions for sealing material, and verifying agreement between engineering surveys and the mine maps.

Fig. 9.12. (*a*) Layout of venturi injectors for controlling chamber air pressure. (*b*) Layout of auxiliary fan for controlling pressure chamber air pressure (after Willett, 1973). A = leakage regulating valve; B = throttle valve.

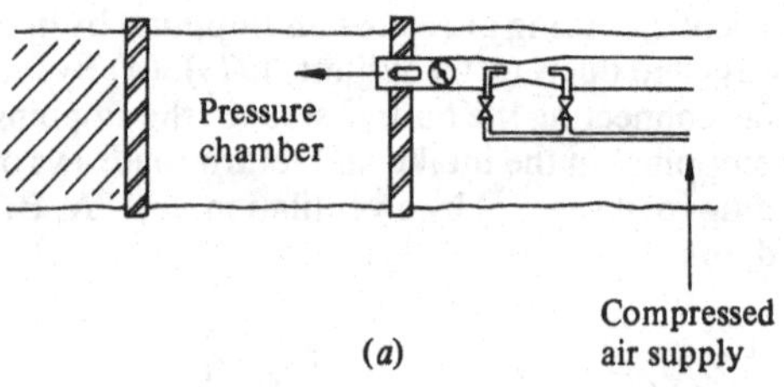

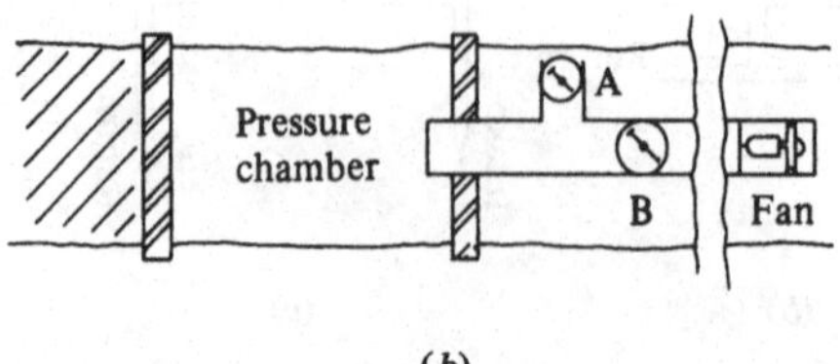

The passageway is then filled with the fly ash bulk sealant. Two adjacent bore holes have been previously drilled to monitor the formation of the gross fly ash seal, using acoustic equipment consisting of a loud speaker and a microphone. As the seal develops, the strength of the signal received at the microphone decreases. This reduction in sound level is a direct measure of the percentage of the mine passage area blocked by the sealant. The electronic signal-processing system includes a technique to eliminate false signals from sound travelling through parallel passageways to the microphone.

At the final stage, a frothy foam topping is added to fill the crater formed by the fly ash and to complete the seal. The foam is formed in place at the bottom of the bore hole. Nitrogen is blown down the inner holes inbye and outbye of the fire zone to cool and smother the fire. Normally, the fly ash is blown down with compressed air or a combination of compressed air and nitrogen in case of danger of an explosion.

The fly ash technique can be modified somewhat to make it capable of impounding water in a fire zone. Modification involves blending equal quantities of fly ash and bentonite to form the bulk of the seal, and blending a thin layer of fly ash and cement as a coating.

9.5 Fire fighting organisation

As the fire hazard is always present, every mine should perfect and maintain its own rescue and fire fighting organisation by:

(1) training rescue crew and other personnel;
(2) preparing a detailed plan of action in which all workers, supervisory officials, rescue crew and management know what they are required to do at the time of a mine fire;
(3) providing necessary fire fighting facilities at suitable places underground as well as providing for their care and maintenance;
(4) providing communications;
(5) providing copies of fire fighting plans at suitable places underground and on the surface;
(6) laying down procedures for conducting fire fighting operations.

9.5.1 Training

The mine should develop a fire fighting organisation by training personnel with special reference to matters such as fire prevention, fire drill, sealing procedure, ventilation control, use of different types of extinguishing agents and methods which may be used in the case of different classes of fires. Personnel should be well aware of the procedures and regulations for the checking of persons leaving and entering the workings, the first aid organisation and facilities, transportation priorities, communication systems, responsibilities and steps to be adopted in case of an emergency. All workers should be made familiar with emergency escapeways or exits from each working section and be instructed on barricading and the system of ventilation of each section.

9.5.2 *Procedures after detection of a fire*

At the appearance of any sign of fire, the worker who notices it should inform the nearest supervisory official as soon as possible. On discovery of an open fire, the person and any others in the vicinity of the fire should make a prompt effort to extinguish it and inform without delay the nearest supervisory official. If the fire cannot be easily and quickly extinguished or if the smoke from the fire is too dense to permit investigation, all workers in the section and other parts of the mine likely to be affected by fire gases or a possible explosion must withdraw to a safe area.

In metal mines which usually have a network of compressed air lines throughout the mine, a stench alarm has been very successfully used for several years. Ethyl mercaptan (C_2H_5SH) and eucalyptus are the best liquid stenches that can be used as warning agents. Ethyl mercaptan has the very disagreeable odour of rotten cabbage while eucalyptus has a characteristic odour which cannot be easily missed. The liquid stench is introduced into the compressed air mains at the surface compressor plant and also sprayed into the downcast shaft. A device which automatically flashes a warning of short flashes followed by long flashes of light indicating the level where an emergency has been reported has also been used in some mines for warning underground personnel.

Evacuation drills conducted once or twice a year using smoke bombs can be used to familiarise all workers with fire procedures, use of escapeways up to the surface, besides throwing light on any deficiencies in the reporting system, evacuation, isolation of fire, etc. Each country has its own prescribed set of regulations about escapeway requirements and procedures for maintaining these in proper form. These regulations should be adhered to strictly.

9.5.3 *Fire fighting facilities*

The efficiency of a mine fire fighting organisation depends on the availability of adequate and proper fire fighting facilities at suitable places underground as well as on their constant maintenance so that they operate reliably at peak operating efficiency at all times. It is advisable for large mines to appoint a safety officer responsible for fire protection, training of fire fighting personnel and education of underground employees in safety.

9.5.4 *Communications*

Every mine should have adequate communication facilities between suitable points underground and the mine surface so that alarms can be sent in emergencies at all times. Telephones are the simplest and the quickest means of communication in mines. They should be provided at the top and bottom of every shaft or mine entry, at each shaft landing, at loading stations and on the intake and return sides of ventilating districts. In large mines with extensive mine workings, it is advantageous to set up a telephone exchange underground which is connected to key points on the surface and to a surface telephone exchange and a central communication office which can be used as the main central office for all operations in case of an emergency. An emergency communication system is invariably installed in mines which is independent of the normal communication system, which has a tendency to be jammed by incoming and outgoing calls during an emergency.

9.5.5 *Plans*

At every mine there must be provided in the control and mine rescue rooms as well as at suitable places underground a complete set of plans and sections of the mine and fire fighting plans showing mine ventilation, water lines with diameters, hydrants and water pressure, location of fire and hose stations, fire trucks, fire doors, stone dust barriers, preparatory fire stoppings, fire-proof zones, telephones, and compressed air, pumping, and electrical systems.

9.5.6 *Conducting fire fighting operations*

With fires which are not serious, the manager or in his absence the official in charge should formulate a plan for the safe conduct of the work in consultation with experienced colleagues who are familiar with local conditions and are well-versed in the procedures following mine fires. The plan should provide for the removal of endangered workers if they have not been removed or warned earlier, the strategy of fire fighting, the selection of suitable sites and the type of stoppings and their order of erection. Some exploratory work by the mine rescue or gas mask crew may be necessary to throw some light on the situation. The plan must be strictly followed with no essential deviation from it. The manager should direct the fire fighting operations from a control room on the surface. If the manager is of the opinion that his presence underground is necessary at times, he should keep the control room in his absence under the charge of a responsible senior mine official.

With large fires, emergency underground and surface organisations should be set up to fight them.

9.6 Recovery of sealed-off areas in mines

The re-opening of an area which has been sealed off owing to an open fire or spontaneous combustion is an arduous and hazardous task. There is no one procedure of recovery which would satisfy all conditions and each case has to be considered separately.

The recovery of a sealed-off area should be undertaken only when conditions inside the area are safe for those engaged in the recovery work and will not cause additional damage to the mine. A certain amount of time must elapse before the conditions become favourable. The time of opening of a sealed-off area depends on the following factors (Ramlu, 1975).

1. Extent and intensity of fire at the time of sealing

The extent and intensity of fire at the time of sealing influence the reduction of oxygen inside a sealed-off area if there is no leakage of air into it. A large amount of burning material reduces the oxygen content quickly but lengthens the time required for cooling off.

2. Nature of adjacent strata

If the adjacent strata are combustible they retain heat for a long time, even after the oxygen content has fallen below the limit when combustion ceases.

3. Airtightness of stoppings and the enclosed area

If there is some leakage of air into the sealed area through the stoppings and the adjacent strata either due to barometric and temperature changes or to differences in the ventilation pressure between different parts of the area, it takes more time for the extinguishment of the fire.

4. Composition of the fire gases

A systematic record of the variation of carbon dioxide, carbon monoxide, oxygen and methane with time in the sealed atmosphere can be used to assess the stage and status of the fire. There will be a rapid decrease in oxygen and a corresponding increase in methane. The amounts of carbon monoxide and carbon dioxide will probably reach a maximum at an early stage and then begin to decrease.

Flames normally disappear when the oxygen concentration is maintained below 12 %. However, heated coals, wood and rock can remain incandescent or smoulder indefinitely when the oxygen concentration is as low as 1 %.

The CO produced/O_2 deficiency* ratio (Graham's ratio) has been found to be a more useful index than carbon dioxide, carbon monoxide and oxygen values as it is practically independent of dilution by the atmospheric air. When this ratio has fallen to the normal value of the district or mine, one can infer that the fire has subsided towards extinction.

The other indices used to ascertain the state of a fire are (1) the CO_2 produced/O_2 deficiency ratio (Young's ratio) and (2) the CO_2/blackdamp† + combustibles ratio (Willett's ratio). If there are no extraneous sources of carbon dioxide other than the fire, and when these ratios approach a constant value, this is a good indication that the activity of the fire has greatly reduced.

A low oxygen concentration and steady low indices can be considered as guides for the re-opening of a sealed area although they cannot provide a guarantee that the fire is out, or, if out, will not re-ignite. The management should prepare itself for this eventuality.

The possible explosion hazard upon admission of air can be determined from Coward triangle (Appendix IV).

9.6.1 Methods for recovery of sealed-off areas

For the successful recovery of any sealed-off fire area, certain preparatory measures should be undertaken. These include assembling the rescue and recovery crew, withdrawal of all workers from the mine other than those required in connection with the recovery operations, stationing a person at the main fan to see that it continues to run efficiently, cutting off electric power from the section of the mine in which a fire is sealed off and from the return airways to be utilised for carrying the fire gases, making necessary adjustments in ventilation, and stone dusting all roadways leading to and from the fire area.

For recovering a sealed-off area, one of the following methods may be employed: (1) recovery by reventilation; (2) recovery by air-locking.

1. Recovery by reventilation

This method is employed when a fire is known to be small in extent and when it is definitely known from the evidence collected that the fire has

* $0.265N_2 - O_2$.

† Excess $N_2 + CO_2$.

been extinguished and that no flammable mixtures will be formed on admission of air. The recovery can be carried out either by direct ventilation or by auxiliary ventilation of the area to be recovered.

When recovering by direct ventilation, an air-lock is constructed preferably near the intake stopping. A fully equipped rescue brigade then breaches the intake stopping and enters the area to observe conditions, take temperature readings and collect air samples. If from the observations and examinations thus made it is concluded that conditions are favourable for re-opening, the necessary adjustments in ventilation are made so that the fire gases may be led directly into the main return. Next, the return stopping is broken by the rescue brigade and fresh air is coursed through the area by opening the air-lock. The fresh air current should be sufficiently strong so that none of the fire gases is left behind and the content of the combustible gases in the main return as well as in the airways carrying the gases to the area is kept below the lower flammable limit. On reaching the fire seat it should be watered down, and if smoke begins to be formed it should be tightly sealed off.

When there are falls in the roadways in the area being recovered which cannot be cleared or when there is only one communicating roadway and direct ventilation is not practicable, recourse is taken to auxiliary ventilation, using a brattice line or ducting with or without a mechanical ventilator. The recovery brigade advances systematically away from the fresh air base by extending the brattice line or ducting.

2. *Recovery by air-locking*

This method of recovery is employed when a sealed-off area is extensive and the fire seat is inaccessible. The area is recovered in successive blocks or stages by means of air-locks. An air-lock consists of two brick walls one and a half to two bricks thick, 5 m apart, equipped with airtight wooden or steel doors which have glass windows for watching the advancing rescue brigade inside. It is erected at a convenient distance outbye of the intake stopping. Within the air-lock all the equipment and materials necessary for the erection of the next air-lock are stored; water and compressed air lines are brought into it. After the air-lock has been dried, the rescue brigade with another in reserve enters the area through the air-lock, breaches the stopping and inspects up to a point where the next air-lock is to be erected, a distance generally lying between 50 and 150 m depending upon the difficulties encountered. The brigade should observe general conditions, collect air samples, take readings of temperature and humidity, and measure for the materials required for the next air-lock.

After the next air-lock is erected, the one previously erected is broken and the area between the air-locks is ventilated by auxiliary ventilation, adequately supported and fitted with the necessary roadway fittings. It is only after the completion of these auxiliary operations that the brigade advances further and by making such advances in successive stages recovers the entire area. It is necessary during recovery operations that the oxygen content of the fire gases be kept as low as possible at all times by limiting air leakage into the area. As leakage of air through the doors of the air-lock takes place when they are opened, the doors should be opened a maximum of twice in a period of 4 hours. To keep back gases and heat during the erection of an air-lock, a temporary brattice stopping or screen is generally erected just beyond the proposed site of an air-lock.

Recovery by use of air-locks is a very costly operation especially when complicated mine workings exist, as a large number of air-locks is required to be constructed and the workers need to wear respiratory equipment.

9.7 References

Bickel, K. L. (1977). Five automatic fire protection systems for mobile mining equipment: their design and inmine reliability testing. In *Metal mine fire protection research, Proceedings of the Bureau of Mines technology transfer seminar*, Tucson, Arizona, 18 March 1977, pp. 14–17. USBM IC 8752. Washington: US Bureau of Mines.

Greene, P. (1973). Designing safety into underground mining equipment – fire suppression on face machinery. *Min. Cong. J.*, **59**, 49–54.

Haven, Y. and Hattingh, C. J. J. (1981). The mobile gas analysis laboratories of the Chamber of Mines of South Africa. *J. Mine Vent. Soc. S. Afr.*, **34**, 51–6.

Johnson, G. A. (1976). Development of mine shaft fire and smoke protection prototype system. *Skilling's Min. Rev.*, **65**, No. 11, 1 and 6–7.

MacKenzie-Wood, A. P. and Ellis, C. G. (1981). Use of a mobile gas laboratory at mine fires. In *Proceedings of the symposium on ignitions, explosions and fires*, University of Wollongong, Wollongong, ed. A. J. Hargraves, pp. 15.1–15.9. Melbourne: Australasian Institute of Mining and Metallurgy.

Pomroy, W. H. (1977). Fire suppressants for mobile mining equipment. In *Metal mine fire protection research, Proceedings of the Bureau of Mines technology transfer seminar*, Tucson, Arizona, 18 March 1977, pp. 18–23. USBM IC 8752. Washington: US Bureau of Mines.

Ramlu, M. A. (1975). *Mine fires, explosions, rescue, recovery and inundations*. Kharagpur, India: Mukherjee Publishers. 314 pp.

Willett, H. L. (1973). Address at the seminar on mine fires on 12th November 1973 at Brisbane, Queensland, Australia. In *Proceedings of the seminar on mine fires*, Brisbane, Australia, pp. 5–29. Melbourne: Australasian Institute of Mining and Metallurgy.

10

Ventilation planning and computer monitoring of environments

10.1 Ventilation planning

First of all, any engineer responsible for designing a ventilation system for a mine has to be familiar with the mine ventilation standards and the occupational safety and health regulations of the state/country for the proposed operation. It is also important to realise that mining is a dynamic, ever-changing process requiring frequent changes, and some flexibility is necessary.

To maintain adequate ventilation throughout the life of a mine, careful ventilation planning in advance is necessary so that a sufficient number of appropriate sized roadways is driven and the correct ventilation plant is selected. Mine ventilation planning involves consideration of two principal factors, namely (1) the total volume flow rate of air required by the mine, its satisfactory and economic distribution and (2) the pressure required by the mine fan(s). After estimating these parameters, the ventilation plant can be designed to meet the requirements.

The mine ventilation plan depends mainly upon the mining method in use and controls the fresh air supply to the faces of all working areas and consequently the safety and health of the workers. Ventilation planning is done for different stages in the development and exploitation of the ore body. The different stages may be:

(1) for the period of entry into the deposit, the development of the mine and up to the first production of ore;
(2) for the 5-year period after the start of production of ore;
(3) for the 15-year period after the start of production of ore; and
(4) for the 25-year period after the start of production of ore.

Inadequate planning in advance has many times given difficulties when the workings have advanced considerable distances from the shafts, when diesel equipment has been introduced into the mine, when deeper levels have been worked, etc.

Development plans for each stage of the mine life are obtained from the mine planning department. The plans show the projected workings of the mine, motor houses, pump houses, locomotive garages, underground workshops, crusher stations, fuel stores, rock handling facilities, material handling/transport roadways, development workings, etc. which will have to be ventilated. Frequently when the ventilation planning is done, modifications such as the driving of additional airways and the increasing of the cross-sections of some airways are warranted in the mine planning. An ill-conceived ventilation plan can adversely affect the cost of ventilation or involve costly rearrangement of the ventilation system.

It is also important to analyse the effect of the proposed mine ventilation system on mine fire control. Or, conversely, the ventilation planning engineer has to appreciate what effect a mine fire may have upon the proposed mine ventilation system (section 9.1.6, Combating fires).

When longwall mining methods are used to extract coal, the ventilation system can influence spontaneous combustion. In the case of bord and pillar workings, leakage can occur through fractures in the coal surrounding stoppings, developing spontaneous combustion in this coal. Pressure differentials obviously require consideration (sections 9.2.2, Factors contributing to spontaneous combustion, and 9.2.3, Prevention of spontaneous combustion).

In a copper pyrite mine, there is a possibility of sulphide dust explosions (section 8.4, Sulphide dust explosions). Dust-laden air from development workings should not be allowed to pass through stopes where the dust could be ignited by the stope blasts.

10.1.1 Volume flow rate of air required for the ventilation of a mine

The main air pollutants to be considered in evaluating the air volume flow rate requirements are methane (in the case of coal mines particularly), blasting fumes, diesel engine exhaust, airborne dust, radon gas (in the case of uranium mines particularly) and heat (in the case of deep mines particularly). Information regarding the emission of these pollutants has been given earlier in the book (3.5.1, Gases from blasting; 3.6.1, Exhaust emission from diesel engines; 4.5, Emission of gas into mine workings; 5.3, Sources of dust; 6.3, Radon gas emission and sources; and 7.3.1, Sources of heat in mines). Using appropriate standards, air volume flow rate requirements are calculated for each and every working place (3.5.2, Control of gases from blasting; 3.6.2, Control of exhaust emission; 4.6, Control of methane emission in mines; 5.4, Control of dust; 6.4, Control of radiation; and 7.4, Control of mine climate). After determining the air volume flow rates for all the working places, the airflows in various airways, shafts, fan drift(s), etc. are built up taking into account the requirements for motor houses, etc. and for leakage at stoppings, air crossings, doors, the surface, etc. Expansion of air in the upcast shaft should also be taken into account to determine the fan volume flow rate.

Some general air volume flow rate requirements are as follows:

Motor and pump houses	2.5 m^3/s
Diesel locomotive garages	0.07 m^3/s per kW
Crusher stations	25 m^3/s
Development workings	0.15 m^3/s per m^2 of excavated area

Transient uncontrolled air losses vary from as low as about 15 % to as high as 90 %. Some of the reported values (%) are as follows:

Bord and pillar coal mines	40–70
Room and pillar non-coal mines	15–90
Sub-level longhole stope mines	20–30
Shrinkage stope mines	20–30
Horizontal cut and fill metal mines	10–60
Sub-level cave mines	20–30
Block cave mines	40–50

The volume flow rate of air has a major effect on the size requirement of shafts and airways. This aspect has been discussed in section 2.6, Economic size of airways. The optimum conditions can be expressed in terms of a velocity range for a given type of shaft or airway. The general ranges of air velocity (in m/s) in various types of primary airways are as follows:

Exhaust shafts (concreted)	15–20
(rock section)	10–15
Production shafts	8–10
Service shafts	5–8
Fan drifts	12–14
Main airways and haulage levels	3–8
Main exhaust airways	10–15
Conveyor roadways, declines	3–5

South African studies have shown that if air velocities between 7.5 and 12.5 m/s are avoided in upcast shafts, the water condensed in the upcast air will not accumulate and contribute to fan stall and water drainage problems.

The air velocities should be calculated in all working places, airways, etc. to check whether they are in a reasonable range. For example, in the case of a longwall face, the lower and upper limits of the face velocity are about 1 and 5 m/s so that the ventilation is not sluggish or uncomfortably high, worsening dust conditions. A minimum velocity of about 1.0 m/s is required to minimise stratification in mine workings and the resultant build-up of methane and diesel exhaust emissions.

10.1.2 Pressure requirements for the fan(s)

After deciding the volume flow rate of air required and its distribution, the pressure required on each face, and on each roadway, intakes and returns are calculated and marked on the plan. Addition of these pressures gives the total pressures required for the whole of the underground ventilation circuits, and to the highest of these, the pressure absorbed in the shafts is added. In this way, the total ventilation pressure is determined. Natural ventilation pressure is deducted from the total ventilation pressure to determine the fan pressure.

Atkinson's equation is used to calculate the pressure losses due to friction (section 2.2.4, Atkinson's equation). Some typical values of the friction factor have been given in Tables 2.1 and 2.2 (p. 24 and p. 25). Pressure losses due to change in direction, cross-sectional area, the presence of an obstruction, entrance or exit should also be calculated for each section of the mine layout. Loss factors for various items have been given in Table 2.3, p. 27.

10.1.3 Distribution of mine air

The resistance of a mine as a whole depends on the layout of the airways as well as on the resistance of each airway. The distribution of the mine air depends upon the ventilation resistance of the workings. In parallel circuits, regulators or booster fans may have to be used to distribute satisfactorily the volume flow rate (section 2.3.2, Parallel circuits). Compound circuits are difficult to solve and programs are available for this purpose (section 2.3.4, Ventilation network analysis, and Appendix I, Mine ventilation network analysis program in BASIC for microcomputers).

10.2 Computer monitoring of environments

Developments in micro-electronics have permitted the transfer of information from a large number of remotely placed points at a quite acceptable cost in mining. The development of a variety of sensors to monitor variables at a fast rate has called for the use of a computer to log the information continuously, process it at a fast rate and present it in a form easily interpretable to enable quick and correct decisions to be made about the state of the environment in any location in a mine. Developments, particularly in the last five years, have resulted in the application of a number of remote monitoring, computer-aided systems in a number of mines throughout the world. This, at the same time, has been almost a primary condition for efficient production techniques to be introduced in very large mines.

Basically there are two monitoring systems applied in mines. In the first, a sample is transferred to the surface through a system of tubes and analysed at a central station (tube bundle system). A schematic diagram of this system is shown in Fig. 10.1. In the second, a sample is analysed on the spot and the results of the analysis are transmitted to the surface and recorded appropriately. An example of such a system for use in ventilation monitoring is given in Fig. 10.2. Based upon these two methods, several computer systems have been developed (Morris and Gray, 1977; Thomas and Chandler, 1978; McPherson, 1979; Ribour and Stenning, 1983; Welsh, 1983; Mine Safety Appliances (MSA), 1984).

10.2.1 Requirements of a remote monitoring system

The requirements of an environmental system depend upon the parameter to be monitored, the importance of the parameter and location of the point with respect to the production process. Frequency of monitoring is dependent upon the purpose of a particular installation. For

Fig. 10.1. Schematic diagram of a tube bundle system (after Fink and Adler, 1975). Solid lines: sample lines; dashed lines: electrical lines. 1, Dust filter; 2, water trap; 3, flame arresters; 4, junction boxes; 5, single line; 6, small harness; 7, large harness; 8, standardisation gases; 9, air; 10, control; 11, timing controls; 12, CO analyser; 13, oxygen analyser; 14, other analysers; 15, recorders; 16, display cabinet.

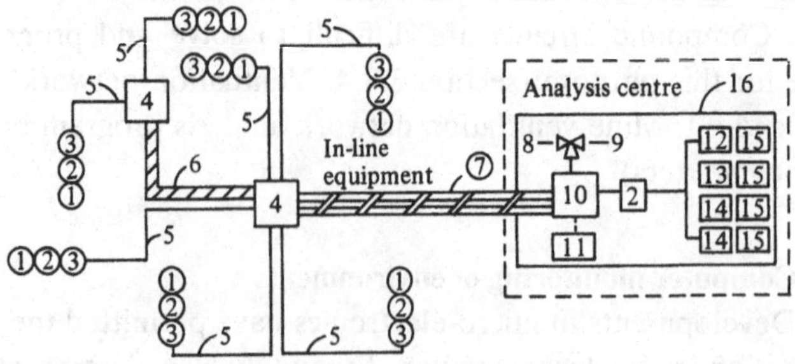

example, where concentration values are changing rapidly, it is essential that the sensor has a fast response and that data transmission is fast and in accordance with the rate of change of the variable. When monitoring methane levels at the face, the monitoring system must have a fast response so that the equipment can be switched off as soon as the methane levels exceed the prescribed statutory requirement limits. On the other hand, with methane monitoring of return roadways at remotely located points, it is enough if the frequency of monitoring is about 10 to 20 per hour. Similarly, dust monitors placed at the face must monitor and transmit the dust levels continuously. For those monitors placed at remote points, the data transmission frequency can be reduced to 10 per hour.

Fig. 10.2. Block diagram of an automated monitoring and control system. A: analogue signal; B: digital signal; A/D: analogue to digital converter; D/A: digital to analogue converter (after Hormozdi, 1979). 1, Pressure; 2, velocity; 3, temperature; 4, carbon dioxide; 5, methane; 6, regulator sensor.

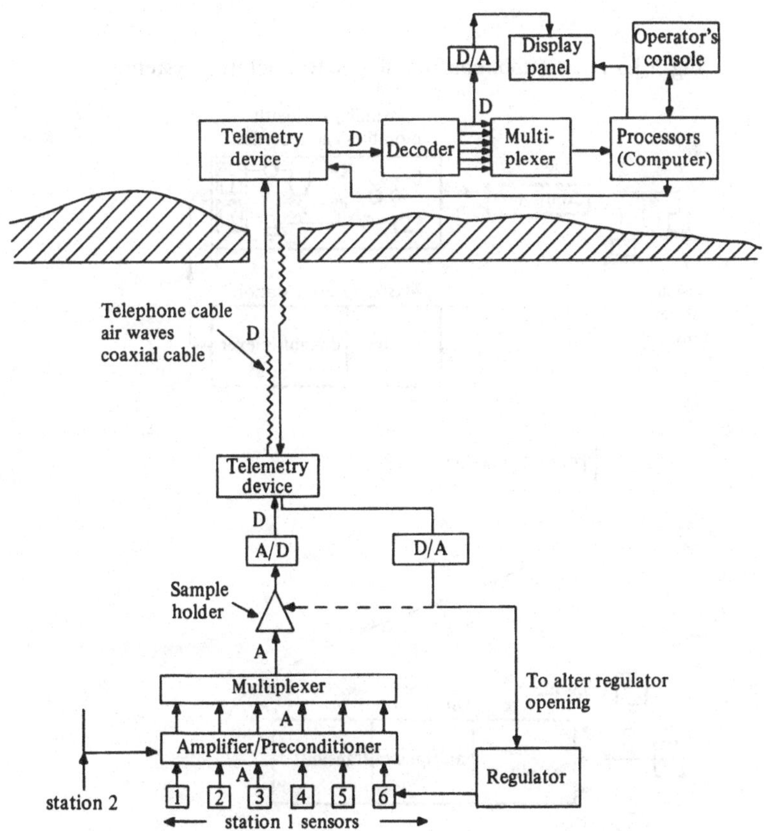

10.2.2 *Elements of a remote monitoring system*

The basic elements of a remote monitoring system are:

(1) a sensing element;
(2) a data transmission system; and
(3) a data recording system.

A schematic diagram of a system is shown in Fig. 10.3. The sensing element or the head to be used for a particular installation depends upon the parameter to be monitored. The data transmission system is dependent upon the type of the signal output from the sensing head, the number of sensing heads and the distance the data are to be transmitted. The recording system is invariably either analogue, such as a chart recorder, or digital, such as a digital computer.

While considering a computer monitoring system, the following points should be kept in view.

(1) Is the system to be coupled to switch on or off production equipment or is it to be used solely for data collection for later analysis?

Fig. 10.3 The components of a simple telemetering system.

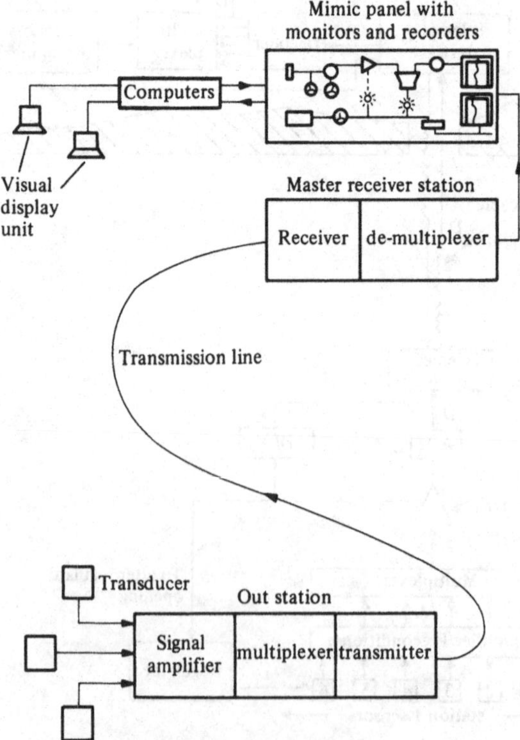

(2) Is it possible to collect the information from a single point/parameter accurately to control a process/equipment or will it need more than one point/parameter to control the process/equipment?
(3) Is the monitoring point fixed in space and time or will it have to be moved periodically? If so what will be the frequency of movement?
(4) What form(s) of signal(s) will be required to control process equipment, warn operators, or inform management?
(5) In decision making, is the history of a parameter needed or is it sufficient to accept a specific measurement?
(6) What is the importance of a particular parameter? Will it need to be monitored continuously, intermittently at certain pre-set intervals? Will it need to be monitored more (or less) frequently if the conditions or its value changes?
(7) Should the monitoring system be independent of the mine power supply system and be capable of transmitting information even when the mine has been completely shut off?

It is not difficult to realise that a modern mine monitoring system should rely on a large, powerful, digital data storage system supported by a continuous visual display of some of the main parameters on a chart recorder.

10.2.3 Sensing elements of a mine environmental monitoring system

Mine environment requires the following elements to be monitored:

(1) velocity of airflow;
(2) percentage of the various gases such as methane, carbon dioxide, carbon monoxide (particularly in the case of fires), radon (in uranium mines), SO_2 (sulphide minerals);
(3) barometric pressures in roadways and on the surface, pressure difference behind stoppings, between intake and return airways or behind sealed areas;
(4) temperature and humidity;
(5) respirable dust;
(6) mine fires;
(7) mine fans;
(8) water supply system.

The principles used in the design of the various sensors have been described earlier in various chapters (Mine ventilation engineering (Chapter 2), Mine gases and their control (Chapter 3), Dust in mine air and its control (Chapter 5), Mine radiation and its control (Chapter 6), Mine climate and its control (Chapter 7), Mine explosions and their control (Chapter 8) and Mine fires and their control (Chapter 9)) and readers are referred to these chapters for detailed study.

The air velocity sensors utilised in mines largely make use of the thermal measuring process (Germany) or an impeller principle (Germany, France and the UK), although the use of vortex separation monitors to measure velocities have been successfully developed in the USA, and pressure sensors in Germany. Correct placement of the units in a roadway or a duct is very important.

Monitoring of various gases (CH_4, CO_2, CO) is usually done by sensors based on the infra-red (Germany), catalytic combustion (USA) or thermal conductivity (France and the UK) principle. All these systems have been found to be quite reliable. For radiation control purposes, instruments relying either on gross alpha or beta counting are used. Infra-red gas monitoring is available for nitric oxide although the application of this is limited to underground diesel workshops, blasting studies, etc. Monitoring of oxygen is still done using a flame safety lamp, electrochemical indicator cells and paramagnetic oxygen differential analysers. These units are not as yet available for remote monitoring.

Sealed units are used for measuring barometric pressures underground and on the surface and by comparison give the pressure differential between the upcast shaft and any point underground. Pressure differential units are used between intakes and returns, at the stoppings and sealed areas. These units make use of either sealed diaphragms with strain gauges bonded to the diaphragm or capacitors for sensing the pressure.

Humidity monitoring underground often presents problems and no reliable sensors have yet been developed for remote monitoring although a number of handheld units have been developed (the USA and Finland).

Equipment is now available for continuous monitoring of respirable dust levels based upon light scattering (Mining Research and Development Establishment, UK, and Bergbau-Forschung, Germany). The Bergbau-Forschung FMA 2 dust station can be used to start/stop or initiate switching functions for dust removal, spraying equipment, etc.

The sensing elements employed in mine fires depend upon the type of fire. Open fires use either heat or smoke sensors, and concealed fires are invariably detected by the presence of combustion gases (CO, CO_2, H_2, the higher hydrocarbon C_2H_6). The best method, which is independent of the type of fire, is still the detection of gases. Carbon monoxide can be detected in the very early stages and infra-red analysers are best suited for this purpose. Metal oxide semiconductors (ZnO_2 or SnO_2) have been developed which can detect CO at 0.1 ppm, but these are highly sensitive to reducing gases such as methane.

Remote control of mine fans invariably means stopping one fan and starting a second fan. The output of a fan can be adjusted by changing the blade position using hydraulic actuators. Monitoring of a fan in normal running includes differential pressure measurements, air volume flow rate measurements, measurement of the percentage of various gases and such parameters as load on the fan motor, etc. (using power meters), temperature of the windings and bearings (using resistance thermometers) and fan vibration (using vibrometers).

Mine water is monitored to detect any sudden inrush of water from unexpected sources. When mine water is used for the fire extinguishing system, the monitoring includes pressure, flow rate, water usage, and detection of bursting of and leakage from the pipes. Water into and out of the mine is monitored to establish water flow patterns. In surface mines, reinjection of water into remote areas away from the pit is monitored for control purposes. Water quality monitoring includes pH values and salt content and is essential in certain areas from the environmental point of view (acid mine drainage) and equipment maintenance.

The accuracy of any measurement depends not only upon the type of sensor, maintenance and transmission losses, but very greatly on the correctness of installation. The air velocity and percentage of gases in a roadway are not constant, particularly when the distance of the measuring point from the source of gas is not large. In every case, it is essential to position the sensor in such a way that it accurately reads the value (e.g. maximum or mean) of the parameter to be measured. When placing sensors in mine roadways, the roadway section should be streamlined by placing a brattice cloth on the sides of the roadway for a minimum length of 10 times the width of the roadway, especially when high velocities exist in the roadway. The distance at which good mixing of the gas takes place depends upon the air velocity and the shape of the roadway. In areas with high turbulence, mixing is much better but in areas with low turbulence, layering in the roof and sides may be found up to many hundreds of metres away from the source of gas emission.

When placing flow detectors in pipes, care should be taken that these are not placed at locations close to bends, valves, T-pieces and at reducers. The sensor head must be placed at the centre of the pipe and the data corrected for any errors in measurement as a result of the location of the sensor.

10.2.4 *Data transmission systems*

Remote data transmission of a large amount of data is done mostly by using two basic systems: (1) analogue current transmission and (2) multiplexing. The analogue current transmission system uses currents from 4 to 20 mA. The current output of the sensor system is proportional to the value of the parameter measured. The disadvantage of this system is that a separate line is required for each channel and the distance to which the signal can be efficiently transmitted is limited (~2 km). As such this is used only in very limited cases.

The multiplexing system is most commonly used in large installations. In this system, the signals are combined, or multiplexed, and sent together along a single line. The carrier signal has a different form to that of the transducer signal and the information is abstracted by changing either the amplitude or the frequency of the carrier signal by an amount proportional to the magnitude of the transducer signal. The latter technique, called frequency modulation, is most common as it is less susceptible to electrical interference (Fig. 10.4).

There are two basic methods used in multiplexing: (1) frequency division multiplexing (FDM) and (2) time division multiplexing (TDM). In FDM, each transducer has its own encoder with an assigned frequency. If the amplitude of the transducer changes, the magnitude of the frequency transmitted by the encoder also changes. Thus the frequency of the carrier signal deviates from the central frequency by an amount

Fig. 10.4. A varying transducer signal with its frequency-modulated carrier signal. The amplitude of the carrier signal remains constant (after Atkins, 1982).

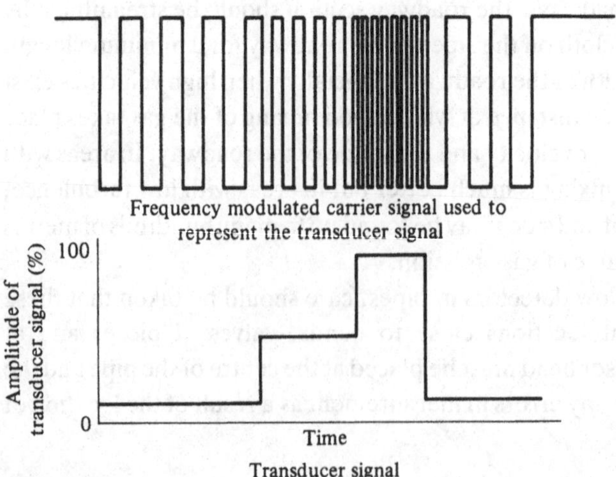

proportional to the transducer output. Using different central frequencies for different channels, the output from a number of channels can be transmitted over a single telephone line.

In TDM, the frequency of each transducer signal is kept constant while the signals for all the transducers are transmitted for short periods at regular intervals and not all at the same time (Fig. 10.5). This method is less expensive than FDM, particularly when a large number of channels are required, but has complex circuitry and is sensitive to electrical interference. Fig. 10.6 gives a comparison of costs of the FDM and TDM systems. All signals are multiplexed at the transmitting station and are transmitted through a twisted pair of telephone lines (5 kHz). On the surface these are de-multiplexed in the case of TDM or filtered through a filtering system and processed to resurrect the signal.

10.2.5 Displaying and recording of information

On arrival the information can be displayed on a mimic panel which represents a block diagram of the system. The signal for each monitor is inserted into the block diagram at the correct position to give on/off information by warning lights or displayed on a chart recorder.

Fig. 10.5. Time division multiplexing with five channels. The frequency of the carrier signal for each channel is the same. The five channels are repeated in sequence over and over again. There is normally a small gap between the carrier signal for each channel and a larger gap at the end of each group or frame (after Atkins, 1982). f_1 = the frequency of transmission.

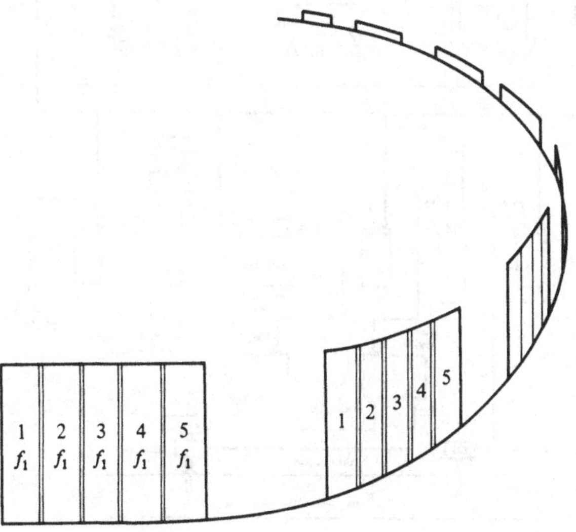

Fig. 10.6. Cost structure of frequency division multiplex (FDM) and time division multiplex (TDM) systems (after Olaf, 1979).

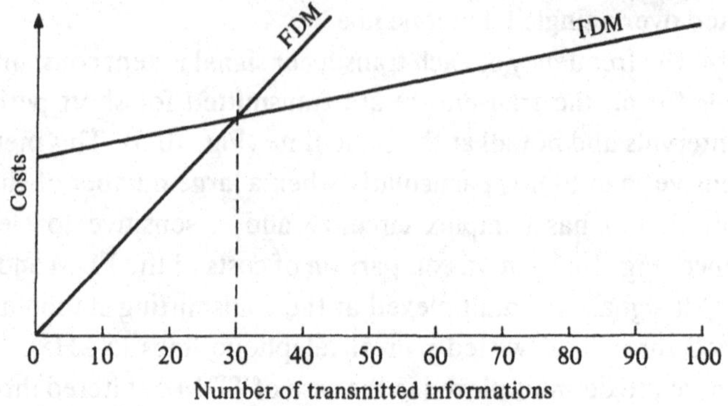

Fig. 10.8. A schematic diagram of the monitoring system for ventilation and methane gas drainage lines used at the West Cliff colliery, Australia (after Fraser *et al.*, 1984).

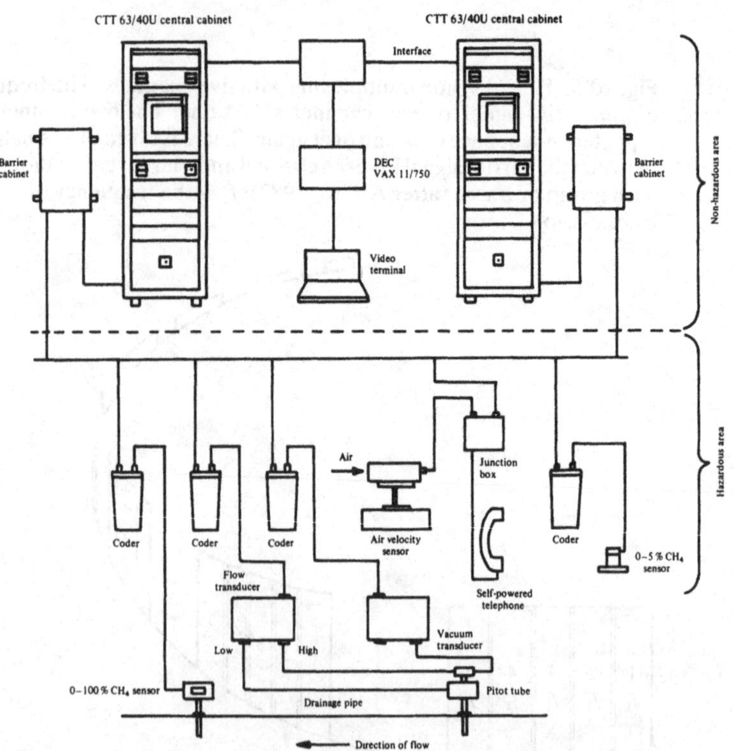

Fig. 10.7 General layout of the telemetering system and the computer interfaces used at the West Cliff colliery, Australia (after Fraser *et al.* 1984). C = colour; BW = black and white

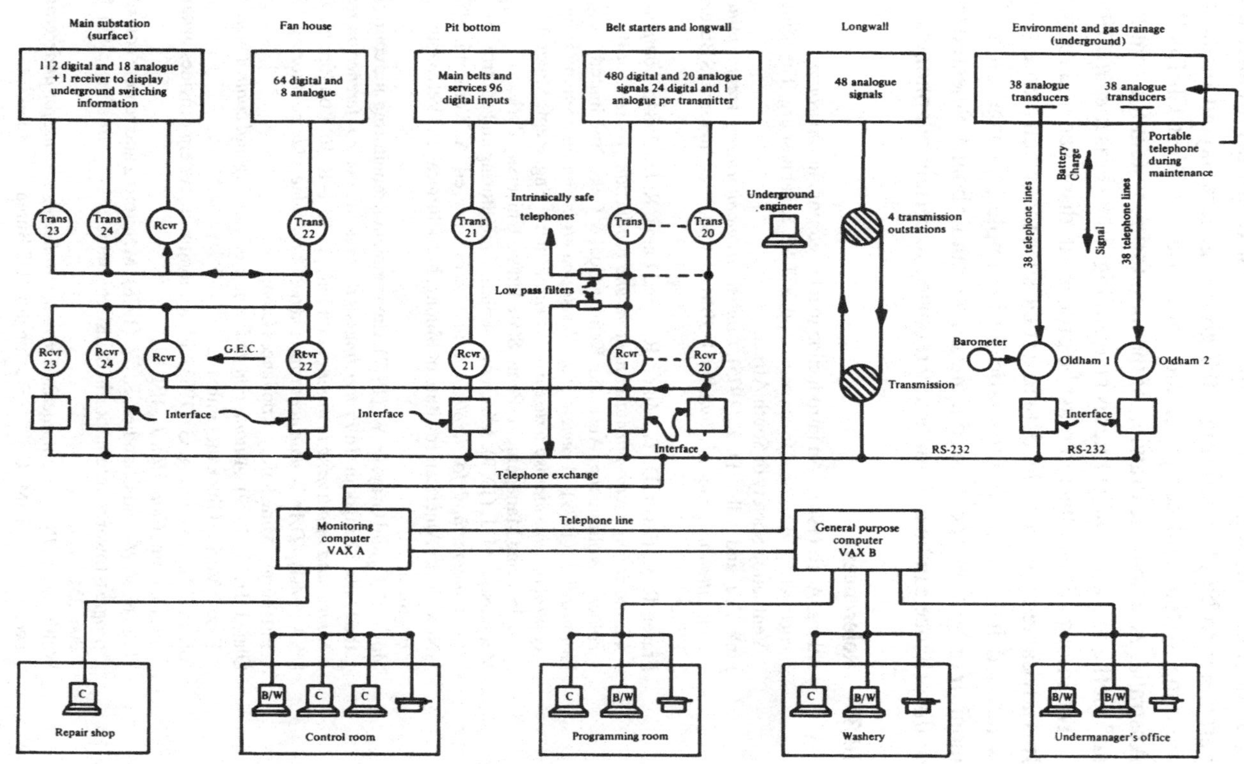

Signals can also be fed to a magnetic tape recorder or an on-line digital computer using appropriate interfaces and stored in the memory of the computer. This real-time storage and fast retrieval system also enables additional information about the specific cause of any disturbance in any parameter to be obtained. Fig. 10.7 gives a general layout of the telemetering system and computer interfaces used at West Cliff colliery, Australia. A schematic drawing of the monitoring system for ventilation and methane gas drainage lines is given in Fig. 10.8 (Fraser *et al.*, 1984).

The information collected must be processed, displayed and reported. It is advisable to process the data to a stage such that when a critical stage is reached it is reported automatically in simple language on a video monitor or by a set of coloured lights on the mimic board. Selected levels or disturbance can be automatically reported to certain telephones.

10.3 References

Atkins, A. R. (1982). An introduction to telemetering. In *Environmental engineering in South African mines*, pp. 877–92. Johannesburg: The Mine Ventilation Society of South Africa.

Fink, Z. J. and Adler, D. T. (1975). *Continuous monitoring system for mine gas concentrations using tube bundles*, USBM RI 8060. Washington: US Bureau of Mines.

Fraser, R. J., Eager, D. T., Miller, R. J. and Lama, R. D. (1984). *Computer monitoring of environmental conditions and production processes at West Cliff colliery*. Wollongong, Australia: Kembla Coal & Coke Pty Limited.

Hormozdi, I. (1979). *A review of the developments in automatic monitoring and control of mine atmospheric environment.* Mineral Eng. Rept, Mineral Engineering Management, Penn. State Univ., University Park.

McPherson, M. J. (1979). Mine environmental monitoring and control. In *Computer methods for the 80s in the mineral industry*, ed. A. Weiss, pp. 705–16. New York: American Institute of Mining, Metallurgical and Petroleum Engineers.

Morris, I. H. and Gray, G. W. (1977). Environmental monitoring and control in the United Kingdom. In *Proceedings of the international conference on remote control and monitoring in mining*, London, pp. 10.1–10.16. London: NCB.

MSA (1984). *'DAN' – data acquisition network for mine wide monitoring.* Pittsburgh: Mine Safety Appliances Company.

Olaf, J. H. E. (1979). *Automation and remote monitoring and control in mines.* Essen: Verlag Gluckauf GmbH.

Ribour, J. and Stenning, G. (1983). Mine monitoring system provides instant information. *Can. Min. J.*, **104**, 91–4.

Thomas, V. M. and Chandler, K. W. (1978). Monitoring and remote control – progress towards automation in British coal mines, 1977. *Min. Engr.*, **136**, No. 206, 251–62.

Welsh, J. H. (1983). *Computerised remote monitoring systems for underground coal mines*, USBM IC 8944. Washington: US Bureau of Mines.

11

Mine water and its control

Although only 3 % of the total water of the world occurs as fresh water locked up in the ground, it is sometimes a major cause of trouble in mines. The presence of water in significant quantities affects the choice of mining method, type and capacity of the dewatering system and determines the steps to be taken to deal with it. Large quantities of water when present may need special attention in the form of dewatering of the area, or the leaving of special barriers to contain water inflow into the workings, or adopting such measures as freezing and cementation techniques to stop flow of water into mine workings. Large-scale dewatering causes ground subsidence and this change in ground level together with subsidence due to mining can result in a change in the pattern of flow of surface and subsurface ground water. The presence of water in sulphide ore bodies compounds mining problems even at ambient temperatures (5 to 35 °C). Mine water can be corrosive, resulting in damage to mining equipment, support systems, ventilation and transport equipment, and mineral processing facilities both on the surface and underground. Corrosion prevention measures may have to be adopted or may need the use of special materials like stainless steel. Special procedures to deal with the problem, such as water neutralisation, protective dressing lubricants, paints, etc. may be needed. The presence of water influences the stability of openings depending upon the sensitivity of the rock. Floor heave and roof falls are common in mudstones and shale in underground coal mines even if only small quantities of water are present. The slope stability of high walls and spoil piles in open pit mines is primarily governed by the presence of pore pressure in these materials.

11.1 Hydrological cycle

The oceans of the world contain 97 % of the total water in the world. The evaporation of water from the oceans and the surface of the earth and its return back to earth (Fig. 11.1) is called the hydrological cycle and is measured over 12 months (November to October). Only a small part of the water evaporating from the ocean reaches the land mass and is returned back into the ocean through the river systems. Part of it evaporates straight into the atmosphere from the land surface or from vegetation. A certain amount of water seeps down to large depths and is stored. Over the history of the earth almost 23 % of the total fresh water (or 0.75 % of total water) has accumulated as ground water (9 % at depths of up to 800 m and 14 % at depths greater than 800 m).

The values of the various components of the hydrological cycle vary greatly from one region to another or from one locality to another. Some areas have very high rainfalls but most of it flows away into the river systems. Others have much less rain but may be prone to flooding of the surface, raising its moisture level and allowing greater permeation into the deeper levels of the land mass.

The part of the hydrological cycle that pertains to the precipitation on the land mass has basically three major components:

(1) infiltration into soil and percolation into deeper levels in the ground resulting in ground water storage;
(2) surface runoff and interflow from the soil into streams;
(3) evaporation of soil moisture and uptake by vegetation.

Fig. 11.1. Hydrological cycle and water balance.

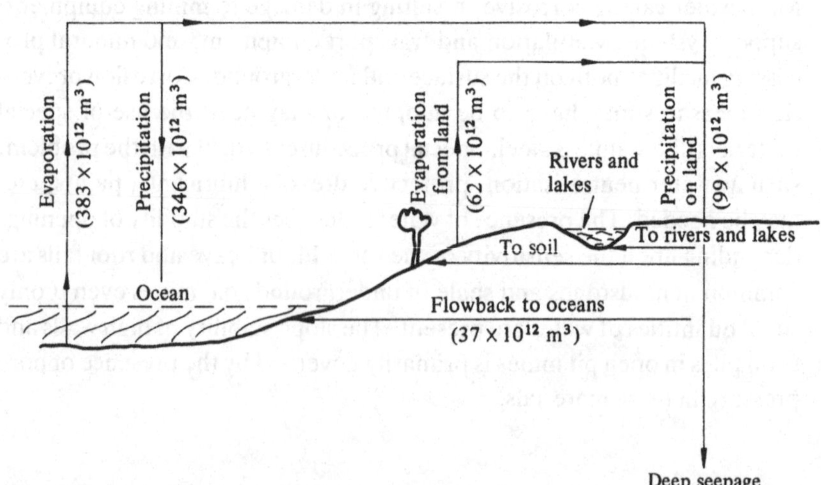

There is no clearcut line between the three components. Some of the ground water storage may evaporate and part of it may flow away slowly into streams particularly from soil and water exposures. The ground water may flow away from or into the area of precipitation depending upon the storage capacity of the rocks (soils) and the general geology of the area. The response times of the various processes of the land phase of the hydrological cycle vary widely. The time lag between the beginning of a storm and the start of the hydrological cycle increases exponentially with the movement of water related to the depth of infiltration (Fig. 11.2) and is related to the rate of precipitation.

The hydrological balance of the area is measured in terms of m^3 of water and is described by the equation

$$H = S + E + R - R' \tag{11.1}$$

where H = rainfall;
S = surface runoff;
E = evaporation;
R = retained water; and
R' = water released from the earlier retention.

Water loss, L, is given by

$$L = E + (R - R') \tag{11.2}$$

and

$$H = S + L. \tag{11.3}$$

Fig. 11.2. Idealised form of the land phase component of the hydrological cycle during a storm for a given rate of precipitation.

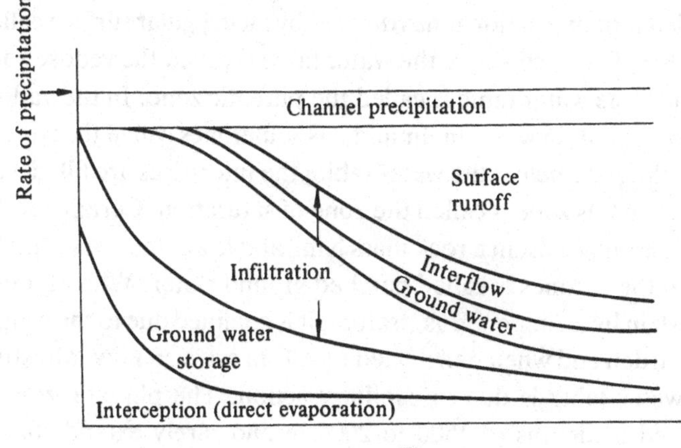

If observations are carried out over a long period of time, then

$$R = R'$$

and therefore

$$L = E$$

or

$$E = H - S.$$

A land mass has a positive water balance, which means greater precipitation than evaporation, and oceans have a negative water balance, which means greater evaporation than precipitation. These two imbalances are equalised by water flow from the rivers into the oceans.

Large-scale mining, requiring the pumping out of large quantities of retained water from the ground, upsets the hydrological balance and also changes the environment by changing the vegetation.

Mining operations are concerned with two types of water inflows: (1) surface precipitation or rain water and surface runoff; (2) underground or subsurface water. The estimation of surface precipitation is based upon long-term rainfall data available from the Department of Meteorology for the area. Information on surface channels and their hydraulic characteristics (stream flow hydrographs and hydraulic routing), soil moisture (saturated and unsaturated as a function of the annual precipitation cycle), evaporation and infiltration into underground rocks is required to estimate the make of water in an area where an open pit or an underground mine is to be opened.

11.1.1 Subsurface water

Fig. 11.3 gives a schematic cross-section of the earth's crust along a coast line with a classification of the subsurface water. The subsurface can be divided into two major zones divided by an irregular surface called the water table. The zone above the water table is called the vadose zone and that below the water table is called the phreatic zone. In the vadose zone, soil pores contain water or air and it is sometimes called the zone of aeration. In the zone below the water table, the interstices are filled with water and hence this zone is called the zone of saturation. Certain locally saturated zones may exist in a rock mass lying above another water table. The water in these zones is called perched ground water. When ground water is overlain by an impervious stratum, it is confined due to the weight of the overburden and when penetrated results in artesian flow. Moisture above the water table is due to capillary action. The phreatic zone is usually limited to depths of 1500 to 2000 m and rarely extends below

3000 m. At these depths, the porosity of rocks is so small that little water is found.

From the mining point of view, it is the water in the phreatic zone that is of great concern. Water in the vadose zone is usually very small and does not present any problems. Perched ground water in certain cases can, however, lead to a slope stability problem in open pit operations.

11.1.2 Presence of water in soils and rocks

Water may be present in soils and rocks as gravity water moving in the larger pores, cracks and joint systems, and as capillary water in the smaller pores, a hygroscopic moisture adhering in the form of a thin layer to grains. Hygroscopic moisture is held by molecular attraction and forms a very small portion of the total water content. It is normally not removable under usual climatic and drainage conditions. Both gravity and capillary water are of concern in mining situations.

Rock formations which contain water and can transmit it from one point to another are called aquifers and those which contain water but cannot transmit it rapidly are called aquicludes. A formation which has no interconnected pore system and cannot hold or transmit water is known as aquifuge.

Fig. 11.3. Schematic cross-section of the earth's crust along a coast line. (*a*) Without wells along the coast; (*b*) with wells along the coast. Y = the difference in elevation between the water table and sea level along the coast line.

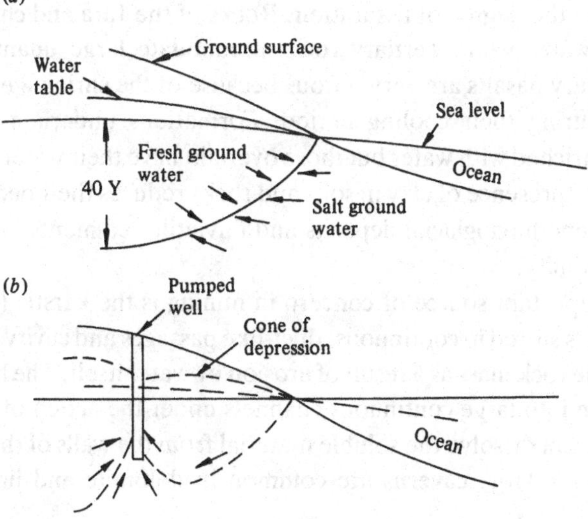

The water accumulation and transport capacity of a formation is dependent upon its porosity. Porosity is defined as the ratio of pore volume to the total volume and is of two types. Primary porosity or the original porosity is defined as that existing at the time of the rock's formation and consolidation. Secondary porosity is that resulting from tectonic stresses causing fracturing and solution channels opening out the flow paths. Well-sorted, coarse-grained, sedimentary deposits have a high primary porosity compared to poorly sorted, fine-grained deposits. Deposition of mineral matter within the pores and consolidation decreases porosity. Tectonic stresses resulting in the fracturing of rocks, weathering along joint planes and removal by solution results in high secondary porosity. Porosity determines both the water holding and the transmission capacity (permeability) of a rock formation.

Assessment of the hydrological conditions of an area where mines are to be opened requires information about the porosity and permeability of the rock as well as knowledge of the other water sources (open channels, ponds, lakes and seas) in the area. The zones of high and low permeability must be defined, along with the general geology of the area. For example, a permeable layer when lying in an anticlinal formation may bear little or no water into an excavation when intersected, but may deliver large quantities when lying in a synclinal formation under a water table. More water is expected to percolate through a disturbed zone when connected to a surface reservoir.

Volcanic rocks have very low primary porosity, but cracks and joints and bedding planes are the main channels of movement of water in these zones. Their permeability therefore will be highly directional and will depend upon the degree of fissuration. Rocks of the Jura and chalk eras have little water while Tertiary rocks accumulate large quantities of water. Tertiary basalts are very porous because of the shrinkage fissures produced during their cooling period. Formations underlain by clay layers are enriched with water but those overlain have their water content reduced. The presence of clay in soils and rocks reduces their permeability. Pleistocene fluvioglacial deposits and fluviatile sediments have very high permeability.

A very important source of concern in mining is the karstic (cavern) water which is stored in continuous, duct-like passages and cavity systems formed in the rock mass as a result of erosion by water itself. The hair-like cracks widen into large continuous channels under the action of flowing water which can dissolve the soluble material from the walls of the crack. For this reason large caverns are common in dolomite and limestone mountains.

11.1.3 Quality of water

The quality of water is expressed in terms of the quantity and types of salts dissolved in it. The importance of these factors is owing to the following reasons:

(1) damage to the drainage plant in terms of the corrosion of the equipment;
(2) harmful effects on the mining equipment used for winning and transportation, etc.;
(3) damage to the support systems used in mines;
(4) environmental aspects of pumping large quantities out into the general drainage system of the area.

Acidic water formed as a result of oxidation of sulphide ores or from peaty areas or rocks containing gypsum are highly corrosive. Clays containing pyrites when exposed to atmospheric moisture give rise to the same problem. Concrete tunnel linings and steel support systems become corroded under the action of these waters. Carbonic acid and particularly free carbon dioxide can leach cement from the concrete and have the capacity to dissolve limestone and dolomite. Sulphates such as sodium sulphate (Na_2SO_4), gypsum and anhydrite ($CaSO_4$) and Epsom salts ($MgSO_4$) are highly detrimental to support systems. The sodium chloride present in sea water is not as destructive as calcium and magnesium salts.

In general, corrosion increases with decrease in pH value and whenever the pH drops to below 6.5, an investigation should be carried out.

11.1.4 Sources of ground water

Almost all ground water is of meteoric origin derived from precipitation at one time or another. Water present in the rocks during their formation and trapped in them (connate water) is frequently highly saline. Juvenile water formed chemically within the earth's crust is brought to the surface along with intrusive rocks. Both connate and juvenile water are of no great consequence in mining operations except that these waters do sometimes contain undesirable minerals.

Water precipitation reaches the ground water by infiltration and percolation and recharges the existing ground water in areas where the water table is high. Highly permeable soils and cavernous rocks lead to a high recharge rate. In areas where the water table is very deep (tens of metres), little or no charge can be expected due to direct percolation. In such areas, seepage from lakes and streams is the only source of deep water recharge. Deep-seated areas may be recharged by underground flow through highly porous aquifers (e.g. coarse gravels, etc.).

The total quantity of water that is capable of percolation into the ground water storage system is usually limited to 10 to 30 % of the total precipitation. Various factors influence it. Slow and consistent precipitation spread around the year, flat topography, dense vegetation, high porosity of the soil, low temperature and high humidity lead to a higher percolation rate.

Ground water recharge may also take place when the mine site is close to the sea or lakes. Salt sea water has a density of about 1.025 g/cm^3 and therefore an imbalance occurs in the flow of water and in the cone of depression (Fig. 11.3) when a pumped well is established close to the sea. An inverted cone of salt water rises into the fresh water. The inverted cone rises up at a rate of 40 m/m of drawdown of fresh water.

11.2 Movement of ground water

The movement of ground water in soils and rocks can be fairly accurately represented by Darcy's law:

$$Q = Aki \tag{11.4}$$

where Q = volume flow rate;
A = cross-sectional area through which the flow takes place;
i = hydraulic gradient; and
k = coefficient of permeability (hydraulic conductivity) having the same units as velocity.

11.2.1 Measurement of permeability

The permeability of the formation may be estimated either by laboratory measurements or by field measurements. Laboratory measurements can be used only when fairly homogeneous material (soils) are present. In general, laboratory values are lower than field values.

Laboratory method

For coarse-grained soils, the constant head permeability test (Fig. 11.4*a*) is used. The specimen is contained in a perspex cylinder of cross-sectional area A. A steady vertical flow of water, under a constant total head, is maintained through the specimen and the volume flow rate of water, Q, is measured. The hydraulic gradient is determined from the tappings from the cylinder. From Darcy's law,

$$k = \frac{Q}{A} \times \frac{l}{h}. \tag{11.5}$$

Prior to running the test, a vacuum is applied to the specimen to obtain full saturation.

For fine-grained soils, the falling head test (Fig. 11.4*b*) is used. A standpipe is connected to the top of the cylinder. The water permeates into a reservoir of constant level. The standpipe is filled with water and a measurement is made of the time, t_1, for the water level to fall from h_0 to h_1. At any intermediate time t the water level in the standpipe is h and its rate of change $-(\mathrm{d}h/\mathrm{d}t)$. At time t the difference in total head between the top and bottom of the specimen is h. Then from Darcy's law,

$$-a\frac{\mathrm{d}h}{\mathrm{d}t} = Ak\frac{h}{l} \tag{11.6}$$

where a = internal area of standpipe.

$$\therefore -a\int_{h_0}^{h_1}\frac{\mathrm{d}h}{h} = \frac{Ak}{l}\int_0^{t_1}\mathrm{d}t$$

$$\therefore k = \frac{al}{At_1}\ln\frac{h_0}{h_1}.$$

Fig. 11.4. Laboratory permeability tests for soils: (*a*) constant head test for coarse-grained soils; (*b*) falling head test for fine-grained soils. A = cross-sectional area of perspex cylinder; Q = volume flow rate of water; h = head of water; a = internal area of the standpipe; l = length of sample

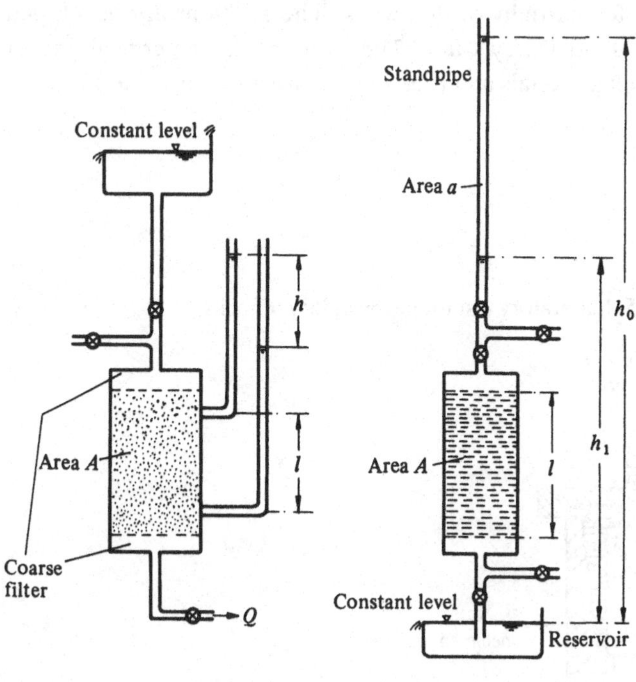

In the case of rocks, pressure gradients of up to 1 in 1000 are used in the laboratory, whereas gradients of only 1 in 10 are usual *in situ* to obtain a measurable rate of percolation. A typical specimen is a cylinder of 60 mm in diameter and a length varying up to 150 mm. The rock specimen is encapsulated in epoxy resin, with the object of preventing leakage along the external cylindrical face of the specimen. The apparatus is shown in Fig. 11.5. This test cannot be applied to very impervious rocks, the permeability limit being about 10^{-8} cm/s (Lama and Vutukuri, 1978).

Field method

Earlier field methods made use of fluorescein dye, which was thrown into the aquifer at the upstream side and the time taken for it to move to the downstream side estimated. Fluorescein can be detected with the naked eye at concentrations of 0.02 ppm, or 0.002 ppm under ultraviolet light.

Modern methods make use of pumping tests using the principles of well hydraulics. The method consists of pumping water from a well at a constant rate and determining the change in water level at various observation holes placed at varying distances from the pumped well. Table 11.1 gives the test configuration for the pump well and observation holes into the aquifer, for a well drilled to the full depth of the aquifer (ideal wells) or for partially drilled wells. The relationships developed here are based upon Darcy's law. The coefficients of permeability of various soils, rock materials and rock masses are given in Table 11.2.

Fig. 11.5. Laboratory test for permeability of rocks.

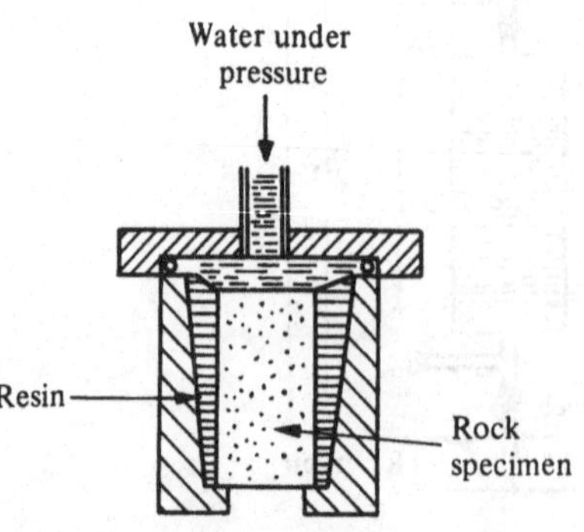

Table 11.1. *(a) Calculations for determining the coefficient of permeability for aquifers not in contact with surface waters*

Illustration	Equation	Remarks
	(*a*) Holes drilled to the full depth of the aquifer (ideal well)	k = coefficient of permeability or hydraulic conductivity, m/s Q = pumping rate, m^3/s, dimensions in figures are in m
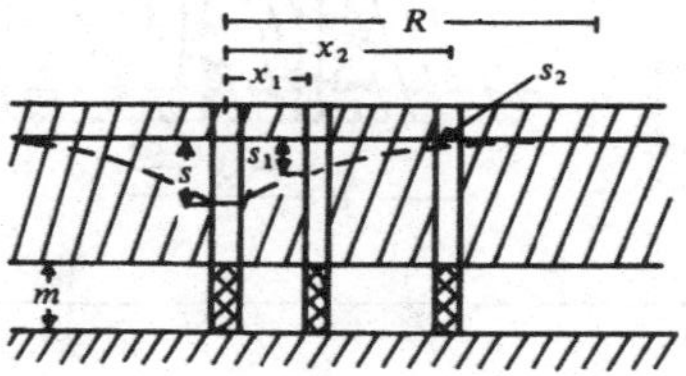	$k = \frac{0.366Q \log \frac{x_2}{x_1}}{m(s_1 - s_2)}$	Two observation holes
	$k = \frac{0.366Q \log \frac{x_1}{r}}{m(s_1 - s_2)}$	One observation hole
	$k = \frac{0.366Q \log \frac{R}{r}}{ms}$	No observation hole
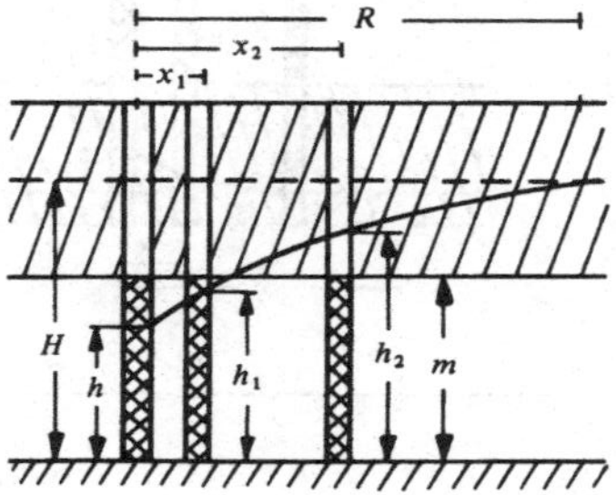	$k = \frac{0.733Q \log \frac{x_2}{x_1}}{(2h_2 - m)m - h_1^2}$	Two observation holes where the water table drops below the aquifer surface during tests
	$k = \frac{0.733Q \log \frac{R}{r}}{(2H - m)m - h^2}$	No observation hole where the water table drops below the aquifer surface during tests

(Table continued)

(*b*) Holes not drilled to the full depth of the aquifer

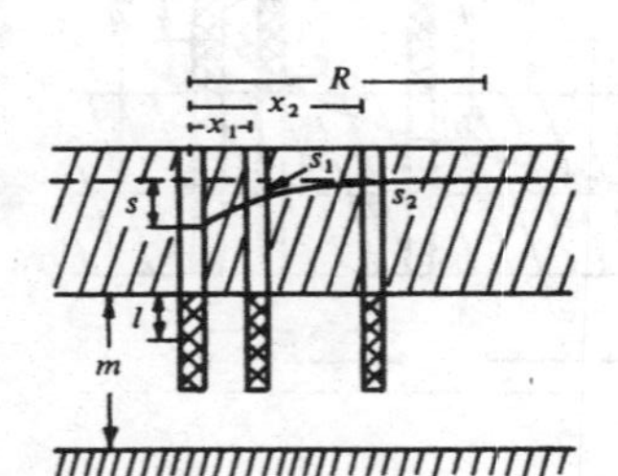	$k = \dfrac{0.366Q \log \dfrac{x_2}{x_1}}{m(s_1 - s_2)} \times \dfrac{1}{b}$	Two observation holes
	$k = \dfrac{0.366Q \log \dfrac{x_1}{r}}{m(s - s_1)} \times \dfrac{l}{b}$	One observation hole
	$k = \dfrac{0.366Q \log \dfrac{R}{r}}{ms} \times \dfrac{1}{b}$	No observation hole $b = \sqrt{\dfrac{l}{m}} \times \sqrt[4]{\dfrac{2m - l}{m}}$
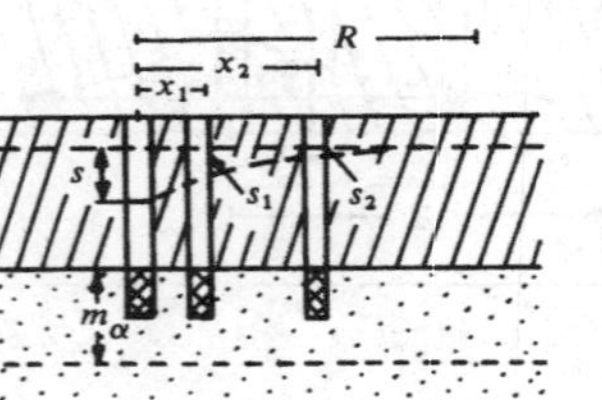	$k = \dfrac{0.366Q \log \dfrac{x_2}{x_1}}{m_\alpha(s_2 - s_1)} \times \dfrac{1}{b}$	Two observation holes
	$k = \dfrac{0.366Q \log \dfrac{R}{r}}{m_\alpha s} \times \dfrac{1}{b}$	No observation hole $b = \sqrt{\dfrac{l}{m_\alpha}} \times \sqrt[4]{\dfrac{2m_\alpha - l}{m_\alpha}}$

$\dfrac{s}{s+l}$	0.2	0.3	0.5	0.8	1.0
$\dfrac{m_\alpha}{s+l}$	1.3	1.5	1.7	1.85	2.0

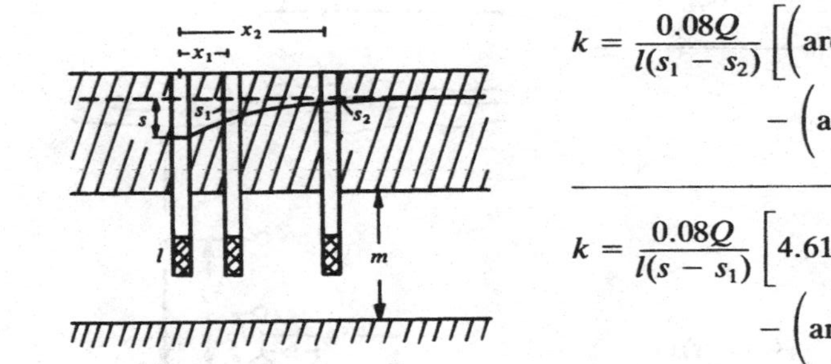

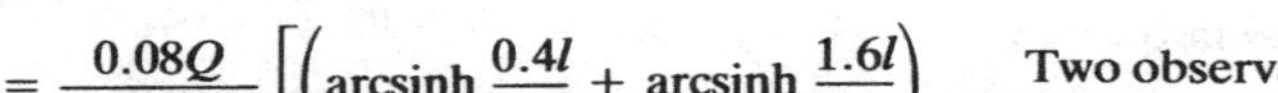

	$$k = \frac{0.08Q}{l(s_1 - s_2)}\left[\left(\operatorname{arcsinh}\frac{0.4l}{x_1} + \operatorname{arcsinh}\frac{1.6l}{x_1}\right) - \left(\operatorname{arcsinh}\frac{0.4l}{x_2} + \operatorname{arcsinh}\frac{1.6l}{x_2}\right)\right]$$	Two observation holes $l < 0.3m$ $x_2 < 0.3m$
	$$k = \frac{0.08Q}{l(s - s_1)}\left[4.61 \log\frac{1.6l}{r} - \left(\operatorname{arcsinh}\frac{0.4l}{x_1} + \operatorname{arcsinh}\frac{1.6l}{x_1}\right)\right]$$	One observation hole $l < 0.3m$ $x_1 < 0.3m$

Table 11.1. *(b) Calculations for determining the coefficient of permeability for aquifers in contact with surface waters*

Illustration	Equation	Remarks
	(*a*) Holes drilled to the full depth of the aquifer (ideal wells)	
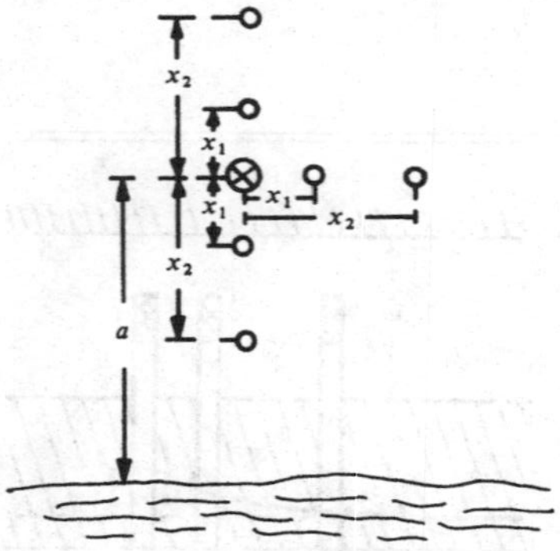	$k = \dfrac{0.366Q}{m(s_1 - s_2)} \log \dfrac{x_2(2a \pm x_1)}{x_1(2a \pm x_2)}$	Two observation holes placed perpendicular to the bank of the surface water '−' when the observation hole is between the pumping well and the surface water '+' when away from the surface water
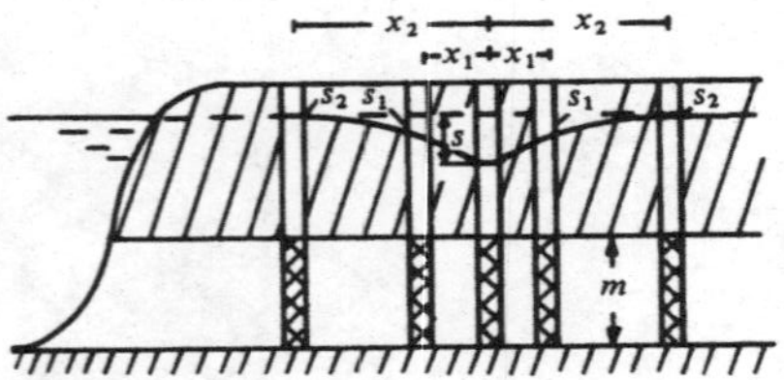	$k = \dfrac{0.366Q}{ms_1} \log \dfrac{2a \pm x_1}{x_1}$	One observation hole '−' when the observation hole is between the pumping well and the surface water '+' when away from the surface water

<table>
<tr><td rowspan="3">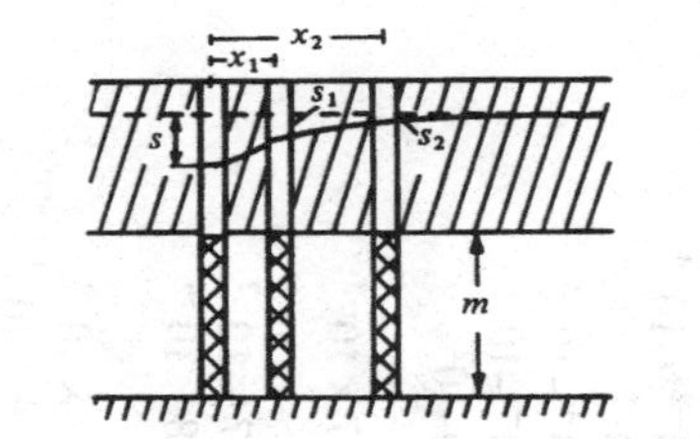</td><td>$$k = \frac{0.366Q}{m(s_1 - s_2)}\left(0.5 \log \frac{4a^2 + x_1^2}{4a^2 + x_2^2} + \log \frac{x_2}{x_1}\right)$$</td><td>Two observation holes placed parallel to the bank of the surface water</td></tr>
<tr><td>$$k = \frac{0.366Q}{ms} \log \frac{\sqrt{4a^2 + x_1^2}}{x_1}$$</td><td>One observation hole placed parallel to the bank of the surface water</td></tr>
<tr><td>$$k = \frac{0.366Q \log \frac{2a}{r}}{ms}$$</td><td>No observation hole</td></tr>
<tr><td></td><td>(b) Holes not drilled to the full depth of the aquifer</td><td></td></tr>
<tr><td rowspan="3"></td><td>$$k = \frac{0.16Q}{ls_1}\left(\operatorname{arcsinh} \frac{l}{x_1} - \operatorname{arcsinh} \frac{l}{2a \pm x_1}\right)$$</td><td>One observation hole placed perpendicular to the bank of the surface water
$a > m$
$x_1 < 0.3m$</td></tr>
<tr><td>$$k = \frac{0.16Q}{ls}\left(\operatorname{arcsinh} \frac{l}{x_1} - \operatorname{arcsinh} \frac{l}{\sqrt{4a^2 + x_1^2}}\right)$$</td><td>One observation hole parallel to the bank of the surface water
'−' when the observation hole is between the pumping well and the surface water
'+' when away from the surface water
$x_1 < 0.3m$
$a > m$</td></tr>
<tr><td>$$k = \frac{0.16Q}{ls}\left(2.3 \log \frac{1.32l}{r} - \frac{l}{2a}\right)$$</td><td>No observation hole</td></tr>
</table>

Table 11.2. *Coefficients of permeability (to water) of various soils, rock materials and rock masses (after Serafim, 1968; Loofbourow, 1973; and Sharp et al., 1977)*

	Permeability, k (cm/s)
Soils (intact permeability)	
Gravel	$1-10^2$
Clean sands	10^{-3}–1
Clayey sands	10^{-6}–10^{-3}
Clays	10^{-9}–10^{-6}
Rock material (laboratory determination)	
Sandstone (Cretaceous Flysch)	10^{-8} to 10^{-10}
Siltstone (Cretaceous Flysch)	10^{-8} to 10^{-9}
Granite	5×10^{-11} to 2×10^{-10}
Slate	7×10^{-11} to 1.6×10^{-10}
Breccia	4.6×10^{-10}
Calcite	7×10^{-10} to 9.3×10^{-8}
Limestone	7×10^{-10} to 1.2×10^{-7}
Dolomite	4.6×10^{-9} to 1.2×10^{-8}
Sandstone	1.6×10^{-7} to 1.2×10^{-5}
Hard mudstone	6×10^{-7} to 2×10^{-6}
Black schists (fissured)	10^{-4} to 3×10^{-4}
Fine-grained sandstone	2×10^{-7}
Oolitic rock	1.3×10^{-6}
Bradfort sandstone	2.2×10^{-5} to 6×10^{-7}
Glenrose sandstone	1.5×10^{-3} to 1.3×10^{-4}
Altered granite	0.6 to 1.5×10^{-5}
Limestone, 2 % porosity	8.5×10^{-8}
Limestone, 16 % porosity	10^{-4}
Silty sandstone	2×10^{-6}
Sandstone, 29 % porosity	2×10^{-3}
Granite	5×10^{-9}
Slate	10^{-10}
Rock mass (in-situ determinations)	
Arterite migmatites	3.3×10^{-3}
Chloritised arteries and shales	0.7×10^{-2}
Gneiss	1.2×10^{-3} to 1.9×10^{-3}
Pegmatoid granite	0.6×10^{-3}
Lignite layer	1.7×10^{-2} to 23.9×10^{-2}
Sandstone	10^{-2}
Mudstone	10^{-4}
Oocene limestone	10^{-2} to 10^{-4}
With clay filled joints	10^{-5}
Moderately jointed	10^{-4}–10^{-2}
Well jointed	10^{-2}–10
Heavily fractured	10–10^2
Clay shale or dense rock with tight fractures, considered impermeable in most excavations	9.7×10^{-8}
Dense rock, few tight fractures, approximate lower limit for oil production	9.7×10^{-7}
Dense rock, 0.015 cm fracture each 0.1 m^2	4.8×10^{-4}
Silt or clay, silt, fine sand. Few water wells in less permeable ground	9.7×10^{-4}
Dense rock with high fracture permeability	19.4×10^{-4}
Clean sand, medium and coarse (0.25 and 1.0 mm)	0.48
Clean gravel (70 % larger than 2.0 mm)	1.2

Testing for permeability of joints

In most rock masses, ground water flow occurs largely through the joints. The test for conductivity of a joint is shown in Fig. 11.6 (Lama and Vutukuri, 1978). This test is known as the packer test. The volume flow rate of water, Q, that filtrates from A to B along the joint aa′, considered of uniform height e, is given by:

$$Q = \frac{\pi e^2}{12\eta} \times \frac{p_1 - p_2}{\ln \dfrac{d}{r_0}} \qquad (11.7)$$

where d is the distance between points A and B in Fig. 11.6, r_0 is the radius of the bore hole, η is the viscosity of water and p_1 and p_2 are the water pressures at the joint level.

11.2.2 Forecasting of inflow of water in mines

The forecasting of inflow of water in mines is a very complex problem and no general method is available which can be used in all situations. Rational forecasting methods, however, must take into consideration both manmade conditions and natural geological environments. The sources that deliver water into mines are as follows.

(1) Flow of water into the mine excavations (shafts, drifts and other entries) driven through water-bearing formations in virgin areas.

(2) Sudden inrush of water from seas, lakes, rivers, swamps, dams, etc. when these are suddenly penetrated as a result of mining subsidence.

Fig. 11.6. Conductivity test of joint aa′. p_1, p_2 = pressure of water at the joint level; d = distance over which water flow rate is measured; e = height of joint.

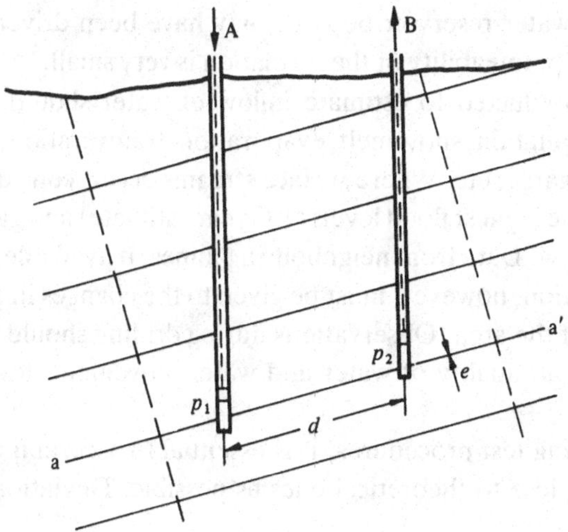

(3) Inrush of karstic water from cavernous rocks during drivage of excavations after blasting.
(4) Penetration of barriers between excavations and old workings of the same mine or a neighbouring underground or open pit mine.
(5) Penetration of certain geological structures such as fault zones especially when they intersect reservoirs placed on the rise side of mine excavations.
(6) Breakage of manmade conduits.
(7) Insufficiently drained hydraulic fill areas.
(8) Inrush of water from overlying or underlying waterlogged worked seams or waterlogged beds as a result of fractures penetrating these areas.
(9) Surface water from exploration bore holes.
(10) Water brought into the mine for use underground, e.g. drilling, dust control, hydraulic fill, etc.

It is therefore important to collect information on all the possible sources of water in and surrounding the mine holdings and mark them clearly on the plan. The maximum rate of delivery and the water reservoir capacity of these sources should be determined along with other data as to the extent of interconnection between them and other sources placed far away from the mine property. Monitoring the water reservoir capacities of the sources in cases of large fluctuations is essential.

Estimation of the water reservoir capacity of the various sources depends upon the geology (simple or very complex), hydrological data (recharge, discharge, storage and permeability of the structure) and the effect of the excavations on the surrounding rock mass. Hydrological investigations must be conducted very carefully, since results may be quite deceptive. A test hole drilled for hydrological investigations may not intersect the water reservoir body or may have been driven in the direction in which permeability of the formation is very small.

Observations conducted to estimate inflow of water should include topography, precipitation, snow melt, evaporation–transpiration, runoff, recharge and discharge, etc. Where surface streams occur, consideration must be given to the highest flood levels (100 year estimate) and gains and losses in stream flow. Data from neighbouring mines, if available, is very useful. Consideration, however, must be given to the changes in geology and topography of the area. Observations during drilling should include the temperature and quality of water and water movement, levels and pressures.

While establishing test procedures, it is essential to ascertain that the conditions are as close to theoretical ones as possible. Deviation of the observations from theoretical calculations and the dispersion of data may mean that either conditions are changing, input information is not correct or the theory is not applicable.

Investigations must be undertaken at the earliest stages of exploration. It is only possible at this stage to undertake a study of alternative mining methods or prepare plans to combat high water inflow.

Space does not permit the description of the various techniques applied to estimate water inflow. In estimation, both hydrologists and hydrogeologists must be included. The various steps that are required to make an estimate are as follows.

(1) Delineate expected waterlogged areas.
(2) Estimate mean and maximum water reservoir capacity under various conditions, keeping in view precipitation, runoff, evaporation and transpiration.
(3) Estimate what part of the surface runoff can be directed away from the mining area so that it does not enter the mine under changed surface topography due to surface subsidence or surface cracking.
(4) Estimate changes in the water inflow pattern as a result of changes in mine extension and depths. It is important to keep in mind that flow in extensively worked mines will not only be radial and horizontal but also vertical in direction.

Inflow of water into an excavation can be calculated using Darcy's law, assuming an arithmetic mean value for the coefficient of permeability. In calculating water inflow it is essential that the boundary conditions of each of the mining excavations placed in or close to the aquifer region are established accurately. Flow into the mine excavation may be laminar, turbulent or both and these flow conditions must be established. With time the location of working places changes. Therefore calculations must be done for the various stages of the mine both in space and time. This method permits establishment of the drainage capacity of the mine pumping plant.

It is essential that the information on water make is updated on a regular basis, predictions checked and appropriate corrections incorporated into the drainage plant requirements of the mine. It must be remembered that taking precautions to avoid inflows of water into the mine, particularly sudden inrushes, is the cheapest way to deal with it.

11.2.3 Instrumentation in ground water control

The most commonly used instruments in ground water control are piezometers, which can be classified as follows: open tube, closed tube, mechanical diaphragm and electrical. The main factor to be considered in selecting a piezometer for a particular installation is the time lag, i.e. the time taken for the pressure in the system to reach equilibrium after a pressure change.

An open-tube or standpipe piezometer consists of a perforated tip for water to enter or leave which is sealed into a section of a bore hole, as shown in Fig. 11.7. The standpipe passing through the seals allows the water level (which is a measure of the pressure at the bottom of the tube) to be measured by means of a water level indicator (Fig. 11.8). The response time of this piezometer installation is sufficient for most applications. Since the standpipe diameter is small, a number of them can be

Fig. 11.7. Typical standpipe piezometer installation. Variations should be made to suit local conditions and to utilise local materials (after Hoek and Bray, 1974).

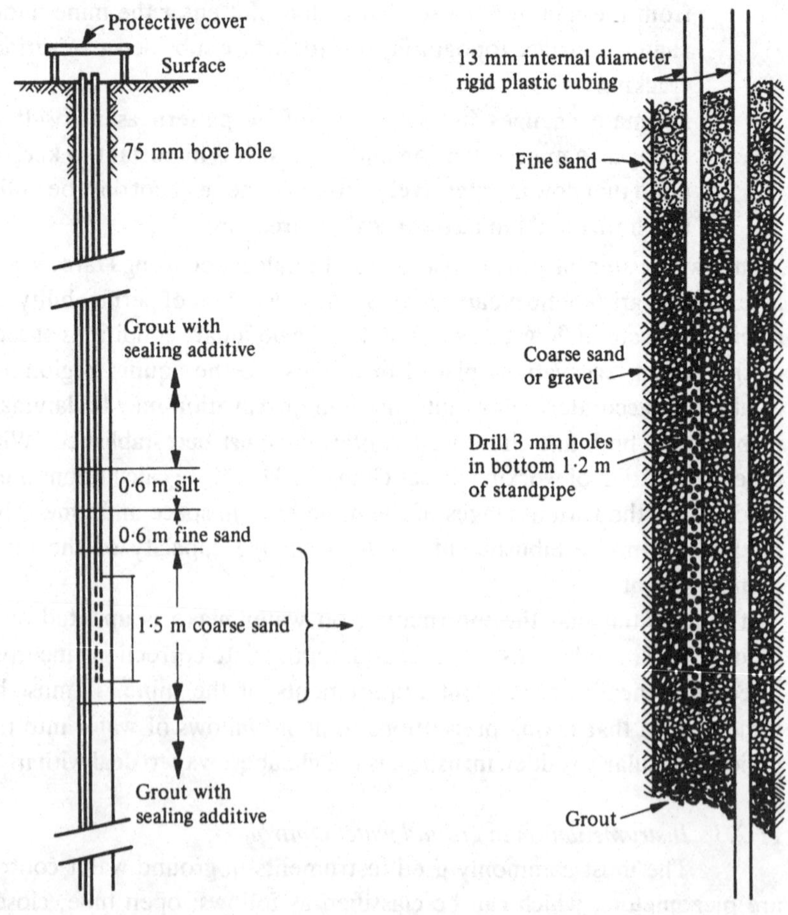

installed in the same hole to determine water pressures in different sections of the bore hole. This advantage is useful when water flow is confined to certain layers within a rock mass.

When the time lag requirement is short, the other types of piezometers are necessary. A closed-tube piezometer is filled with de-aired water, resulting in an improved time lag, and is suitable for the measurement of small water pressures. The principal characteristics of this type of piezometer are shown in Fig. 11.9.

The mechanical diaphragm piezometers are of two types, one utilising air and the other oil (Fig. 11.10). The water pressure is measured by a balancing air or oil pressure acting against a diaphragm. These piezometers respond very quickly.

In electrical piezometers, electrical strain gauges attached to the diaphragm are used to measure the deflection of the diaphragm due to water pressure (Fig. 11.11). These piezometers respond almost instantaneously and are ideal for measuring the water pressure within the test cavity during a pumping test.

11.2.4 Ground water flow analysis

Ground water flow can be represented by a pattern of flow lines and equipotentials called a flow net (Fig. 11.12). The potential is defined

Fig. 11.8. A simple probe for water level detection (after Sharp *et al.*, 1977).

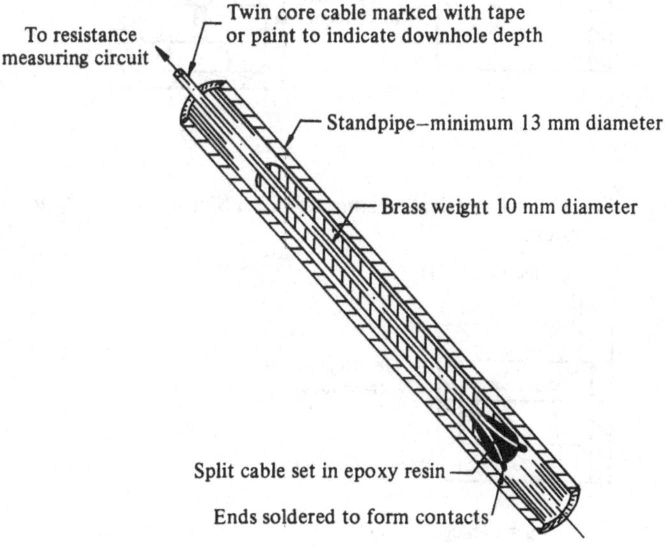

To resistance measuring circuit

Twin core cable marked with tape or paint to indicate downhole depth

Standpipe–minimum 13 mm diameter

Brass weight 10 mm diameter

Split cable set in epoxy resin

Ends soldered to form contacts

Fig. 11.9. Closed-tube piezometer (after Sharp *et al.*, 1977). u = pressure in the system; γ_m = unit weight of mercury; γ_w = unit weight of water; H_m = height of mercury; H_w = height of water.

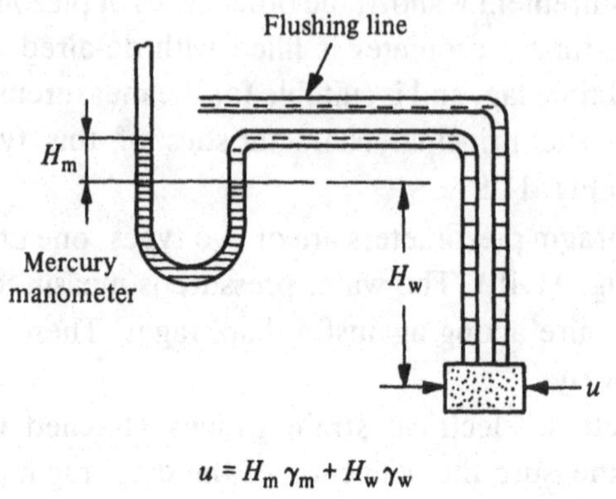

Fig. 11.10. Mechanical diaphragm piezometers (after Sharp *et al.*, 1977). (*a*) Water pressure against the diaphragm seals off the outlet. P is increased until the valve opens, then $u = P$. (*b*) Water pressure against the diaphragm opens the ball check valve. P is increased until the valve closes, then $u = P$; u = pressure in the system.

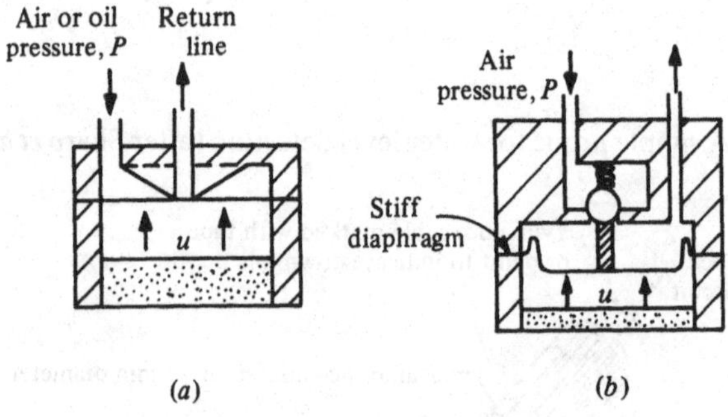

Fig. 11.11. Electrical piezometer (after Sharp *et al.*, 1977). u = pressure in system.

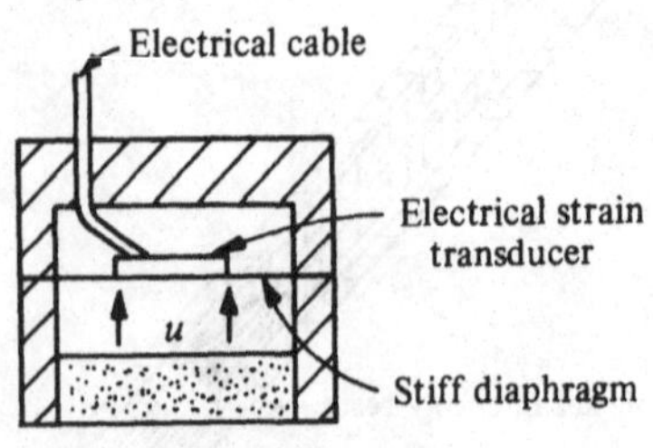

as the elevation at a given point plus the pressure expressed as head of water and influences ground water flow. Ground water moves from high to low potential. There is no flow along equipotentials.

For the construction of a flow net the following methods can be used: (1) graphical sketching; (2) electrical resistance analogues; (3) numerical analysis by computer.

Sharp, Ley and Sage (1977) have given details about these methods. First, the boundary conditions are defined. These are the limits to which flow occurs and are represented by the phreatic surface and the physical boundaries of the area being considered. The position of the phreatic surface is normally estimated from piezometric measurements.

To estimate ground water flow, the flow net parameters are used in the following equation:

$$Q = k \times h_L \times \frac{N_f}{N_e} \tag{11.8}$$

where Q = flow rate;
k = coefficient of permeability;
h_L = total head difference or head loss;
N_f = number of flow paths; and
N_e = number of equipotential drops.

Fig. 11.12. Flow net for seepage through a slope (after Sharp *et al.*, 1977). Hydraulic potential at point X = elevation of X + pressure head ($Z_X + h_p$). h_p is determined from the intersection of equipotential through point X and ground water table.

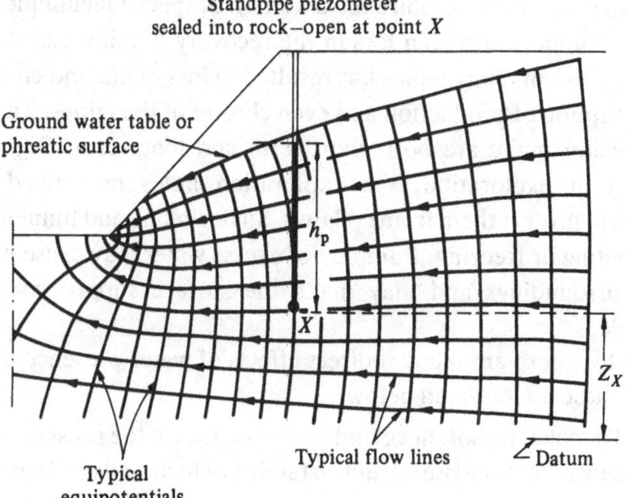

11.2.5 Use of tracers

The direction of movement of subsurface water can be best indicated using tracers which can be easily introduced into the water at one point and monitored at other points. The properties of a good tracer are that it should be easy to recognise in dilution, is non-toxic, should not be adsorbed by various rocks and bacteria, and has a high solubility. The most commonly used tracers are dyes, salts and radioactive isotopes. The most common dye tracer is fluorescein. It is easily detectable by eye (0.02 ppm), and under ultraviolet light can be detected at a concentration of 0.002 ppm.

Many salts are easily detected either by chemical methods or by electric potential (dextrose, sodium chloride). Some of them, however, can be adsorbed by rocks and attacked by bacteria.

Medium-life radioactive isotopes can be detected in very small concentrations (10^{-18} mol), depending upon background interference. The equipment used is a Geiger counter which is easy to operate and objective. Certain isotopes are adsorbed by ion-exchange reaction and permission must be obtained from the appropriate authority before use.

Tracer monitoring can also help establish quantitatively the movement of water, dilution, recirculation of water and establishment of preferred channels of movement.

11.3 Effects of water in mines

The effects of the presence of water in mines is twofold. Firstly, it adds direct costs to mining as a result of pumping requirements. In certain mines the cost of pumping may be a principal item. The presence of excessive water may require the adoption of special techniques during drivage of tunnels, or mean loss in the recovery of minerals. A sudden inrush of water in many mines has resulted in loss of life and equipment, the interruption of production and even closure of the mines. The effects of high water make are both short term and long term. Right at the beginning of exploration, the exploration holes may need special measures to protect them from collapse. Shaft sinking and tunnelling may need grouting or freezing. Pumping of excess water may cause pollution of the surroundings and may make the surface supply systems unacceptable.

Secondly, there are many indirect effects of water presence in mines. Some of these are outlined below.

(1) Damage to roof, floor and walls of mines in the presence of rocks sensitive to moisture, and instability of high walls and spoil piles.

(2) Harmful effects on the working conditions in mines, particularly in deep mines where the temperatures are high. It may require extraction of excessive moisture from the air and lowering the temperature.
(3) Uncomfortable conditions due to water dripping on the miners' faces, reducing the efficiency of workers.
(4) Higher costs of maintenance of equipment, higher wear and rust and lower life of equipment (ropes, belts, tyres, etc.).
(5) High water inflows result in the washing of fines from belts, chutes and storage bins, resulting in higher cleaning costs.
(6) Problems associated with explosives when charge holes get filled with water. Charge holes can collapse under the effect of water.
(7) High moisture content of the ore may require drying before shipping. It is serious in cold climates where it can freeze in ships, transport wagons or storage bunkers and require blasting.
(8) Pumping of water from the mine property may lead to excessive subsidence, resulting in changes in make in water wells placed near the mine property and bringing about changes in water quality which may be unacceptable to the urban authorities.

11.4 Control of water in mines

The shafts and excavations should be located in the best ground possible. Water at or near the surface should be diverted or drained. Shafts should be lined with impervious linings. The permeability of the rock mass should be reduced. Pumping should be used to dewater the bedrock if necessary.

11.4.1 Open pit mines

Surface water results in erosion of pit slopes, haulage roadways and drainage ditches, erosion fan deposition, softening of haulage roadways and frost heave and glaciation in winter. Subsurface water can result in reduction of soil and rock shear strength, instability of pit slopes requiring flatter slopes, increase in blasting costs, heave of pit slopes and floor bottoms and seepage of slopes resulting in erosion and glaciation.

The basic dewatering methods include drainage ditches at the surface (Fig. 11.13) and the bottom (Fig. 11.14) of the mine, horizontal drains (Fig. 11.15), vertical wells drilled from the surface (Fig. 11.16) or benches or pit bottom, dewatering shafts and galleries for pre-drainage and a combination of one or more of these methods.

Fig. 11.13. Dewatering by means of surface ditches (after Straskraba, 1979).

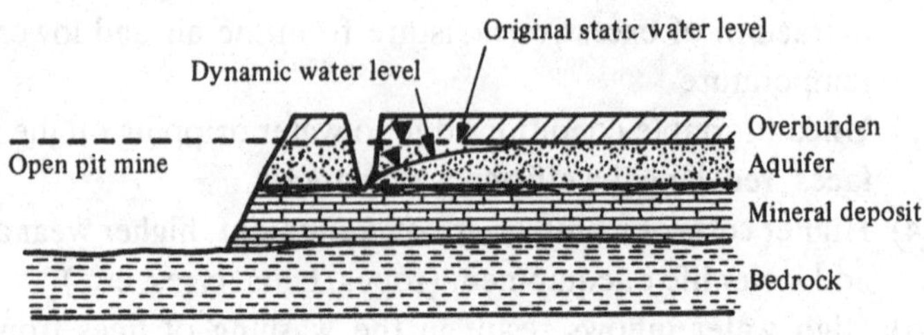

Fig. 11.14. Dewatering by means of bottom pit ditches (after Straskraba, 1979).

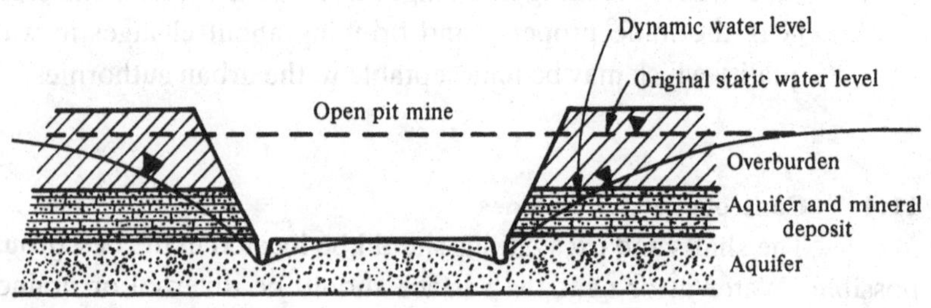

Fig. 11.15. Dewatering by means of horizontal drains (after Straskraba, 1979).

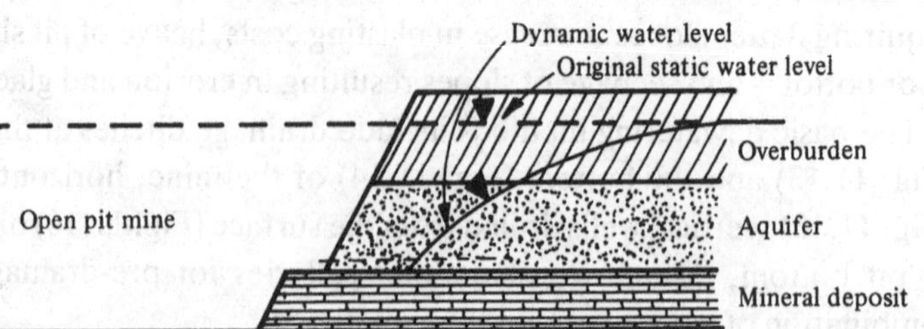

Surface ditches are applied to drain unconfined aquifers whereas bottom ditches are used to drain confined aquifers. Bottom ditches are more difficult to maintain than surface ditches.

Horizontal drains are used to dewater an aquifer affecting the slope stability. A typical horizontal drain consists of 40 to 50 mm of PVC slotted pipe installed inside drill holes at an angle of a 1 to 5° upslope. Installation and maintenance costs are low. There are no running costs. The main disadvantage is that they cannot be installed before completion of the excavation. Drilling of the holes in certain rocks may create problems requiring the use of special techniques.

Vertical wells drilled from the surface use pump, gravity or artesian flow discharge. The main advantages are the possibility of dewatering ahead of excavation, no interference with the mining operation and the

Fig. 11.16. Dewatering by means of vertical wells from the surface. (*a*) The depth of the aquifer is less than the depth of the open pit mine; (*b*) the depth of the aquifer is more than the depth of the open pit mine. (After Straskraba, 1979.)

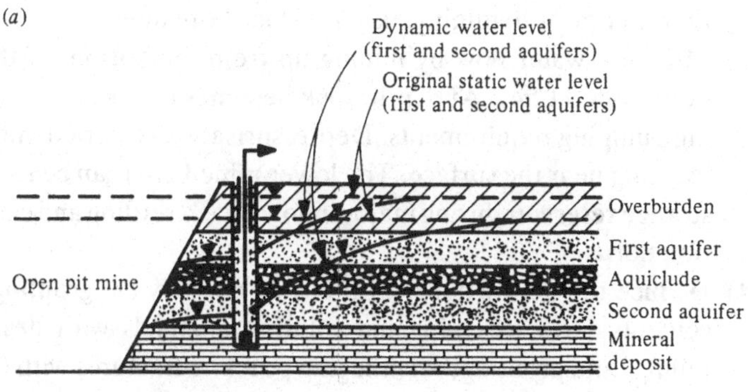

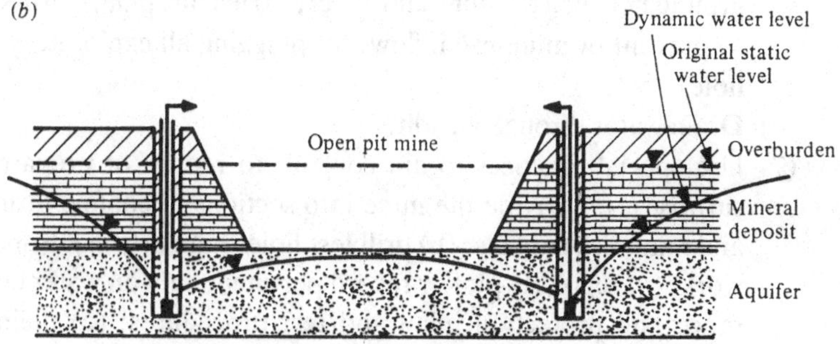

possibility of draining several aquifers with a single well. The main disadvantages are the cost of the energy, the limited drawdown, and the inability to dewater more impervious aquifers effectively.

Vertical wells drilled from the benches or pit bottom have the advantage of reduced pumping costs because of the lower head of pressure. The disadvantages include: (1) postponement of drilling until the site has been excavated; (2) more difficult access to drilling locations; and (3) interference with mining operations. To prevent water seepage from aquifers located below the pit floor, pressure relief wells can be drilled at the pit bottom.

Dewatering shafts and galleries are driven only in complicated hydrogeological conditions where other dewatering methods are not effective. Table 11.3 gives a summary of several examples (see also Figs. 11.17–11.23).

11.4.2 Underground mines

The most common approaches to controlling water in underground minds are to

(1) Divert or intercept surface water.
(2) Dewater prior to mining, using surface bore holes.
(3) Minimise water flow by mining up from the bottom of the ore body (Fig. 11.24). At depth, rock becomes less pervious, reducing pumping requirements. Depressurisation is carried out when working near the surface. The lower mined area can be used as a storage reservoir during major inflow or as a settling and clarification horizon.
(4) Reduce the permeability of the rock mass by (a) grouting with cement slurry; (b) plugging solution channels with desanded tailings and grouting; (c) plugging pores or fractures with clay.
(5) Protect the working areas from inflow by (a) leaving enough solid ground between the mine and water; (b) leaving pillars on fissures to prevent or minimise inflow; (c) plugging all exploratory drill holes.
(6) Drain water through an adit.
(7) Use special practices to aid control, to reduce or prepare for surprises. (a) Divide the mine into sections to confine inrushes and minimise damage; (b) drill test holes regularly in advance of work; (c) regularly plot or record pertinent data on water occurrence and protection, plan and test procedures; (d) maintain materials, tools and trained crew ready to carry out protective procedures.

Table 11.3. *Summary of open pit stabilisation by drainage (after Brawner, 1979)*

Location	Problem	Solution
Cyprus Anvil Mine Faro, Yukon	Surface inflow from Faro Creek causing erosion in summer and glaciation in winter. Subsurface seepage into the mine causing unstable pit slopes, increased blasting costs, glaciation on slopes, increased haul and maintenance, ice build-up in truck boxes and shovel buckets. Overly wet ore required stockpiling for drainage. Excess stockpile time lead to oxidation.	Diverted creek and placed water-proof lining along high permeability zones. 1830 m of horizontal drains were installed with 38 mm o.d. slotted plastic pipe. Average drawdown ranged from 2.1–4.3 m. Daily volumes of water average about 380 m^3. Collector sump in bedrock constructed with pumps and 15–20 cm polyethylene pipes used to remove water from pit. A steamer is used to thaw frozen pipelines.
Pine Point Mines Pine Point, NWT	Extensive seepage into numerous open pits causes difficult and hazardous mining conditions and high explosives costs. Seepage is from aquifers that continuously recharge from mountains to the south. Sinkholes are frequently encountered in the open pit areas.	Installation of deep wells (120–150 m) and pumps around the perimeter of the open pits. Design based on pumping tests and analysis methods of Cooper and Jacob and of Thiem. Well design allows for 10 % loss due to well collapse and 80 % effective pumping rate. Well locations selected away from sinkholes. Wells 37.5 cm diameter with no casing or screens at depth. Dewatering costs represent 16 % of direct mining costs. Volume is 22 700 m^3/day (Fig. 11.17).
Bougainville Copper Mine Papua and New Guinea	Very high rainfall region combined with severe earthquake potential would require expensive flat pit slopes to maintain stability unless the water pressures in the rock slopes could be minimised. The rock contained high fracture frequency with a majority of steeply dipping joints.	Development of drainage adits below the ore body developed an extensive zone of depressurisation near the adits. Flows increased when semi-impervious zones were intersected, i.e. clay-filled faults. Field permeability values ranged from 10^{-2} to 10^{-5} cm/s.

(*Table continued*)

Table 11.3—*Continued*

Location	Problem	Solution
Twin Buttes Mine Arizona	Two large slide zones developed in the pit over a vertical depth of some 180 m. The movement was largely influenced by high water pressures in the slope. These pressures varied due to numerous clay-filled faults and shears in the slope.	An underground adit 1040 m long was mined and 7770 m of drain holes were drilled from the adit to increase the effective drain radius. PVC slotted pipe was installed in the drain holes. The water pressure in monitoring piezometers dropped an average of 67 %. The overall permeability averaged 10^{-6} cm/s. If a risk factor of 12 % is acceptable the average slope angle can be increased from 27° to 35°.
Jeffrey Mine Canadian Johns Manville Quebec	Granular soils overlying the bedrock provided a major ground water supply to develop water pressures in the pit slopes. This led to a major slide involving about 20 000 000 t of rock. The slide intersected the ore skipway and came within 25 m of the primary crusher building. The instability was aggravated by leakage of water from town service mains and mining at the toe of the slope. Considerable water infiltrated the tension cracks.	Horizontal drains were installed from the skipway bridge area. Drain holes were drilled upward from a horizontal adit to intersect the failure zone. The adit was used for previous mining and exploration. Tension cracks were filled in. Horizontal drains were installed in the granular layers in the upper over-burden slopes to reduce recharge into the rock. Surface interception and diversion ditches were developed (Figs. 11.18 and 11.19).
Iron Ore Company Quebec	High precipitation, very cold winter climate, variable strength rock with considerable folding and faulting led to numerous slope stability problems. The high water table resulted in a high moisture content of the ore and very wet haulage access with high maintenance of roads and trucks.	Surface drainage was installed around the pit to control surface water. 47 in pit and perimeter wells (38 cm diameter) have been installed to lower the water table. Special precautions are required to protect the pipe lines from freezing. Problems still exist with operating delays due to blasting, relocation of pipe lines and inspection of pumps.

Konin and Turow Mines Poland	A shallow water table existed where 40 to 150 m of overburden required removal to expose the coal measures. Depressurisation of the overburden, coal and underlying strata was required to maintain slope stability and prevent pit bottom heave.	The zone proposed for initial mining was dewatered by deep wells and drainage galleries in the coal horizon. As mining progresses, wells are installed ahead of mining and from the galleries. These latter wells drain by gravity as pressure relief wells. Horizontal drains are installed from the pit slopes. Dewatering trenches and pumping stations are developed in the pit bottom. Pumping systems are designed for 120 % capacity (Fig. 11.20).
Rhenish Lignite Mining District West Germany	Water-bearing sand zones exist between multi-layered coal horizons. The lignite extends to a depth of over 600 m. Some major faults exist in the deposits which usually act as barriers to ground water flow. The topwall aquifers in the pit area must be completely dewatered and the footwall aquifers sufficiently depressurised to maintain stability and control heave. Continued monitoring of ground water is essential each year. About 25 000 m of monitoring holes are drilled. Local depressurisation of sand pockets is required ahead of the dredgers to control slope blowouts.	Ground water budgets to evaluate parameters, yields and boundary conditions are developed. Design diagrams based on Dupuit-Thiem and Sichardt are used for design of gravel-packed dewatering wells. One- and two-dimensional numerical aquifer models have been developed. Because of high transmissivities and large areal extent of the acquifer, gravity discharge into vertical tube wells is used with high capacity submersible motor pumps to lift the inflow. Well depths extend to 500 m. Drilling diameters range from 1.2–1.8 m with well screens and inner casings of 30–80 cm. Reverse circulation air injection drilling is used. Vacuum dewatering is used to lower the water table ahead of excavators in the slopes in areas of low hydraulic conductivity (Fig. 11.21).

(*Table continued*)

Table 11.3—*Continued*

Location	Problem	Solution
Morwell Open Cut Brown Coal Project Australia	Two water-bearing aquifers are located below a coal seam up to 120 m thick. The water pressure below the base of the pit must be reduced to control pit bottom heave and development of zone of tension near the toe of the pit batters. High water pressures in the slopes partially induced by surface watering to minimise fire potential must be reduced to reduce risk of batter movement. Water pressure reduction to improve stability has increased the effective vertical stress in the coal which is slightly compressible. This has resulted in differential settlement of up to 1.5 m near the open cut and some settlement up to 16 km distance.	Wells have been developed into both aquifers to reduce the water pressure to control pit bottom heave and horizontal movement at the toe of the batters. Piezometers have been installed as control installations. Control elevations have been established at each piezometer to assist in determining the required pumping rates. To minimise differential settlement near the open cut, depressurisation wells can be located to develop uniform levels of vertical effective stress. The stability of the batters has been maintained by extensive installation of horizontal drains up to 185 m long (Figs. 11.22 & 11.23).
Swift Agricultural Chemicals Phosphate Mine, Florida	The major problems to maximise phosphate recovery are pit slope stability, ground water control and influence on spoil volume to matrix recovery. In cohesive soil the existence of water pressure increases the potential for sliding. In cohesionless soil the slopes tend to flow. Surface water in the pit base can cover the base and make spotting of the bucket for efficient matrix recovery impossible.	Drilling and aquifer water pressure testing revealed that a deep aquifer in the Avon Park formation (approx. 210 m below surface) had a lower piezometric top elevation than a shallow Hawthorne aquifer immediately below the matrix to be excavated. Connector wells joining the surface sediments and Hawthorne to the deeper Avon Park formation have resulted in a vertical flow by gravity reducing the water levels in the upper formations some 13 m. This has allowed prestripping with scrapers, steeper pit slopes and increased matrix recovery.

Western Phosphate Field, Idaho	Water resources within the phosphate field exist in complex ground water and surface flow systems. The water has the potential to hamper mining operations by pit flooding and pit and waste dump stability. It is desirable to be able to predict in advance of mining the input of water on the mining programme.	Extensive use of stream flow gain–loss monitoring indicated that certain geological formations supported ground water flow systems while others did not. Structural geological features have a significant effect on the development of ground water and surface water flow systems. Major surface drainages are generally parallel to fold or fault structures. Geological sections for various combinations of formation dip, smooth or broken ridges, valley or ridge location were developed to allow a prediction of ground water systems at proposed mine sites.

Fig. 11.17. Typical well and ditch arrangement at the Pine Point open pit mine, Northwest Territories (after Durston, 1979).

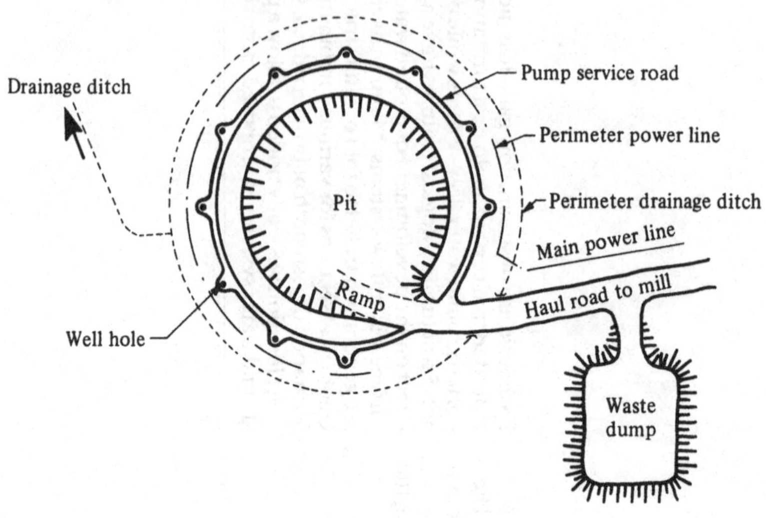

Fig. 11.18. Cross-section through axis of slide – Jeffrey mine, Quebec (after Sharp, 1979).

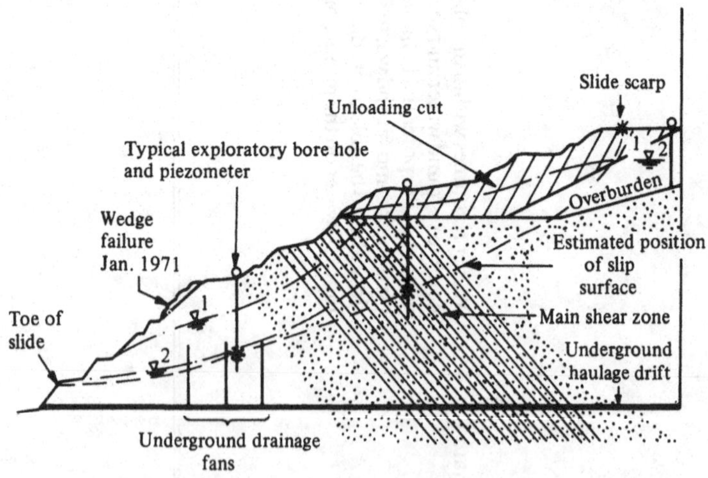

Fig. 11.19. Cross-section through an underground drift showing fanned drainage holes – Jeffrey mine, Quebec (after Sharp, 1979).

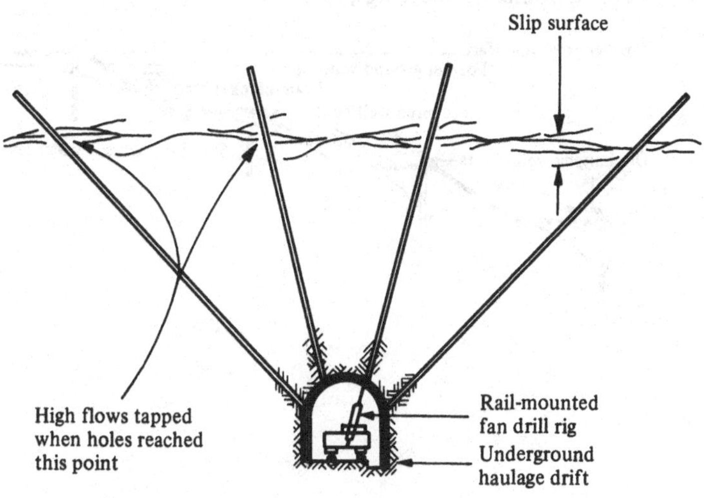

Fig. 11.20. Schematic diagram of the dewatering system at the Konin surface lignite mine, Poland (after Libicki, 1979). gwt = ground water table.

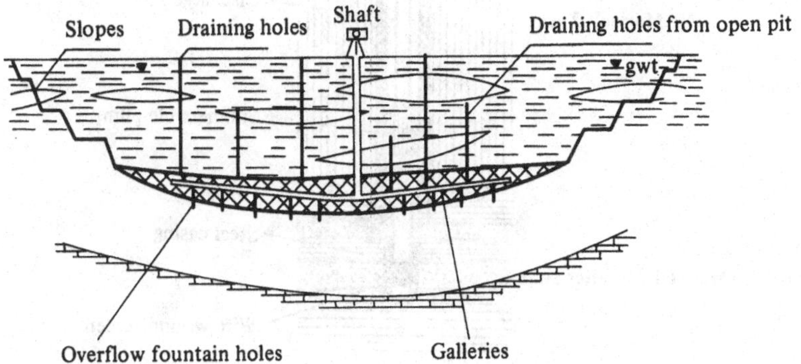

Fig. 11.21. Stabilisation of ground water level by vacuum dewatering measures at the Rhenish lignite mining district in West Germany (after Boehm, Schneider and Voigt, 1979).

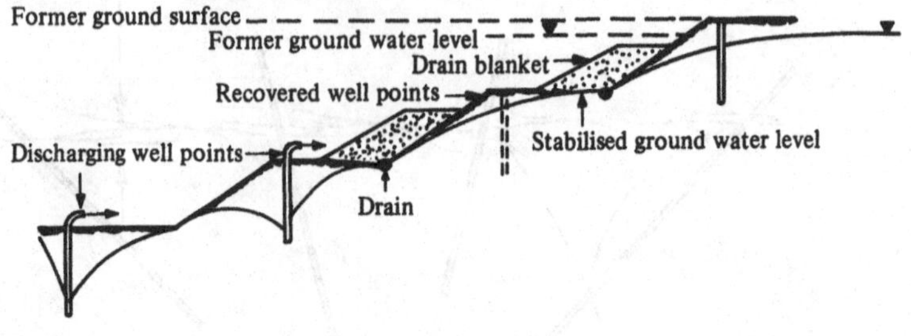

Fig. 11.22. The Morwell 2 aquifer pumping bore at the Morwell open cut brown coal project in Victoria, Australia (after Fraser and Pitt, 1979).

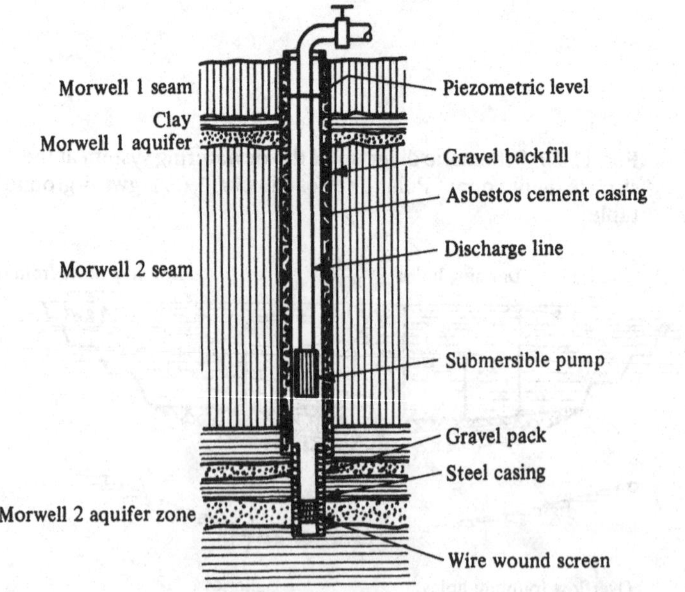

Fig. 11.23. Location of bores at the Morwell open cut brown coal project, Victoria, Australia (after Fraser and Pitt, 1979).

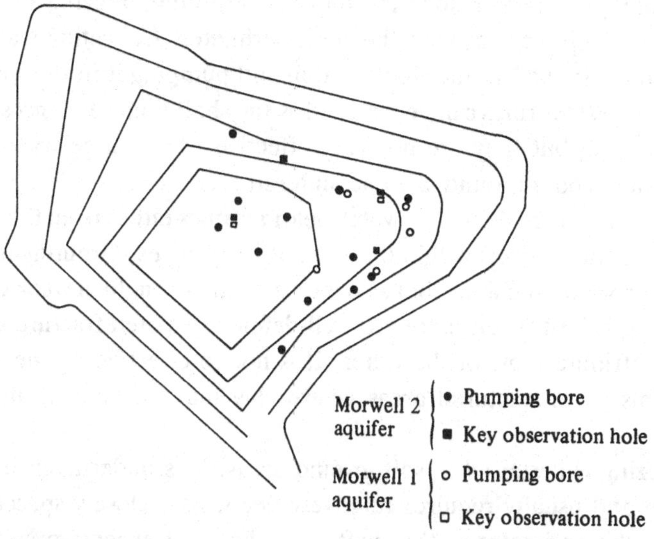

Fig. 11.24. First mining carried out in the tightest ground to minimise water inflow, drawdown and total pumping (after Loofbourow, 1979).

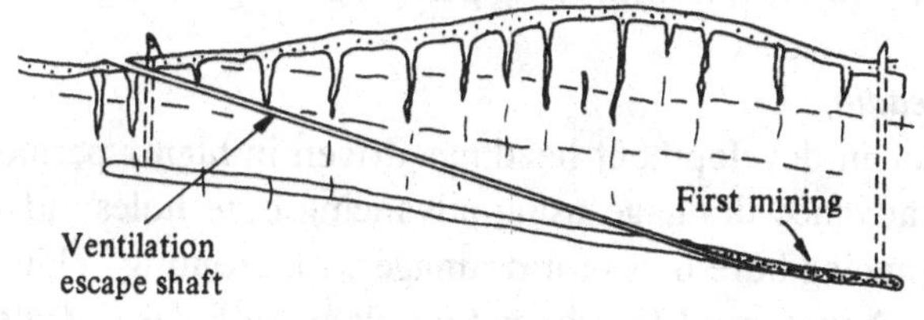

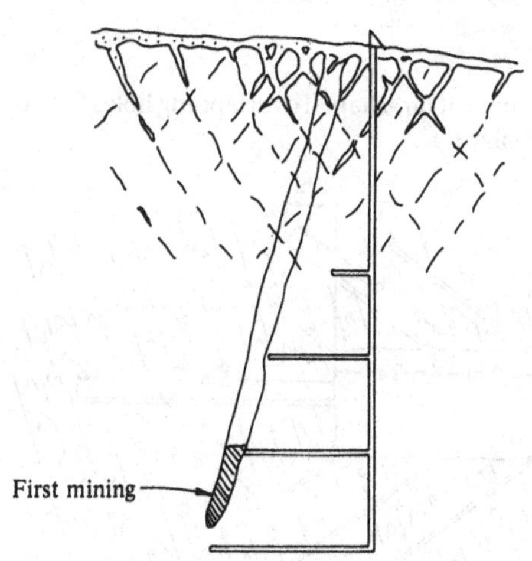

Shaft sinking

The common systems of controlling water while sinking shafts include sump pumping, installation of water rings, grouting, freezing and pumping from deep wells around the shaft perimeter. Collecting water that flows into the shaft at the shaft bottom and pumping it to the surface is common. Water rings can be installed as the shaft lining advances during shaft sinking but they are not very effective where large water flows, especially in poor ground, are encountered.

Grouting can be done into water-bearing zones either from the surface or from various levels within the shaft as it advances. Grouting not only reduces rock permeability but also increases the strength of weak ground. Grouting is best when there are well-defined, isolated fracture systems that contribute most of the water. It is not so effective in fine-grained materials or in fractured areas where clay may be present along the fractures.

Freezing is practised in soft ground areas. It is undertaken from the surface and usually requires relatively deep, very closely spaced holes around the perimeter of the shaft. It is the most expensive method of water control.

Deep dewatering wells can be used to reduce water inflow rates. They are often used in conjunction with sump pumping and grouting.

Development headings

Control of water in development headings driven in highly permeable rocks requires advance drainage using advancing bore holes, advance grouting or advancing bore holes for drainage with grouting behind the working places. A system of fan-shaped overlapping holes is drilled at suitable intervals (2–5 m) ahead of the face with an advance cover of 1 m (Fig. 11.25).

Fig. 11.25. A scheme of fan-shaped overlapping holes for a development heading to control water.

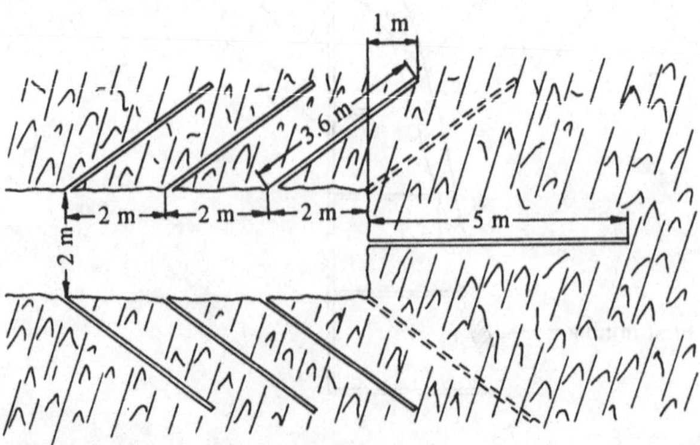

Pressurisation of the excavation with compressed air has been used quite successfully in driving several tunnels. Air pressure keeps the water away from the excavation. This method requires special dams and a depressurisation chamber for people working in environments under high air pressures. Because of the long acclimatisation time involved for the workers moving in and out of the pressurised working face, the technique gives very low productivity.

While driving development workings in areas where a sudden inrush of water may be expected, it is advisable to drill a hole in advance of the heading with equipment which can control the inflow of water and has the facility to measure water pressure (e.g. Burnside safety boring apparatus). With standard drilling equipment, a minimum of 5 m of grouted length of pipe should be installed in the face and all drilling conducted through it. Suitable valves are provided so that overflow of water can be controlled if the bore hole hits a high water pressure area. Specially constructed watertight dams are built which can be easily closed to stop flooding of the whole mine in case of a sudden inflow of water (Fig. 11.26).

When structures containing water are expected during mining, it is advisable to stop development 40 to 60 m before the anticipated position of the structures and drill a bore hole in advance to tap the structure and drain it. Headings driven to intersect these structures should be driven at right angles to the strike of the structure. An intersection of the development heading at a low angle to the long axis of these structures should be avoided.

When several headings are being driven, it is advisable to stagger the development, with one heading leading the others by 40 to 60 m. The leading heading should be placed on the dip side. Suitable isolation doors should be placed between the various headings such that these can be easily closed and protected from an inrush of water.

An estimate, as accurate as possible, must be made of the reservoir pressure and capacity when approached (old workings, natural reservoirs). Vertical sections should be prepared to estimate level differences and evaluate the pressure head. Plans of old mine workings are helpful but these may not be accurate. Provision of a sump to hold water or a connection to an older part of the mine (an abandoned area) to divert the water is helpful when a sudden inrush of water in a developing heading is expected.

Old workings when approached may also contain obnoxious gases. Good ventilation of the area is essential to deal with this emergency. Self rescuers should be provided and regular sampling of mine air should be conducted. The advance of the headings should be regularly surveyed and plotted and final penetration should be done by a rescue team.

As far as possible development headings should be driven at an inclination to the rise with drainage bore holes placed on the sides having a gradient of not less than 0.75 %. Ditches can be lined, particularly when these are in soft floors.

Many secondary pumps should be provided to cater for sudden plant failure and requirements of maintenance of pumps and sumps.

Secondary extraction

Control of water during secondary extraction where full extraction is envisaged usually requires large-scale drainage of the whole area. Overlying beds or workings under waterlogged areas, dams, river beds or under the sea require special precautions. If complete drainage of the reservoir

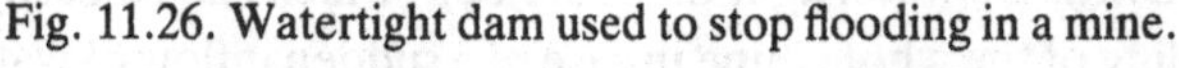

Fig. 11.26. Watertight dam used to stop flooding in a mine.

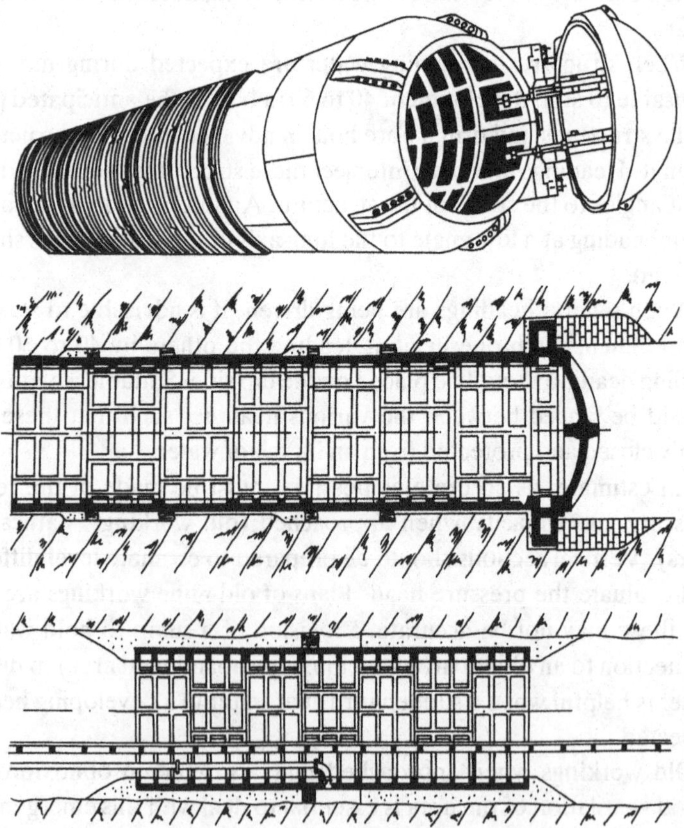

is not possible, partial extraction or full or partial packing of the goaf may have to be adopted.

Several countries have outlined conditions for mining under dams, seas and lakes. In Nova Scotia, it is required that a cover of 30 m must exist between the sea bed and an opening and a cover of 54 m between the secondary extraction level and the sea bed. In the UK, the National Coal Board stipulates a minimum cover of 105 m for longwall and 60 m for bord and pillar workings (minimum 45 m for carboniferous rocks) and a tensile strain of not more than 0.01. In Newfoundland, the minimum cover required is 66 m. In steeply dipping deposits, the conditions applicable for mining under seas may be more stringent.

Monitoring of the base of the sea using a sounding technique is useful in establishing the sea floor level.

The area must be divided into panels which can be easily isolated with watertight dams after extraction has been completed in each panel. Transport roadways for workers and material and escape roadways for the area should be placed on the rise side of the area to protect them from accidental flooding and provide time in case of a mishap for withdrawal of persons.

Water inflow into the area should be monitored on a regular basis to establish flow pattern. Any steadily increasing rate of flow or any sudden change in the flow pattern is a sign of danger. The chemistry of the surface water in the waterlogged area and that of the underground water should be monitored to ascertain if there is any change which may suggest water sources other than those anticipated. While mining under the sea, an increase in the salt water underground is a sign of inflow of water from the sea. The use of a tracer is a very useful technique to establish quantity and source of water make.

Salt mines

The flow of fresh water into a salt mine results in the easily soluble salts being dissolved increasing the width and length of cracks and creating sudden collapses of the mine excavations. However, if the amount of water entering the mine can be eliminated or if the flow can be controlled then the movement of the concentrated salt solution can lead to the precipitation of salt and healing of the crack system. This system of mining must be based upon the use of solid barriers and bulkhead doors.

Monitoring of water intake and the concentration of salt solutions, particularly using tracer elements, is very helpful in controls. The movement of roof and floor should be limited so that cracks do not develop which may lead to excessive infiltration of fresh waters. If any cracks develop which may lead to excessive water circulation, cementation of the barriers can be resorted to, using mixtures of magnesium oxide and magnesium chloride, or other grouting materials.

Selection and use of materials for the building of dams in salt mines is important. Dams must be built using sealed tubes or concrete and clinker bricks with magnesium oxide as a cementing material.

Acid mine drainage

Sulphide minerals in the broken rock mass in surface and underground mines become oxidised and disintegrate when exposed to atmospheric oxygen and humidity. The disintegration exposes new surfaces such that oxidation continues to proceed further inside the mineral particles. Surface and underground water circulating through the rock mass dissolves the salts so formed, hydrolyses them and produces acidic waters. The pumping of these acidic waters from mines (acid mine drainage) or natural seepage has resulted in the pollution of the environment in coal mining areas of the world (particularly in the eastern USA), killing vegetation and aquatic life.

The chemistry of the reaction is very complex, but a simplified version can be given as follows:

$$\underset{\text{(pyrite)}}{FeS_2} + H_2O + 7O \rightarrow FeSO_4 + H_2SO_4$$

This first reaction proceeds slowly depending upon temperature, pH of the water and the properties of the pyrite. In the presence of further oxygen and iron-oxidising bacteria, the iron sulphate ($FeSO_4$) is further oxidised to ferric sulphate ($Fe_2(SO_4)_3$) as follows:

$$2FeSO_4 + O + H_2SO_4 \rightarrow Fe_2(SO_4)_3 + H_2O$$

The hydrolysis of ferric sulphate precipitates ferric hydroxide and sulphuric acid.

$$Fe_2(SO_4)_3 + 6H_2O \xrightarrow{pH=5.5} 2Fe(OH)_3 + 3H_2SO_4$$

The oxidation of pyrite is an exothermic reaction producing heat. A rise in temperature increases the reaction rate (every 10° rise in temperature results in an increase of the reaction rate by 10 %). The presence of water is essential to the reaction. Acid resulting from the reaction dissolves mineral matter (e.g. limestone, dolomite, clays, etc.), producing a high salt content and carbon dioxide gas.

Control. Control of acid drainage is therefore fourfold:

(1) avoid water infiltration into a broken rock mass;
(2) stop inflow of fresh air;

(3) separate sulphide areas and store sulphide-bearing refuse in areas inaccessible to water;
(4) neutralise acid mine water.

The avoidance of water infiltration into a rock mass is the best method. Water should be collected at the source using drains, pipes, etc. and directed to the sumps. The sumps should be located as close as possible to the water source and continuously pumped out. The maximum length of surface drains and ditches should be limited to 500 m. Water should not be allowed to filter through broken rock mass or goaf areas, but positive steps should be taken to collect water by a peripheral drainage system.

Tailings and refuse should be compacted at disposal sites to minimise infiltration of water and to reduce oxidation. Suitable distribution of particle size is essential to achieve maximum compaction.

Submergence of all materials containing acid-forming minerals in stagnant waters is the best method of control in open pit mines, underground goaf areas and old mines. It is essential, however, to ensure that there is no outflow of acid water and no inflow of fresh water into the areas. Mine areas on elevated positions which can permit gravity drainage should be so planned that on closure they can be sealed off effectively. The size of safety pillars, stoppings, impervious curtains and bulkheads should be designed to withstand both ground movement and water heads. The use of grouting to prevent movement of water, surface dams and the reconstruction of channels to ensure that the site to be abandoned remains fully submerged with no inflow and outflow of water are recommended. Highly toxic materials can be buried and sealed.

Treatment of acid mine drainage is essential not only while handling water in and around the mine, but also before disposal into the surface drainage system. Complete treatment methods are designed to neutralise acidity, remove ferric salts and dispose of sludge.

Lime (CaO) or limestone ($CaCO_3$) is most commonly used for neutralisation. The chemistry of neutralisation is given by

$$Fe_2(SO_4)_3 + 6H_2O \xrightarrow{pH=5.5} 2Fe(OH)_3 + 3H_2SO_4$$
$$CaCO_3 + H_2SO_4 \xrightarrow{pH=7} CaSO_4 + H_2O + CO_2$$
$$FeSO_4 + CaCO_3 \rightarrow CaSO_4 + FeCO_3$$
$$FeCO_3 + H_2O \xrightarrow{pH=9.5} Fe(OH)_2 + CO_2$$

The precipitation of ferrous hydroxide does not occur below a pH value of 9.5 but the reaction of $CaCO_3$ and H_2SO_4 is stopped at higher pH values. As such, after the initial reaction on the surfaces of calcium carbonate particles, they become inactive. Also, coating the larger particles of calcium carbonate with oxides of iron and calcium sulphate results in loss of the neutralising power. The efficiency of their utilisation can be increased by using finer limestone particles. In case water is to be pumped out, treatment can be done in a rotating drum working as an autogenous grinder where carbon dioxide and ferrous oxide can be removed.

Limestone and lime in combination can be used. Limestone is used in the first stage to remove ferric oxide at a pH $\leqslant 5$ and then lime is added to raise the pH to 9.5, when ferrous oxide can be removed.

Demineralisation processes using ion-exchange resins, or passing the water through a membrane, using the principle of reverse osmosis where the dissolved particles can be removed, are effective enough methods to give drinking quality water. However, the cost of these methods is prohibitive under normal conditions.

Before pumping acid mine drainage into the surface stream system, it is essential to take samples for analysis so that the mineral content of the water does not exceed the legal limits.

Cementation and grouting

The flow of water into excavations or from the reservoirs can be reduced by cementing or grouting the rock surrounding it. The process of cementation or grouting relies on plugging the fractures and reducing the overall porosity of the system by filling the holes present with an impervious material, usually cement but also certain chemicals and organic resins. Cementation or grouting also increases the strength of the rock and thereby reduces erosion. Grouting is also used in plugging exploratory drill holes, as concrete plugs for dams, etc.

The method involves drilling a set of holes at suitable intervals and pumping the grout material into the holes sequentially according to a desired pattern, making sure that the grout enters the fracture spaces but does not cause additional fractures as a result of the hydraulic pressures developed in the impervious rock space. In general, grouting pressures are below the gravitational stresses and are kept under constant surveillance during the various stages of grouting. Various types of tracers can be used to monitor grout flow. Control procedures include drilling after grouting to assess the success of grouting and pre- and post-permeability testing.

The most commonly used grout materials are: (1) clays; (2) Portland cement in combination with sodium silicate, bentonite and early setting additives like sodium chloride; and (3) chemical grouts in single- or multiple-component systems.

Clays are commonly used where the porosity is large but mostly consisting of smaller size pores in weak grounds. Colloidal clay can be used to fill pores as small as 0.01 mm. Clay grouts, however, remain in suspension for a very long time but are not sensitive to deformation and ground movement.

Portland cement slurries (5 or 10 parts of water + 1 part cement) can be used in comparatively competent rocks with a pore and fracture system of 1 mm. The penetration rate of cement slurries can be improved by pre-treatment with sodium silicate or bentonite slurries. Bentonite, however, decreases the strength of the cement grout. A minimum cement grout density of 1.5 g/cm^3 is required for the cement particles to bind together.

Chemical grouts include acrylamides, chrome lignins, resorcinol formaldehydes, etc. These have a very low viscosity and can be designed to set almost suddenly after a given time. These are of very high strength and can stand high gradients (300 kPa/m).

Clay and cement grouts are the cheapest, while chemical resin grouts are almost 10 to 100 times costlier. Chemical grouts are used only to seal small areas or after treatment with cement grouts.

The equipment used for pumping grouts consists of reciprocating simplex or duplex grout-filled slush pumps fed by gravity from agitator tanks containing the slurries. When two component chemicals are used, two separate positive displacement pumps with a predetermined stroke are used so that a definite amount of each component is delivered. The two components are mixed at the collar of the hole or at the discharge point.

Dams and stoppings

There are two types of dams generally used in mines: (1) filtration dams and (2) water dams, bulkheads and plugs.

Filtration dams. Filtration dams are of light construction designed to stop movement of suspended particles and thereby allow clean (filtered) water to pass through them. Such dams are made of wooden props with a filtering membrane (hessian cloth, jute mat) placed on the upstream side (Fig. 11.27). Because of the low water pressure which these have to stand, they are very thin. The wedging of props into the roof, sides and floor or occasionally a brick wall support is normally sufficient. There are no special requirements for the sites of such dams.

Water dams, bulkheads and plugs. Preparatory stoppings which are built so that they can be converted into bulkheads, plugs and dams to stand high water pressures are part of the mining methods in mines likely to experience high water inflows. The selection of the site for the location of these dams is very important. Sites should be selected so that they are in stable ground which is not affected by high rock pressures developed as a result of mining. The ground should be, as far as possible, free from large-scale fracturing and away from faults, shear zones, dykes, etc. which can be a source of high leakage. Certain sites, if possible, can be cement grouted to reduce their permeability. Water leakage invariably occurs through the rock rather than through a properly constructed dam.

Preparatory stopping for the dam can be prepared containing steel bulkhead doors (Fig. 11.26). The space between the steel tubings and the excavation can be grouted using the injection technique. Such preparatory bulkhead doors can be placed at the entrance of each panel and at the development headings and can be quickly closed if a sudden inrush of water should occur.

Permanent fully closed dams are constructed with cement concrete (Fig. 11.28) and have the shape of a serrated roller with an arched or a conical body. The resistance of the dam to movement is obtained from the sides of the tunnel forming the serrated abutments or the frictional reaction of the dams and the walls. The thickness of the concrete dam can be calculated using the following relationships.

Cylindrical arch dams: In the design of these dams the shear strength of the abutment of the arch must not be exceeded. The thickness of dam, $a = (R - r)$ and can be calculated as follows (Fig. 11.29):

$$a = \frac{rp}{\tau - p} = \frac{Rp}{\tau} \tag{11.9}$$

where r = internal radius of the dam, m;
p = water pressure expected, MPa;
τ = shear strength of the rock mass, MPa; and
R = external radius, m

If the water pressure is very large and the shear strength of the rock mass is low, then the dam can be built with a serrated surface with a number of walls (n) each having a width of a_n such that

$$a = na_n$$

and

$$a_n = \frac{rp}{n\tau - p} = \frac{Rp}{n\tau} \tag{11.10}$$

Spherical dams: In the case of high pressures, instead of cylindrical arch dams, spherical arch dams are designed. The geometrical centre of the dam is determined and all brick or concrete work is so placed that elements lie along the radius projected from the centre. The dam can be

Fig. 11.27. Two types of filtration dams.

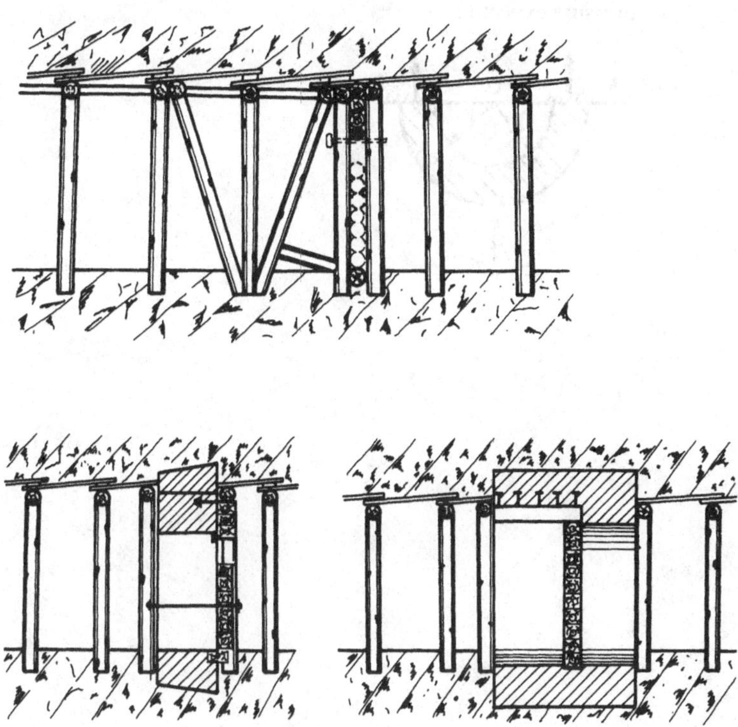

Fig. 11.28. Two types of permanent dams (after Lesiecki, 1949). a = the thickness of the dam; r = internal radius of the dam; R = external radius of the dam; b = width of the tunnel; h = height of the tunnel.

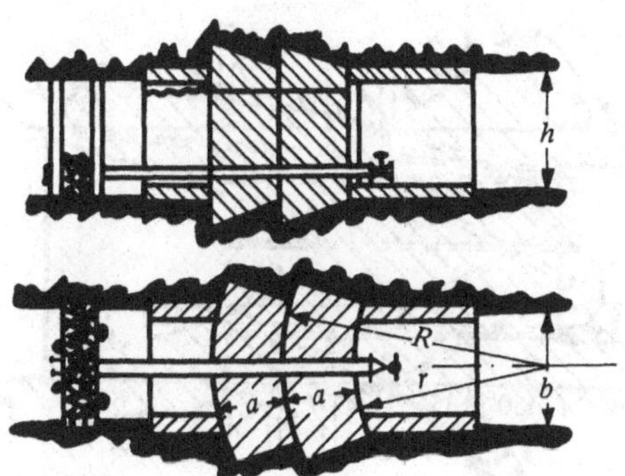

Fig. 11.29. Nomenclature for arch dams. a = the thickness of the dam; r = internal radius of the dam; R = external radius of the dam; p = water pressure expected.

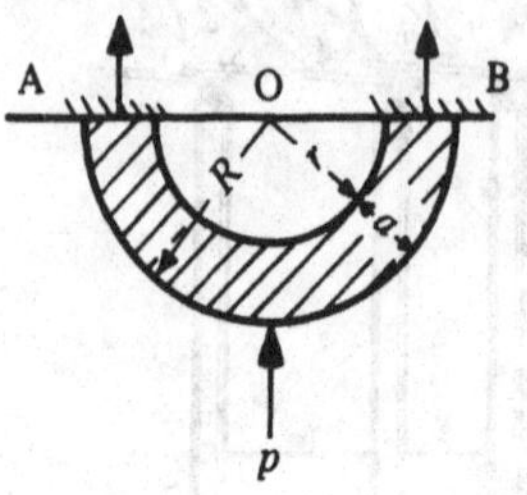

Fig. 11.30. A cylindrical dam (after Trembecki, 1972).

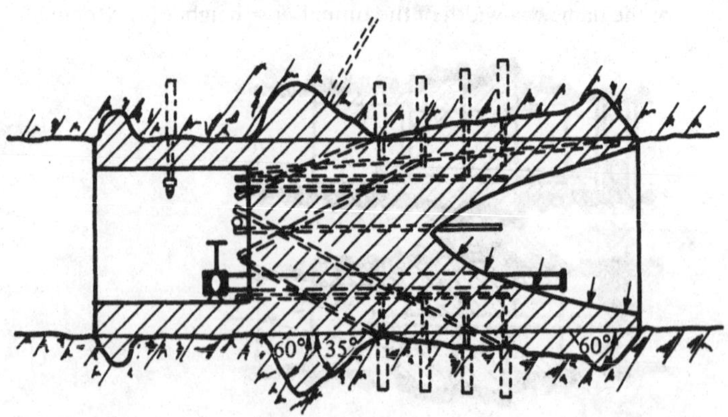

constructed in two or three segments as the cylindrical arch dam. The thickness of the dam can be calculated from the relationship

$$a = r \times \sqrt{\frac{\tau}{\tau - p} - 1} \tag{11.11}$$

where r = radius of the internal spherical surface of the dam, m;
τ = shear strength, MPa; and
p = water pressure, MPa.

Since the water pressure acts uniformly in spherical dams, the thickness of the dam works out to be much smaller than that of cylindrical arch dams.

Cylindrical dams: Preparation of inserts to act as bearing surfaces is not required where there are hard strong rocks and where good shear strength between concrete and rock interface can be achieved. The building of such dams is easy and much faster. In such cases, cylindrical dams can be built (Fig. 11.30). The thickness of the dam can be calculated using the relationship

$$b = \frac{Ap}{s\tau} \tag{11.12}$$

where A = cross-sectional area of the excavation, m^2;
p = water pressure, MPa;
s = excavation perimeter, m; and
τ = shear strength of rock and concrete interface.

The dam is usually made thinner in the middle which allows a positive pressure on the wall of the dam, increasing the frictional resistance and strength of the dams.

All dams are provided with drain pipes so that water can be drained till the cement concrete obtains the desired strength (28 days). A steel pipe with a bulkhead is provided as a means of exit for people after the completion of the dam.

11.4.3 Pumps and pumping

The main methods of disposing of mine water are drain tunnels and pumps. Drain tunnels can be used in mountainous areas where rainfalls are heavy. They can be economical especially if they are also used for ore transport and ventilation. They are also useful in the case of flooding.

In the case of pumps, the general layout includes sump, pump and delivery mechanism.

Sumps

The mine water is collected in reservoirs, called 'sumps' from which it is pumped to the surface. These are usually located at the shaft bottom whereas sumps in the upper levels are excavated near the shaft. In deep mines the water is sometimes pumped in stages, each sump having its own pump.

A sump acts as a reservoir from which the pumps draw water as necessary. Pumps can be run at an economic load for part of the day, often in the night shift to take advantage of cheap off-peak rates. A sump also provides storage in case of inundation or increase in quantity from a feeder or in case of pump repairs. It also provides an opportunity for abrasive solid matter in the water to settle down and allows for addition of chemicals to neutralise acids, etc. if required. Sumps should be divided for cleaning out. The minimum storage should be two days normal water make.

When sumps are used as settlers, their design is important. There are two types of settlers, the horizontal flow type and the vertical flow type. In horizontal flow settlers, the incoming water enters at one end of a long sump and discharges at the other end over a battery of overflow lip launders (Fig. 11.31). The overflow is then channelled along a collecting launder into the clear water sump ahead of the pumps. The pattern of sludge deposition in the horizontal flow settler is such that the coarse particles come out of suspension in the first few metres, rapidly building up sludge at the inflowing end. Since the flow is local through the body of the settler, little further settling occurs until the water reaches the collecting end. A secondary build-up of sludge of finer particles occurs under the overflow launders, i.e. in the region of almost wholly vertical flow. As the sludge build-up grows and extends into the settler from the end walls, the efficiency of the settler decreases. As a result, progressively larger particles are carried over into the clear water sump. When the size and quantity of solid material entering the clear water sump becomes excessive, the settler requires cleaning. If the settler is not cleaned in

Fig. 11.31. Section through a horizontal flow settler showing actual flow pattern and typical sludge pile growth (after Goninan, Grudzinskas and Fleming, 1968).

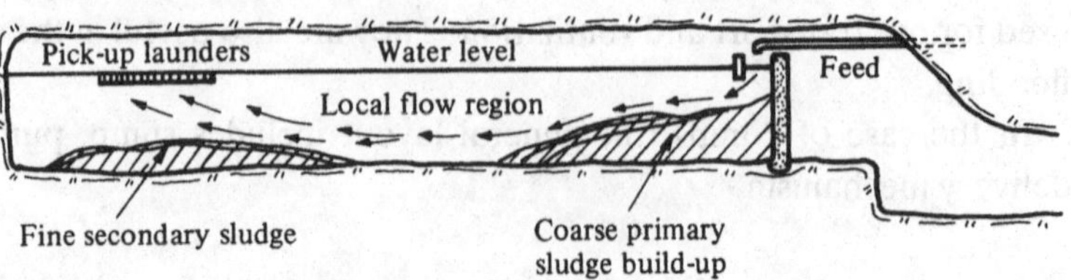

time, the clear water sump becomes partially filled with sludge which can reach the pumps, causing accelerated wear and damage.

When cleaning the settler, the inflowing water is bypassed to feed directly into the back of the clear water sump. The water above the sludge in the settler is then decanted into the clear water sump after the level in the clear water sump has been pumped down as low as possible.

The mud near the outer settler wall is then agitated into a slurry with high-pressure air and water jets. It is then either gravitated to a stope on a lower level or fed to a plunger pump which lifts it to a higher level.

To clean the clear water sump, the water levels in both the settler and clear water sump are lowered as far as possible. Mine water is then run into the settler while the clear water sump is being cleaned.

Vertical flow settlers are superior. Two types are shown in Fig. 11.32 (de Villiers, 1961). In design (*a*) the inflow water is taken to below the bed of settling particles behind a wall. The excavation width governs the length of the horizontal flow. A more stable condition can be obtained with design (*b*). In this case, the length of the horizontal flow is reduced to only half the width of the excavation at the bottom of the divergent upward flow portion. This opening must not be made too narrow as this would trap the settling particles in the upward flow portion and create unstable conditions. In both cases, the settled sludge must not be allowed to accumulate to such an extent that the openings at the bottom of the walls become restricted, as this would result in settled particles becoming re-entrained.

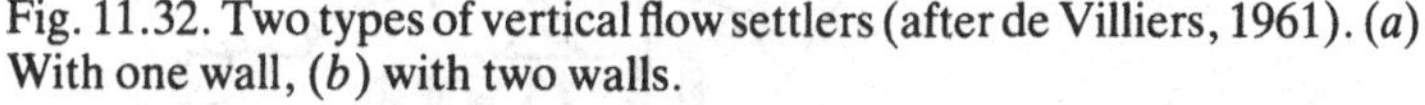

Fig. 11.32. Two types of vertical flow settlers (after de Villiers, 1961). (*a*) With one wall, (*b*) with two walls.

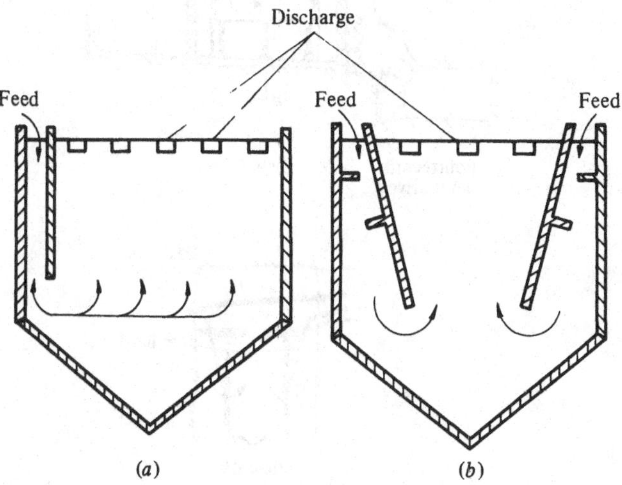

An arrangement of the settler installation using design (b) is shown in Fig. 11.33. Here, there are two settler systems operating side by side, each being fed from an eight-way distributor and discharging into separate clean water dams.

Fig. 11.33. Arrangement of settler installation on the fourteenth level at the Browne shaft, Broken Hill (after Goninan *et al.*, 1968).

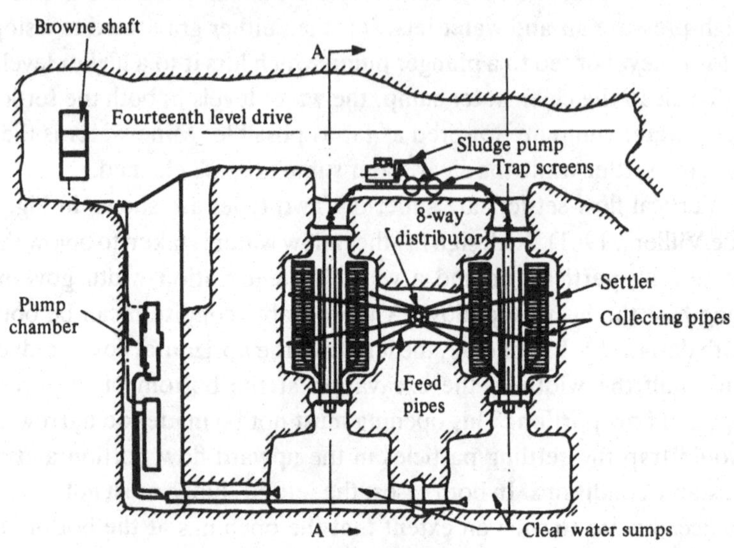

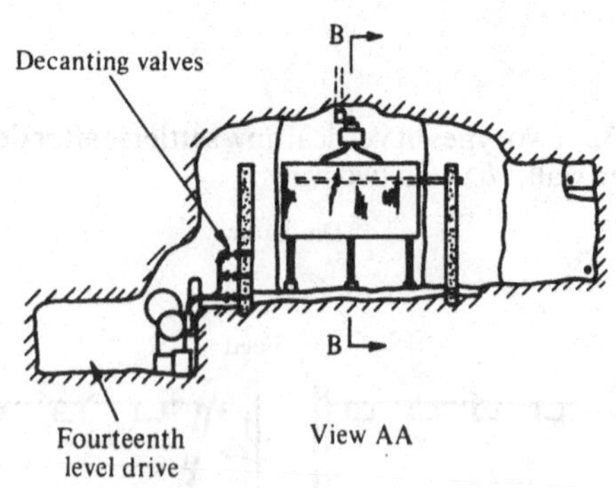

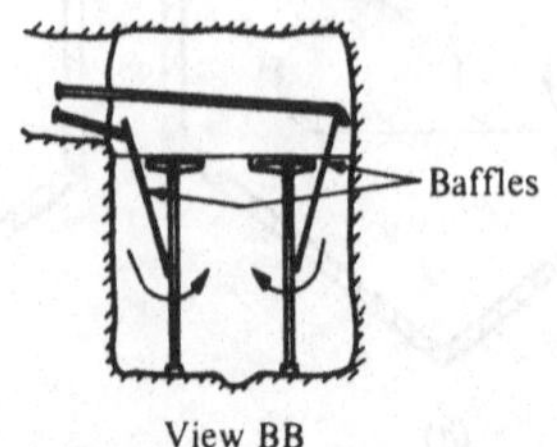

Pumps

Pumps may be classified into two groups: rotodynamic and positive displacement. Rotodynamic pumps include both axial and centrifugal pumps. The most commonly used pump in the mining industry is the centrifugal or radial flow type (Fig. 11.34). It may be single or double suction, or single or multi-stage. A centrifugal pump employs an impeller with fixed blades (vanes) mounted on a rotating shaft and housed in a suitably shaped fixed casing (diffuser). The pumping action is produced by the pump applying moment to the water handled, water motion through the pump at the same time producing a pressure reduction on the inlet side, or suction. The water enters at a central 'eye' in the impeller and is then flung outwards by centrifugal force to the periphery. Water is ejected from the impeller into diffuser chambers, arranged, both as to direction and gradually increasing section, so as to achieve conversion of energy while directing the water finally into the delivery pipe, or in multi-stage pumps, into the eye of the succeeding impeller.

A multi-stage pump has a number of impellers on one shaft in a multiple casing assembly. The number of stages may be limited by shaft length and the deflections which can be tolerated. The stages may be manufactured as units and assembled to the required number for the particular duty. End covers incorporate inlet and outlet ports. Each stage consists of impeller, guide passages and return passages within a chamber which discharges the water into the next stage. Multi-stage pumps having

Fig. 11.34. Centrifugal pump with diffuser. Water is ejected from the impeller into the diffuser chambers, arranged both as to direction and gradually increasing section so as to achieve conversion of energy, while directing the water ultimately into the delivery pipe, or in multi-stage pumps, into the eye of the succeeding impeller.

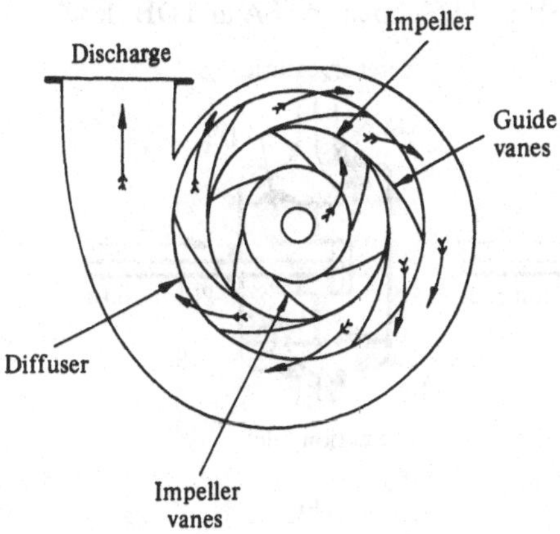

single-entry impellers incur considerable axial end thrust towards the suction end, because the impeller surface on the entry side is smaller than that on the outlet side. To carry the end thrust of all the impellers, the delivery cover generally contains a hydraulic balancing disc keyed to the shaft. End thrust can also be balanced by incorporating an even number of stages and by suitable arrangement of flow passages.

Positive displacement pumps include both reciprocating pumps and diaphragm pumps. They derive their pumping action from stroke motions applied to a working chamber.

In a reciprocating pump, the working chamber or cylinder is alternately filled and emptied by forcing in and drawing out the water by mechanical action. It consists of two main parts: a power end and a water end. The power end converts the rotatory motion of the driver (electric motor or diesel engine) to the reciprocating motion required by the water end. Output is in the form of sine wave pulsations. To smoothe the delivery, either two (at 180°) or three (at 120°) cylinders are used with overlapping strokes. The pump may be single acting, when one side of the piston is utilised for pumping, or double acting, when both sides are used.

In the water end, the sliding, oscillating moving member within a cylinder is either a piston or a plunger. The cylinder also has a suction and a discharge valve.

Piston pumps are either single or double acting and Fig. 11.35 illustrates the water end of a double-acting piston pump. The cylinder has two separate water chambers, each with its own suction and discharge valves, with the piston operating through the centre of both. As the piston moves

Fig. 11.35. Double acting piston pump. In (*a*) the piston is moving to the right; suction valve A (SA) and discharge valve B (DB) open. Suction valve B (SB) and discharge valve A (DA) close. In (*b*) the piston is moving to the left; SB and DA open and SA and DB close.

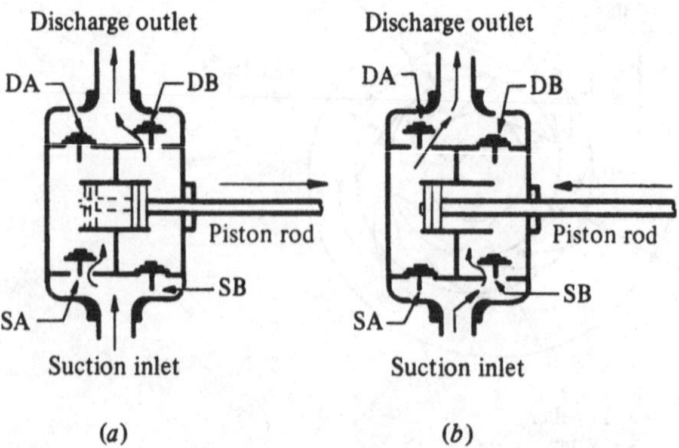

to the right (Fig. 11.35*a*), water is forced out of the right-hand chamber into the discharge pipe through the discharge valve DB, and at the same time the suction valve SA opens, filling the left-hand chamber. On the reverse stroke (Fig. 11.35*b*), water is forced out of the left-hand chamber through DA, and SB opens, filling the right-hand chamber.

The water end of the plunger pump may consist of two or more single-acting plungers, individual or monoblock water cylinders, and a suction and a discharge valve for each cylinder.

Reciprocating pumps are employed to pump muddy or corrosive mine water and slurries. However, special measures must be taken to protect the sealing system from abrasive slurry particles. Plunger-type pumps inject clean water in front of the packing on the suction stroke to prevent contact between the slurry and the packing materials.

In the case of the Mitsubishi–Mars pump (Fig. 11.36), oil is interposed between the pump and the slurry. A low-speed reciprocating, duplex oil pump displaces the pumped slurry into a vertical oil/slurry chamber, and the slurry does not contact the pump piston. However, the suction and discharge valves operate in the slurry.

In diaphragm pumps, a flexible diaphragm is oscillated to provide suction and discharge through non-return valves and these are of two types: (1) mechanically operated and (2) hydraulically operated. In a mechanically operated pump, a flexible diaphragm is clamped in a shallow chamber and oscillated by a mechanical movement connected to the centre of the diaphragm. In a hydraulically operated pump, pressure pulsations are initiated in a fluid chamber connected to one side of the diaphragm. Diaphragm pumps can be driven by an electric motor through a crank mechanism or by directly acting compressed air. Variable speed can be achieved by altering either the oscillation frequency or the stroke.

Fig. 11.36. Simplified diagram of the Mitsubishi–Mars pump.

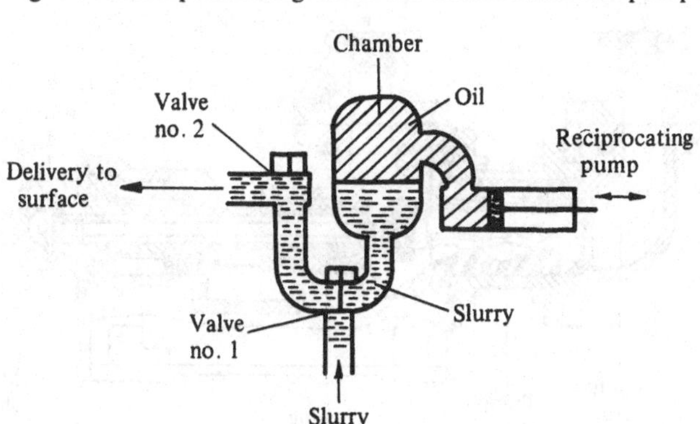

These pumps are suitable for handling difficult slurries with large and abrasive particles from shaft sumps or excavations in mines and cope with medium delivery heads and volume flow rates. They have good suction lift capability and are self-priming, being able to run dry without damage.

A single screw/flexible stator pump (a Mono pump – Fig. 11.37) is also used in mines to handle gritty or corrosive mine water, slurry, etc. It consists of a helical metal rotor rotating within a resilient stator. The stator has the form of a double internal helix and the rotor does not fit perfectly into the stator, so that the rotor motion transmits a continuous seal through the pumping element. A universal joint between the rotor and drive unit causes the eccentric motion of the rotor. The pump is self-priming and capable of suction lifts of up to 8 m of water. There are no valves and abrasive wear is minimum. A bypass is used to provide sufficient water for lubricating the stator when the pump is drawing air, or 'snoring'.

Submersible pumps. Centrifugal pumps may be submerged in the water to be pumped by mounting the motor above the water level and connecting it to the pump through a vertical drive shaft. Fully submersible pumps are also available with the casing and impeller assembly coupled to a special motor designed for operation underwater. For high heads and large capacities, multi-stage submersibles are used. The advantages of submersible pumps are: (1) no special pump chamber or foundations are required, as the pump is suspended or stands in the sump; (2) the sudden rise in water level which could flood a pump chamber is of no concern; (3) control is simple as no priming is required; (4) no maintenance or supervision is required, as overhaul is required only at long intervals; (5)

Fig. 11.37. Diagram of a Mono pump.

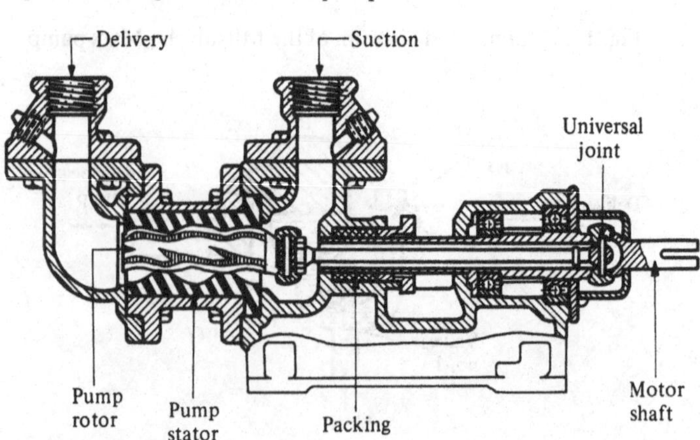

the water in the sump is not subject to any significant aeration or turbulence which can lead to the formation of insoluble impurities causing the build-up of deposits in the pump.

Pump characteristics. In the case of centrifugal pumps, the head generated by each impeller is a function of the diameter and rotational speed. Typical characteristic curves are given in Fig. 11.38. These curves are very similar to the centrifugal fan characteristic curves except that the pumps have a suction lift curve also. The discharge head is the same as the pressure. The discharge head/volume flow rate curve shows the volume flow rate expected against various discharge heads. The power curve indicates the power consumed at any point on the discharge head/volume flow rate curve. The efficiency curve shows the percentage efficiency to be expected at any point on the discharge head/volume flow rate curve. The suction lift curve shows the suction lift the pump can manage at any point on the discharge head/volume flow rate curve.

In the case of reciprocating pumps, the volume flow rate is constant regardless of the delivered head, and is affected only by speed changes, but this is limited by the amount of water which can physically enter the cylinder. If this is not sufficient, cavitation is caused by the breaks in water flow, and bubbles filled with vapour collapse when the pressure is increased, causing noise, vibration, possible damage to the casing walls, valve pitting or metal fatigue.

Comparison of centrifugal and reciprocating pumps. The advantages of centrifugal pumps are:

(1) their higher speed results in lower size and cost;

Fig. 11.38. Typical characteristics of centrifugal pumps.

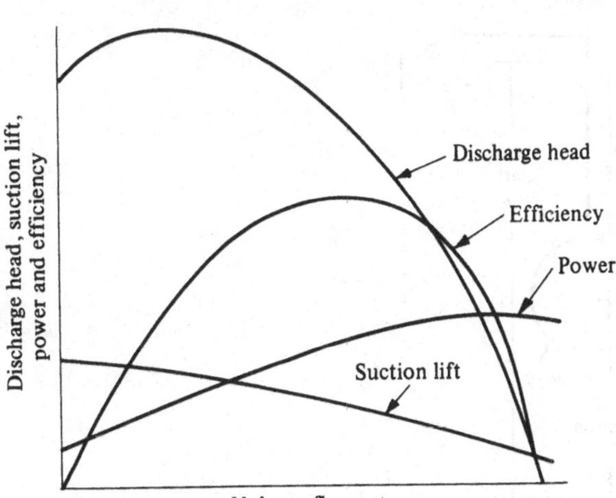

(2) the water delivery is steady and uniform, free from pressure fluctuations;
(3) they can (a) operate on minimum flow rate without exceeding a predetermined delivery head, (b) operate on maximum flow rate without exceeding a predetermined power demand and (c) be designed to meet several duties;
(4) no gearing is required as the high speeds are suitable for direct drives with consequent smooth and quiet running;
(5) the absence of vibration and comparatively inexpensive foundations.

The advantages of reciprocating pumps are:

(1) within their field of applicability they have higher efficiencies;
(2) they are applicable to variable delivery heads without speed adjustment – useful when the head is uncertain;
(3) they are self-priming and therefore able to handle a certain amount of air without pump failure.

Centrifugal pumps are used for duties varying in volume flow rate from the smallest to the largest, the upper limit being dictated only by the ability to manufacture and to transport (up to, say, 4.5 million litres/per minute and a maximum stage head of 600 m). Reciprocating pumps are selected for duties where volume flow rate is low and delivery head high (e.g. 10 litres/s, 1500 m).

Pump duty. The main duties are: (1) rate of flow required, i.e. litres/s which are required to be pumped; (2) the total head which the pump must

Fig. 11.39. Terms used in assessing heads in a suction lift installation.

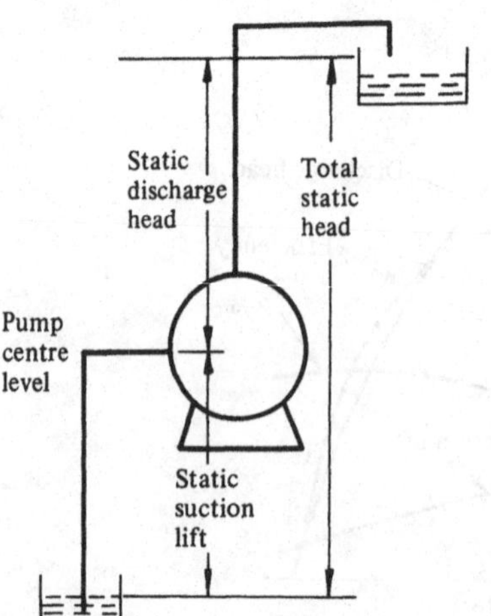

generate when pumping the required flow rate; and (3) the type of mine water.

Volume flow rate depends upon the volume and mode of water inflow, seasonal fluctuations and the selected load factor of the pump. (It is important to ensure that the suction lift capacity exceeds the volume flow rate of the pump.)

The total head requirement depends upon the vertical distance through which water is to be pumped and the friction losses in the associated pipework. It is usual to treat the inlet or suction side of the pump pipework separately from the delivery or pressure side. The reason for separating these calculations is to ensure that the suction lift is within the capability of the pump to be selected.

In a suction lift installation (Fig. 11.39), the total head, H, against which the pump will be required to operate is as follows:

$$H = h_d + h_s + f_d + f_s \tag{11.13}$$

where h_d = static discharge head;
h_s = static suction lift;
f_d = friction head in discharge (i.e. the head required to overcome friction in the pipe, valves, fittings, turns, etc. in the discharge system); and
f_s = friction head in suction (i.e. the head required to overcome friction in the suction system).

When the pump is under a suction head installation (Fig. 11.40),

$$H = h_d - h_s + f_d + f_s. \tag{11.14}$$

Fig. 11.40. Terms used in assessing heads in a suction head installation.

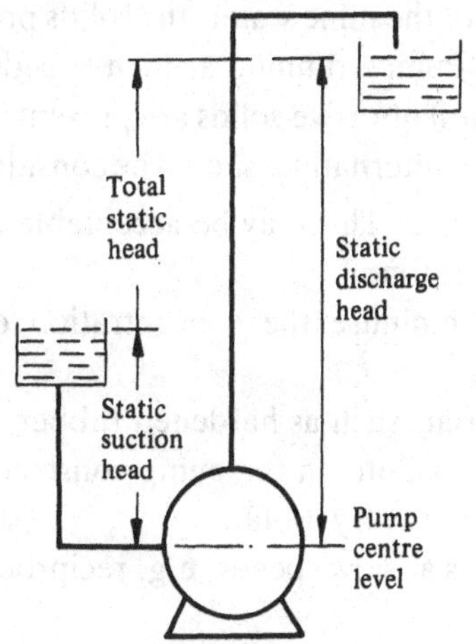

The head is usually expressed in metres of water (head in metres = pressure in Pa/(density in kg/m^3 × 9.8)).

The friction head is generally of the form KQ^2. The value of the loss factor, K, is dependent upon the individual source of loss in the system. The head required to overcome friction in the pipe can be calculated using Darcy's equation:

$$f_d \text{ or } f_s = \frac{4fLV^2}{2gD}$$

$$= \frac{64fLQ^2}{2 \times 9.87 \times g \times D^5} \quad (11.15)$$

where f_d or f_s = friction head in discharge or suction, m;
f = friction factor determined from Moody's diagram (Fig. 2.11, p. 22), dimensionless;
L = pipe length, m;
Q = volume flow rate, m^3/s;
g = gravitational acceleration, m/s^2; and
D = pipe diameter, m.

Equivalent friction heads for pipe fittings are as follows:

Foot valve and strainer	90 × pipe diameter in m
Sluice valve, fully open	6 × pipe diameter in m
Reflux valve, hinged type	50 × pipe diameter in m
Globe valve, fully open	150 × pipe diameter in m
Bends (radius 3 to 5 × pipe diameter)	10 × pipe diameter in m
Round elbows	30 × pipe diameter in m
Sharp elbows and tees	90 × pipe diameter in m

In addition to the chemical nature of the mine water, the solids present in it should be taken into account. Pumps running at high speeds can experience considerable rates of wear if abrasive solids are present in the mine water. In such a situation, various alternatives should be considered.

(1) Accept the cost of maintenance. This may be acceptable if the system is to be used for a short time.
(2) Use sumps to remove or minimise the concentration of all offending material.
(3) Use specially selected materials such as hardened rubber, non-ferrous metals, hardened steels, etc. in the pump construction. Easy change of parts or linings is very useful.
(4) Select a pump which operates at slow speeds, e.g. reciprocating pump.

Selection. After establishing the volume flow rate, the total head to be generated by the pump and the suction lift for the installation, the characteristic curves of the pumps are referred to in order to choose a pump that suits all three of these requirements.

The characteristic curves of the pumps are given either (1) at constant speed with several head–volume flow rate curves for different impeller diameters or (2) at varying speeds for a particular diameter impeller. In both the cases, the pump efficiency, power absorbed by the pump and the suction lift for the volume flow rate range of the pump are also given.

If the required volume flow rate, total head and suction lift are plotted on the characteristic curves, the speed and input power required can be determined. The input power is usually given for clear water (i.e. density = 1000 kg/m^3). If the density of the mine water is more, more power will be required.

After establishing the speed and power required, a suitable motor is selected.

The fan laws given in Chapter 2 are equally applicable to pumps.

Pipe systems. The suction pipe should be at least as large in diameter as the pump inlet connection and should be kept as short and as straight as possible. When a pipe larger than the pump inlet is used, an eccentric reducer should be fitted to reduce the possibility of air pockets. If the pump is to operate on suction lift, it is best for the pipe to rise continuously to the pump. Discharge pipe lines should be designed to minimise losses. If necessary, long radius bends should be used. At or near a centrifugal pump discharge a 'non-return' valve should be used to prevent water from running back through the pump when it stops, otherwise the pump will rotate in a reverse direction and may damage the pump or its motor. Valves should also be installed in such a way that the pump can be isolated for maintenance purposes.

Cavitation. Cavitation is the local vaporisation of the water. It occurs when the absolute value of pressure falls below the vapour pressure of the water, and can be seen as small bubbles in the body of the water. When such bubbles move to a region of higher pressure, they collapse violently. Cavitation results in a loss of performance and can cause damage if allowed to persist.

Pressure surges and water hammer. These can occur whenever the velocity in a pipe is changed from one steady state to another, e.g. instantaneous starting and stopping of the pump, instantaneous opening and closing of a valve in the pipe line. Water hammer may cause noise and vibration in the system, piping to burst, or a pump casing to crack. The more common methods of control include the use of air vessels and accumulators, increased pump inertia, slower valve or gate operation, air admission valves, pressure relief valves and bypass valves.

Pipe diameter. In long pipe lines, friction becomes a major part of the total head and should be kept to an economic minimum by choosing an appropriate diameter. Velocities of about 2 m/s are usually found to be optimum when pumping water over long distances.

11.5 References

Boehm, B., Schneider, D. and Voigt, R. (1979). Dewatering techniques for Rheinbraun's open pit lignite mines. In *Proceedings of the 1st international mine drainage symposium*, Denver, Colorado, May, ed. G. O. Argall, Jr and C. O. Brawner, pp. 437–62. San Francisco: Miller Freeman Publications Inc.

Brawner, C. O. (1979). General report on mine drainage. In *Proceedings of the 1st international mine drainage symposium*, Denver, Colorado, May, ed. G. O. Argall, Jr and C. O. Brawner, pp. 33–53. San Francisco: Miller Freeman Publications Inc.

Durston, K. J. (1979). Open pit dewatering at Pine Point. In *Proceedings of the 1st international mine drainage symposium*, Denver, Colorado, May, ed. G. O. Argall, Jr and C. O. Brawner, pp. 275–303. San Francisco: Miller Freeman Publications Inc.

Fraser, C. J. and Pitt, H. A. (1979). Artesian dewatering operations at Morwell open cut. In *Proceedings of the 1st international mine drainage symposium*, Denver, Colorado, May, ed. G. O. Argall, Jr and C. O. Brawner, pp. 362–82. San Francisco: Miller Freeman Publications Inc.

Goninan, E. K., Grudzinskas, V. and Fleming, M. R. (1968). The development of mine dewatering installations at Broken Hill South Limited. In *Broken Hill Mines – 1968*, ed. M. Radmanovich and J. T. Woodcock, pp. 361–74. Melbourne: The Australasian Institute of Mining and Metallurgy.

Hoek, E. and Bray, J. (1974). *Rock slope engineering*. London: The Institution of Mining and Metallurgy. 309 pp.

Lama, R. D. and Vutukuri, V. S. (1978). *Handbook on mechanical properties of rocks*, vol. IV. Clausthal: Trans Tech Publications.

Lesiecki, W. (1949). *Odwadnianie Wyrobisk. Gornictwo*, Tom XII. Katowice: Instytut Weglowy. 281 pp.

Libicki, J. S. (1979). The design of mine dewatering systems in Poland. In *Proceedings of the 1st international mine drainage symposium*, Denver, Colorado, May, ed. G. O. Argall, Jr and C. O. Brawner, pp. 304–23. San Francisco: Miller Freeman Publications Inc.

Loofbourow, R. L. (1973). Ground water and ground-water control. In *SME mining engineering handbook*, vol. 2, ed. I. A. Given, pp. 26-1–26-55. New York: SME of the American Institute of Mining, Metallurgical and Petroleum Engineers.

Loofbourow, R. L. (1979). Controlling mine water. In *Proceedings of the 1st international mine drainage symposium*, Denver, Colorado, May, ed. G. O. Argall, Jr and C. O. Brawner, pp. 507–27. San Francisco: Miller Freeman Publications Inc.

Serafim, J. L. (1968). Influence of interstitial water on the behaviour of rock masses. In *Rock mechanics in engineering practice*, ed. K. G. Stagg and O. C. Zienkiewicz, pp. 55–97. London: Wiley.

Sharp, J. C. (1979). Drainage used to control movements of a large rock slide in Canada. In *Proceedings of the 1st international mine drainage symposium*, Denver, Colorado, May, ed. G. O. Argall, Jr and C. O. Brawner, pp. 423–36. San Francisco: Miller Freeman Publications Inc.

Sharp, J. C., Ley, G. M. M. and Sage, R. (1977). *Pit slope manual*, Chapter 4, Groundwater, Canada Centre for Mineral and Energy Technology Report 77-13. Ottawa: Printing and Publishing Supply and Services Canada. 240 pp.

Straskraba, V. (1979). Some technical aspects of open pit mine dewatering. In *Proceedings of the 1st international mine drainage symposium*, Denver, Colorado, May, ed. G. O. Argall, Jr and C. O. Brawner, pp. 481–91. San Francisco: Miller Freeman Publications Inc.

Trembecki, A. (1972). *Zagrozenia Wodne w Gornictwie*. Katowice: Wydawnictwo 'Slask'. 496 pp.

de Villiers, J. W. (1961). An investigation into the design of underground settlers. *J. S. Afr. Inst. Min. Met.*, **61**, 501–21.

12

Noise and its control in mines

A vibrating source produces vibrations into the medium in which it is placed. These vibrations are propagated as waves in the form of pressure variations and are termed acoustic waves. If they fall within the range capable of exciting the sense of hearing, they are called sound waves. An acoustic wave travels in a given medium at a constant velocity. When the level of sound becomes objectionable it is called noise. Sound becomes noise when:

(1) the sound level is so high that hearing can be permanently impaired;
(2) it interferes with normal work;
(3) it disturbs or prevents sleep.

The main sources of noise are unbalanced rotating or reciprocating machines, impacts, friction, alternating electromagnetic effects, air turbulence from fast rotating parts, fluid movement within ducts and through orifices, explosions or the sudden release of gas under pressure.

12.1 Properties of vibrations

As sound waves propagate through a medium the particles of the medium move up and down or sideways around their mean position while the wave moves forward in the direction of vibration of the particles (longitudinal wave) or at right angles to the direction of vibration of the particles (transverse wave). The minimum time interval after which the vibrating particle repeats its cycle of vibration around its mean position is called the time period. For a sinusoidal wave, the relationship between frequency and the time period is given by

$$f = \frac{1}{T} \tag{12.1}$$

or

$$f = \frac{\omega}{2\pi} \tag{12.2}$$

where f = frequency of the propagating wave, Hz;
T = time period, s; and
ω = angular velocity, radians/s.

The sinusoidal vibration can be represented by a vector rotating at a constant speed of ω (Fig. 12.1). The vertical displacement of this vector is given by

$$x = u \times \sin \omega t \tag{12.3}$$

$$x_{max} = u \tag{12.4}$$

where x = vertical displacement of the particle at a given time, m; and
$x_{max} = u$ = maximum displacement of the particle, called the amplitude of vibration, m.

The particle velocity, v, of this sinusoidal vibration is related to the displacement, u, by the relationship

$$v = \frac{du}{dt} = u \times \omega \times \cos \omega t \tag{12.5}$$

$$v_{max} = u \times \omega = \frac{2\pi u}{T} = 2\pi f u \tag{12.6}$$

where v = particle velocity at a given time t, m/s; and
v_{max} = maximum particle velocity, m/s.

The particle acceleration, a, is given by

$$a = \frac{d^2u}{dt^2} = -u\omega^2 \sin \omega t \tag{12.7}$$

Fig. 12.1. Sinusoidal vibration. ω = angular velocity; u = maximum displacement of the particle; T = time period.

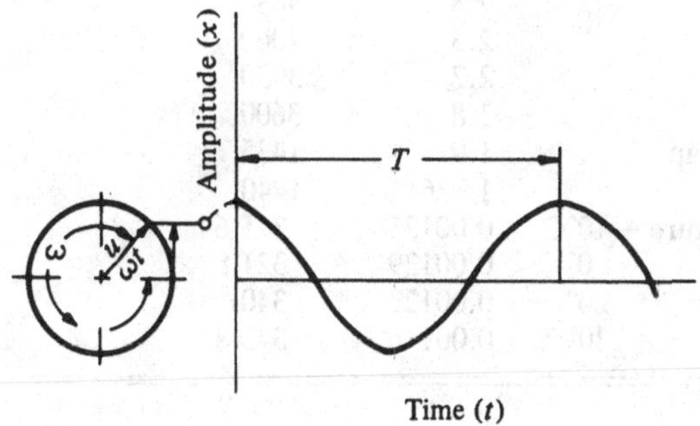

$$a_{max} = \frac{4\pi^2 u}{T^2} = 4\pi^2 f^2 u \tag{12.8}$$

where a = acceleration at a given time, t, m/s^2;
a_{max} = maximum acceleration of the particle, m/s^2; and
T = time period, s.

Besides maximum amplitude, maximum velocity and maximum acceleration, the terms effective amplitude, effective velocity and effective acceleration are also used and they are described by the following equations:

$$u_{eff} = \frac{x_{max}}{\sqrt{2}} \tag{12.9}$$

$$v_{eff} = \frac{v_{max}}{\sqrt{2}} \tag{12.10}$$

$$a_{eff} = \frac{a_{max}}{\sqrt{2}}. \tag{12.11}$$

Acoustic vibrations travel in the medium at a velocity which is a function of the medium. In liquids and gases the movement of the particles is in the direction of travel of the wave as the medium cannot sustain high shearing. As a result, only longitudinal waves can propagate through gases and liquids. The movement of particles in solids is more complex and these can transmit both longitudinal and transverse waves. The velocities of acoustic waves in various media are given in Table 12.1.

Table 12.1. *Acoustic velocity in different media*

Medium	Density (g/cm^3)	Velocity of acoustic waves (m/s)
Aluminium	2.7	5200
Glass	2.5	5190
Iron	7.9	5100
Steel	7.8	4880
Concrete	2.3	4000
Granite	2.7	3950
Brick	1.8	3600
Water from tap	1.0	1435
Sea water	1.026	1440
Air temperature – 10°C	0.00135	325.6
0°	0.00129	321.8
20°	0.00122	340
30°	0.00116	343.8

The vibrations travel through the medium a certain distance λ, called the wavelength of the vibrations in the time period T, such that

$$\lambda = cT = \frac{c}{f} \tag{12.12}$$

where λ = wavelength of the vibration, m; and
c = velocity of vibration, m/s.

Thus when a vibration source is placed in a medium and it produces vibrations of a certain frequency, the wavelength of these vibrations depends upon the properties of the medium.

The human ear does not respond to all frequencies. Acoustic vibrations in the range of 16 to 20 000 Hz (usually 16 to 16 000 Hz) are capable of being heard by humans and hence this range of acoustic waves are called the audible range and the waves lying in this range are called sound waves. Also, the human ear cannot distinguish all frequencies within the audible range with equal facility. Sounds within 1000 to 4000 Hz are most easily heard. The ability to hear higher frequencies declines with age.

12.1.1 Acoustic pressure or sound pressure

The propagation of a wave through air results in changes in pressure which are related to amplitude of the vibrations of the particles. Fig. 12.2 shows a longitudinal wave where the particles are vibrating

Fig. 12.2. (*a*) Displacement of individual particles; (*b*) compression and rarefaction of the medium corresponding to the individual particle; (*c*) change in acoustic pressure.

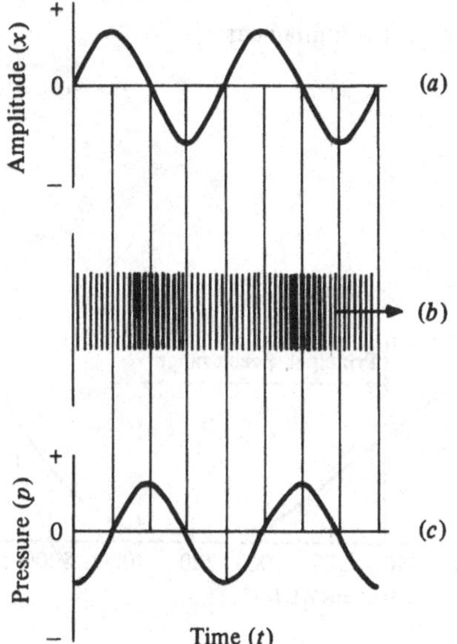

around their mean position in a direction parallel to the direction of the propagation of the wave. As a result of the vibrations of particles, regions develop where the density of the particles increases and decreases with time, giving rise to variations in pressure with respect to the barometric pressure where there are no sound vibrations. The human ear responds to a change in sound pressure which lies approximately in the range of 0.00002 to 20 Pa. The faintest sound that can be heard is termed the threshold of audibility. This threshold depends upon the frequency and is given in Fig. 12.3. The sound pressure at which physical discomfort becomes apparent is the threshold of feeling or audible pain. The ratio of these two thresholds is about 10^6 to 1 and it is inconvenient to use a linear unit to cover such wide limits. Consequently a logarithmic unit is normally used.

The sound pressure level, L_p, is expressed in dimensionless form as the ratio between two sound energies. These are proportional to the squares of the sound pressures, p. The basic unit is called a bel (B).

$$L_p = \log \left(\frac{p}{p_r}\right)^2. \tag{12.13}$$

A more convenient unit is, however, the decibel (dB).

$$L_p = 10 \log \left(\frac{p}{p_r}\right)^2 = 20 \log \left(\frac{p}{p_r}\right) \tag{12.14}$$

where p = the actual pressure level, Pa; and
p_r = the reference pressure level, Pa.

Fig. 12.3. The auditory field for the human ear.

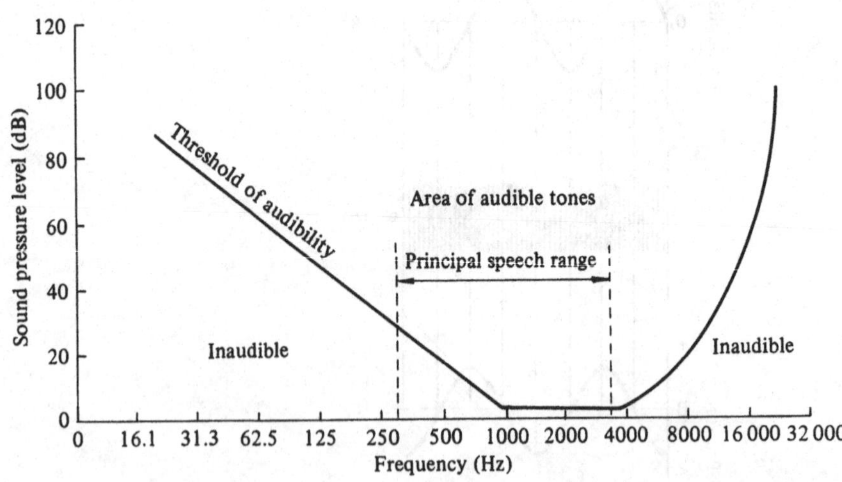

The 'threshold of hearing', representing the lowest sound audible to the ear, is taken as the reference pressure, p_r, and has a value of 2×10^{-5} Pa. Then

$$L_p = 20 \log \left(\frac{p}{2 \times 10^{-5}} \right), \text{ dB.} \tag{12.15}$$

The reference pressure of 2×10^{-5} Pa represents the reference level of 0 dB.

12.1.2 Sound intensity and power

Sound intensity is a measure of the amount of energy flowing through a unit area in a unit time and is given by the equation

$$I = \frac{p^2}{wc} \tag{12.16}$$

where I = intensity, W/m^2;
p = sound pressure, Pa;
w = density of the medium, 1.2 kg/m^3; and
c = velocity of sound, m/s.

Sound intensity levels and sound pressure levels are related to each other as follows:

$$L_I = L_p = 10 \log \left(\frac{p}{p_r} \right)^2 = 10 \log \left(\frac{I}{I_r} \right), \text{ dB.} \tag{12.17}$$

where I_r = reference intensity, W/m^2.

Thus to determine the acoustic power of a source, it is required to measure the pressure level at some distance from the source, on the assumption that sound travels from the source in all directions. Thus for a source suspended in space, the acoustic power, N, (W) of the source is given by

$$\begin{aligned} N &= I \times S \\ &= I \times 4\pi r^2 \end{aligned} \tag{12.18}$$

where N = acoustic power of the source, W;
I = intensity, W/m^2;
S = surface area of the sphere, m^2; and
r = distance at which the sound pressure intensity is measured, m.

For the surface area of hemispherical shape, i.e. the source placed on a level ground,

$$N = 2\pi r^2 \times I. \tag{12.19}$$

Once again the sound power level L_N, is to be related to the reference level. The reference level for acoustic power, N_r, is 10^{-12} W which corresponds to the lowest intensity the ear can normally detect. Therefore the acoustic power level or sound power level (in dB) as it is sometimes called can be given by

$$L_N = \log\left(\frac{N}{N_r}\right) = \log\left(\frac{N}{10^{-12}}\right)$$

$$= 10 \log\left(\frac{N}{10^{-12}}\right). \tag{12.20}$$

The acoustic power and the sound intensity of some typical sources are given in Table 12.2.

The acoustic power of any equipment does not change materially for any change in its location and hence can be assumed to be constant unless some internal changes are made to it. However, the total sound pressure level experienced in a given system may change depending upon the resonance effects of the equipment in relation to the other structures placed around or next to it.

12.1.3 Directional characteristics of emission of noise

Invariably the equipment producing the noise is very complex, consisting of a number of mechanical systems working with variable frequency and a method of transmission of power (rotary, percussive or longitudinal), and the spatial distribution of sound pressure level is not uniform. To define their spatial characteristics, tests have to be conducted on an imaginary spherical surface of radius r, where the sound pressure is measured and the mean value is calculated using the relationship

$$p_{av} = \left[\frac{p_1^2 + p_2^2 + \cdots + p_n^2}{n}\right]^{1/2} \tag{12.21}$$

Table 12.2. *The acoustic power and sound power level of some typical sources*

Source	Acoustic power (W)	Sound power level (dB)
Jet plane	10^4	160
Pneumatic equipment	1	120
Tractor	10^{-1}	110
Piano	2×10^{-2}	103
Human voice during normal talking	2×10^{-5}	73
Small electric bell	2×10^{-8}	43
Push sound	2×10^{-9}	30

and

$$L_{p_{av}} = 10 \log \left[\frac{p_1^2 + p_2^2 + \cdots + p_n^2}{np_r^2} \right]. \tag{12.22}$$

where p_{av} = the average pressure, Pa;
$p_1, p_2, p_3 \ldots p_n$ = the pressures measured at various positions, Pa;
n = number of positions;
$L_{p_{av}}$ = the average pressure level, dB; and
p_r = the reference pressure, Pa.

If the difference between the individual values is less than 5 dB, then the source can be said to have spherical characteristics. If the difference is more than 5 dB, the directional coefficient of the acoustic field is calculated from the relationship

$$\phi_a = \frac{p_a^2}{p_{av}^2} \tag{12.23}$$

where ϕ_a = coefficient of acoustic field in direction a;
p_a = pressure in direction a, Pa; and p_{av} = average pressure, Pa

and the acoustic field factor is given by

$$G_a = 10 \log \phi_a. \tag{12.24}$$

Thus, for a source with spherical characteristics, $\phi_a = 1$ and $G_a = 0$. When $G_a \neq 0$, then the shape of the surface of equal sound pressure levels becomes elongated in one or more directions.

12.1.4 Summation of sound pressure levels

The scale of the sound pressure levels is logarithmic. Therefore the sound pressure levels from two independent sources cannot be added arithmetically.

If the sound pressures of two sources are p_1 and p_2 Pa respectively, then the combined sound pressure level in dB due to the two sources can be calculated as follows:

$$L_p = 10 \log \left(\frac{p_1^2 + p_2^2}{2 \times 10^{-5}} \right). \tag{12.25}$$

Alternatively, if the sound pressure levels of the two sources are known, then the combined sound pressure level can be calculated using the values given in Table 12.3. Thus if the sound pressure level of a source is 85 dB and the other source is 81 dB, then the combined noise from the two sources will be 85 + f(85 − 81) = 85 + f(4) = 85 + 1.5 = 86.5 dB. When the difference between the two sources is zero (say, 80 dB machines) as may be the case with two identical machines, the increase in sound pressure level will be 3 dB. In other words, adding another similar

Table 12.3. *Addition of unequal sound levels*

Difference between dB values of two sources	Add this value to the higher dB value
0	3
1	2.5
2	2
3	2
4	1.5
5	1
6	1
7	1
8	0.5
9	0.5
10	0
>10	0

Fig. 12.4. (*a*) Addition of two unequal sound levels; (*b*) addition of equal sound levels.

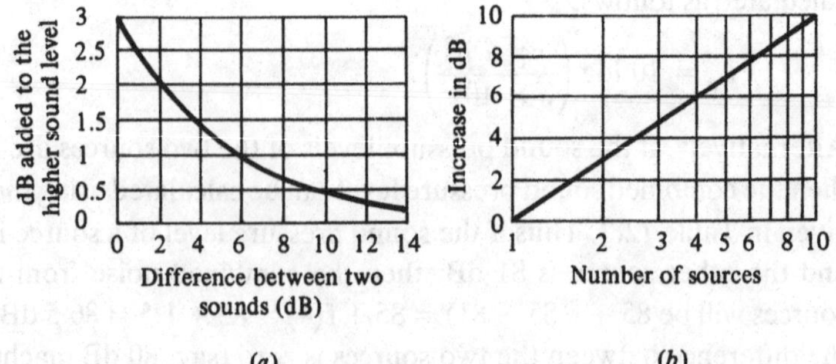

machine increases the sound pressure level by 3 dB. At this stage, if another machine is added then the sound pressure level will increase by a further 2 dB, i.e. the total sound pressure level will be (83 + f(83 – 80)) = 83 + 2 = 85. If a fourth machine is added, then the new sound pressure level will be (85 + f(85 – 80)) = 85 + f(5) = 85 + 1 = 86. A simplified method of adding unequal and equal sound levels is given in Fig. 12.4.

As seen from Table 12.3, the addition of 3 dB represents a doubling of the sound pressure level but does not of course sound two times as loud to the human ear.

The human ear does not respond with equal sensitivity to all frequencies within the audible range. Because of this, the sound pressure level is not an accurate indicator of loudness or intensity of sound. The unit of loudness level is the phon, which is derived by a subjective comparison with a 1000 Hz reference tone. When the reference tone and the sound under consideration are judged to be equally loud, the loudness level of the sound in phons is numerically equal to the sound pressure level of the reference tone. Fig. 12.5 shows equal loudness contours.

12.1.5 Effect of distance from the source

A spherical point source emits sound equally in all directions. As indicated earlier, the sound intensity as measured at the surface of a

Fig. 12.5. Equal loudness contours (in phons) using a frequency of 1000 Hz as a base.

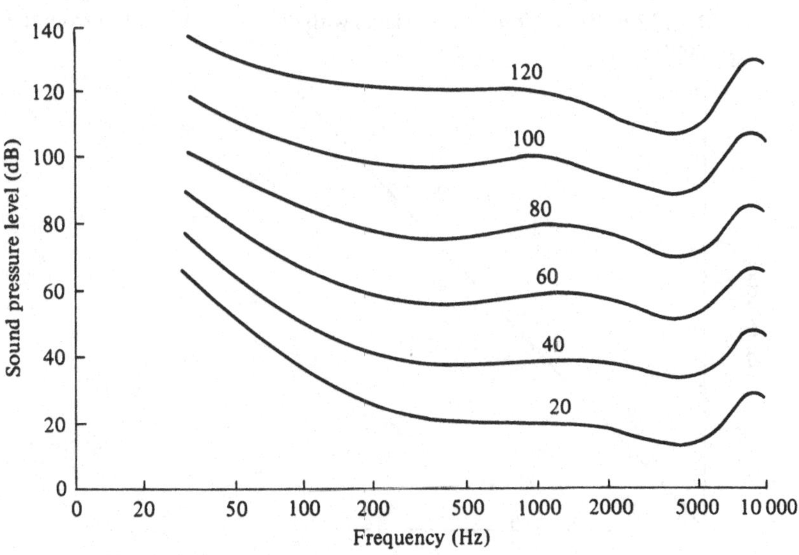

sphere is inversely proportional to the square of the distance:

$$\frac{I_1}{I_2} = \left(\frac{r_2}{r_1}\right)^2. \tag{12.26}$$

where I_1 = sound intensity at a distance r_1 (m) from a point source, W/m^2; and

I_2 = sound intensity at a distance r_2 (m) from the same source, W/m^2.

Thus, if the intensity of the sound level, L_I, in dB is known at one point, the intensity of sound level at another point can be calculated using the relationship

$$L_{I_2} = L_{I_1} - 20 \log\left(\frac{r_2}{r_1}\right) \tag{12.27}$$

and

$$\Delta L_I = 20 \log\left(\frac{r_2}{r_1}\right).$$

Thus, for example, if the sound pressure level at 10 m from a source is 118 dB (L_{p_1}), then the sound pressure level at 80 m from the source (L_{p_2}) shall be

$$L_{p_2} = L_{p_1} - 20 \log\left(\frac{r_2}{r_1}\right)$$

$$= 118 - 20 \log\left(\frac{80}{10}\right) = 118 - 20 \times 0.9 = 100 \text{ dB}.$$

Fig. 12.6. Reduction in sound level with distance beyond 1 m from the source.

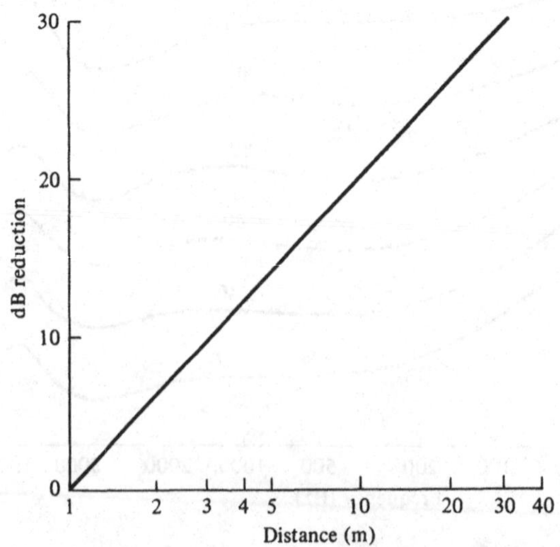

Fig. 12.6 shows the reduction in sound pressure level at distances beyond 1 m from the source.

12.1.6 *Acoustic frequency spectrum*

The acoustic frequency of the various sources varies within wide limits. The various sound frequencies emitted do not have the same sound pressure levels. The analysis of sound levels emitted by a source therefore must be conducted for the full audible range of 16 to 20 000 Hz. In practice, it is sufficient to analyse between 40 and 8000 Hz. This analysis is conducted by measuring the sound pressure levels within well-defined bands. The widest bands used in frequency analysis are called octave bands. An octave band is defined as one which has the ratio of upper frequency, f_2, to the lower frequency, f_1, equal to two, i.e. $f_2/f_1 = 2$. A half octave has $f_2/f_1 = \sqrt{2}$ and a third octave has $f_2/f_1 = \sqrt[3]{2}$. The band is defined by its mid-band frequency, given by the geometric mean of the upper and lower limits, i.e. $f = \sqrt{f_1 f_2}$ Hz. Table 12.4 gives the octave band and the mid-band frequencies which are normally used in analysis. The bands may be narrowed using half an octave, one-third of an octave or may be even narrower.

In the case of a continuous sound (white sound), when the sound pressure levels of the various bands are measured, the total sound pressure level, $L_{p_{total}}$, of the source for the whole of the audible sound spectrum is given by

$$L_{p_{total}} = 10 \log (10^{0.1L_1} + 10^{0.1L_2} + \cdots + 10^{0.1L_n}) \qquad (12.28)$$

where $L_1, L_2, \ldots L_n$ = sound pressure levels of various bands, dB. The sound pressure level for a narrower band is smaller. For example, if $L_{p_{total}}$ for the frequencies $\Delta f = f_2 - f_1 = 8000 - 40 = 7960$ is $L_{p_{7960}}$ dB, then the sound pressure level for frequencies $\Delta f = 100$ (in dB) is given by

$$L_{p_{100}} = L_{p_{7960}} - 10 \log \frac{7960}{100}.$$

When measurements are done within a given octave, then

$$\begin{aligned} L_{p_{oct}} &= 3 + L_{p_{\frac{1}{2}oct}} \\ &= 5 + L_{p_{\frac{1}{3}oct}} \end{aligned} \qquad (12.29)$$

where $L_{p_{\frac{1}{2}oct}}$ and $L_{p_{\frac{1}{3}oct}}$ are the sound levels measured at the one-half and one-third octave frequencies, and $L_{p_{oct}}$ is the sound level at one octave frequency.

12.1.7 *Acoustic absorption*

When a sound wave hits a physical surface it is partially reflected and partially absorbed; the greater the rigidity of the surface, the greater

Table 12.4. *Octave bands and mid-band frequencies normally used in the analysis of sound levels emitted by a source (after Howe, 1982)*

Band limits (Hz)	22		44		88		176		353		707		1414		2829		5650		11 300		22 500
Mid-band frequency (Hz)		31.5		63		125		250		500		1000		2000		4000		8000		16 000	

Normally used: 63–4000 Hz

the reflection. The ratio of the energy absorbed to the incident energy is called the coefficient of absorption, α, and is given by the relationship

$$\alpha = \frac{I_2}{I_c} \tag{12.30}$$

where $I_2 = I_c - I_1$;
I_c = incident energy;
I_1 = reflected energy; and
I_2 = absorbed energy.

The acoustic absorption, A, in m^2 of a surface of area, S, is given by

$$A = \alpha S. \tag{12.31}$$

In an enclosed space the acoustic absorption (A) of a number of different surfaces and objects is given by the relationship

$$\begin{aligned} A &= \alpha_1 S_1 + \alpha_2 S_2 + \cdots + \alpha_n S_n \\ &\quad + n_1\alpha'_1 + n_2\alpha'_2 + \cdots + n_m\alpha'_m \\ &= \sum_{i=1}^{n} \alpha_i S_i + \sum_{i=1}^{m} n_i\alpha'_i \end{aligned} \tag{12.32}$$

where $\alpha_{1,2\ldots,n}$ = absorption coefficients of the various surfaces, roof, floor, etc.;
$S_{1,2\ldots,n}$ = surface areas of the respective surfaces;
$\alpha'_{1,2\ldots,m}$ = absorption coefficients of the various objects in the enclosed space; and
$n_{1,2\ldots,m}$ = number of respective objects with absorption coefficients of $\alpha'_1, \alpha'_2, \ldots \alpha'_m$.

The value of A for a given enclosed environment can be calculated by measuring the time for the noise level from a steady-state high level to drop by 60 dB. This time T is called the reverberation time and is given by

$$T = \frac{0.161V}{S\alpha_{av}} \tag{12.33}$$

where V = volume of the closed environment, m^3;
S = total surface area of the enclosure; and
α_{av} = average absorption coefficient of the enclosure.

Concrete and other rock surfaces have absorption coefficients of about 0.05–0.1, while expanded polyurethane can have an absorption coefficient of 0.98. Absorption coefficients are also dependent upon the frequency, and the value increases with higher frequencies. Frequencies less than 250 Hz are much more difficult to absorb. The absorption coefficients of rock for different frequencies are given in Table 12.5.

Table 12.5. *Absorption coefficients of rock surfaces in underground excavations (after Howe, 1982)*

Mid-band frequency (Hz)	63	125	250	500	1000	2000	4000	8000
Absorption coefficient	0.05	0.10	0.13	0.14	0.15	0.16	0.16	0.16

Table 12.6. *Attenuation in dB/m of 760 mm diameter ducts (after Howe, 1982)*

Mid-band frequency (Hz)	63	125	250	500	1000	2000	4000	8000
Steel ducts	0.08	0.13	0.17	0.21	0.25	0.28	0.30	0.31
Flexible ducts	0.15	0.25	0.55	0.75	0.95	1.05	1.15	1.25

Table 12.7. *Reduction in sound pressure levels between the inside and outside of a duct (1 m from the duct) (after Howe, 1982)*

Mid-band frequency (Hz)	63	125	250	500	1000	2000	4000	8000
760 mm steel duct (dB)	30	18	28	32	23	22	27	29
760 mm flexible duct (dB)	23	11	8	7	10	16	21	22

Ducts used underground for ventilation radiate noise to the surroundings. Duct surfaces can bend and absorb sound energy. The absorption is dependent both on the material of construction of the duct and the duct diameter. Attenuation in dB per metre length of the duct is given in Table 12.6. The differences between sound pressure levels inside and outside at 1 m distance are given in Table 12.7.

When equipment is placed in an elongated environment, e.g. a tunnel or a development heading, it is not only the direct sound but also the reflected sound reaches an area. The method of transmission is given in Fig. 12.7. The drop in sound level as a function of distance can be calculated by the equation

$$dB = 12.6 \frac{P}{A} x^{0.8} \alpha^{1.4} \tag{12.34}$$

where dB = sound level reduction over distance x, m–dB;
P = perimeter of the excavation, m;
A = cross-sectional area of the excavation, m^2;
x = distance from the source, m; and
α = absorption coefficient (see Table 12.5).

Fig. 12.7. Stages of sound transmission in an excavation (after Howe, 1982).

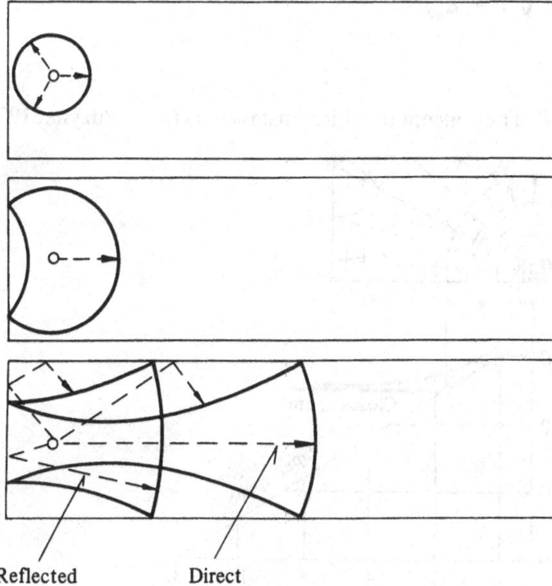

12.1.8 *Critical distance*

In an enclosed environment the reflected waves add to the sound pressure levels. At very close distances to the source, the sound pressure level drops with distance as in an open space, but at longer distances, reflected waves add to the sound levels and the sound levels remain constant. This is schematically explained in Fig. 12.8. Thus for a given environment, there is a critical distance, r_g, from the source at which the equilibrium between the direct and the reflected (indirect) noise pressure levels is obtained. The value of r_g can be calculated by using the following equations. For a spherical source

$$r_g \simeq 0.14\sqrt{A}. \tag{12.35}$$

For a half-spherical source

$$r_g \simeq 0.2\sqrt{A}. \tag{12.36}$$

where A = cross-sectional area of the excavation, m^2.

If the average absorption coefficient of all surfaces (roof, floor, equipment, etc.) is α_{av}, then the exact value of r_g is given by the following equations. For a spherical source

$$r_g = 0.14\sqrt{\frac{S_a\alpha_{av}}{1-\alpha_{av}}} = 0.14\sqrt{R}. \tag{12.37}$$

For a half-spherical source

$$r_g = 0.2\sqrt{\frac{S_a\alpha_{av}}{1-\alpha_{av}}} = 0.2\sqrt{R}. \tag{12.38}$$

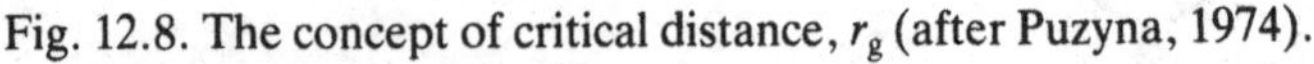

Fig. 12.8. The concept of critical distance, r_g (after Puzyna, 1974).

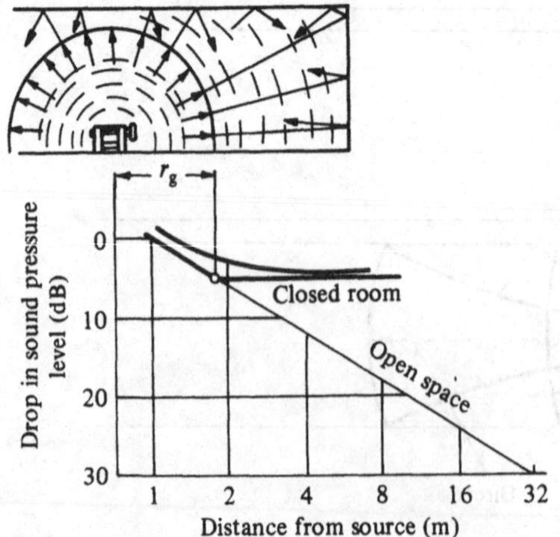

where S_a = sum total of all the surfaces, m^2; and
R = acoustic constant of the room, m^2.

For example, if some equipment is placed in a room of 5 m × 4 m × 4 m, with an average acoustic absorption coefficient of 0.2, then $S_a = 2 \times (5 \times 4 + 5 \times 4 + 4 \times 4) = 112\ \text{m}^2$. Therefore $A = S_a\alpha_{av} = 112 \times 0.2 = 22.4\ \text{m}^2$. From equation 12.35, $r_g \simeq 0.14\sqrt{22.4} \simeq 0.6$ m, or the exact value is given by equation (12.37):

$$r_g = 0.14\sqrt{\frac{22.4}{1 - 0.2}} = 0.74\ \text{m}.$$

It is therefore obvious that when equipment is placed in an open space the sound pressure level will drop as a function of distance (equation 12.27) but when this equipment is placed in an enclosed environment, then the minimum sound pressure level will be that which is achieved at a distance r_g in the enclosed environment.

12.2 Measurement of sound

While making noise measurements it is essential to understand the working of the equipment. Certain machines, for example a diesel engine, will tend to give a more or less uniform sound intensity level, and in such a case it is sufficient to define the maximum and minimum sound intensity levels. In other cases the sound level may not remain constant, for example a shuttle car operating between the face and a conveyor. In such a case it is essential to define not only the minimum and maximum levels but also the time periods during which they occur and how the minimum and maximum levels vary.

In general, noise is measured with a sound level meter which consists of a microphone, amplifier, weighting networks (or filters) and an indicating meter (or recording instrument) which registers the sound level in decibels. When making measurements it is essential to understand the characteristics of the sound level meters. Fig. 12.9 gives the directional characteristics of the Bruel and Kjaer sound meter with a capacitor microphone as a function of the angle of placement of the instrument in relation to the sounds. It is therefore essential that all measurements are conducted by placing the microphone at 0° to the direction of measurements. Here the microphone has a sensitivity which is almost linear from 0 to 10 000 Hz while at other angles, it is linear only up to 1000 Hz.

Modern instruments give direct sound pressure levels in dB and do correct automatically for any non-linearity in the sensitivity. However, where this facility is not available, and measurements are made using microphones and recording the data onto tape recorders, the sensitivity of the microphone, and the sensitivity of the tape recording unit both during recording and playback as a function of the frequency must be analysed and any variations must be taken into account.

Standard sound measuring units make allowance for the frequency dependent response of the ear (Fig. 12.5) by using filters (weighting networks) which attenuate the levels in the different frequencies and thereby give direct readings which reflect the sensitivity of the ear. Since the amount of attenuation is dependent upon the sound level, the three networks (A, B and C) usually employed have frequency characteristic responses simulating 40, 70 and 100 dB equal loudness contours. Fig. 12.10 gives the normalised characteristics of the three weighting networks.

In making measurements it is essential that the observer stands at least 0.5 m from the instrument so that no reflected waves affect the measurements. Measurement errors including those due to instruments usually do not exceed 2 dB. The cause of errors exceeding this limit should be carefully investigated.

When making measurements at a place of work, the microphone is set up 1.6 m from the floor level for operators working in a standing position or 0.6 m for workers working in a sitting position.

Fig. 12.9. Directional characteristics of the Bruel and Kjaer sound meter (after Puzyna, 1974).

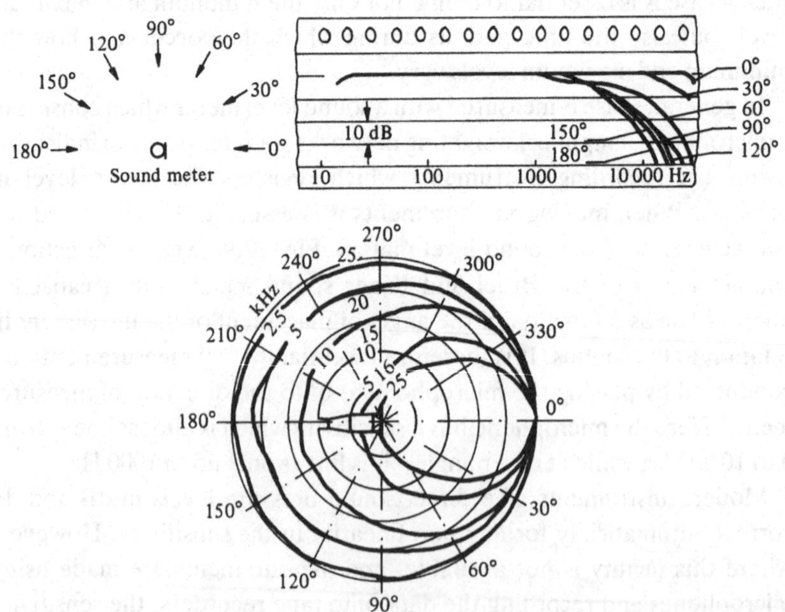

Experience has shown that the A weighting correlates well with subjective noise ratings even at high levels and consequently it is in general use whenever a measurement corresponding to a subjective reaction to a noise is required. Recent research has shown that the B weighting more nearly equates hearing losses than does A and that no weighting overcompensates. The C weighting lies between the B weighting and no weighting and is suitable for sound pressure levels over 85 dB.

The use of the A weighting only in hearing conservation results in an error of the order of 2 dB. In mining, the frequency spectrum tends to fall. The mean sound levels in the 2000 and 4000 Hz bands are lower than the mean sound levels in the 250 and 500 Hz frequencies. The use of the A scale weighting network therefore results in a small underestimation of the damaging effect of noise.

The weighting network allows a single value to be used to describe the environment but gives no information on the frequency.

When recording sound level meter readings it is important to state which, if any, weighting network was used, e.g. 65 dB(A).

Noises may be characterised not only by loudness, but also by tonal range and duration. All three of these factors can be important in assessing the problems which a particular noise can cause.

Where it is desirable to investigate the tonal characteristics of a noise, the sound level meter can be used with octave or one-third octave band pass filters for frequency analysis (Table 12.4).

12.2.1 Open area measurements

The positions of sound-measuring instruments for spherical and hemispherical propagation are given in Fig. 12.11. The distance at which the measuring instrument is placed should not be less than twice the

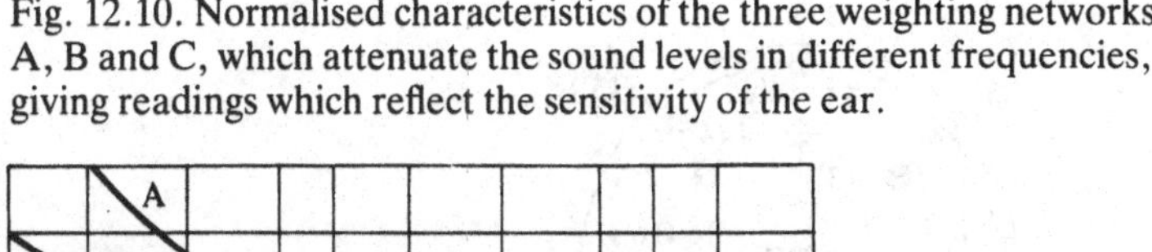

Fig. 12.10. Normalised characteristics of the three weighting networks A, B and C, which attenuate the sound levels in different frequencies, giving readings which reflect the sensitivity of the ear.

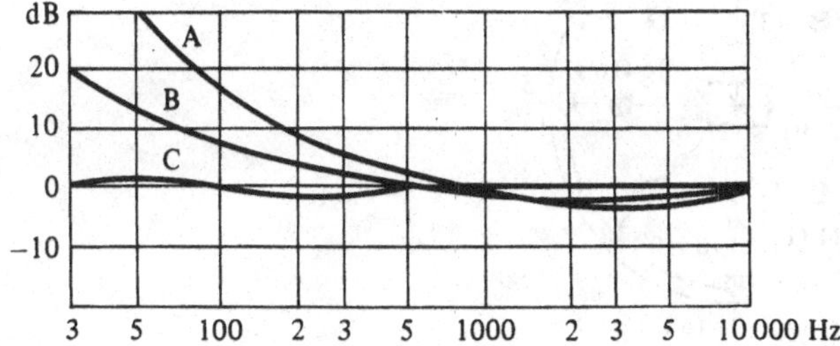

maximum dimensions of the source. If the difference in noise levels measured exceeds 5 dB, then the measurements must be repeated.

Measurements can be converted to a standard (1 m distance) for comparing the acoustic power of various equipment using the equation

$$L_N = L_{av} + 10 \log \frac{S}{S_r} \text{ dB} \tag{12.39}$$

where L_N = sound power level;
L_{av} = average value of sound pressure levels measured at the surface S; and
S_r = surface of 1 m^2.

Fig. 12.11. Positions of a sound measuring instrument for (*a*) hemispherical propagation; (*b*) spherical propagation (after Puzyna, 1974). 1–16 represent positions above the source; 1′–16′ represent positions below the source.

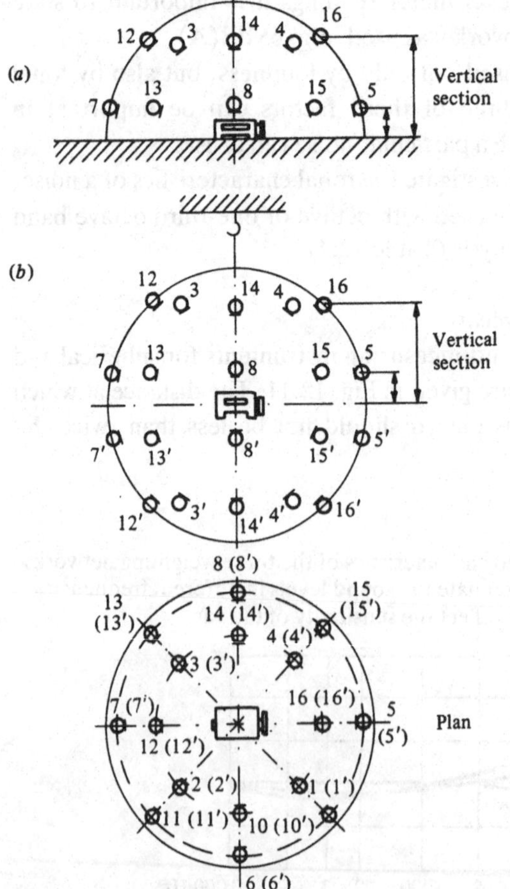

Thus, for example, if the measurements are taken 5 m away from a machine of dimensions 1 m × 1.5 m × 1.2 m, and the average sound pressure level for a number of measurements is 100 dB(A), then the acoustic power (N) of the machine is calculated as follows:

$$L_N = 100 + 10 \log \frac{2\pi 5^2}{1}$$

$$= 121.98 \simeq 122 \text{ dB(A)}.$$

From equation (12.20)

$$122 = 10 \log \left(\frac{N}{N_r}\right).$$

Since the reference level for acoustic power, $N_r = 10^{-12}$ W

$$\log \left(\frac{N}{10^{-12}}\right) = 12.2$$

$$N = 1.58 \text{ W}.$$

12.2.2 Measurements in an enclosed environment with dissipated noise

In the case of enclosed environments with dissipated (diffused) noise, particularly in a small room where the coefficient of acoustic absorption is small ($\alpha < 0.1$), at least five measurements should be taken at a distance $r > 1$ m ($r > r_g$, the critical distance), (equation 12.36) from the geometric centre of the equipment. At the same time the location of the measuring instrument must be at least 1 m away from the walls, roof, floor and the equipment and at least 2 m from the corners of the room. The acoustic power level, L_N, of the equipment can be calculated thus (in dB):

$$L_N = L_{av} + 10 \log \frac{A}{A_r} + 6 \qquad (12.40)$$

where L_{av} = average value of the noise pressure level measured at five points, dB;
A = acoustic absorption of the room, m^2 (equation 12.33); and
A_r = acoustic absorption for reference, 1 m^2.

12.2.3 Measurements in large rooms

In large rooms it is advisable to 'make' measurements of sound pressure levels of equipment which has elongated shapes at a constant distance from the external surface of equipment in two planes (vertical and horizontal – Fig. 12.12). The minimum distance between the point of measurement, walls and equipment surface should be about 1 m and from the corners 2 m. The distance from the floor, h, should be $H/2$ (see Fig. 12.12). At least five measurements must be taken.

In the case of such a measurement for a hemispherical source, the sound pressure level, L_N, can be calculated by using the equation

$$L_N = L_{av} + 20 \log \frac{r}{r_a} + 10 \log 2\pi \qquad (12.41)$$

where L_{av} = average value of sound pressure level measurements, dB;
r_a = radius of the weighted half sphere, m,
$= \sqrt{\dfrac{ab + ac}{2}}$ where a, b and c are the dimensions in Fig. 12.12; and
r = distance from the enclosed surface of the equipment (1 or 3 or 10 m) as dictated by the experimental conditions ($r = d$ in Fig. 12.12), m.

12.3 Effects of noise

12.3.1 *The human ear*

The ear is a delicate structure that turns physical vibration in the audible range into electrical impulses or messages to the brain. Its structure is illustrated in Fig. 12.13. Sound waves in the air cause minute variations of pressure in fluid in a non-elastic coiled tube, the cochlea,

Fig. 12.12. Measurement of sound pressure levels in large rooms (after Puzyna, 1974). h = distance of measuring equipment above the floor.

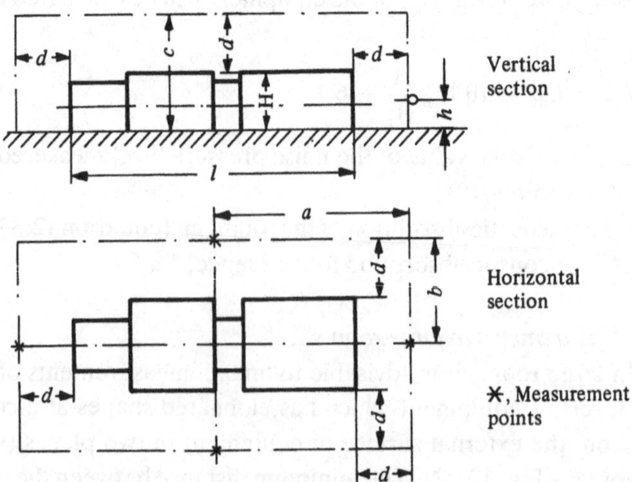

which contains the organ of Corti, along which are about 24 000 tiny hairs. These hairs all respond to different pressure variations and together send an electric message to the brain.

With age, the ability to hear high frequencies is reduced. This is due to the natural reduction in the number of hairs on the organ of Corti. Since the understanding of speech depends very much on the distinguishing of consonants, and consonants generate high frequencies, the elderly tend to become hard of hearing.

Exposure to high sound levels for an excessive time has a similar effect, causing a permanent reduction in acuity in the higher frequencies, often with a marked reduction of hearing around 4000 Hz. This reduction is called the permanent threshold shift. Noise also causes these hairs to become 'numb' for a day or two, this reaction being referred to as the temporary threshold shift.

A very loud sound may rupture the ear drum membrane. This generally heals, but it may leave a hole, resulting in defective hearing. The term 'deaf' refers to this partial loss of hearing. To be 'stone deaf' is to hear nothing at all.

Using audiometry techniques, it is possible to assess hearing loss due to exposure to high sound levels during employment. Exposure during leisure time may complicate matters. Complications also arise owing to a number of factors such as heredity, age, disease and poisons.

Fig. 12.13. Schematic diagram of the ear.

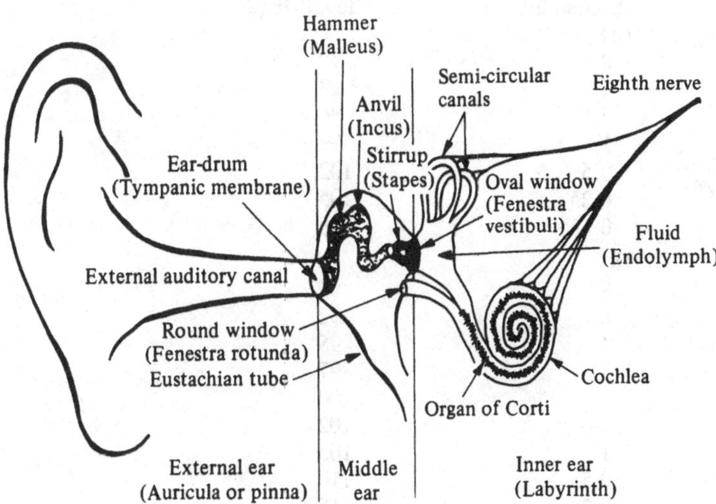

Table 12.8. *Critical sound pressure levels and their effects on humans (after Puzyna, 1974)*

L_p (dB(A))	Type of work	Human reaction	Communication
100	Working more than 5 hours per day without protection	Hearing damage	Difficult
90	General mechanical equipment, workshops, etc.	Affects nervous system	
80	General erection sites		Needs raised voice
70	Equipment assembly requiring precision		
60	Laboratory work, offices, etc.	Affects mental state	Normal voice
50	Work requiring special attention		
40	Work requiring extreme attention		

Table 12.9. *(a) UK Department of Employment code of practice and (b) maximum sound pressure levels set by the Occupational Safety and Health Act in the USA*

Maximum exposure (hours/day)	Sound pressure level(dB(A))
(*a*)	
8	90
4	93
2	96
1	99
0.5	102
0.25	105
0.125	108
(*b*)	
8	90
6	92
4	95
3	97
2	100
1.5	102
1	105
0.5	110
⩽0.25	115

12.3.2 *Acceptable levels*

The effect of noise on humans is psychological as well as leading to possible hearing damage. It also affects the quality and precision of work. Table 12.8 gives the critical sound pressure levels and their effects on humans.

Table 12.9 shows the code of practice issued by the UK Department of Employment, and USA's Occupational Safety and Health Act. The UK code recommends no one with unprotected ears shall be exposed to a sound level above 135 dB or to an impact noise exceeding 150 dB, and that no part of the body be exposed to continuous noise exceeding 150 dB.

According to the Occupational Safety and Health Act in the USA, sound levels can be allowed to rise higher than those in the UK code before reducing the exposure time by a half. In no case should a worker be exposed to a continuous sound level of more than 115 dB(A), nor to impulsive or impact noise in excess of 140 dB(C).

When the daily exposure is composed of periods of noise exposure at different levels ($c_1, c_2 \ldots$) the combined effect should be considered, i.e. the sum of the following fractions must not exceed unity.

$$\frac{c_1}{t_1} + \frac{c_2}{t_2} + \cdots + \frac{c_n}{t_n} \leqslant 1.00 \qquad (12.42)$$

where c_n = the total time of exposure at a specified noise level; and
t_n = the permissible time at that exposure.

Variable noise which varies at intervals of more than 1 s is called impact noise. The maximum number of impacts for any given peak sound pressure level is given in Fig. 12.14.

Fig. 12.14. Suggested limits for impact noise (after Howe, 1982).

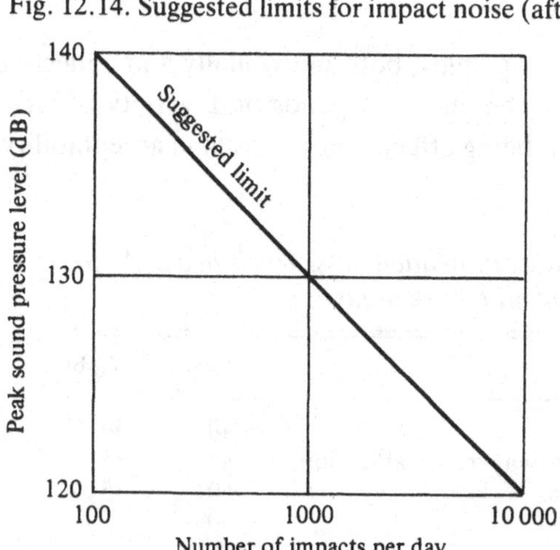

Aside from the hearing damage that may result from loud noise, ease of communication underground is very important. Speech interference occurs between 500 and 2000 Hz at levels below hearing damage level. The permissible noise level will depend upon the distance between the communicators. At 80 dB, the minimum distance is 0.5 m. For each doubling of the distance, the noise rating will have to drop by 5 dB.

Another method adopted for calculating equivalent exposure sound pressure level for an 8 hour shift in places of variable sound levels and duration is to use the following equations:

$$L_{eq} = \frac{\log F}{0.1} + 90 \qquad (12.43)$$

where $F = \Sigma f$ and

$$f = \frac{t}{8} \text{ antilog } [0.1(L - 90)]$$

where L_{eq} = equivalent continuous sound level;
F = total exposure factor;
f = fractional exposure factor;
t = time period in hours; and
L = sound level, dB.

The equivalent exposure sound pressure level should not exceed the prescribed limits given in Table 12.9. Table 12.10 gives some of the recommended levels in neighbourhood areas and work places. An increase in noise level of 10 dB(A) in the neighbourhood as a result of introducing a new source usually leads to complaints. A rise of 5 dB(A) may be considered as marginal.

12.4 Noise control

Several basic principles, both individually and collectively, are used to control noise. The choice depends on a variety of factors, the primary considerations being effectiveness, cost and acceptability.

Table 12.10. *Recommended noise levels in dB(A) in neighbourhood and work areas*

Situation	Day	Night
Country areas	40	30
Suburban areas away from traffic routes	45	35
Busy urban areas	50	35
Design offices	40	
Precision workshops	50	

Undoubtedly, the right time to consider the noise levels of machines and plant equipment is at the design stage. Good functional design often helps to reduce noise levels.

Noise can be controlled by (1) preventing it from being generated; (2) confining it by isolating the source; (3) modulating it by using air silencers, resilient mounting, etc.; (4) isolating the whole machine or plant from people; (5) isolating people from the main source and protecting them by wearing ear defenders and finally; (6) reducing the time that persons are exposed (alternatively, making the noise at selected times only).

Prevention

Machine vibrations are the source of most noise. Fast rotating parts in perfect balance and properly aligned and rigidly mounted bearings produce less vibration. Generally, a slow-running machine makes less noise than a fast-moving one. Wear increases vibration and therefore produces more noise.

Electric motors are quieter than compressed air or internal combustion engines. Care is needed to avoid electromagnetic vibration in all electrical equipment. Air turbulence can be reduced by smoothing the flow as far as possible.

The Helmoltz expansion box, sometimes combined with a lining of sound-absorbent material, remains the commonest way of controlling the explosive noise from internal combustion engines, compressed air equipment and blowers.

Impact noise can be minimised by not allowing parts to bang or clatter. If this is absolutely necessary, either or both of the contact surfaces may be of resilient material.

Isolation

Within a machine, but away from the source, much can be done by building in sound-absorbent barriers. The water-cooling jacket of an internal combustion engine dampens both explosive and vibration noise. Properly designed resilient mountings not only control the vibration of the source but prevent it from passing to surrounding structures or foundations. Brick walls around the machines can also attenuate noise, for example, a 25 cm brick wall attenuates noise by about 45 dB. Double-glazed windows can be used where visual control is essential.

Automation or remote control can help to isolate equipment and allows removal of the operator to a remote area.

An operator confined in a chamber with reflecting surfaces and a noisy process will experience a sound level some 5 to 10 dB higher than that in free space or in a chamber lined with absorbent material.

Modulation
Noise can be reflected, absorbed and diffracted. A smooth, hard surface, e.g. glazed tiles, reflects noise and a suitably curved surface disperses it. Any fabric absorbs noise, reducing both transmission and reflection, the most effective being blocks or sheets of glass-wool fibre firmly pressed together. A masonry wall or earth pile forms a substantial barrier. Also a steel plate with resilient material, e.g. a rubber sheet, acts as a barrier. At the edge of a barrier, diffraction takes place causing sound energy to penetrate into geometric shadows (Fig. 12.15).

Limiting exposure
When noise levels are extreme, reduced operator exposure is the only answer. The operator can visit the noisy operation only when required. A roster of duties can be prepared such that several people are exposed for a short time instead of one person for an extended time.

When considering noise with respect to neighbours, noisy operations should be confined to the day time, if at all possible.

Ear defenders
The use of ear defenders is the ultimate means of isolating a person from noise. There are three main kinds of defenders: ear muffs, ear plugs and

Fig. 12.15. Effects of screen location on the noise experienced by a receiver. θ = angle of diffraction.

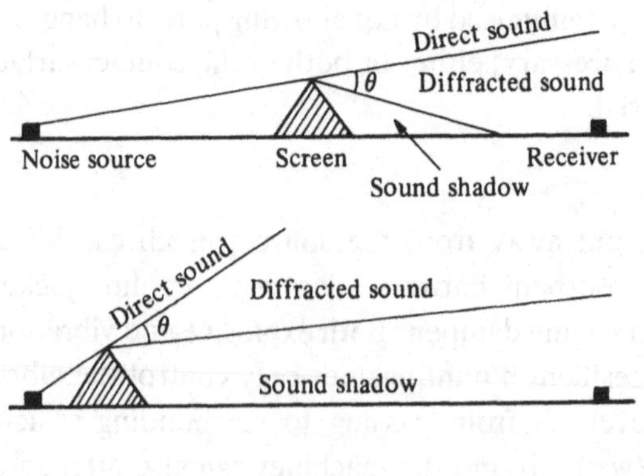

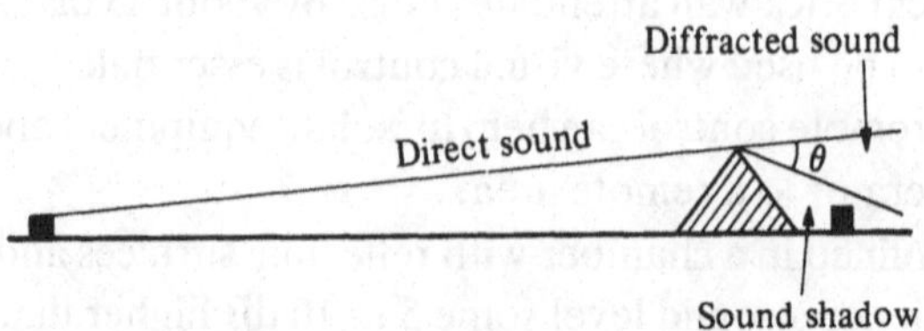

acoustic wool (or glass down). Ear muffs are best, being the most effective and hygienic if they are not reused. Plugs and acoustic wool are generally more acceptable. Both muffs and plugs require frequent attention. In practice, employees often dislike wearing ear defenders of any kind and may do so only under pressure. When worn for several hours, defenders can give the wearer a sense of isolation and may affect his sense of orientation.

Industrial relations

When employees complain of noise that is known to be at a high level, they need a sympathetic reception. The employer should do all that is possible to reduce the noise and the employees should be provided with ear defenders to use.

12.5 Sources of noise and its control

Noise in open pits and underground mines is related to the drilling, cutting, blasting, treatment and transfer of mineral. Ventilation ducts can be a source of high noise levels and so can transfer points. Table 12.11 gives some idea of the noise levels associated with various equipment and locations.

Pneumatic percussive drilling machines

Measurements made in headings where single machines are in use give sound levels of between 108 and 118 dB(A) with an L_{eq} (equation 12.43) of about 100 to 109 dB(A) at the operator's ear level. Rock hardness, length of drill rod exposed and water feed affect sound levels. Air exhausting from the machine generates the dominant sound level. To

Table 12.11. *Noise levels occurring in mines*

Noise level (dB)	Type of work
87–95	Working longwall faces; drilling of holes for blasting; transport of coal by chain conveyors
90–98	Pit tops with ore transport system
90–100	Compressor
94–95	Power generator stations
95–100	Sorting screens
96–98	Blind shafts; ore transfer points
98–100	Unloading and loading of mine cars into cages; skip winding places
105–112	Development headings using pneumatic equipment
100–120	Exhaust of main radial fans

control noise levels, plastic-hood-type silencers are used and air can be remotely exhausted through flexible pipes. Excessive shrouding, however, can cause condensation and freezing of moisture. With proper care, the noise level of the hammer on the drill steel would obtrude at about 99 dB(A). Plastic-covered drill rods reduce noise, but cuttings will not clear. The use of sharp bits, the delivery of air to the machines at optimal pressure, and proper maintenance of equipment can help to reduce the sound levels by 10 dB(A).

Mobile loaders and transporters

These are frequently powered by diesel engines, and sound levels at the operator's position can reach 105 dB(A) at full load and drop to 89 dB(A) on a light load, with an equivalent continuous sound level (L_{eq}) above 90 dB(A). Improved engine, mounting and exhaust silencer design can help reduce sound levels, although the latter may lead to some loss in power.

Pneumatically powered rocker shovels

These present special problems since the addition of exhaust silencers and ducts causes freezing up and blockage of the orifice, with resultant loss of power and ultimate stoppage. Sound levels up to 118 dB(A) have been recorded and, however short the bursts of activity, an L_{eq} in excess of 90 dB(A) results. The use of electrohydraulically operated machines appears to offer a better solution.

Underground fans

Noise levels from 90 to 100 dB(A) have been measured near underground fans; but, normally, workers are exposed to this only when passing by and as such high sound levels are only transient. Where workers are affected, investigations should be made to see if a new site can be arranged. Straight-through and pod-type silencers can be used; the latter offer some resistance to flow and both become fouled and lose effectiveness with time. Although a centrifugal fan is generally quieter than an axial flow fan for the same duty, a mixed-flow fan has given encouraging results.

Mobile plant equipment

Mobile plant equipment on dirt disposal tips can cause residents to complain. Ambient levels on wet and windy occasions generally mask plant noise, but on a dry, still day (or night) manmade sounds may obtrude. An L_{eq} of 90 to 100 dB(A) is normal.

These problems can be resolved by a combination of improved engineering, good public relations, good sense and the cooperation of drivers. Personal idiosyncrasies can cause one person, and not another, to be affected and complain. It is important to show that concern is felt. The problem should be discussed with complainants in their own homes: if they understand the problem, they may be more tolerant. Noise at night should be avoided.

Main ventilation fans

Generally, sound directly from the fan motor house and air leaks from air-lock doors obtrude above normal sound from the evasee chimney. Simple precautions include absorbent silencers, isolation walls, etc.

12.6 References

Howe, M. J. (1982). Noise control. In *Environmental engineering in South African mines*, pp. 923–51. Johannesburg: The Mine Ventilation Society of South Africa.

Puzyna, C. (1974). *Zwalczanie Halasu W Przemysle*. Warszawa: Wydawnictwa Naukowo-Techniczne.

13

Mine illumination

Illumination of the work place is one of the primary factors concerning the safety, productivity and morale of the workers. An extensive amount of work has been done in the area of illumination of factories, offices, streets and houses, but very little attention has been paid to the underground mining industry in the last decade. One of the main reasons has been the fact that the large-scale electrification and introduction of electric illumination in mines after the second world war, along with battery operated personal lamps, has almost eliminated the mining diseases associated with poor lighting. Battery operated personal cap lamps have almost eliminated carbide lamps and oil safety lamps as a source of lighting in mines. Incandescent lamps and fluorescent lamps are the most commonly used sources of light in fixed and mobile units in mines. The oil safety lamp is used in underground mines only as a secondary method of testing for methane. It can also be used for detecting oxygen deficiency and giving an approximate measure of carbon dioxide.

Illumination in a mine has its own problems. Working places in mines are spread over large distances both vertically and laterally. Except for a few locations where activity continues for long periods over the life of a mine, the places of work continually move. At the same time, in many mines, particularly metal mines requiring blasting, permanent installations for lighting at the face are most impractical because of high costs. Poor natural reflectance of rocks, particularly coal, and absorption of light because of dust, smoke and fog make the achievement of good lighting very difficult. In spite of all this, the effect of better illumination on safety and productivity are now well recognised. Better illumination is becoming more important with remotely operated equipment. As a result, lighting standards have been developed in various countries and are being introduced on an increasing scale in mines.

13.1 Properties of light

Light is electromagnetic in nature and can travel through space at a velocity of 3×10^8 m/s and possesses measurable characteristics such as frequency and wavelength, which are related to each other by the equation

$$c = f\lambda \qquad (13.1)$$

where c = velocity of light, 3×10^8 m/s;
f = frequency, Hz; and
λ = wavelength, m.

The visible portion of the electromagnetic spectrum lies in a very narrow band with wavelengths lying between 350 nm* (8.6×10^{14} Hz) and 700 nm (4.3×10^{14} Hz). Light with a wavelength below 350 nm is called ultraviolet light and that with a wavelength above 700 nm is called infra-red light.

The quantity of light is measured by photometers. The four basic units of light are: intensity of light or luminous intensity; luminous flux; illuminance; and luminance.

The unit measure of intensity of light or luminous intensity (I) is called a candela (cd), which is one-sixtieth of the intensity of light emitted by monochromatic source radiation with a frequency of 540×10^{12} Hz (wavelength 555 nm) (black body radiation) from a source of 1 cm^2 at a temperature of 2042 K (the freezing point of platinum). This unit of intensity of light is very close to that of a standard candle, the definition of which has been so common since the eighteenth century. One candela is equal to one lumen per steradian†.

Luminous flux (Φ) is the rate of flow of light. Thus, from a source of one candela, the luminous flux passing through an area of 1 m^2 at a distance of 1 m is called a lumen (lm). Thus 1 lumen = 1 candela × 1 steradian (cd · sr). Thus a source of one spherical candela emits 12.57 lm. In general practice, the intensity of a light source is measured in Watts. A 40 W incandescent lamp has an output of about 400 lumens. Similarly, a 40 W fluorescent lamp has an output of 2400 lumens.

As the distance of a surface from the point source of light increases, and assuming no absorption of light as it travels through a medium, the flux per unit area will decrease and is thus inversely proportional to the square of the distance. This is called the inverse square law. The luminous flux falling per unit area is known as the illumination value or illuminance (E).

* nm – nanometer, i.e. 10^{-9} m.

† A steradian (sr) is a measure of the unit solid angle at the centre of a sphere of unit radius formed by a unit area at the surface of the sphere. Since the surface area of the sphere is proportional to 4π, there are 4π or 12.57 steradians in a sphere.

The unit of illuminance is lux (lx) or lm/m²

$$E = \frac{I}{d^2} \tag{13.2}$$

where E = illuminance, lm/m² (lx);
I = luminous intensity of the source, cd; and
d = distance of the surface from the source, m.

Luminance is defined as the luminous flux per unit of projected area and unit solid angle either leaving a surface at a given point and direction or arriving at a given point on a surface from a given direction. Alternatively, it is the luminous intensity of any surface in a given direction per unit of projected area (cd/m²). However, it must be kept in mind that sources of light in practice do not emit light equally in all directions. It depends upon the construction of the light source, the type of fittings, the reflector shape, etc.

Another factor which has to be kept in mind is that the surface on which the light is falling may not be at right angles to the direction of travel of the light. When the angle between the incident light and the normal to the surface to be illuminated is θ, then the surface area to be illuminated increases. Thus, by inclining the plane of the surface by an angle θ (Fig. 13.1), the surface area has increased from A to $A/\cos\theta$ and hence the illuminance must drop proportionally such that if the illuminance at surface a was lx, then at the surface b it will be $lx \times \cos\theta$. This is known as the cosine correction.

Fig. 13.1. Illustration of cosine law of illumination. a = surface normal to incident light; b = surface whose normal is inclined at an angle, θ, to incident light; A = area of surface normal to incident light; θ = angle between the incident light and normal to the surface to be illuminated.

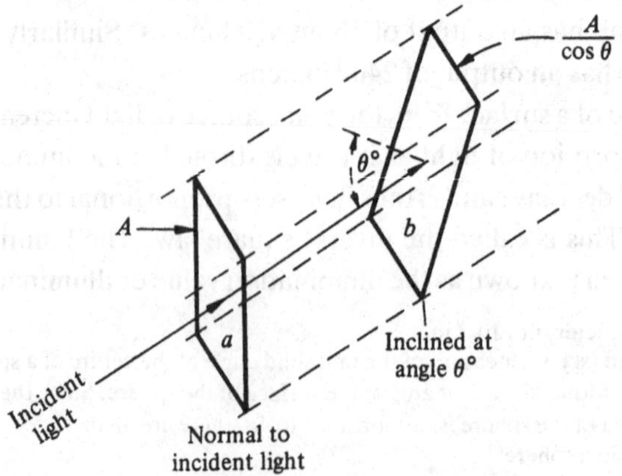

What the human eye sees is not the light falling upon an object but in fact the light which is reflected by the object and enters the eye. Thus the brightness of the surface depends not only upon the intensity of light striking it, but also on the light reflected by it. Reflectance (ρ) is the ratio of the total reflected luminous flux to the total incident luminous flux. The reflectance is obtained by measuring the incident light (illuminance) and the reflected light (luminance) at right angles to the surface. It is calculated from the following equation:

$$L = \frac{E\rho}{\pi} \tag{13.3}$$

where L = luminance, cd/m^2;
E = illuminance, lx; and
π = constant for diffuse reflection.

Table 13.1 summarises the various photometric units.

13.1.1 Brightness and glare

The subjective interpretation of the measured physical quantity known as luminance is brightness. When brightness causes discomfort, it is called glare. The discomfort experienced is a function of the adaptation of the eye and depends upon a number of factors, some of which are:

(1) the intensity of the luminance (task brightness);
(2) the size of the bright area in the field of view;
(3) the position of the luminance relative to the eye;
(4) the contrast between the luminance of the task and its surroundings;
(5) the time of exposure.

Table 13.1. *Photometric units and their definitions*

Physical quantity	Symbol	Name of unit	How defined	Abbreviation for unit
Luminous intensity	I	Candela	Basic SI unit	cd
Solid angle	ω or Ω	Steradian	Mathematically defined	sr
Luminous flux	Φ	Lumen	cd · sr	lm
Illuminance	E	Lux	lm/m^2	lx
Luminance	L	Candela per square metre	cd/m^2	cd/m^2
Reflectance	ρ	No unit	$\frac{L\pi}{E}$	None

A light source of high luminance may or may not appear to be bright. For example, a cap lamp underground causes glare, but on the surface where the eyes are adapted to a much higher level of luminance, it does not appear bright. Similarly, a sudden change in the illuminance level from, say, 0.5 lx to 100 lx will force the pupils of the eyes to contract and can cause discomfort through glare or strain. Glare decreases with increase in the angle between the glare source and the line of sight and with increase in the distance from the glare source (thereby decreasing the effectiveness of the source). Glare can originate from general lighting or from local lighting. Tungsten filament lights with clear bulbs have a small area but a high intensity of light which can cause glare.

The following steps can be taken to avoid glare in mines:

(1) Use a large number of small sources rather than a small number of high-luminance sources.
(2) Mount luminaires (light fittings) out of the field of view of the common work place.
(3) Screening, shielding or covering with diffusing plates or filters greatly reduces glare.
(4) Since glare is also due to sudden changes of illuminance, it can be reduced by more uniform lighting.
(5) Educate workers not to shine cap lamps into other workers' eyes when travelling in cages or passing instructions. (Note: a cap lamp must be lit at all times.)
(6) Use proper and correct lighting and avoid specular materials such as metallic paint on machines or wall rock and choose a flat paint when possible.
(7) Keep the background and surround luminances high. Only one-thirtieth of the task luminance is necessary.

13.2 Light sources in mines

The introduction of the Davy safety lamp in 1815 and its subsequent modifications provided a safe source of light in mines. However, the introduction of the electric cap lamp was the real breakthrough in underground lighting and this proved to be the major factor which almost eliminated nystagmus. At present, efforts in lighting are being directed towards providing a quality illumination (i.e. no glare, effective contrast between task and background, diffuse sources, etc.) which is best for the safety, performance and morale of miners at a minimum cost.

The oil safety lamp, although still used in mines as a standby unit for the testing of methane, has been almost virtually replaced by the electric cap lamp in all major mining countries. In any new mines, electric cap lamps are accepted as the standard for personal illumination.

Fixed lighting and lighting on mobile equipment consist of tungsten filament lamps, tungsten halogen lamps, fluorescent lamps, mercury vapour lamps, metal halide lamps and low- and high-pressure sodium lamps.

Light sources in mines can be divided into two main categories: (1) general illumination lamps; (2) personal lamps.

13.2.1 General illumination lamps

The source of power used in the production of light is electricity, which is converted into light either by incandescence or photoluminescence. Incandescence is the emission of visible radiation produced by raising the temperature of a body (filament) to a high level by passing an electric current through it. Examples of such lamps are tungsten filament lamps and quartz halogen lamps. Photoluminescence is the light emitted when an electron of a neutral gas which has been knocked out of its orbit reverts back to its place. The energy emitted by it may be in the visible spectrum, or in the ultraviolet range and converted into visible light by exciting phosphorescence and fluorescence in certain materials. Examples of such lamps are fluorescent lamps and high-intensity discharge lamps such as mercury, metal halide and high-pressure sodium lamps.

Tungsten filament lamps

The filament of a general service incandescent lamp employs tungsten wire wound as a single, double or triple coil enclosed in a sealed glass bulb filled with a mixture of nitrogen and argon in varying amounts, depending upon the size of the lamp. The power of the lamp may vary from 0.1 W to 20 kW with an operating temperature range of 2500 to 3400 K. The efficacy of these lamps varies between 10 and 15 lm/W and they have a rated life of 750 to 1000 hours. Lamps working on 120 V are more efficient than 240 V lamps and the highest efficacy is obtained between 12 and 20 V. The efficacy is very much dependent upon the voltage fluctuations. A 10 % change in applied voltage can cause a 70 % change in lumen output.

Ordinary tungsten filament lamps are not designed for use under conditions subject to vibrations. Special shock-resistant filament lamps (25 to 500 W) are available with an extra support and should be used in places such as belt structures, chutes, grizzlies, etc. where vibrations occur. Safety lamps with a silicon-coated bulb can withstand higher shocks and remain intact after impact. Cap lamps and certain high-wattage lamps are filled with krypton. Krypton gas has a higher vapour pressure than either nitrogen or argon and conducts heat much more slowly and hence increases efficacy by some 10 %. High-wattage lamps are provided with a filament shield to eliminate glare.

Tungsten halogen lamps contain a suitable quantity of halogen gases. They have a higher efficacy and increased life over ordinary gas-filled tungsten lamps but they are expensive. They operate at a higher filament temperature and their concentrated light source can cause glare. They are very compact and give excellent light control.

The source luminance of tungsten filament lamps is high, ranging from 10^7 for clear to 10^5 cd/m^2 for frosted lamps. Tungsten filament lamps, although very cheap, have a low efficacy (10 to 12 %) and are liable to damage due to shock. These lamps are not safe in an explosive atmosphere and therefore require flameproof protection for use in gassy mines.

Fluorescent lamps

A fluorescent lamp is an electric-discharge lamp consisting of electrodes placed at either end of a long tube containing mercury vapour. When a suitable potential difference is maintained between the electrodes, the mercury vapour emits largely ultraviolet radiation which is converted into visible light as this impinges upon molecules of certain silicates, tungstates, borates and phosphates. By selecting the appropriate mixture of powders for coating the tube, a natural daylight colour can be obtained. Fluorescent lamps are available in diameters from 15 to 50 mm, in straight or circular tubes with a range of 4 to 65 W and in lengths of up to 2.5 m, and their life ranges from 500 to 30 000 hours. The source luminance of these lamps ranges from 5×10^4 to 2×10^5 cd/m^2 and their efficacy varies between 35 and 85 lm/W. They possess much less inherent glare owing to their large area and low luminance characteristic, and because of their natural daylight characteristics they are most often used in mines.

Low-pressure sodium lamps

Low-pressure sodium lamps are becoming more common these days because of their high efficacy. These lamps consist of a borate-coated, U-shaped tube under very low gas pressure with small amounts of inert gases such as helium and neon, and operate at 235 °C. In order to equalise temperatures the gas tube is enclosed inside an outer glass tube. These sources have very low brightness and a working efficacy of up to 150 lm/W. They maintain their initial lumen output and are not affected by voltage changes or variations in ambient conditions. The other advantage of sodium vapour lamps is that yellow light is not greatly affected by refraction and with an output in the 589.4 nm region it is close to the peak response of the human eye (555 nm). They are excellent for floodlighting but give a very poor colour response. In some countries, they are not approved for use in coal mines.

High-intensity discharge lamps
High-intensity discharge (HID) lamps consist of gaseous discharge tubes which operate at pressures and current densities sufficient to create desired quantities of radiation within their arcs. Since these lamps are all gas discharge lamps they have negative resistance characteristics and hence need ballasts to supply the necessary starting voltage and to limit current after the start. Restrike time depends upon the type of lamp and the fixture, as the gases must cool to a point when the arc will strike again. The three categories of HID lamps used in mines are:

(1) mercury vapour lamps;
(2) metal halide lamps; and
(3) high-pressure sodium lamps.

Mercury vapour lamps. In these lamps, collision between electrons and mercury atoms produces the mercury spectral lines (404.7, 435.8, 546.1, 577.9 and 700 nm). Using various fluorescent powders, colour improvement is possible which also results in less glare. The size of the arc tube varies between 25 and 100 mm in length and 10 and 25 mm in diameter and they are marketed in sizes from 50 to 100 W with a life of 10 000 to 24 000 hours. The efficacy varies between 35 and 60 lm/W. Mercury vapour lamps are commonly used on mobile equipment in mines because they do not have a filament which can break. The disadvantages include long warm-up periods (7 to 9 minutes), long restart time (10 minutes), cold-weather starting requiring a higher voltage and special ballasts, and the outside possibility of skin burns and eye inflammation due to excess ultraviolet radiation.

Metal halide lamps. These lamps are filled with certain halides such as iodides of sodium, thallium and indium. These additives produce red and yellow light which with the blue and yellow of the mercury arc give a better colour quality and a higher efficacy (50 % more for the same wattage). However, these lamps require a higher open circuit voltage and thereby have a lower operational life and a higher restart time. They are very expensive and are susceptible to voltage drop. As such they are not very common in mines.

High-pressure sodium lamps. Although they work on the same principle as other HID lamps, they have a different construction where sodium, mercury and xenon vapours are maintained in a translucent ceramic arc tube at high temperature and pressure (but lower than atmospheric pressure). These lamps have a high efficacy of 110 to 130 lm/W and a life of 20 000 to 24 000 hours. They produce much less ultraviolet radiation. They need a high striking voltage of up to 480 A.C. and thus require a special ballast to start the arc. These lamps are becoming common in permanently mounted lighting installations in mines. In some countries, they are not approved for use in coal mines.

Table 13.2 gives a comparison of the various sources of light. Fig. 13.2 gives a comparison of efficacy figures.

Compressed air lamps

Where electric power cannot be supplied, a number of small, self-contained luminaires can be made available, operated from a small, compressed air turbine generator. The size of the luminaires may vary from 15 to 80 W and each contains either a mercury vapour or an incandescent lamp. The exit pressure from the turbine is allowed to circulate in the lamp chamber of the luminaire for cooling purposes. This also serves as a safety valve in the event of a cover glass breakage, so shutting power to the turbine.

These luminaires are often used in remote drilling sites and in development areas. They have high capital and operating costs and are noisy in operation.

Fig. 13.2. Efficacy ranges of light sources in mines (after Trotter, 1982).

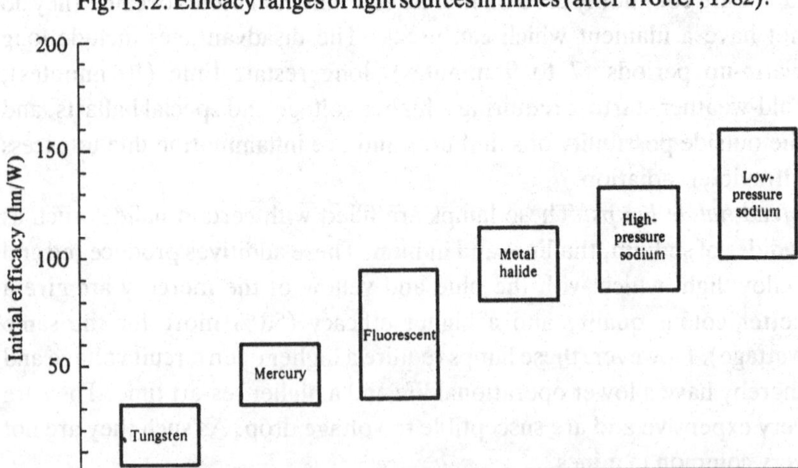

Table 13.2. *Comparison of light sources in mines (after Trotter, 1982)*

Type of source	Approximate luminance (cd/m^2, clear bulb)	Average rated life (hours)	Capital cost rank (1 = costly)	DC source	Warm-up time (minutes)	Restrike time (minutes)	Colour appearance and rendition
Tungsten filament	10^5 to 10^7	750 to 1000	7	Yes	—	—	Warm, white to yellow, excellent rendition
Tungsten halogen	2×10^7	5[a] to 2000	4	Yes	—	—	Warm, white, slight yellow, excellent rendition
Fluorescent	5×10^4 to 2×10^5	500[b] to 30 000	3	Yes	—	—	Warm, white, excellent rendition
Mercury vapour	10^5 to 10^6	16 000 to 24 000	6	Yes with limitations	7–9	3–10	Cool, bluish, average rendition
Metal halide	5×10^6	10 000 to 20 000	2	Yes with limitations	2–5	10–20[c]	Cool, blue-white, good rendition
High-pressure sodium	10^7	12 000 to 24 000	1	Not advised	3–4	0.5–1	Warm, golden colour, fair rendition
Low-pressure sodium	10^5	10 000 to 18 000	5	Not advised	7–9	1–2	Warm, amber, poor rendition

[a] This is not general service lamp.
[b] The 500 hour lamp is not a normal source. The 30 000 hour lamp is a cold cathode lamp – not suitable for mines.
[c] Less if special circuits are used.

Lamp reflectors and luminaire modules

A luminaire module consists of a lamp, reflector, lens and an enclosure. The purpose of reflectors within luminaire modules are to direct the light and distribute it and also to protect the source. The factors which influence the choice of a luminaire and reflector combination are:

(1) type of distribution required – concentrated or dispersed;
(2) environmental conditions – robustness, vibration-proof, dustproof, humidity-proof, etc.;
(3) size of the lamp and the area to be illuminated;
(4) ease of maintenance.

Reflectors are used to distribute light in a specific area and to cut off the light distribution in the directions not required. The most common types of reflectors are those of parabolic or elliptical shape. Reflector systems may also require suitable lenses to locate the light accurately. Parabolic reflectors produce a narrow beam of light if the filament is located at the focal point of the reflector. Some special bulbs are manufactured with integral aluminised parabolic and elliptical reflectors which distribute their light without the aid of added reflectors or lenses. These are commonly called PAR bulbs.

Elliptical reflectors ideally concentrate the beam at a point some 50 mm in front of the cover glass and then broaden it into a beam pattern. This is sometimes known as the x-principle and normally prevents the beam from fanning. Such luminaires are recommended at areas such as belt transfer points and other places which are monitored by closed-circuit television and are positioned such that no light enters the monitor directly from the source.

Several luminaires have been developed which contain more than one light source (usually two to four fluorescent lamps), allowing more flexibility for the lighting designer. These modules are generally provided with diffuser guards to cut glare. For use in hazardous conditions, flameproof luminaires are required. The essential requirements of a flameproof luminaire are as follows:

(1) It should be proof against the build-up of flammable vapours within an enclosure.
(2) It should be proof against a possible internal explosion transferring to the outside atmosphere.
(3) The surface temperature of the exterior should be low so as not to cause burns or the ignition of methane. The maximum temperature should not exceed 150 °C.

13.2.2 Personal lamps

The personal lamps now used in mines are almost invariably electric cap lamps, although in some mines electric hand lamps are still in use. Flame safety lamps are used in some old mines and by the supervisory staff as part of the statutory requirements.

Flame safety lamp

A modern flame safety lamp is based upon the same principle as the Davy safety lamp, namely it is equipped with wire gauzes surrounding the flame which cool down the hot gases sufficiently so as not to ignite any methane surrounding it, provided enough care is taken when in use. A modern flame safety lamp also includes such features as the lowering and raising of the wick and the ability to relight. Some lamps have a total luminous output which almost matches that of electric cap lamps although, of course, the intensity is considerably less.

Electric cap lamp

An electric cap lamp is the single obligatory source of lighting in underground coal mines. The lamp has three main parts, the headpiece, the cable and the battery. The headpiece clips on to the cap of the miner while the battery is worn on the miner's belt. The total mass of the lamp varies from 1.8 to 2.5 kg.

The most common lamps use tungsten filament bulbs with one krypton-filled bulb plus an emergency low-output bulb or a two-filament bulb rated at between 0.75 and 1.2 A at 4 V. The life of the bulb is from 600 to 800 hours. The headpiece can have a beam intensity which is greater than 2000 cd and can be provided with a focussing device which can be adjusted by means of a special outside-operated cam mechanism for the provision of a narrow (specular), medium (soft) or broad (diffused) beam characteristic with beam-centre intensities of 2200, 700 and 200 cd, respectively. (Other cap lamps are provided with a choice of reflectors.) With 73 % reflection, this permits visibility at 20, 10 and 6 m with the narrow, medium and broad beams, respectively (Bell, 1983*a*).

In most cases, lead acid (Australia, the UK and France) or cadmium alkaline (Germany) batteries are used. Lead acid batteries have a rating of up to 1.2 A at 4 V which provides a useful life of 13 hours. Lead acid batteries have two cells while alkaline batteries used in Germany have two or three cells. The battery is housed in a polycarbonate container which is shock-proof. A fuse is incorporated in the battery top when it is to be used in gassy mines. A lock ensures that it can only be opened by an authorised person in a safe location.

With lead acid batteries, special automatic charging racks are provided on the surface and when the headpiece is located on a charging frame and turned by 180°, battery charging automatically starts. Charging is required to be done for a minimum of 16 hours. The life of the battery is about three years.

Certain other types of cap lamps such as tungsten halogen lamps using 6 V nickel–cadmium alkaline cells have been developed by the Mining Industry Research Organisation of Canada and the US Bureau of Mines. These lamps have undergone extensive testing and have a higher efficacy, lower mass and occupy a smaller space.

13.3 Methods of illumination in underground mines

In underground lighting installations and in cases of lighting units mounted on mobile equipment, it is essential that all lamps are ruggedly constructed to resist vibrations. Fluorescent lamps can be operated on alternating or direct currents. Thermoplastic mouldings are impact-resistant and can be used underground in fixed lighting systems. Flameproof mountings are required in gassy mines.

The method of providing illumination in mines is dependent upon the location of the tasks to be illuminated, the need for constant (24 hour) illumination, the mobility of the area to be lit and the size of the area. Basically, there are two methods of illumination: (1) fixed luminaires and (2) mobile luminaires.

13.3.1 Fixed illumination systems

Places which require fixed illumination are those which have to be illuminated around the clock. These include machine chambers, underground workshops, main pump stations, small shaft hoisting chambers mounted on blind shafts, personnel carrier stations, loading and unloading stations, important transfer points, main transport arteries, etc.

For network illumination using fixed sources, tungsten filament, discharge or fluorescent light sources are most commonly used. In high chambers, mercury discharge lamps are preferable. High-pressure sodium lamps are used in some countries. Fluorescent lamps are used at pit bottoms, transport roadways, intersections, etc. where the height is less than 3.5 m and where high activity is expected. The length of the fluorescent tubes may vary from 0.6 to 1.5 m and two tubes are usually employed in each luminaire. Mercury lamps over 125 W and incandescent lamps over 150 W should not be suspended below 2 m above the floor level.

Most of the places requiring 24 hour illumination are located at quite some distance from each other and as such a small transformer of the 1.6 to 10 kVA range may be used when the maximum distance does not exceed 500 m. In large areas such as main transport roadways, three-phase transformers are generally placed every 800 to 1000 m.

In places where continuous 24 hour illumination is not required, switches are provided so that lights are switched on only during working periods.

13.3.2 Mobile illumination systems

Movable illumination systems are needed in places where the working front continually advances. Such places include the bottoms of shafts during sinking, and advancing faces. Movable working platforms suspended 15 to 40 m above the bottom of a shaft under sinking provide a base for the electric power supply for the equipment used in the shaft sinking. This platform can be used for lighting the shaft using two to three mercury vapour lamps of 250 to 400 W mounted on reflectors. The lamps are protected from blast damage by a 90° tilt. This method permits illumination of the shaft bottom by providing about 120 lx (Peretiatkowicz, 1982).

It is important that all lamps within 20 m from face in the case of coal headings and 30 m from the face in the case of stone headings are removed or lowered and protected before blasting.

Illumination of longwalls

Self-advancing supports are provided with intrinsically safe or flameproof fluorescent tube luminaires energised from an alternate two-phase and three-phase main line electricity supply. These lamps are mounted under the roof bars in the case of thick seams or inbetween the props in the case of thin seams.

In the UK, a typical luminaire consists of a miniature 8 or 13 W fluorescent tube. These are placed 1 m apart and this spacing permits low glare and a reasonable illumination over the walkway.

Certain installations use miniature fluorescent tubes operating at 20 kHz, supplied from a constant-current, intrinsically safe source.

When the seam thickness is greater than 2 m, mercury vapour lamps of 80 to 125 W are used in some countries although they have many disadvantages. Use of built-in ballast lamps fed with stud cables from a main line permits illumination of thick seams. The voltage supply may vary from 100 to 125 V depending upon the system.

The number of lamps depends upon the intensity of the individual unit and the thickness of the seam. The spacing between the individual units may vary from 1.5 to 6 m. Each or every second support is usually provided with a lighting unit.

Individual lighting of mining machines

Underground mining machines and transporters are equipped with headlights having sealed beam lamps. Quartz halogen lamps have been adopted in some countries. The operation may be 6, 12, 24 or 36 V with up to a 500 W intensity. Mercury vapour lamps are additionally specified for moving machines because of their higher efficacy and longer life. Certain machines are provided with machine-mounted luminaires. All lamps mounted on mining machines must be ruggedly constructed to resist vibrations.

13.4 Lighting in opencut mines

Opencut mines have usually two main methods of illumination. Each machine is equipped with its own task lighting luminaires using non-adjustable headlights and a fixed floodlight system or a combination of one or the other of these with additional luminaires for the surrounding area. In the case of large machines such as draglines, the positioning of the lights should be in such a way so that the operator can position the bucket at exactly the right point, scoop in the material and move to the discharge point. It is therefore essential that good three-dimensional illumination is provided. Lighting provisions around machines should be so designed that an area of about 6 m radiating from the machine and all working areas beneath the main frame, exterior walkways and platforms are illuminated. Fig. 13.3 shows the arrangement of lighting equipment on the boom and cab of a typical large dragline.

The three-dimensional illumination of such an installation needs a computer design to work out the positions of the luminaires coupled with proper photometric surveys (Hottinger, Faux and Yantz, 1982). Table 13.3 gives the recommended lighting levels and methods of measurement for large machines.

Drilling for blasting, surface reclamation sites, spoil pile bulldozer operations, etc. in opencut mines require fixed floodlighting besides the individual equipment being provided in the form of their own headlights.

Fig. 13.3. Dragline boom and cab floodlights (after Bell, 1983*a*). Numbers are distances in metres.

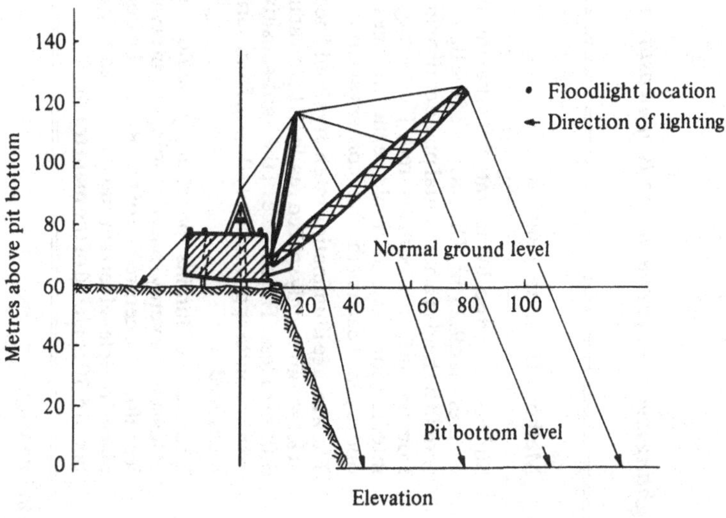

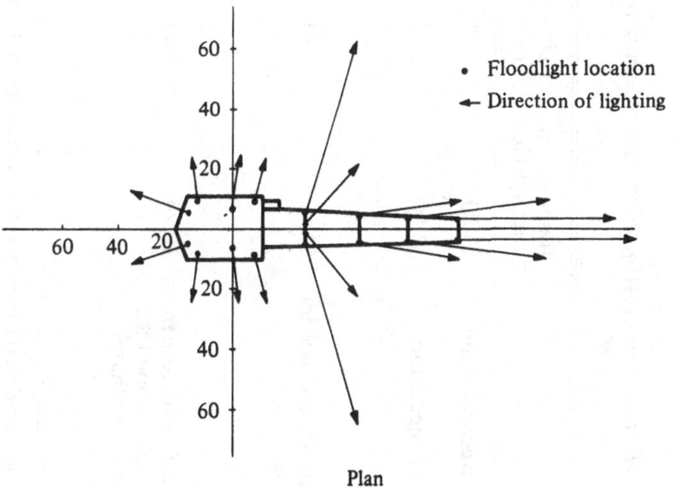

Table 13.3. *US Mine Safety and Health Administration recommended lighting levels in lux and methods of measurement for large machines (after Hottinger et al., 1982)*

Location	Illumination intensity (lx)	Uniformity ratio[a] (UR)	Method of measurement
(1) All areas 6.10 m in all directions from the main frame	53.80	10/1	Surface should be divided into fields having an area not greater than 9.29 m^2. The surface of the light-sensitive cell should be held in a horizontal plane approximately 0.76 m from the surface to be measured. The average illumination intensity for that field should be the average of five readings; one taken at each of the four corners and one at the centre
(2) All work or travel areas beneath the main frame	53.80	10/1	
(3) Exterior walkways, catwalk and ladders on main frame and boom	53.80	10/1	The average illumination intensity should be the average of at least four evenly spaced readings taken at intervals of 3.05 m or less. The surface of the light-sensitive cell should be horizontal approximately 0.76 m from surface to be measured
(4) Area beneath the boom from 6.10 m from the main frame to the furthest point the equipment can excavate or discharge material	10.76	10/1	The average illumination intensity should be the average of at least four evenly spaced readings at intervals of 3.05 m or less along the centre line of the boom. The boom should be placed over level terrain and the first reading should be taken 6.10 m ahead of the main frame

[a] Uniformity ratio should be calculated as the average to minimum illumination intensity.

Table 13.4. *Mine lighting (illuminance) standards in some countries (after Bell, 1983a)*

	Lighting location and minimum recommendation in lux (lm/m^2)						
	Shaft bottom	Houses (engine, pumping, etc.)	Main roads (haulage)	Loading points	Underground offices	Locomotive repair area/ inspection area	Coal faces
Country							
Australia	20	20		20	100		
Belgium	20	25	10	20			
Canada	50	50	20	20	270	270	
Czechoslovakia	15	20	5	20			5
East Germany	30	80	15	40			
West Germany	30	40	15	40			
Hungary	60	20	2	40		20	10
Poland	50(s)	50(s)	2(s)	15(s)		50/100	2(s)
United Kingdom	70	30		30	60	50/150	
United States							15(s)
Surface locations	100	100	100	100	400	400	

Note: Where a blank space exists for some locations, this does not necessarily signify that there is no lighting at these points but rather that no official recommendation has been made.
'(s)' = statutory.

13.5 Standards for mine lighting

Guidelines for mine lighting standards have been suggested for underground lighting in coal mines by Technical Committee 4.10 of the International Commission on Lighting (CIE). The following is a summary of the important guidelines with respect to luminance:

> Where traffic is light and mechanisation is minimal but general lighting is desirable because of safety considerations: 0.05 cd/m^2
>
> Where mechanised equipment normally operates: 0.2 cd/m^2
>
> Underground chambers where precision work is not performed: 10 cd/m^2
>
> Underground chambers where precision work is being carried out: 20 cd/m^2

The mine lighting standards of some countries are given in Table 13.4. Many countries define illuminance values in their standards because these are easy to measure. In the USA, however, recommended values are in luminance. For example, 0.2 cd/m^2 is prescribed in the USA for all areas where mechanised work is to be conducted. Assuming 0.05 as the diffuse reflectance, an illuminance of 12.6 lx is required to produce a luminance of 0.2 cd/m^2.

13.6 Photometry in mines

Measurement of luminance and illuminance or illumination values requires the use of light sensors (photometers) that have a spectral sensitivity curve similar to that of the human eye. These photometers work on the principle that light falling upon the sensing head is converted into electrical energy which is then processed and related to the light level. The most commonly employed light sensor is a selenium photovoltaic cell (the selenium photometer was developed by Evans Electroselenium Ltd of England). Certain other photometers employ silicon photodiodes (Tectronix Inc. of the USA). Selenium photometers are more sensitive than photometers employing silicon photodiodes, which require amplification. Certain photometers do not measure light but only indicate whether or not a certain level of light has been exceeded. These include the 'go/no go' photometer developed by the General Energy Development Corp. of the USA.

Photographic photometry based upon the principle that the amount of blackening of a negative is proportional to the luminance of the photographed object has been used to evaluate the illumination of mine tunnels (Lewin and Bell, 1968). Visual photometers are used in many cases. The basic principle of these photometers is that the luminance of the area being measured is compared to that of a standard test plate of known luminance (Macbeth illuminometer and the Salford Electrical Instruments photometer).

The photometer to be used depends upon the type of measurement to be conducted. Photometers employing selenium photovoltaic cells and silicon photodiodes measure the light received by the surface, i.e. the incident light (illuminance survey). Photographic photometry and the go/no go photometer measure the light given off or scattered by the surface, i.e. the reflected light (luminance measurements). Visual photometers can measure illuminance, luminance and reflectance but require batteries to operate them. Therefore they are not used in coal mines.

13.6.1 Illuminance measurements

There are various techniques used in illuminance measurements. The most common of these are: (*a*) planar measurements; (*b*) maximum value measurement. These results are used to draw isolux contours which can help to determine the total light flux, design an appropriate lighting installation and calculate the efficiency of the system.

In planar measurements, a surface grid is established and measurements are taken at the centre of the square, along grid lines or at the junction of intersecting lines. The grid distance depends upon the size of the surface and the uniformity of illuminance. Usually a grid of 1 m intervals is recommended.

Maximum value measurements are conducted by pointing the photometer towards the source. The reading is resolved normal to the plane and this component is assumed to be the illuminance at the point of measurement.

13.6.2 Luminance measurements

In luminance measurements, the instrument is aimed at the surface. The surface is divided into a grid with 0.6 m intervals and the distance from the photometer to the surface is so chosen that the area covered in each measurement is not greater than 325 cm^2 in the case of enclosed obstructed places and 0.46 m^2 in an unobstructed area (US Mine Safety and Health Administration recommendations). Values are averaged by taking at least four uniformly spaced measurements in each grid square.

13.6.3 Reflectance measurements

The knowledge of reflectance of the surface is essential in the planning of illuminance or luminance measurements. Light falling on a surface may be specularly reflected, diffusely reflected or both (Fig. 13.4) depending upon the surface. In field measurements where conditions can change with time and where specimens cannot be collected for measurements in the laboratory, the methods employed are reflected incident light comparison and standard chips comparison.

In the reflected incident light comparison method, the photometer is placed against the surface and readings are taken for the illuminance (E_2) striking the surface. The meter is then turned around and drawn back about 0.1 m from the surface to measure the illuminance (E_1) emitted by the surface. The reflectance ρ is then calculated from

$$\rho = \frac{E_1}{E_2}. \tag{13.4}$$

Fig. 13.4. Types of reflectance (after Trotter, 1982). (*a*) Specular reflectance; (*b*) diffuse reflectance; (*c*) diffuse specular reflectance.

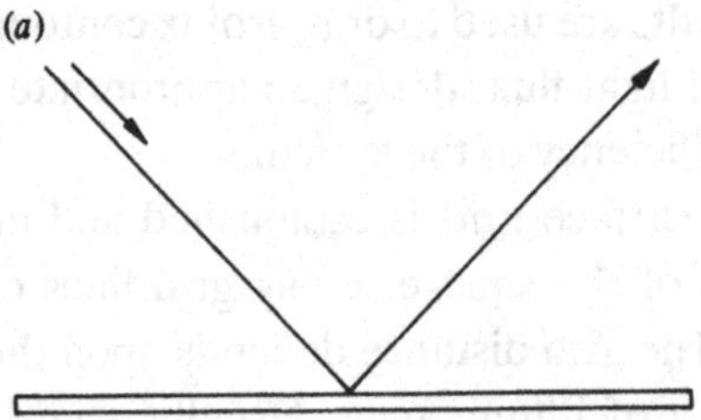

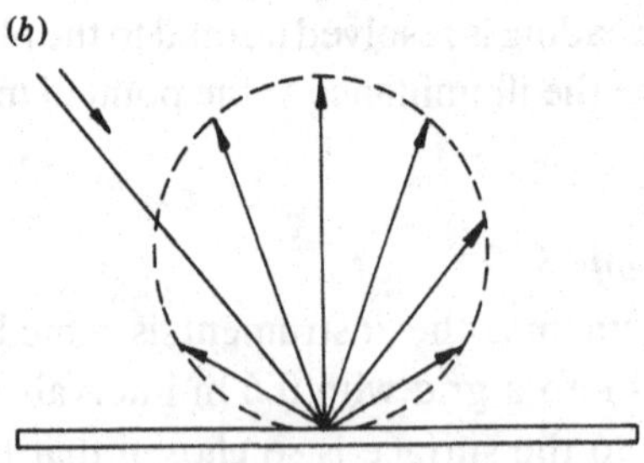

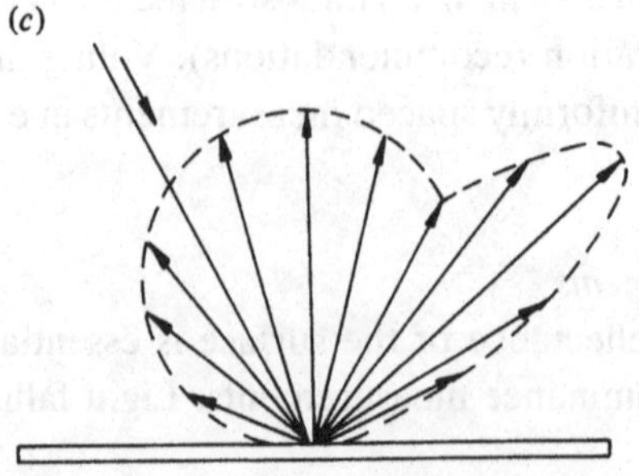

This method, however, suffers from the disadvantage that light can either be blocked, or reflected light from other points can enter the meter.

In the standard chips comparison method, neutral chips of known reflectance are placed on the object and the chip that most closely matches the sample is compared to the sample surface (Fig. 13.5) (Macbeth, 1949). A mask is placed over the chip, and the chip, the sample and the mask can all be observed simultaneously. Reflectance values of a number of ores and other materials are given in Fig. 13.6.

13.7 Design of lighting systems in mines

13.7.1 *Illuminance calculations*

Consider a light fitting F mounted at a height h m above the working place (e.g. the floor of mine roadway) (Fig. 13.7). Let the intensity of the source in a downward vertical direction be I_0. This intensity (candelas) divided by the square of the height (h) – the inverse square law of illumination – gives the illumination value or illuminance (E) on the floor below the fitting (lm/m^2). Therefore

$$\frac{I_0}{h^2} = E_0. \tag{13.5}$$

Similar values are obtainable for points 1, 2, 3, etc. (at various distances from the fitting). For example

$$\frac{I_0}{l_1^2} = E_1 \text{ (max)} \tag{13.6}$$

$$E_\theta \text{ (planar)} = E_1 \text{ (max)} \times \cos\theta. \tag{13.7}$$

Fig. 13.5. Determination of reflectance of a surface by comparing chips of known reflectances.

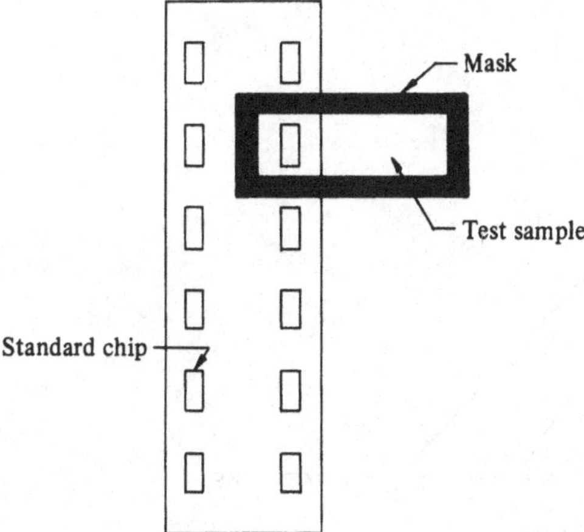

Fig. 13.6. Experimental results of reflectance values of various ores and other materials (after Trotter, 1982).

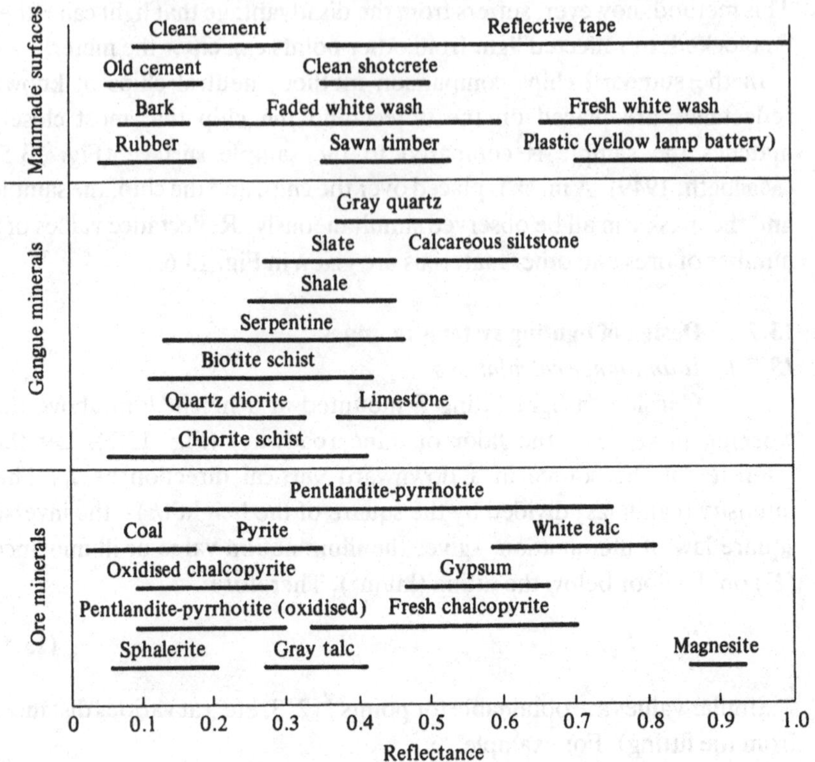

Fig. 13.7. Illuminance on a horizontal plane. F = light fitting; point 0 = point vertically beneath the light source; points 1, 2, 3, 4 . . . = points at the working place at various distances from the light source; θ = angle between the vertical and a line from the light source to point 1; l_1 = the distance from the source to point 1.

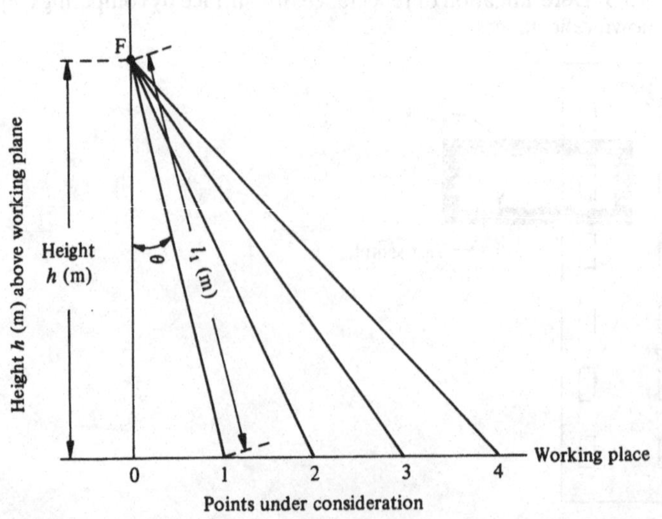

To simplify calculations of E_θ (planar) the following formula is normally used:

$$l_1 = \frac{h}{\cos\theta} \tag{13.8}$$

$$\therefore E_\theta = \frac{I_0}{l_1^2} \times \cos\theta \tag{13.9}$$

$$= \frac{I_0}{(h/\cos\theta)^2} \times \cos\theta \tag{13.10}$$

$$= \frac{I_0}{h^2} \times \cos^3\theta \tag{13.11}$$

$$= E_0 \cos^3\theta. \tag{13.12}$$

where I_0 = the intensity of the source, cd;
l_1 = the distance from the source to point 1, m (see Fig. 13.7);
E_1 (max) = maximum illumination at point 1, lm/m^2;
E_θ (planar) = illumination on the horizontal plane at point 1, lm/m^2; and
θ = angle between the vertical and a line from the source to the measurement point 1, degrees.

Thus for a given point in any plane, the illuminance can be calculated and hence for a given standard of illuminance, the spacing can be determined.

13.7.2 *Luminance calculations*

In luminance calculations, the reflectance must be taken into account. However, in an enclosed system, internal reflection from the roof, floor and walls act as a light source and this process continues to infinity. If the uniform reflectance of the rock surface is ρ and the number of lumens given off by the source are ϕ, then the total number of lumens reflected by the rock surfaces (lu) can be given by

$$lu = \phi \times \rho(1 + \rho + \rho^2 + \cdots + \rho^\infty). \tag{13.13}$$

For $\rho < 1$,

$$lu = \frac{\phi\rho}{(1-\rho)}. \tag{13.14}$$

For $\rho = 0.3$, $lu = 0.43\phi$; for $\rho = 0.7$, $lu = 2.3\phi$. Thus by changing the rock reflectance a great improvement in luminance can be achieved. However, for lower reflectance values, the multiple reflection does not offer much improvement.

Since $\rho = \pi L/E$ (equation 13.3) equation (13.11) can be rewritten to include the effect of reflection as follows:

$$L_\theta = \frac{I_0 \cos^3 \theta \times \rho}{\pi h^2 (1 - \rho)}. \qquad (13.15)$$

where L_θ = luminance at point 1, cd/m^2;
I_0 = the intensity of the source, cd;
θ = the angle between the vertical and a line from the source to point 1, degrees;
ρ = reflectance of the rock surface; and
h = height above the working plane.

For values of ρ less than 2,

$$L_\theta = \frac{I_0 \cos^3 \theta \times \rho}{\pi h^2}. \qquad (13.16)$$

Thus if the prescribed standards of luminance values are known, then the optimal spacing between the lamps can be determined.

In underground mines the walls of the excavation and the roof illumination are as important as the ground surface illumination, and the method of calculation of the luminance is the same. The sources can be provided with reflectors which direct the light in the required direction.

Nomograms can be prepared for a given geometry of the excavations which can help in the planning of lighting installations (Young, Potts and Bell, 1955).

13.8 References

Bell, W. B. (1983*a*). A feasibility study into the lighting requirements of the Australian coal mining industry. In *20th session of Internationale de l'Eclairage*, Amsterdam. Paris: Bureau Central CIE.

Bell, W. B. (1983*b*). Mine lighting – a search for standards. *Lighting in Australia*, **3**, 11–14.

Hottinger, D. D., Faux, K. R. and Yantz, R. D. (1982). Illuminating large surface mining machines, problems and solutions. In *2nd international mine lighting conference of the International Commission on Illumination (CIE)*, ed. K. L. Whitehead and W. H. Lewis, pp. 271–304. USBM IC 8886. Washington: US Bureau of Mines.

Lewin, I. and Bell, W. B. (1968). Luminance measurement by photographic photometry. *Illum. Engr.*, **63**, No. 11, 582–9.

Macbeth, N. (1949). Munsell value scales for judging reflectance. *Illum. Engr.*, **44**, 106–8.

Peretiatkowicz, A. (1982). Mine underground illumination in Poland. In *2nd international mine lighting conference of the International Commission on Illumination (CIE)*, ed. K. L. Whitehead and W. H. Lewis, pp. 23–31. USBM IC 8886. Washington: US Bureau of Mines.

Trotter, D. A. (1982). *The lighting of underground mines*. Clausthal-Zellerfeld: Trans Tech Publications. 201 pp.

Young, W., Potts, E. W. and Bell, W. B. (1955). A lumen method of design for mine lighting installations. In *13th session of Commission Internationale de l'Eclairage*, Zürich, ed. N. A. Halbertsma and J. L. Jansen, vol. II, paper G-Y. Paris: Bureau Central CIE.

APPENDIX I

Mine ventilation network analysis program in BASIC for microcomputers

This program is based on the one given by Y. J. Wang and L. W. Saperstein (1970) in their paper entitled Computer-aided solution of complex ventilation networks, *Trans. Soc. Min. Engrs AIME*, vol. 247, pp. 238–50. It is written in BASIC, the most common language of microcomputers.

Notes

(1) The facilities available include free-splitting and fixed-volume flow rate branches, internal and external fans, and natural ventilation pressures.

(2) A fan may be placed in any branch of the network, except those with fixed-volume flow rates.

(3) All the branches leading to and away from a junction should not be specified by fans or fixed-volume flow rates.

(4) Node selection: Working from the mine ventilation plan, all nodes (junctions) are numbered. The order is insignificant, but no number may be used more than once. Any number from 1 to 185 may be used. To use a number higher than 185, the dimension of JC (in line 10 of the program listing, p. 460) has to be changed accordingly. The surface is taken as a common node so that all branches which connect to the surface use the same node number.

(5) The program can handle up to 100 branches, 65 meshes, 6 fans, 8 points for each fan characteristic, 15 fixed-volume flow rate branches, and 15 branches with natural ventilation pressures. It requires 32 K RAM. If storage space is available, the dimensions (in line 10 of program listing, p. 460) can be changed to increase the quantities as follows:

Branches:	B, R, Q, OU, HH
Nodes:	J1, J2
Meshes:	ME, SN
Fans:	ND, C, FQ
Fan characteristic points:	X, FX
Fixed volume flow rate branches:	DD
Branches with natural ventilation pressure:	RR

Nomenclature

The following list defines all the variables in the program that are necessary for proper input and output.

DD(I): air volume flow rate assigned to fixed-volume flow rate branch I, m^3/s;

E: error, the desired precision of solution, m^3/s;

FF: friction factor, Ns^2/m^4;

FI: number of branches with fixed-volume flow rate;

FX(I): fan pressure for the point I on a fan characteristic, Pa;

H: pressure loss, Pa, height of rectangular airway, m;

I: branch number;

IT: maximum number of iterations to be performed;

J, J1, J2: J for junction or node. J1 and J2 indicate the two nodes which define a branch; positive airflow is taken from J1 to J2;

L: length of airway, m;

MJ: maximum number of junctions that is introduced in the data;

NB: number of branches;

NF: number of fans;

NJ: number of junctions actually involved in the solution; must be less than or equal to MJ;

NM: number of meshes;

NP: number of points used to designate one fan characteristic;

NV: number of branches with natural ventilation pressure;

Q: volume flow rate, m^3/s;

R: resistance of airway, Ns^2/m^8;

RR(I): natural ventilation pressure associated with branch I, Pa;

SN(I): summation of natural ventilation pressure for mesh I, Pa;

W: width of rectangular airway, m;

X(I): volume flow rate for the point I on a fan characteristic, m^3/s.

Input

The computer asks questions for the input data. The one restriction on the input of airway characteristics is that all branches with a fixed-volume flow rate must be input first, then all branches with fans, and finally the general branches. The order of input within each group is of no concern. The order in which the nodes are input for each airway gives the direction of the airway. Except for the fixed-volume flow rate branches and those with fans, the directions of the airways are of little concern because the computer will simply give a minus sign to a volume flow rate if it flows opposite to the given direction.

For each branch with natural ventilation pressure, two fields of data are to be input: (1) branch number and (2) natural ventilation pressure. If the pressure acts opposite to the direction implied for the airway (input earlier), a minus sign must be employed for natural ventilation pressure.

Output

'COEFFICIENTS' refer to the coefficients of the polynomial equation which is generated to reproduce the fan characteristic. The general form of the equation is

$$y = a + bx + cx^2 + \cdots$$

In problems having fixed-volume flow rate branches, a call for positive pressure loss would require a booster fan; negative pressure loss would require a regulator.

Examples

A small ventilation network is given in Fig. 1. The details of the airways are:

Branch	J1	J2	R (Ns²/m⁸)
1	3	6	2.793
2	6	1	0.279
3	1	2	1.117
4	2	3	13.406
5	4	3	2.234
6	1	4	12.848
7	4	5	20.109
8	2	5	3.352
9	5	6	49.156

Solve the network given:

1a. Two fans are located in branches 1 and 2. The characteristics of fan 1 (in branch 1) are:

Volume flow rate, m^3/s	Pressure (Pa)
9.44	1280
28.32	1280

The characteristics of fan 2 (in branch 2) are:

Volume flow rate, m^3/s	Pressure (Pa)
14.16	2490
23.60	2190
33.04	1690
42.48	1000

1b. Branches 1 and 2 have natural ventilation pressures of 75 Pa and 50 Pa respectively.

2a. One fan with the following characteristics is located in branch 2:

Volume flow rate, m^3/s	Pressure (Pa)
14.16	2490
23.60	2190
33.04	1690
42.48	1000

2b. Branches 1 and 2 have natural ventilation pressures of 75 Pa and 50 Pa respectively.

2c. Branch 1 is a fixed-volume flow rate branch with a requirement of 23.59 m^3/s.

3. Branches 1 and 2 have natural ventilation pressures of 1360 Pa and 2020 Pa respectively.

Fig. 1. An example of a ventilation network with node numbers (in parentheses) and branch numbers. The arrows show assigned flow directions.

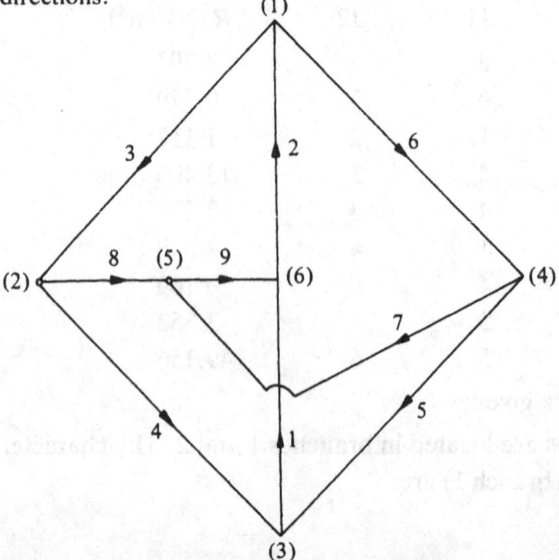

```
INPUT DATA FOR EXAMPLE 1

TITLE OF THE PROBLEM=?
EXAMPLE 1

NUMBER OF BRANCHES=?
9

NUMBER OF JUNCTIONS=?
6

MAXIMUM NUMBER OF JUNCTIONS=?
6

NUMBER OF FANS=?
2

MAXIMUM NUMBER OF ITERATIONS=?
100

NUMBER OF BRANCHES WITH NATURAL VENTILATION PRESSURE=?
2

NUMBER OF BRANCHES WITH FIXED VOLUME FLOW RATE=?
0

ERROR=?
.0001

DO YOU WANT PRINTOUT OF AIRWAY DETAILS ON PRINTER? - ANS. Y/N ---
Y

DO YOU WANT PRINTOUT OF MESH DETAILS ON PRINTER? - ANS. Y/N ---
Y

DO YOU WANT TO INPUT R's OR FF's, H's, W's AND L's? - ANS. R/F --
-
R

J1 1?
3
J2 1?
6
R1?
2.793
J1 2?
6
J2 2?
1
R 2?
.279
J1 3?
1
J2 3?
2
R 3?
1.117
J1 4?
2
J2 4?
3
R 4?
13.406
J1 5?
4
J2 5?
3
R 5?
2.234
J1 6?
1
J2 6?
4
R 6?
12.848
J1 7?
4
J2 7?
5
R 7?
20.109
J1 8?
2
```

```
J2 8?
5
R 8?
3.352
J1 9?
5
J2 9?
6
R 9?
49.156

NOW YOU ENTER DATA IN REGARD TO BRANCHES WITH NATURAL
VENTILATION PRESSURE

BRANCH NUMBER?
1
NATURAL VENTILATION PRESSURE IN BRANCH 1?
75
BRANCH NUMBER?
2
NATURAL VENTILATION PRESSURE IN BRANCH 2?
50

NOW YOU ENTER DATA REGARDING FANS

NUMBER OF POINTS FOR FAN 1?
2
X1?
9.44
FX1?
1280
X2?
28.32
FX2?
1280
NUMBER OF POINTS FOR FAN 2?
4
X1?
14.16
FX1?
2490
X2?
23.6
FX2?
2190
X3?
33.04
FX3?
1690
X4?
42.48
FX4?
1000

OUTPUT FOR EXAMPLE 1

EXAMPLE 1


NUMBER OF BRANCHES= 9
NUMBER OF JUNCTIONS= 6
MAXIMUM NUMBER OF JUNCTIONS= 6
NUMBER OF MESHES= 4
NUMBER OF FANS= 2
MAXIMUM NUMBER OF ITERATIONS= 100
NUMBER OF BRANCHES WITH NATURAL VENTILATION PRESSURE= 2
NUMBER OF BRANCHES WITH FIXED VOLUME FLOW RATE= 0
ERROR= .0001

AIRWAY    J1      J2        R

  1        3       6        2.793
  2        6       1         .279
  3        1       2        1.117
  4        2       3       13.406
  5        4       3        2.234
  6        1       4       12.848
  7        4       5       20.109
  8        2       5        3.352
  9        5       6       49.156
```

BRANCH	NVP
1	75
2	50

MESH	NO. OF AIRWAYS	SUMNVP	AIRWAY NUMBERS
1	6	75	1
			-9
			-8
			-3
			6
			5
2	4	50	2
			3
			8
			9
3	4	0	7
			-8
			-3
			6
4	4	0	4
			-5
			-6
			3

TOTAL 18

FAN	J1	J2
1	3	6

POINT	X	FX
1	9.44	1280
2	28.32	1280

COEFFICIENTS

1280
0

FAN	J1	J2
2	6	1

POINT	X	FX
1	14.16	2490
2	23.6	2190
3	33.04	1690
4	42.48	1000

COEFFICIENTS

2577.875
9.3220339
-1.0941091

ITERATIONS 28

BRANCH	J1	J2	R	Q	H
1	3	6	2.793	23.59	1554.467
2	6	1	.279	28.31	223.618
3	1	2	1.117	18.87	397.912
4	2	3	13.406	9.44	1193.804
5	4	3	2.234	14.15	447.604
6	1	4	12.848	9.44	1144.112
7	4	5	20.109	- 4.72	- 447.661
8	2	5	3.352	9.44	298.551
9	5	6	49.156	4.72	1094.782

FAN DATA

FAN	J1	J2	Q	FQ
1	3	6	23.59	1280.000
2	6	1	28.31	1964.865

```
INPUT DATA FOR EXAMPLE 2

TITLE OF THE PROBLEM=?
EXAMPLE 2

NUMBER OF BRANCHES=?
9

NUMBER OF JUNCTIONS=?
6

MAXIMUM NUMBER OF JUNCTIONS=?
6

NUMBER OF FANS=?
1

MAXIMUM NUMBER OF ITERATIONS=?
100

NUMBER OF BRANCHES WITH NATURAL VENTILATION PRESSURE=?
2

NUMBER OF BRANCHES WITH FIXED VOLUME FLOW RATE=?
1

ERROR=?
.0001

DO YOU WANT PRINTOUT OF AIRWAY DETAILS ON PRINTER? - ANS. Y/N ---
N

DO YOU WANT PRINTOUT OF MESH DETAILS ON PRINTER? - ANS. Y/N ---
Y

DO YOU WANT TO INPUT R's OR FF's, H's, W's AND L's? - ANS. R/F --
-
R

J1 1?
3
J2 1?
6
R1?
2.793
J1 2?
6
J2 2?
1
R 2?
.279
J1 3?
1
J2 3?
2
R 3?
1.117
J1 4?
2
J2 4?
3
R 4?
13.406
J1 5?
4
J2 5?
3
R 5?
2.234
J1 6?
1
J2 6?
4
R 6?
12.848
J1 7?
4
J2 7?
5
```

```
R 7?
20.109
J1 8?
2
J2 8?
5
R 8?
3.352
J1 9?
5
J2 9?
6
R 9?
49.156

NOW YOU ENTER DATA IN REGARD TO BRANCHES WITH NATURAL
VENTILATION PRESSURE

BRANCH NUMBER?
1
NATURAL VENTILATION PRESSURE IN BRANCH 1?
75
BRANCH NUMBER?
2
NATURAL VENTILATION PRESSURE IN BRANCH 2?
50

NOW YOU ENTER DATA REGARDING FANS

NUMBER OF POINTS FOR FAN 1?
4
X1?
14.16
FX1?
2490
X2?
23.6
FX2?
2190
X3?
33.04
FX3?
1690
X4?
42.48
FX4?
1000

NOW YOU ENTER DATA REGARDING FIXED VOLUME FLOW RATE BRANCHES

VOLUME FLOW RATE ASSIGNED TO FIXED VOLUME FLOW RATE BRANCH 1?
23.59
```

```
OUTPUT FOR EXAMPLE 2

EXAMPLE 2

NUMBER OF BRANCHES= 9
NUMBER OF JUNCTIONS= 6
MAXIMUM NUMBER OF JUNCTIONS= 6
NUMBER OF MESHES= 4
NUMBER OF FANS= 1
MAXIMUM NUMBER OF ITERATIONS= 100
NUMBER OF BRANCHES WITH NATURAL VENTILATION PRESSURE= 2
NUMBER OF BRANCHES WITH FIXED VOLUME FLOW RATE= 1
ERROR= .0001


BRANCH   NVP

 1        75
 2        50
```

MESH	NO. OF AIRWAYS	SUMNVP	AIRWAY NUMBERS
1	6	75	1 -9 -8 -3 6 5
2	4	50	2 3 8 9
3	4	0	7 -8 -3 6
4	4	0	4 -5 -6 3

TOTAL 18

FAN	J1	J2
1	6	1

POINT	X	FX
1	14.16	2490
2	23.6	2190
3	33.04	1690
4	42.48	1000

COEFFICIENTS

2577.875
9.3220339
-1.0941091

F.Q. BRANCH	Q
1	23.59

ITERATIONS 10

BRANCH	J1	J2	R	Q	H
1	3	6	2.793	23.59	1554.271
2	6	1	.279	28.31	223.599
3	1	2	1.117	18.87	397.881
4	2	3	13.406	9.44	1193.659
5	4	3	2.234	14.15	447.547
6	1	4	12.848	9.44	1143.993
7	4	5	20.109	- 4.72	- 447.582
8	2	5	3.352	9.44	298.541
9	5	6	49.156	4.72	1094.898

HEAD LOSS TO BE ADJUSTED IN FIXED VOLUME FLOW RATE BRANCH/BRANCHES

BRANCH	J1	J2	R	Q	SUMH
1	3	6	2.793	23.59	1279.491

FAN DATA

FAN	J1	J2	Q	FQ
1	6	1	28.31	1964.928

```
INPUT DATA FOR EXAMPLE 3

TITLE OF THE PROBLEM=?
EXAMPLE 3

NUMBER OF BRANCHES=?
9

NUMBER OF JUNCTIONS=?
6

MAXIMUM NUMBER OF JUNCTIONS=?
6

NUMBER OF FANS=?
0

MAXIMUM NUMBER OF ITERATIONS=?
100

NUMBER OF BRANCHES WITH NATURAL VENTILATION PRESSURE=?
2

NUMBER OF BRANCHES WITH FIXED VOLUME FLOW RATE=?
0

ERROR=?
.0001

DO YOU WANT PRINTOUT OF AIRWAY DETAILS ON PRINTER? - ANS. Y/N ---
N

DO YOU WANT PRINTOUT OF MESH DETAILS ON PRINTER? - ANS. Y/N ---
N

DO YOU WANT TO INPUT R's OR FF's, H's, W's AND L's? - ANS. R/F --
-
R

J1 1?
3
J2 1?
6
R1?
2.793
J1 2?
6
J2 2?
1
R 2?
.279
J1 3?
1
J2 3?
2
R 3?
1.117
J1 4?
2
J2 4?
3
R 4?
13.406
J1 5?
4
J2 5?
3
R 5?
2.234
J1 6?
1
J2 6?
4
R 6?
12.848
J1 7?
4
J2 7?
5
R 7?
20.109
```

```
J1 8?
2
J2 8?
5
R 8?
3.352
J1 9?
5
J2 9?
6
R 9?
49.156

NOW YOU ENTER DATA IN REGARD TO BRANCHES WITH NATURAL
VENTILATION PRESSURE

BRANCH NUMBER?
1
NATURAL VENTILATION PRESSURE IN BRANCH 1?
1360
BRANCH NUMBER?
2
NATURAL VENTILATION PRESSURE IN BRANCH 2?
2020
```

OUTPUT FOR EXAMPLE 3

```
EXAMPLE 3

NUMBER OF BRANCHES= 9
NUMBER OF JUNCTIONS= 6
MAXIMUM NUMBER OF JUNCTIONS= 6
NUMBER OF MESHES= 4
NUMBER OF FANS= 0
MAXIMUM NUMBER OF ITERATIONS= 100
NUMBER OF BRANCHES WITH NATURAL VENTILATION PRESSURE= 2
NUMBER OF BRANCHES WITH FIXED VOLUME FLOW RATE= 0
ERROR= .0001
```

BRANCH	NVP
1	1360
2	2020

ITERATIONS 13

BRANCH	J1	J2	R	Q	H
1	3	6	2.793	23.63	1559.218
2	6	1	.279	28.35	224.273
3	1	2	1.117	18.90	399.066
4	2	3	13.406	9.45	1197.441
5	4	3	2.234	14.18	448.975
6	1	4	12.848	9.45	1147.533
7	4	5	20.109	- 4.73	- 449.098
8	2	5	3.352	9.45	299.372
9	5	6	49.156	4.72	1097.291

PROGRAM LISTING

```
5 PUT31
10 DIM NA(765),JC(185),B(100),R(100),Q(100),RR(15),OU(100),J1(75),J2(75),ME(65)
SN(65),ND(6),DD(15),X(8),FX(8),W(6),Z(6),S(6),AL(6),BE(6),SI(6),C(6,6),P(7,20),
Q(6),HH(100)
15 REM IDENTIFICATION OF THE PROBLEM
20 INPUT("TITLE OF THE PROBLEM=?")A$
25 REM INPUT PARAMETERS
30 INPUT("NUMBER OF BRANCHES=?")NB
40 INPUT("NUMBER OF JUNCTIONS=?")NJ
50 INPUT("MAXIMUM NUMBER OF JUNCTIONS=?")MJ
60 INPUT("NUMBER OF FANS=?")NF
```

```
70 INPUT("MAXIMUM NUMBER OF ITERATIONS=?")IT
80 INPUT("NUMBER OF BRANCHES WITH NATURAL VENTILATION PRESSURE=?")NV
90 INPUT("NUMBER OF BRANCHES WITH FIXED VOLUME FLOW RATE=?")FI
100 INPUT("ERROR=?")E
108 INPUT("DO YOU WANT PRINTOUT OF AIRWAY DETAILS ON PRINTER? - ANS. Y/N ---") E
$
109 INPUT("DO YOU WANT PRINTOUT OF MESH DETAILS ON PRINTER? - ANS. Y/N ---") D$
110 NM=NB-NJ+1
120 PRINT#8,A$:PRINT#8:PRINT#8
130 PRINT#8,"NUMBER OF BRANCHES=";NB
131 PRINT#8,"NUMBER OF JUNCTIONS=";NJ
132 PRINT#8,"MAXIMUM NUMBER OF JUNCTIONS=";MJ
133 PRINT#8,"NUMBER OF MESHES=";NM
134 PRINT#8,"NUMBER OF FANS=";NF
135 PRINT#8,"MAXIMUM NUMBER OF ITERATIONS=";IT
136 PRINT#8,"NUMBER OF BRANCHES WITH NATURAL VENTILATION PRESSURE=";NV
137 PRINT#8,"NUMBER OF BRANCHES WITH FIXED VOLUME FLOW RATE=";FI
138 PRINT#8,"ERROR=";E:PRINT#8:PRINT#8
139 REM INPUT BRANCH PARAMETERS
140 INPUT("DO YOU WANT TO INPUT R's OR FF's, H's, W's AND L's? - ANS. R/F ---")B
$
150 IFB$="F"THEN 260
160 IF E$="Y" THEN PRINT#8,CHR$(27);CHR$(68);CHR$(10);CHR$(20);CHR$(33);CHR$(0);
"AIRWAY";CHR$(137);"J1";CHR$(137);"J2";CHR$(137);"R"
170 IF E$="Y" THEN PRINT#8
180 FOR I= 1 TO NB
190 PRINT"J1";I;:INPUT J1(I)
200 PRINT"J2";I;:INPUT J2(I)
210 PRINT"R";I;:INPUT R(I)
220 B(I)=I
230 RR(I)=R(I)
240 IF E$="Y" THEN PRINT#8,CHR$(27);CHR$(68);CHR$(10);CHR$(20);CHR$(30);CHR$(0);
I;CHR$(137);J1(I);CHR$(137);J2(I);CHR$(137);R(I)[3.3]
250 NEXTI:PRINT#8:PRINT#8:GOTO380
260 IF E$="Y" THEN PRINT#8,CHR$(27);CHR$(68);CHR$(10);CHR$(20);CHR$(33);CHR$(43)
;CHR$(51);CHR$(61);CHR$(72);CHR$(0);"AIRWAY";CHR$(137);"J1";CHR$(137);"J2";CHR$(
137);"R";CHR$(137);"FF";CHR$(137);"H";CHR$(137);"W";CHR$(137);"L"
265 PRINT#8
270 FOR I= 1 TO NB
273 PRINT"J1";I;:INPUTJ1(I)
276 PRINT"J2";I;:INPUTJ2(I)
280 PRINT"FF";I;:INPUTFF
290 PRINT"HEIGHT";I;:INPUT H
300 PRINT"WIDTH";I;:INPUTW
310 PRINT"LENGTH";I;:INPUT L
320 B(I)=I
330 T=H*W
340 R(I)=FF*(H+W)*2*L/(T*T*T)
350 RR(I)=R(I)
360 IF E$="Y" THEN PRINT#8,CHR$(27);CHR$(68);CHR$(10);CHR$(20);CHR$(30);CHR$(40)
;CHR$(50);CHR$(60);CHR$(70);CHR$(0);I;CHR$(137);J1(I);CHR$(137);J2(I);CHR$(137);
R(I)[2.3];CHR$(137);FF[1.4];CHR$(137);H[1.1];CHR$(137);W[1.1];CHR$(137);L[7.1]
370 NEXT I
375 PRINT#8:PRINT#8
377 REM GENERAL BRANCHES ARRANGED IN DECREASING ORDER OF RESISTANCE
380 IS=FI+NF+1
390 IE=NB-1
400 JE=IE
410 L=0
420 FOR I= IS TO IE
430 FOR J= IS TO JE
440 IF RR(J+1)-RR(J)<=0 THEN 520
450 T=RR(J)
460 RR(J)=RR(J+1)
470 RR(J+1)=T
480 T=B(J)
490 B(J)=B(J+1)
500 B(J+1)=T
510 L=1
520 NEXT J
530 IF L<=0 THEN 560
540 JE=JE-1
550 NEXT I
555 REM BASIC BRANCHES DETERMINED
560 FOR I= 1 TO MJ
570 JC(I)=0
580 NEXT I
590 I=NB+1
600 L=0
610 N=0
620 FOR IJ=IS TO NB
```

```
622 I=I-1
624 OU(I)=0
625 K=B(I)
626 JA=J1(K)
628 JB=J2(K)
629 IFJC(JA)-JC(JB)<0THEN730
630 IF JC(JA)-JC(JB)=0 THEN 760
640 IF JC(JB)=0 THEN 710
650 JJ=JC(JB)
660 FOR J= 1 TO MJ
670 IF JC(J)-JJ <>0 THEN 690
680 JC(J)=JC(JA)
690 NEXTJ
700 GOTO830
710 JC(JB)=JC(JA)
720 GOTO830
730 IFJC(JA)<>0 THEN 650
740 JC(JA)=JC(JB)
750 GOTO830
760 IFJC(JA)=0THEN 800
770 OU(I)=1
780 N=N+1
790 GOTO830
800 L=L+1
810 JC(JA)=L
820 JC(JB)=L
830 NEXTIJ
840 IF(N+FI+NF-NM)=0THEN870
850 PRINT"NO. OF BASIC BRANCHES=";N:STOP
870 IF(FI+NF)<=0THEN910
880 FORI=1TO(FI+NF)
890 OU(I)=1
900 NEXTI
905 REM MESHES FOUND
910 JK=0
920 JE=0
930 L=0
940 FOR I=1 TO NB
950 IF OU(I) <=0 THEN 1360
960 K=B(I)
970 L=L+1
980 JK=JK+1
990 NA(JK)=K
1000 JA=J1(K)
1010 JB=J2(K)
1020 N=I+1
1030 FORJ=N TO NB
1040 IF OU(J)<>0 THEN 1180
1050 K=B(J)
1060 IF JB-J1(K)<>0THEN 1110
1070 JB=J2(K)
1080 JK=JK+1
1090 NA(JK)=K
1100 GOTO 1150
1110 IFJB-J2(K)<>0 THEN 1180
1120 JB=J1(K)
1130 JK=JK+1
1140 NA(JK)=-K
1150 IF JB-JA=0THEN1300
1160 OU(J)=-1
1170 GOTO1290
1180 NEXTJ
1190 K=ABS(NA(JK))
1200 IFNA(JK)<0THEN1230
1210 JB=J1(K)
1220 GOTO1240
1230 JB=J2(K)
1240 JK=JK-1
1250 IF JK-JE>0THEN1030
1260 K=B(I):PRINT"NO MESH FOUND AT BRANCH";K:STOP
1290 GOTO1030
1300 FOR J= N TO NB
1310 IF OU(J)>=0 THEN 1330
1320 OU(J)=0
1330 NEXT J
1340 ME(L)=JK
1350 JE=JK
1360 NEXT I
1370 FOR I=1 TO NB
1380 RR(I)=0
1385 NEXTI
```

```
1390 IF NV<=0THEN 1460
1395 REM INPUT NATURAL VENTILATION PRESSURES
1405 PRINT"NOW YOU ENTER DATA IN REGARD TO BRANCHES WITH NATURAL VENTILATION PRE
SSURE":PRINT
1406 FOR J=1 TO NV
1410 INPUT("BRANCH NUMBER?")I
1420 PRINT"NATURAL VENTILATION PRESSUURE IN BRANCH";I;:INPUTRR(I)
1430 NEXT J
1435 PRINT#8,"BRANCH","NVP"
1437 PRINT#8
1440 FOR I = 1 TO NB
1445 IF RR(I)=0THEN 1455
1450 PRINT#8,I,RR(I)
1455 NEXT I
1456 PRINT#8:PRINT#8
1458 REM MESH TABLE PRINTED
1460 PRINT"MESH","NO. OF AIRWAYS","SUMNVP","AIRWAY NUMBERS":PRINT:PRINT
1461 IF D$="Y" THEN PRINT#8,CHR$(27);CHR$(68);CHR$(10);CHR$(30);CHR$(40);CHR$(0)
;"MESH";CHR$(137);"NO. OF AIRWAYS";CHR$(137);"SUMNVP";CHR$(137);"AIRWAY NUMBERS"
:PRINT#8:PRINT#8
1470 JE=0
1480 FOR I = 1 TO NM
1490 JS=JE+1
1500 JE=ME(I)
1510 SN(I)=0
1530 FOR J= JS TO JE
1535 K=ABS(NA(J))
1540 IF NA(J)<0THEN 1570
1550 SN(I)=SN(I)+RR(K)
1560 GOTO1580
1570 SN(I)=SN(I)-RR(K)
1580 NEXT J
1590 L=JE-JS+1
1595 PRINTI,L,SN(I),
1600 IF D$="Y" THEN PRINT#8,CHR$(27);CHR$(68);CHR$(10);CHR$(30);CHR$(40);CHR$(0)
;I;CHR$(137);L;CHR$(137);SN(I),
1610 FORJ=JS TO JE
1611 PRINTTAB(50),NA(J)
1612 IF D$="Y" THEN PRINT#8,CHR$(27);CHR$(68);CHR$(50);CHR$(0);CHR$(137);NA(J)
1614 NEXT J
1615 NEXTI
1620 PRINT"          TOTAL";JE
1625 IF D$="Y" THEN PRINT#8,"     TOTAL";JE :PRINT#8:PRINT#8
1630 IF FI+NF<=0THEN2640
1635 REM FAN CHARACTERISTICS APPROXIMATED
1640 IF NF<=0THEN 2680
1645 PRINT"NOW YOU ENTER DATA REGARDING FANS"
1650 FOR L= 1 TO NF
1660 JJ=L+FI
1667 PRINT
1670 PRINT"NUMBER OF POINTS FOR FAN";L;:INPUT NP
1680 KE=0
1695 PRINT #8,"FAN","J1","J2":PRINT#8
1705 PRINT#8,L,J1(JJ),J2(JJ):PRINT#8
1710 IFKE>1 THEN 2640
1720 FOR I=1 TO NP
1730 PRINT"X";I;:INPUT X(I)
1740 PRINT"FX";I;:INPUT FX(I)
1750 NEXT I
1765 PRINT#8,"POINT","X","FX":PRINT#8
1770 FOR I= 1 TO NP
1785 PRINT#8,I, X(I),FX(I)
1790 NEXT I
1795 PRINT
1798 PRINT#8
1800 DD(JJ)=X(NP)
1810 IF (NP -2) > 0 THEN 1900
1820 ND(L) = 1
1830 N1=2
1840 IF (NP -1) =0 THEN 1870
1850 IF (NP -1) >0 THEN 1892
1860 STOP
1870 C(L,1) = FX(1)
1880 C(L,2)=0
1890 GOTO 2590
1892 IFX(1)-X(2)<>0THEN 1896
1894 STOP
1896 C(L,1)=(FX(1)*X(2)-FX(2)*X(1))/(X(2)-X(1))
1898 C(L,2)=(FX(2)-FX(1))/(X(2)-X(1))
1899 GOTO2590
1900 DS=0
```

```
1910 FOR J= 1 TO NP
1920 DS=DS+FX(J)*FX(J)
1930 P(1,J)=0
1940 P(2,J)=1
1945 NEXT J
1950 W(1)=NP
1960 BE(1)=0
1970 IF(NP-7) <0 THEN 2000
1980 N1=6
1990 GOTO 2010
2000 N1= NP-1
2010 FOR I= 1 TO N1
2020 K=I+1
2030 Z(I)=0
2040 FOR J=1 TO NP
2050 Z(I)=Z(I)+FX(J)*P(K,J)
2060 NEXT J
2070 S(I)=Z(I)/W(I)
2080 DS=DS-S(I)*S(I)*W(I)
2090 SI(I)=DS/(NP-I)
2100 IF SI(I) - (10^-6) <=0 THEN 2310
2110 IF (I-N1) >=0 THEN 2220
2120 AL(I)=0
2130 FOR J= 1 TO NP
2140 AL(I)=AL(I)+X(J)*P(K,J)*P(K,J)
2145 NEXT J
2150 AL(I)=AL(I)/W(I)
2160 W(K)=0
2170 FOR J= 1TO NP
2180 P((K+1),J)=(X(J)-AL(I))*P(K,J)-BE(I)*P(I,J)
2190 W(K)=W(K)+P((K+1),J)*P((K+1),J)
2200 NEXT J
2210 BE(K)=W(K)/W(I)
2215 NEXT I
2220 SM=SI(1)
2230 I=1
2240 FORJ=2 TO N1
2250 IF(SM-SI(J))<=0 THEN 2280
2260 I=J
2270 SM=SI(J)
2280 NEXT J
2310 N1=I
2320 N=N1-1
2330 IFN>0THEN 2380
2340 ND(L)=1
2350 C(L,1)=S(1)
2360 C(L,2)=0
2370 GOTO2590
2380 ND(L)=N
2390 FOR I= 1 TO N1
2400 FOR J = 1 TO I
2410 P(I,J)=0
2420 NEXT J
2430 P(I,I+1)=1
2440 NEXT I
2450 FOR J= 1 TO N
2460 P(1,(J+2))=-P(1,(J+1))*AL(J)-P(1,J)*BE(J)
2470 NEXT J
2480 FOR I=2 TO N
2490 FOR J = I TO N
2500 P(I,(J+2))=P(I-1,J+1)-P(I,J+1)*AL(J)-P(I,J)*BE(J)
2510 NEXT J
2520 NEXT I
2530 FOR I= 1 TO N1
2540 C(L,I)=0
2550 FOR J= I TO N1
2560 C(L,I)=C(L,I)+P(I,J+1)*S(J)
2570 NEXT J
2580 NEXT I
2590 PRINT"COEFFICIENTS":PRINT
2595 PRINT#8,"COEFFICIENTS":PRINT#8
2600 FOR I = 1 TO N1
2610 PRINTC(L,I)
2615 PRINT#8,C(L,I)
2620 NEXT I
2625 PRINT
2628 PRINT#8
2630 NEXT L
2631 ND(L)=ND(KE):N1=ND(KE)+1
2632 DD(JJ)=DD(KE+FI)
```

```
2633 FOR J= 1 TO N1
2634 C(L,J)=C(KE,J)
2635 NEXT J
2638 REM INITIAL VALUES SET FOR HARDY-CROSS ITERATION
2640 IF (NF+FI)>0 THEN 2680
2645 DD(1)=10
2660 IE=1
2670 GOTO 2770
2680 IE=NF+FI
2690 IF FI <=0 THEN 2770
2700 PRINT"NOW YOU ENTER DATA REGARDING FIXED VOLUME FLOW RATE BRANCHES":PRINT
2710 PRINT#8,"F.Q.","Q"
2720 PRINT#8,"BRANCH"
2725 PRINT#8
2730 FOR I= 1 TO FI
2733 PRINT"VOLUME FLOW RATE ASSIGNED TO FIXED VOLUME FLOW RATE BRANCH";I;:INPUT
DD(I)
2735 PRINT#8,I,DD(I)
2736 NEXT I
2738 PRINT#8:PRINT#8
2740 GOTO2770
2770 FOR I= 1 TO NB
2780 Q(I)=0
2790 NEXT I
2792 JE=0
2793 FOR I = 1 TO IE
2794 JS=JE+1
2795 JE=ME(I)
2800 FOR J= JS TO JE
2810 K=ABS(NA(J))
2820 IF NA(J)<0 THEN 2850
2830 Q(K)=Q(K)+DD(I)
2840 GOTO2860
2850 Q(K)=Q(K)-DD(I)
2860 NEXT J
2865 NEXT I
2868 REM ITERATION BEGUN
2870 IB=FI+1
2880 FOR MI=1 TO IT
2885 PRINT"ITERATIONS";MI
2890 IF FI<=0 THEN 2920
2900 JE=ME(FI)
2910 GOTO 2930
2920 JE=0
2930 L=0
2940 SD=0
2950 FOR I= IB TO NM
2960 IM=I-FI
2970 JS=JE+1
2980 JE=ME(I)
2990 SH=-SN(I)
3000 DH=0
3010 DF=0
3020 IF(FI+NF)-I<0 THEN 3140
3025 REM PRESSURES IN FAN BRANCHES FOUND USING FAN CURVE POLYNOMIAL
3030 N=ND(IM)
3031 FQ(IM)=0
3032 IF N=0 THEN 3110
3040 FQ(IM)=C(IM,N+1)*Q(I)
3050 IF (N-1)=0 THEN 3110
3060 J=N
3070 FOR IL = 2 TO N
3080 FQ(IM)= (FQ(IM)+C(IM,J))*Q(I)
3100 J=J-1
3105 NEXT IL
3110 FQ(IM)=FQ(IM)+C(IM,1)
3111 IF(N-1)<=0 THEN 3120
3112 J=N
3113 FOR IJ = 2 TO N
3114 DF=(DF+J*C(IM,(J+1)))*Q(I):J=J-1
3115 NEXT IJ
3120 SH=SH-FQ(IM)
3130 DF=DF+C(IM,2)
3135 REM PRESSURES FOUND USING H = R*Q^2
3140 FOR J=JS TO JE
3150 K = ABS(NA(J))
3160 DS=R(K)*ABS(Q(K))
3170 H=DS*Q(K)
3175 REM NEW INCREMENT FOR Q COMPUTED
3180 DH=DH+DS
```

```
3190 IF NA(J)<0 THEN 3220
3200 SH=SH+H
3210 GOTO 3230
3220 SH=SH-H
3230 NEXT J
3240 DH=DH+DH-DF
3250 IF ABS(DH)-10^-20 <=0 THEN 3360
3260 D=-SH/DH
3265 IF ABS(D)<=10^-20 THEN 3370
3270 FOR J=JS TO JE
3280 K=ABS(NA(J))
3290 IF NA(J)<0 THEN 3320
3300 Q(K)=Q(K)+D
3310 GOTO 3330
3320 Q(K)=Q(K)-D
3330 NEXT J
3340 SD=SD+ABS(D)
3350 IF ABS(D)-E<=0 THEN 3370
3360 L=1
3370 NEXT I
3380 IF L<=0 THEN 3430
3390 NEXT MI
3400 PRINT "ITERATIONS";MI:PRINT:PRINT
3405 PRINT #8,"ITERATIONS";MI:PRINT #8:PRINT#8
3410 PRINT"D STILL GREATER THAN E"
3415 PRINT#8,"D STILL GREATER THAN E":PRINT #8:PRINT#8
3420 GOTO3440
3430 PRINT "ITERATIONS";MI:PRINT:PRINT
3435 PRINT #8,"ITERATIONS";MI:PRINT#8:PRINT#8
3440 PRINT #8,CHR$(27);CHR$(68);CHR$(10);CHR$(20);CHR$(37);CHR$(50);CHR$(65);CHR
$(0);"BRANCH";CHR$(137);"J1";CHR$(137);"J2";CHR$(137);"R";CHR$(137);"Q";CHR$(137
);"H":PRINT#8:PRINT#8
3445 PRINTTAB(1);"BRANCH";TAB(11);"J1";TAB(21);"J2";TAB(38);"R";TAB(49);"Q";TAB(
67);"H"
3450 FOR I = 1 TO NB
3460 HH(I)=R(I)*ABS(Q(I))*Q(I)
3470 PRINT #8,CHR$(27);CHR$(68);CHR$(10);CHR$(20);CHR$(30);CHR$(45);CHR$(50);CHR
$(60);CHR$(0);I;CHR$(137);J1(I);CHR$(137);J2(I);CHR$(137);R(I)[6.3];CHR$(137);Q(
I)[3.2];CHR$(137);HH(I)[5.3]
3475 PRINTTAB(1);I;TAB(11);J1(I);TAB(21);J2(I);TAB(31);R(I)[6.3];TAB(45);Q(I)[3.
2];TAB(61);HH(I)[5.3]
3476 FOR Z=1TO 270
3477 NEXT Z
3480 NEXT I
3485 PRINT #8:PRINT#8
3487 PRINT:PRINT
3490 IF FI<=0 THEN 3660
3495 PRINT #8:PRINT#8
3500 PRINT "HEAD LOSS TO BE ADJUSTED IN FIXED VOLUME FLOW RATE BRANCH/BRANCHES":
PRINT:PRINT
3505 PRINT #8,"HEAD LOSS TO BE ADJUSTED IN FIXED VOLUME FLOW RATE BRANCH/BRANCHE
S":PRINT#8:PRINT#8
3510 PRINT #8,CHR$(27);CHR$(68);CHR$(10);CHR$(20);CHR$(37);CHR$(50);CHR$(68);CHR
$(0);"BRANCH";CHR$(137);"J1";CHR$(137);"J2";CHR$(137);"R";CHR$(137);"Q";CHR$(137
);"SUMH":PRINT#8:PRINT#8
3515 PRINTTAB(1);"BRANCH";TAB(11);"J1";TAB(21);"J2";TAB(38);"R";TAB(49);"Q";TAB(
67);"SUMH"
3520 JE=0
3530 FOR I = 1 TO FI
3540 JS=JE+1
3550 JE=ME(I)
3560 SH(I)=-SN(I)
3570 FOR J=JS TO JE
3580 K=ABS(NA(J))
3600 SH(I)=SH(I)+NA(J)/K*HH(K)
3630 NEXT J
3640 PRINT #8,CHR$(27);CHR$(68);CHR$(10);CHR$(20);CHR$(31);CHR$(45);CHR$(65);CHR
$(0);I;CHR$(137);J1(I);CHR$(137);J2(I);CHR$(137);R(I)[5.3];CHR$(137);Q(I)[3.2];C
HR$(137);SH(I)[5.3]
3645 PRINTTAB(1);I;TAB(11);J1(I);TAB(21);J2(I);TAB(31);R(I)[5.3];TAB(45);Q(I)[3.
2];TAB(61);SH(I)[5.3]
3650 NEXT I
3655 PRINT:PRINT
3657 PRINT#8:PRINT#8
3660 IF NF<=0 THEN 4000
3663 PRINT"FAN DATA":PRINT:PRINT
3666 PRINT#8,"FAN DATA":PRINT#8:PRINT#8
3670 PRINTTAB(1);"FAN";TAB(11);"J1";TAB(21);"J2";TAB(36);"Q";TAB(52);"FQ":PRINT:
PRINT
```

```
3675 PRINT #8,CHR$(27);CHR$(68);CHR$(10);CHR$(20);CHR$(35);CHR$(55);CHR$(0);"FAN
";CHR$(137);"J1";CHR$(137);"J2";CHR$(137);"Q";CHR$(137);"FQ";CHR$(137);PRINT#8:P
RINT#8
3680 FOR I=1 TO NF
3690 PRINTTAB(1);I;TAB(11);J1(I+FI);TAB(21);J2(I+FI);TAB(32);Q(I+FI)[3.2];TAB(46
);FQ(I)[5.3]
3695 PRINT #8,CHR$(27);CHR$(68);CHR$(10);CHR$(20);CHR$(32);CHR$(50);CHR$(0);I;CH
R$(137);J1(I+FI);CHR$(137);J2(I+FI);CHR$(137);Q(I+FI)[3.2];CHR$(137);FQ(I)[5.3]
3700 NEXT I
3705 PRINT #8:PRINT#8
3710 PRINT:PRINT
4000 END
```

APPENDIX II

Programs in BASIC for auxiliary ventilation calculations

Program 1

This program is based on the theory developed for method 1 on p. 60, Mathematical analysis of airflow in ducts with a uniformly distributed leakage, in section 2.5.1, Airflow analysis in leaky ducts.

It has an option to calculate the resistance coefficient of leakage paths using the Woronin formula (equation 2.49, p. 63). If the duct length is known, one of the following is required as input:

(1) volume rate handled by the fan;
(2) pressure developed by the fan;
(3) fan data;
(4) volume rate reaching the face.

For a given minimum face volume rate, the maximum duct length can be calculated requiring one of the following input data:

(1) volume rate handled by the fan;
(2) pressure developed by the fan;
(3) fan data.

If the fan data is input, then details about the fan characteristic curve, number of fans (if more than one, in series or parallel) and natural ventilation pressure are required.

The results, depending upon the input data, include some of the following:

Maximum duct length

Volume rate reaching the face

Volume rate handled by the fan

Volume increase ratio
Pressure drop for leakless duct
Pressure developed by the fan, i.e. pressure drop for leaky duct
Pressure increase ratio
Equivalent resistance
Input power to motor
Duct efficiency

In the calculation of input power to motor, the combined efficiency of fan and motor has been assumed to be 70 %. Duct efficiency is calculated from the pressure increase ratio and the volume increase ratio and this value shows how large a proportion of the power is wasted, but it bears very little relationship to the condition of the duct because the length of the duct is an overriding factor.

The program has a facility for rerunning after changing some of the data.

Examples

(1) A 0.61 m diameter duct 300 m long has an intake volume rate of 1.70 m^3/s and a delivery volume rate of 1.42 m^3/s. Calculate the resistance coefficient of leakage paths per 100 m of duct. The friction factor for the duct is 0.0038 Ns^2/m^4.

(2) If it is required to exhaust 7 m^3/s from the end of a 400 m long, 0.76 m diameter duct, determine what volume rate a fan at the other end will have to handle and what pressure it will have to produce. The friction factor for the duct is 0.0038 Ns^2/m^4. The resistance coefficient of leakage paths per 100 m of duct is 10 000 Ns^2/m^8.

(3) What is the maximum length of 0.61 m diameter duct which will deliver 1.42 m^3/s at its inbye end if a fan with the characteristics given in Table A is connected at the other end? The friction factor for the duct is 0.0038 Ns^2/m^4. The resistance coefficient of leakage paths per 100 m of duct is 10 000 Ns^2/m^8.

Table A. *Fan characteristics*

Q (m^3/s)	p (Pa)
4	1100
5	800
6	500

```
INPUT DATA FOR EXAMPLE 1

Duct diameter, m=?
0.61

Duct friction-factor, Ns2/m4=?
0.0038

Do you want to calculate resistance coefficient of leakage
paths/100 m, Ns2/m8? - Answer yes or no ----
yes

Volume rate handled by the fan, m3/s=?
1.7

Volume rate reaching the face, m3/s=?
1.42

Duct length, m=?
300


  OUTPUT FOR EXAMPLE 1

          AUXILIARY VENTILATION CALCULATIONS

INPUT DATA

Duct diameter, m= .61
Duct friction-factor, Ns^2/m^4= .0038

Calculation for resistance coefficient of leakage paths

Volume rate handled by the fan, m^3/s= 1.7
Volume rate reaching the face, m^3/s= 1.42
Duct length, m= 300

RESULT

Resistance coefficient of leakage paths/100 m, Ns^2/m^8=  9872.23

******************************************************************


INPUT DATA FOR EXAMPLE 2

Duct diameter, m=?
0.76

Duct friction-factor, Ns2/m4=?
0.0038

Do you want to calculate resistance coefficient of leakage
paths/100 m, Ns2/m8? - Answer yes or no ----
no

Resistance coefficient of leakage paths/100 m, Ns2/m8=?
10000

Do you want to calculate maximum duct length, m? - Answer yes or
no ----
no

Duct length, m=?
400

One of the following should be available for input:-

Volume rate handled by the fan, m3/s (QI)

Pressure drop for leaky duct i.e. Pressure developed by the fan,
Pa (PI)

Fan data (FD)

Volume rate reaching the face, m3/s (QO)

Which input data is available?
QO

Volume rate reaching the face, m3/s=?
7
```

```
OUTPUT FOR EXAMPLE 2

AUXILIARY VENTILATION CALCULATIONS
INPUT DATA

Duct diameter, m= .76
Duct friction-factor, Ns^2/m^4= .0038
Resistance coefficient of leakage paths/100 m, Ns^2/m^8= 10000
Duct length, m= 400
Volume rate reaching the face, m^3/s= 7

RESULTS

Volume rate handled by the fan, m^3/s= 8.18
Volume increase ratio= 1.17
Pressure drop for leakless duct, Pa=1904.81
Pressure developed by the fan i.e. Pressure drop for leaky duct, Pa=2154.78
Pressure increase ratio= 1.13
Equivalent resistance, Ns^2/m^8= 32.21
Input power to motor, kW= 25.18
Duct efficiency, %=75.66

******************************************************************

Do you want another run with the data changed? - Answer yes or no
----
no

INPUT DATA FOR EXAMPLE 3

Duct diameter, m=?
0.61

Duct friction-factor, Ns2/m4=?
0.0038

Do you want to calculate resistance coefficient of leakage
paths/100 m, Ns2/m8? - Answer yes or no ----
no

Resistance coefficient of leakage paths/100 m, Ns2/m8=?
10000

Do you want to calculate maximum duct length, m? - Answer yes or
no ----
yes

Volume rate required at the face, m3/s=?
1.42

One of the following should be available for input:-

Volume rate handled by the fan, m3/s (QI)

Pressure drop for leaky duct i.e. Pressure developed by the fan,
Pa (PI)

Fan data (FD)

Which input data is available?
FD

Number of points on fan characteristic curve=?
3

Number of fans=?
1

Natural ventilation pressure, Pa=?
0

Q 1?
4
p 1?
1100
Q 2?
5
p 2?
800
Q 3?
6
p 3?
500
```

OUTPUT FOR EXAMPLE 3

```
        AUXILIARY VENTILATION CALCULATIONS

INPUT DATA

Duct diameter, m= .61
Duct friction-factor, Ns^2/m^4= .0038
Resistance coefficient of leakage paths/100 m, Ns^2/m^8= 10000

Calculation for maximum length of duct

Volume rate required at the face, m^3/s= 1.42
No. of fans= 1
Natural ventilation pressure, Pa= 0
Fan characteristics
Q, m^3/s              p, Pa
 4                    1100
 5                    800
 6                    500

RESULTS

Maximum duct length, m= 951.42
Volume rate reaching the face, m^3/s= 1.42
Volume rate handled by the fan, m^3/s= 3.29
Volume increase ratio= 2.32
Pressure drop for leakless duct, Pa= 559.22
Pressure developed by the fan i.e. Pressure drop for leaky duct, Pa=1313.79
Pressure increase ratio= 2.35
Equivalent resistance, Ns^2/m^8=121.57
Input power to motor, kW=  6.17
Duct efficiency, %=18.38


******************************************************************

    Do you want another run with the data changed? - Answer yes or no
    ----
    no
```

PROGRAM 1 LISTING

```
10 LPRINT"            AUXILIARY VENTILATION CALCULATIONS"
20 LPRINT
30 LPRINT"INPUT DATA"
40 LPRINT
50 INPUT "Duct diameter, m=?",D
60 PI=3.141592654#
70 INPUT "Duct friction-factor, Ns^2/m^4=?",K
80 INPUT "Do you want to calculate resistance coefficient of leakage paths/100 m
, Ns^2/m^8? - Answer yes or no ----",Y$
90 LPRINT"Duct diameter, m=";D
100 LPRINT"Duct friction-factor, Ns^2/m^4=";K
110 IF Y$ ="yes"  THEN 1080
120 INPUT "Resistance coefficient of leakage paths/100 m, Ns^2/m^8=?",R
130 LPRINT"Resistance coefficient of leakage paths/100 m, Ns^2/m^8=";R
140 INPUT "Do you want to calculate maximum duct length, m? - Answer yes or no -
---",X$
150 IF X$ ="yes"  GOTO 190
160 INPUT "Duct length, m=?",L
170 LPRINT"Duct length, m=";L
180 GOTO 250
190 LPRINT:LPRINT"Calculation for maximum length of duct"
200 LPRINT
210 INPUT "Volume rate required at the face, m^3/s=?",V
220 LPRINT"Volume rate required at the face, m^3/s=";V
230 PRINT
240 PRINT
250 PRINT"One of the following should be available for input:-"
260 PRINT"Volume rate handled by the fan, m^3/s (QI)"
270 PRINT"Pressure drop for leaky duct i.e. Pressure developed by the fan, Pa (P
I)"
280 PRINT"Fan data (FD)"
290 IF X$="yes" THEN 330
```

```
300 PRINT"Volume rate reaching the face, m^3/s (QO)"
310 PRINT
320 PRINT
330 INPUT"Which input data is available?",Z$
340 IF Z$="QI"  THEN INPUT "Volume rate handled by the fan, m^3/s=?",Q
350 IF Z$="QI" THEN LPRINT"Volume rate handled by the fan, m^3/s=";Q: GOTO 420
360 IF Z$="QO"  THEN INPUT "Volume rate reaching the face, m^3/s=?",O
370 IF Z$="QO" THEN LPRINT"Volume rate reaching the face, m^3/s=";O: GOTO 420
380 IF Z$="PI"  THEN INPUT "Pressure drop for leaky duct i.e. Pressure developed
 by the fan, Pa=?",P
390 IF Z$="PI" THEN LPRINT"Pressure drop for leaky duct i.e. Pressure developed
by the fan, Pa=";P: GOTO 420
400 IF AA$="yes" THEN 420
410 GOSUB 1290
420 LPRINT
430 LPRINT"RESULTS"
440 LPRINT
450 IF X$="yes" THEN L=100
460 B0=0
470 U=K*PI*D/(PI*D*D/4)^3
480 R=R*10000
490 E=(2*(U/R)^.5)^(2/3)*L
500 F=1.0203344#-.268171*E+1.068933*E*E-.726484*E^3+.258632*E^4-.0309553*E^5+6.9
628E-04*E^6
510 G=1.0088958#-.062632*E+.731745*E*E-.466485*E^3+.180633*E^4-.0288809*E^5+1.55
255E-03*E^6
520 IF Z$="QI"  THEN 600
530 IF Z$="FD"  THEN 910
540 IF Z$="PI"  THEN 1020
550 N=U*L*O*O
560 P=F*N
570 Q=G*O
580 R=R/10000
590 GOTO 640
600 O=Q/G
610 N=U*L*O*O
620 P=F*N
630 R=R/10000
640 IF X$="no" GOTO 680
650 B0=B0 + 1
660 IF B0>1 THEN 1800
670 IF X$="yes" THEN 1740
680 PRINT"Volume rate reaching the face, m^3/s=";:PRINT USING "##.##";O
690 PRINT"Volume rate handled by the fan, m^3/s=";:PRINT USING "##.##";Q
700 PRINT"Volume increase ratio=";:PRINT USING "#.##";G
710 PRINT"Pressure drop for leakless duct, Pa=";:PRINT USING "####.##";N
720 PRINT"Pressure developed by the fan i.e. Pressure drop for leaky duct, Pa=";
:PRINT USING "####.##";P
730 PRINT"Pressure increase ratio=";:PRINT USING "#.##";F
740 PRINT"Equivalent resistance, Ns^2/m^8=";:PRINT USING "####.##";P/Q^2
750 PRINT"Input power to motor, kW=";:PRINT USING "###.##";P*Q/700
760 PRINT"Duct efficiency, %=";:PRINT USING "##.##";1/G/F*100
770 IF Z$<>"QO" THEN LPRINT"Volume rate reaching the face, m^3/s=";:LPRINT USING
 "##.##";O
780 IF Z$<>"QI" THEN LPRINT"Volume rate handled by the fan, m^3/s=";:LPRINT USIN
G "##.##";Q
790 LPRINT"Volume increase ratio=";:LPRINT USING "##.##";G:LPRINT"Pressure drop
for leakless duct, Pa=";:LPRINT USING "####.##";N
800 IF Z$<>"PI" THEN LPRINT"Pressure developed by the fan i.e. Pressure drop for
 leaky duct, Pa=";:LPRINT USING "####.##";P
810 LPRINT"Pressure increase ratio=";:LPRINT USING "##.##";F
820 LPRINT"Equivalent resistance, Ns^2/m^8=";:LPRINT USING "###.##";P/Q^2
830 LPRINT"Input power to motor, kW=";:LPRINT USING "###.##";P*Q/700
840 LPRINT"Duct efficiency, %=";:LPRINT USING "##.##";1/G/F*100
850 LPRINT
860 LPRINT
870 LPRINT"**********************************************************************"
880 LPRINT:INPUT"Do you want another run with the data changed? - Answer yes/no
----",AA$
890 IF AA$="yes" THEN 1880
900 IF AA$="no" THEN END
910 O=10
920 N=U*L*O*O
930 P=F*N
940 Q=G*O
950 H=P/Q/Q
960 Q=(-B-SQR(B*B-4*(A-H)*C))/2/(A-H)
970 O=Q/G
980 N=H*Q*Q/F
990 P=H*Q*Q
1000 R=R/10000
```

```
1010 GOTO 640
1020 N=P/F
1030 O=SQR(N/U/L)
1040 Q=O*G
1050 P=F*N
1060 R=R/10000
1070 GOTO 640
1080 LPRINT
1090 LPRINT"Calculation for resistance coefficient of leakage paths"
1100 LPRINT
1110 INPUT"Volume rate handled by the fan, m^3/s=?",Q
1120 INPUT"Volume rate reaching the face, m^3/s=?",O
1130 INPUT"Duct length, m=?",L
1140 LPRINT"Volume rate handled by the fan, m^3/s=";Q
1150 LPRINT"Volume rate reaching the face, m^3/s=";O
1160 LPRINT"Duct length, m=";L
1170 U=K*PI*D*100/(PI*D*D/4)^3
1180 R=U*((L/100)^3)/(((Q/O)^.5-1)*3)^2
1190 PRINT"Resistance coefficient of leakage paths/100 m, Ns^2/m^8=";:PRINT USIN
G "######.##";R
1200 LPRINT
1210 LPRINT"RESULT"
1220 LPRINT
1230 LPRINT"Resistance coefficient of leakage paths/100 m, Ns^2/m^8=";:LPRINT US
ING "######.##";R
1240 LPRINT
1250 LPRINT"***************************************************************"
1260 LPRINT
1270 LPRINT
1280 END
1290 A=0
1300 B=0
1310 C=0
1320 SX=0
1330 SY=0
1340 X2=0
1350 XY=0
1360 INPUT"Number of points on fan characteristic curve=?",I
1370 INPUT"Number of fans=?",A6
1380 LPRINT"No. of fans=";A6
1390 INPUT"Natural ventilation pressure, Pa=?",B1
1400 LPRINT"Natural ventilation pressure, Pa=";B1
1410 IF A6=1 THEN 1450
1420 INPUT"Are the fans in series or parallel? - Answer s/p-------",W$
1430 IF W$="s" THEN LPRINT"Fans are in series"
1440 IF W$="p" THEN LPRINT"Fans are in parallel"
1450 LPRINT"Fan characteristics"
1460 LPRINT "Q, m^3/s" TAB(20) "p, Pa"
1470 FOR J=1 TO I
1480 PRINT"Q";J;:INPUT M
1490 PRINT"p";J;:INPUT S
1500 LPRINT M TAB(20) S
1510 IF A6=1 THEN 1540
1520 IF W$="s" THEN S=S*A6: GOTO 1540
1530 IF W$="p" THEN M=M*A6
1540 SX=SX+M
1550 SY=SY+S+B1
1560 X2=X2+M^2
1570 XY=XY+(S+B1)*M
1580 A=A+M^3
1590 B=B+M^4
1600 C=C+M^2*(S+B1)
1610 NEXT J
1620 T=X2-SX^2/I
1630 A7=XY-SX*SY/I
1640 A5=A-SX*X2/I
1650 A0=C-X2*SY/I
1660 A1=B-X2^2/I
1670 A2=(A0*T-A7*A5)/(T*A1-(A5)^2)
1680 A3=(A7*A1-A0*A5)/(T*A1-(A5)^2)
1690 A4=SY/I-A3*SX/I-A2*X2/I
1700 A=A2
1710 B=A3
1720 C=A4
1730 RETURN
1740 A8=0
1750 A9=3000
1760 IF ABS(A9-A8)>=1 THEN 1780
1770 GOTO 1850
1780 L=(A8+A9)/2
```

```
1790 GOTO 470
1800 IF (O-V)<=0 THEN 1830
1810 A8=L
1820 GOTO 1760
1830 A9=L
1840 GOTO 1760
1850 PRINT"Maximum duct length, m=";:PRINT USING "####.##";L
1860 LPRINT"Maximum duct length, m=";:LPRINT USING "####.##";L
1870 GOTO 680
1880 LPRINT"Changed data"
1890 LPRINT
1900 PRINT"Which parameter you want to change - Answer"
1910 PRINT"          D for diameter"
1920 PRINT"          K for friction-factor"
1930 PRINT"          R for resistance coefficient of leakage paths"
1940 PRINT"          L for duct length"
1950 PRINT"          QO for volume rate required at the face"
1960 INPUT BB$
1970 IF BB$="D" THEN INPUT"What is the new diameter?",D
1980 IF BB$="D" THEN LPRINT"The new diameter=";D:GOTO 2070
1990 IF BB$="K" THEN INPUT"What is the new friction-factor?",K
2000 IF BB$="K" THEN LPRINT"The new friction-factor=";K:GOTO 2070
2010 IF BB$="R" THEN INPUT"What is the new resistance coefficient of leakage pat
hs?",R
2020 IF BB$="R" THEN  LPRINT"The new resistance coefficient of leakage paths=";R
:GOTO 2070
2030 IF BB$="L" THEN INPUT"What is the new length?",L
2040 IF BB$="L" THEN LPRINT"The new length=";L:GOTO 2070
2050 IF BB$="QO" THEN INPUT"What is the new volume rate required at the face?",V
2060 IF BB$="QO" THEN LPRINT"The new volume rate required at the face=";V
2070 PRINT"Do you want to change any other parameter before rerun of the problem
? ---- Answer y/n------":INPUT CC$
2080 IF CC$="y" THEN 1900
2090 GOTO 400
```

Program 2

This program is based on the theory developed for method 3, on p. 65, treatment of the airflow through a leaky duct as a series–parallel combination of airflows along the duct and through leakage paths, in section 2.5.1, Airflow analysis in leaky ducts.

For a given duct system, the values of diameter, friction factor, resistance coefficient of leakage paths per 100 m of duct, total length and segment length are known. One of the following should be available for input:

Volume rate handled by the fan

Volume rate reaching the face

Fan data

The results, depending upon the input data, include some of the following:

Volume rate handled by the fan

Volume rate reaching the face

Pressure developed by the fan

Equivalent resistance

The program has an option to change the resistance of any segment or leakage path.

Examples

(1) If it is required to exhaust 7 m^3/s from the end of a 400 m long, 0.76 m diameter duct, determine what volume rate a fan at the other end will have to handle and what pressure it will have to produce. The friction factor for the duct is 0.0038 Ns^2/m^4. The resistance coefficient of leakage paths per 100 m of duct is 10 000 Ns^2/m^8. The segment length is 4 m.

(2) A fan with the characteristics given in Table B is connected to one end of a 950 m long 0.61 m diameter duct system. The friction factor for the duct is 0.0038 Ns^2/m^4. The resistance coefficient of leakage paths per 100 m of duct is 10 000 Ns^2/m^8. The segment length is 2.5 m. Calculate:

(*a*) volume rate handled by the fan;
(*b*) volume rate reaching the face;
(*c*) pressure developed by the fan; and
(*d*) equivalent resistance.

Table B. *Fan characteristics*

Q (m^3/s)	p (Pa)
4	1100
5	800
6	500

```
INPUT DATA FOR EXAMPLE 1

Duct diameter, m=?
0.76

Duct friction-factor, Ns2/m4=?
0.0038

Resistance coefficient of leakage paths/100 m, Ns2/m8=?
10000

Duct length, m=?
400

Segment length, m=?
4

One of the following should be available for input:-

Volume rate handled by the fan, m3/s (QI)

Volume rate reaching the face, m3/s (QO)

Fan data (FD)

Which input data is available?
QO

Volume rate reaching the face, m3/s=?
7

DO YOU WANT TO CHANGE ANY DUCT SECTION RESISTANCES?
no

DO YOU WANT TO CHANGE ANY LEAKAGE RESISTANCES?
no
```

```
OUTPUT FOR EXAMPLE 1

                AUXILIARY VENTILATION CALCULATIONS

INPUT DATA

Duct diameter, m= .76
Duct friction-factor, Ns^2/m^4= .0038
Resistance coefficient of leakage paths/100 m, Ns^2/m^8= 10000
Duct length, m= 400
Segment length, m= 4
Volume rate reaching the face, m^3/s= 7

RESULTS

Volume rate reaching the face, m^3/s= 7.00
Volume rate handled by the fan, m^3/s= 8.19
Volume increase ratio= 1.17
Pressure developed by the fan, Pa= 2177.13
Equivalent resistance, Ns^2/m^8=   32.43

*******************************************************

INPUT DATA FOR EXAMPLE 2

Duct diameter, m=?
0.61

Duct friction-factor, Ns2/m4=?
0.0038

Resistance coefficient of leakage paths/100 m, Ns2/m8=?
10000

Duct length, m=?
950

Segment length, m=?
2.5

One of the following should be available for input:-

Volume rate handled by the fan, m3/s (QI)

Volume rate reaching the face, m3/s (QO)

Fan data (FD)

Which input data is available?
FD

Number of points on fan characteristic curve=?
3

Number of fans=?
1

Natural ventilation pressure, Pa=?
0

Q 1?
4
p 1?
1100
Q 2?
5
p 2?
800
Q 3?
6
p 3?
500

DO YOU WANT TO CHANGE ANY DUCT SECTION RESISTANCES?
no

DO YOU WANT TO CHANGE ANY LEAKAGE RESISTANCES?
no
```

```
OUTPUT FOR EXAMPLE 2

              AUXILIARY VENTILATION CALCULATIONS
INPUT DATA

Duct diameter, m= .61
Duct friction-factor, Ns^2/m^4= .0038
Resistance coefficient of leakage paths/100 m, Ns^2/m^8= 10000
Duct length, m= 950
Segment length, m= 2.5
No. of fans= 1
Natural ventilation pressure, Pa= 0
Fan characteristics
Q, m^3/s           p, Pa
 4                  1100
 5                  800
 6                  500

RESULTS

Volume rate handled by the fan, m^3/s= 3.28
Volume rate reaching the face, m^3/s= 1.42
Pressure developed by the fan, Pa=1314.92
Equivalent resistance, Ns^2/m^8=121.96

**************************************************************
```

PROGRAM 2 LISTING

```
10 LPRINT"                    AUXILIARY VENTILATION CALCULATIONS"
20 LPRINT
30 LPRINT"INPUT DATA"
40 LPRINT
50 OPTION BASE 0
60 DIM A(3000)
70 DIM B(3000)
80 DIM C(3000)
90 INPUT"Duct diameter, m=?",DD
100 LPRINT"Duct diameter, m=";DD
110 INPUT"Duct friction-factor, Ns^2/m^4=?",KK
120 LPRINT"Duct friction-factor, Ns^2/m^4=";KK
130 PI=3.141592654#
140 A=KK*PI*DD*100/(PI*DD*DD/4)^3
150 INPUT"Resistance coefficient of leakage paths/100 m, Ns^2/m^8=?",B
160 LPRINT"Resistance coefficient of leakage paths/100 m, Ns^2/m^8=";B
170 INPUT"Duct length, m=?",C
180 LPRINT"Duct length, m=";C
190 INPUT"Segment length, m=?",SL
200 LPRINT"Segment length, m=";SL
210 PRINT
220 PRINT
230 PRINT"One of the following should be available for input:-"
240 PRINT"Volume rate handled by the fan, m^3/s (QI)"
250 PRINT"Volume rate reaching the face, m^3/s (QO)"
260 PRINT"Fan data (FD)"
270 PRINT
280 PRINT
290 INPUT"Which input data is available?",Z$
300 IF Z$="QI" THEN 340
310 IF Z$="QO" THEN 370
320 GOSUB 1110
330 GOTO 400
340 INPUT"Volume rate handled by the fan, m^3/s=?",QI
350 LPRINT"Volume rate handled by the fan, m^3/s=";QI
360 GOTO 400
370 INPUT"Volume rate reaching the face, m^3/s=?",P
380 LPRINT"Volume rate reaching the face, m^3/s=";P
390 QI=10
400 D=C/SL
410 A(D+1)=QI
420 H=A*C/100/D
430 FOR K = 1 TO D
440 B(K) = H
450 NEXT K
460 I=B*10000*D*D/C/C
470 FOR K = 1 TO (D-1)
480 C(K) = I
490 NEXT K
500 INPUT "DO YOU WANT TO CHANGE ANY DUCT SECTION RESISTANCES?",ZZ$
```

```
510 IF ZZ$="yes" THEN 1560
520 INPUT "DO YOU WANT TO CHANGE ANY LEAKAGE RESISTANCES?",YY$
530 IF YY$ = "yes" THEN 1590
540 J=0
550 FOR K=1 TO D-1
560 J=(B(D-K+1)+J)*C(D-K)/(SQR(B(D-K+1)+J)+SQR(C(D-K)))^2
570 A(K)=J
580 NEXT K
590 J=J+B(1)
600 A(D)=J
610 IF Z$="QI" THEN 640
620 IF Z$="QO" THEN 640
630 A(D+1)=(-F-SQR(F*F-4*(E-A(D))*G))/(2*(E-A(D)))
640 FOR K=1 TO D-1
650 L=A(D+K)*SQR(C(K)/(B(K+1)+A(D-1-K)))/(1+SQR(C(K)/(B(K+1)+A(D-1-K))))
660 A(D+1+K)=L
670 NEXT K
680 M=D*H*A(D+D)*A(D+D)
690 N=A(D)*A(D+1)*A(D+1)
700 S=A(D+1)/A(D+D)
710 IF Z$="QO" THEN 900
720 PRINT
730 PRINT"RESULTS"
740 PRINT
750 PRINT"Volume rate handled by the fan, m^3/s=";:PRINT USING "##.##";A(D+1)
760 PRINT"Volume rate reaching the face, m^3/s=";:PRINT USING "##.##";A(D+D)
770 LPRINT
780 LPRINT"RESULTS"
790 LPRINT
800 LPRINT"Volume rate handled by the fan, m^3/s=";:LPRINT USING "##.##";A(D+1)
810 LPRINT"Volume rate reaching the face, m^3/s=";:LPRINT USING "##.##";A(D+D)
820 PRINT"Volume increase ratio=";:PRINT USING "##.##";S
830 PRINT"Pressure developed by the fan, Pa=";:PRINT USING "####.##";N
840 LPRINT"Pressure developed by the fan, Pa=";:LPRINT USING "####.##";N
850 PRINT"Equivalent resistance, Ns^2/m^8=";:PRINT USING "###.##";N/(A(D+1))^2
860 LPRINT"Equivalent resistance, Ns^2/m^8=";:LPRINT USING "###.##";N/(A(D+1))^2
870 LPRINT
880 LPRINT"************************************************************"
890 GOTO 1100
900 R=A(D)*(S*P)^2
910 Q=D*H*P*P
920 LPRINT
930 PRINT
940 LPRINT"RESULTS"
950 PRINT"RESULTS"
960 LPRINT
970 PRINT
980 PRINT"Volume rate reaching the face, m^3/s=";:PRINT USING "##.##";P
990 PRINT"Volume rate handled by the fan, m^3/s=";:PRINT USING "##.##";S*P
1000 PRINT"Volume increase ratio=";:PRINT USING "##.##";S
1010 PRINT"Pressure developed by the fan, Pa=";:PRINT USING "#####.##";R
1020 LPRINT"Volume rate reaching the face, m^3/s=";:LPRINT USING "##.##";P
1030 LPRINT"Volume rate handled by the fan, m^3/s=";:LPRINT USING "##.##";S*P
1040 LPRINT"Volume increase ratio=";:LPRINT USING "##.##";S
1050 LPRINT"Pressure developed by the fan, Pa=";:LPRINT USING "#####.##";R
1060 PRINT"Equivalent resistance, Ns^2/m^8=";:PRINT USING "#####.##";A(D)
1070 LPRINT"Equivalent resistance, Ns^2/m^8=";:LPRINT USING "#####.##";A(D)
1080 LPRINT
1090 LPRINT"************************************************************"
1100 END
1110 AA=0
1120 BB=0
1130 CC=0
1140 SX=0
1150 SY=0
1160 X2=0
1170 XY=0
1180 INPUT"Number of points on fan characteristic curve=?",II
1190 INPUT"Number of fans=?",A6
1200 LPRINT"No. of fans=";A6
1210 INPUT"Natural ventilation pressure, Pa=?",B1
1220 LPRINT"Natural ventilation pressure, Pa=";B1
1230 IF A6=1 THEN 1270
1240 INPUT"Are the fans in series or parallel? - Answer s/p-----",W$
1250 IF W$="s" THEN LPRINT"Fans are in series"
1260 IF W$="p" THEN LPRINT"Fans are in parallel"
1270 LPRINT"Fan characteristics"
1280 LPRINT"Q, m^3/s" TAB(20) "p, Pa"
1290 FOR JJ=1 TO II
1300 PRINT"Q";JJ;:INPUT MM
1310 PRINT"p";JJ;:INPUT SS
```

```
1320 LPRINT MM TAB(20) SS
1330 IF A6=1 THEN 1360
1340 IF W$="s" THEN SS=SS*A6:GOTO 1360
1350 IF W$="p" THEN MM=MM*A6
1360 SX=SX+MM
1370 SY=SY+SS+B1
1380 X2=X2+MM^2
1390 XY=XY+(SS+B1)*MM
1400 AA=AA+MM^3
1410 BB=BB+MM^4
1420 CC=CC+MM^2*(SS+B1)
1430 NEXT JJ
1440 TT=X2-SX^2/II
1450 A7=XY-SX*SY/II
1460 A5=AA-SX*X2/II
1470 A0=CC-X2*SY/II
1480 A1=BB-X2^2/II
1490 A2=(A0*TT-A7*A5)/(TT*A1-(A5)^2)
1500 A3=(A7*A1-A0*A5)/(TT*A1-(A5)^2)
1510 A4=SY/II-A3*SX/II-A2*X2/II
1520 E=A2
1530 F=A3
1540 G=A4
1550 RETURN
1560 INPUT "WHICH SEGMENT?",SEG
1570 INPUT "WHAT IS THE VALUE?",B(SEG)
1580 GOTO 500
1590 INPUT"WHICH SEGMENT?",SEG
1600 INPUT"WHAT IS THE VALUE?",C(SEG)
1610 GOTO 520
```

APPENDIX III

Psychrometrics of air–water vapour mixtures

In solving climatic problems in mines, it is necessary to determine the psychrometric parameters of the air–water vapour mixtures. In a single circuit in which there is no leakage but where changes of temperature, pressure and moisture content take place:

(1) the total mass of air remains constant;

(2) the total mass of the air–water vapour mixture changes by the addition or abstraction of moisture;

(3) the volume varies inversely as the mass of air per unit volume of air–water vapour mixture.

The other relationships are:

$$V_m = V_a = V_v$$
$$t_m = t_a = t_v$$
$$P_m = P_a + P_v$$
$$H_m = H_a + H_v$$

where V = volume;
t = temperature;
P = pressure; and
H = enthalpy (see next section for definition).

The subscripts m, a and v refer to air–water vapour mixture, air and water vapour, respectively.

Definitions

Atmospheric pressure – The actual pressure of the atmosphere and is equal to the sum of the pressures exerted by the air and water vapour in the atmosphere. It is measured with an aneroid barometer.

Dry-bulb temperature – The temperature as indicated by a thermometer with the bulb exposed to the atmosphere but sheltered from excessive amounts of radiation.

Wet-bulb temperature – The temperature as indicated by a thermometer with the bulb covered with a wet muslin or cotton sleeve exposed to an atmospheric current of not less than 3 m/s and shielded from radiation in excess of the atmospheric dry-bulb temperature. Dry- and wet-bulb temperatures are measured by a whirling hygrometer.

Vapour pressure – The partial pressure exerted by the water vapour in air–water vapour mixture.

Saturated vapour pressure – The maximum pressure that can be exerted by the water vapour in an air–water vapour mixture, i.e. the pressure exerted by the water vapour in a saturated air–water vapour mixture.

Relative humidity – The ratio of the actual vapour pressure to the saturation vapour pressure at the dry-bulb temperature and is usually expressed as a percentage.

Moisture content – The actual mass of water vapour per unit mass of air. This is also known as apparent specific humidity.

Dew point temperature – The temperature for which the saturation vapour pressure equals the actual (partial) vapour pressure of the water vapour in an air–water vapour mixture.

Apparent specific volume of air–water vapour mixture – Volume of air–water vapour mixture per unit mass of air.

Apparent density of air–water vapour mixture – Mass of air per unit volume of air–water vapour mixture.

True specific volume of air–water vapour mixture – Volume per unit mass of air–water vapour mixture.

True density of air–water vapour mixture – Mass of air–water vapour mixture per unit volume of air–water vapour mixture.

Enthalpy of air–water vapour mixture – The summation of the sensible heat, latent heat and superheat of an air–water vapour mixture between an arbitrary datum point and the temperature and state under consideration. (Sensible heat is the heat which is added or abstracted from a substance to change its temperature without changing its state. Latent heat is the heat which is added or abstracted from a substance which does not affect its temperature but changes its state from solid to liquid or liquid to vapour or vice versa. Superheat is the heat which raises the temperature of a vapour above the saturation temperature corresponding to the pressure.)

Sigma heat – Enthalpy minus the heat of water at the wet-bulb temperature.

Psychrometric equations

The commonly used psychrometric equations are listed below (Whillier, 1971; Barenbrug, 1974).

(1) Saturated vapour pressure (e_t, Pa) at temperature t (°C):

$$e_t = 610.5 \exp(17.27t/(t + 237.3)).$$

(2) Vapour pressure (e) in the air–water vapour mixture (Pa):

$$e = e_{wb} - 0.000644P(t_{db} - t_{wb})$$

where e_{wb} = saturated vapour pressure at the wet-bulb temperature t_{wb} (Pa);
P = barometric pressure (Pa);
t_{db} = dry-bulb temperature (°C); and
t_{wb} = wet-bulb temperature (°C).

(3) Relative humidity (ϕ, %):

$$\phi = (e/e_{db}) \times 100$$

where e = vapour pressure in the air–water vapour mixture (Pa); and
e_{db} = saturated vapour pressure at the dry-bulb temperature (Pa).

(4) Moisture content (r, g of vapour/kg of air):

$$r = 622e/(P - e)$$

where e = vapour pressure in the air–water vapour mixture (Pa); and
P = barometric pressure (Pa).

(5) Apparent specific volume of air–water vapour mixture (v_a, m^3/kg of air):

$$v_a = (287.045(t_{db} + 273.15))/(P - e)$$

where t_{db} = dry-bulb temperature (°C);
P = barometric pressure (Pa); and
e = vapour pressure in the air–water vapour mixture (Pa).

(6) Apparent density of air–water vapour mixture (w_a, kg of air/m^3 of air–water vapour mixture):

$$w_a = 1/v_a.$$

(7) True specific volume of air–water vapour mixture (v_t, m^3/kg of air–water vapour mixture):

$$v_t = (287.045(t_{db} + 273.15))/(P - 0.378e)$$

where t_{db} = dry-bulb temperature (°C);
P = barometric pressure (Pa); and
e = vapour pressure in the air–water vapour mixture (Pa).

(8) True density of air–water vapour mixture (w_t, kg of air–water vapour mixture/m³):

$w_t = 1/v_t$.

(9) Enthalpy of air–water vapour mixture (H, kJ/kg of air):

$H = 1.005t_{db} + r(2.5016 + 0.0018t_{db})$

where t_{db} = dry-bulb temperature (°C); and
r = moisture content (g of vapour/kg of air).

(10) Sigma heat (S, kJ/kg of air):

$S = H - 0.004187rt_{wb}$

where H = enthalpy of air–water vapour mixture (kJ/kg of air);
r = moisture content (g of vapour/kg of air); and
t_{wb} = wet-bulb temperature (°C).

During any adiabatic saturation process, the sigma heat remains constant (but not the enthalpy) and the use of sigma heat in calculations involving wet heat transfer is more accurate than when the enthalpy is used.

(11) Dew point temperature (t_{dew}, °C):

$t_{dew} = 237.3 \log_e (e/610.5) \div 17.27 - \log_e (e/610.5)$

where e = vapour pressure (Pa).

A program (in BASIC) for the Casio FX 702P is given in Table 1 for psychrometric calculations.

Table 1. *Program for psychrometric calculations*

```
10 INP "P=",A,"DB TEMP=",B,"WB TEMP=",C
20 D=610.5*EXP ((17.27*C)/(C+237.3))
30 E=D-.000644*A*(B-C)
40 F=610.5*EXP ((17.27*B)/(B+237.3))
50 G=E/F
60 H=622*E/(A-E)
70 I=287*(B+273.15)/(A-E)
80 J=1/I
90 K=((A-.378*E)/(287*(B+273.15)))
100 L=2.5016+.0018*B
110 M=1.005*B+H*L
120 N=M-.004187*H*C
125 SET F3
130 PRT "R.H.=";G
140 PRT "MOIS CONT=";H
150 PRT "SPEC VOL=";I
160 PRT "DRY AIR DENS=";J
170 PRT "TRUE DENS=";K
180 PRT "ENTHALPY=";M
190 PRT "SIGMA HEAT=";N
195 SET N
200 END
```

Psychrometric charts

Barenbrug (1974) has given various psychrometric charts for a wide range of barometric pressures (80 kPa to 130 kPa). Whillier (1971) has given three figures for all barometric pressures.

Fig. 2 gives a psychrometric chart for any pressure. To locate a state point on the chart, a wet-bulb line is first drawn with the slope indicated, from the wet-bulb scale that is given along the saturation line. The intersection of this wet-bulb line with the dry-bulb line gives the relative humidity. Moving to the right-hand side of the intersection gives the vapour pressure and by proceeding further to the right to intersect the vertical line at the appropriate barometric pressure, the moisture content is obtained. (Lines of constant enthalpy are not shown on this figure.)

Fig. 3 gives an apparent specific volume chart. To determine the apparent specific volume, the vapour pressure must be known, either from Fig. 2, or approximated by using the abridged version of Fig. 2 that is included in the top left-hand corner of Fig. 3.

Fig. 4 gives a sigma heat chart. The determination of sigma heat requires the knowledge of barometric pressure and wet-bulb temperature.

Calculations

In mine climate problems, it is often necessary to calculate the heat and moisture picked up by the air(–water vapour mixture) around a path. In solving climatic problems, the effects of mixing two volumes of air(–water vapour mixtures) or installing a refrigeration plant or of removing heat sources may be considered. The methods used to study these effects are either psychrometric charts or computer programs.

First, the psychrometric parameters of the air(–water vapour mixture) are determined either from psychrometric charts or by using computer programs. Between any two points the heat added or lost by the air(–water vapour mixture) is calculated by multiplying the change in sigma heat of the air(–water vapour mixture) (kJ/kg of air) by the mass flow rate of air(–water vapour mixture) (kg of air/s) and is in units of kJ/s. The moisture pick-up is calculated by multiplying the change in moisture content (g of vapour/kg of air) by the mass flow rate of air(–water vapour mixture) (kg of air/s) and is in units of g/s. To determine the effect of refrigeration, the cooling power of the refrigeration plant (kW) divided by the mass flow rate of the air(–water vapour mixture) is subtracted from the sigma heat of the air(–water vapour mixture). The temperature of the air(–water vapour mixture) is then found from the new sigma heat, and if there is no condensation the original moisture content is used.

Fig. 2. Psychrometric chart for any pressure (after Whillier, 1971).

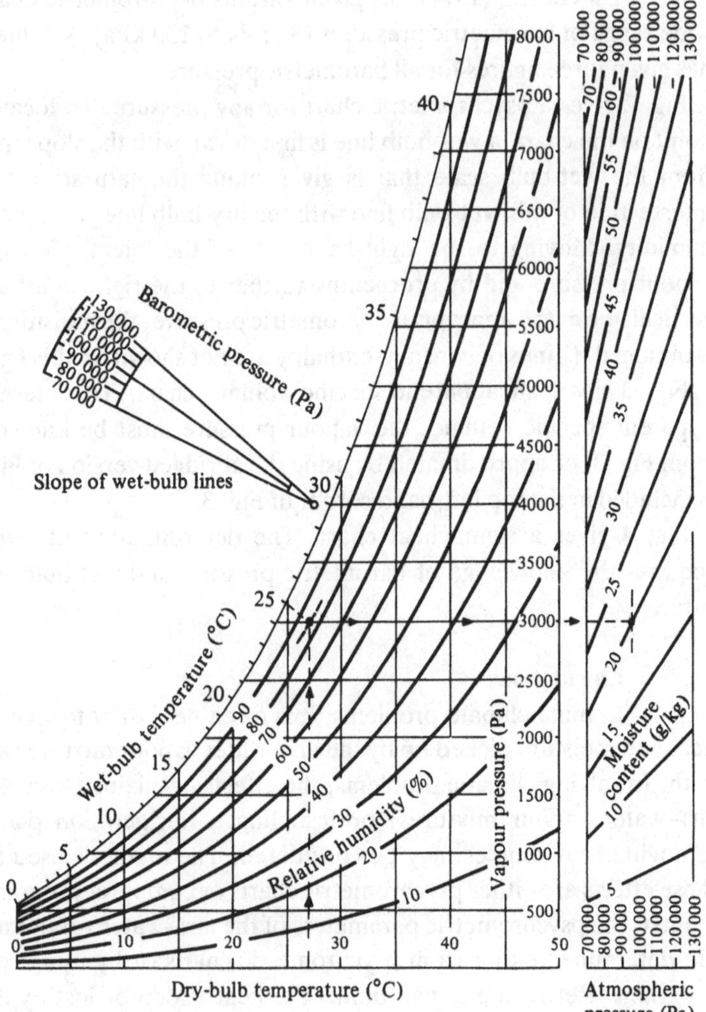

Fig. 3. Apparent specific volume chart (after Whillier, 1971). R.H = relative humidity.

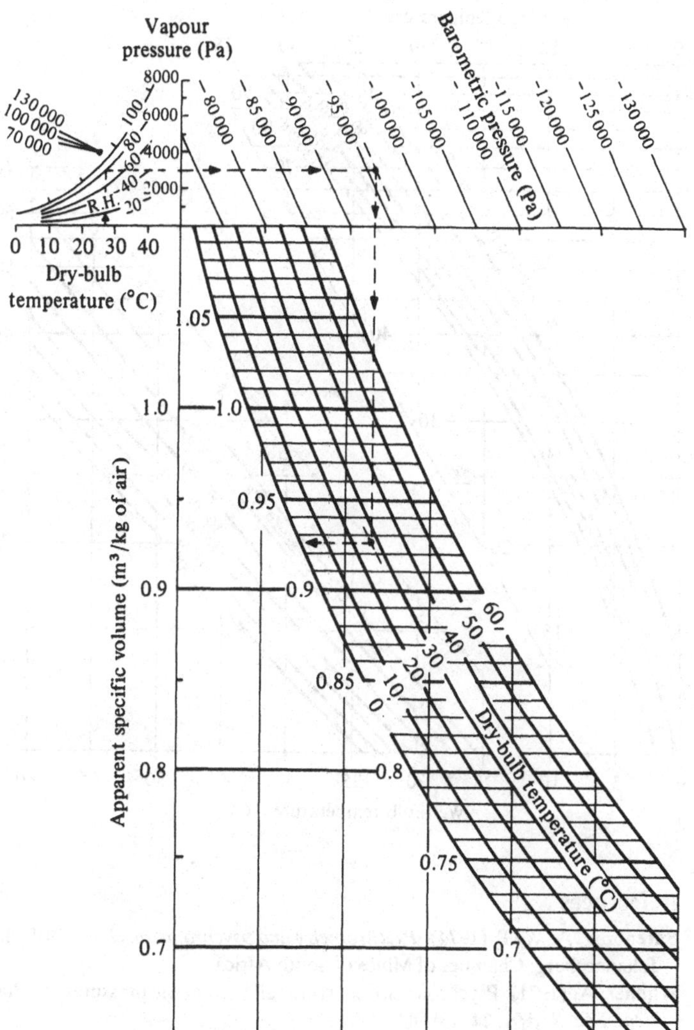

Fig. 4. Sigma heat chart (after Whillier, 1971).

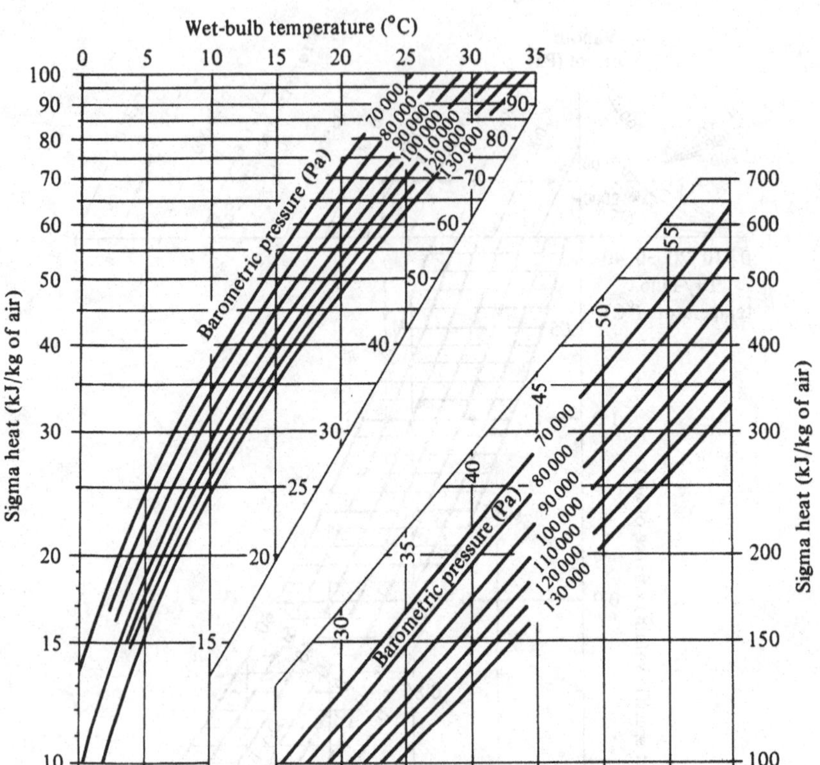

References

Barenbrug, A. W. T. (1974). *Psychrometry and psychrometric charts*, 3rd edn. Johannesburg: Chamber of Mines of South Africa.

Whillier, A. (1971). Psychrometric charts for all barometric pressures. *J. Mine Vent. Soc. S. Afr.*, **24**, 138–43.

APPENDIX IV

Explosibility of mine atmospheres and fire gases

The main combustible gas found in coal mines is methane. It has explosive properties when its concentration ranges from 5.3 to 14 % of the air volume. Mixtures below and above this range will burn if ignited; mixtures within the range are capable of giving violent explosions. The other combustible gases found in mines are:

Carbon monoxide	12.5 to 74 %	Explosive range
Hydrogen sulphide	4 to 44 %	Explosive range

Underground mine fires contaminate the mine air with a variety of combustible gases. These gases are formed by the combustion of various hydrocarbons and may accumulate because of insufficient ventilation or ventilation failure. These combustible gases are dangerous because there is the possibility of an explosive mixture being formed and if this comes into contact with fire, an explosion. If rescue work is in progress or if fire fighting teams are underground, the effects of an explosion can be disastrous.

Once a fire has progressed beyond fire fighting control, it is necessary to seal off part of the mine to exclude air and thus extinguish the fire. When the seals are in place, explosive mixtures may build up in the fire area because the combustible gases are not being swept away. On the other hand, the oxygen concentration will decrease through combustion processes until it becomes too low to support combustion. To avoid death or injury, all personnel are withdrawn from the mine.

Air samples taken from a fire area are evaluated for explosibility in order to maintain the safety of those involved in rescue and recovery operations. The two methods used widely are based upon (1) the Coward triangle (Hughes and Raybould, 1960) and (2) the US Bureau of Mine's explosibility diagram (Zabetakis, Stahl and Watson, 1959). (Smith and

Nugent (1977) have written a program for use with the Hewlett-Packard Model 65 programmable calculator to do the required calculations in regard to the use of this explosibility diagram.) Both the methods require the plotting of diagrams and they become cumbersome if a large number of assessments of various sampling locations over a period of time are required. This may make the approach of a dangerous situation more difficult to predict. To overcome these difficulties and for rapid assessment, Ellicott (1981) has further developed the Coward triangle method with the aid of a handheld programmable calculator (Hewlett-Packard Models 67 and 41C) which does not require the drawing of diagrams to assess the explosibility of gas mixtures. Simple Cartesian coordinates representing the explosive potential of gas mixtures are obtained, which may readily be plotted to ascertain trends with time. The details of this method are given here.

Fig. 5 shows the Coward triangle for methane. The y-axis is calibrated from 0 to 20.93 % oxygen (this is equivalent to 0 to 100 % air). The x-axis is calibrated from 0 to 100 % methane. Point A represents pure air, B represents pure methane and the line AB represents mixtures of methane with air. Point P represents the lower explosive limit and point Q the upper explosive limit of methane in air. All points below the line APQB

Fig. 5. The Coward triangle for methane. A – pure air; B – pure methane; P – lower explosive limit; Q – upper explosive limit; R – nose limit; S – air-free analysis of nose limit mixture; X – methane–air mixture; P′ – lower explosive limit for mixture X; Q′ – upper explosive limit for mixture X; a – point on line joining X and A; b – point on line joining X and B; 0 – point on line joining X and origin.

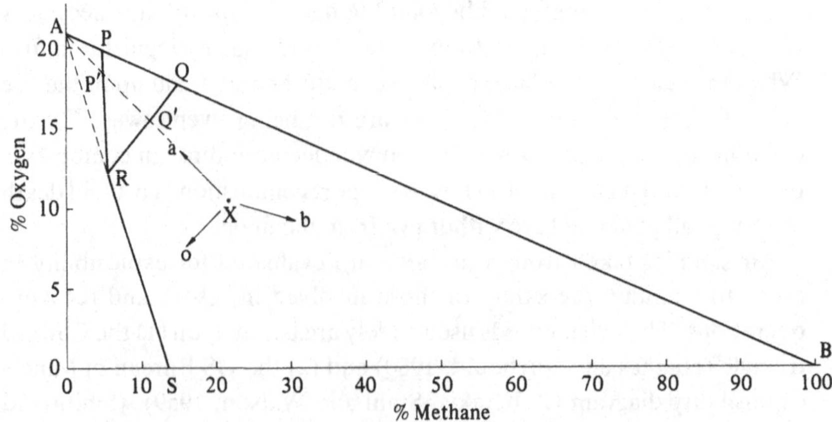

represent mixtures of methane, air and excess nitrogen*. The line PR is the line of the lower limits of explosibility and QR the line of the higher limits. R is commonly termed the nose limit (point). No mixture containing less oxygen than R can explode. The air-free analysis of the nose limit is represented by point S.

In drawing this diagram, points P and Q are first plotted from the lower and upper explosive limits of the methane–air mixture, then point R from the methane and oxygen contents at the nose limit for the mixture of methane, air and excess nitrogen (Tables 2 and 3). Line SR is then drawn such that if continued it would pass through point A. The composition point for the sample X is then plotted from its methane and oxygen contents.

Three regions are of interest in the diagram. If the sample point lies in the area PQR it is explosive and could cause an explosion. Sample points lying in region BQRS will become explosive when mixed with air. They contain so much methane and perhaps excess nitrogen that they are not explosive; however, they would form explosive mixtures when mixed with air. Sample points lying in the region APRSO are non-explosive. They contain either so little methane or so much excess nitrogen that they cannot be ignited 'as is'; furthermore, they will not form explosive mixtures when mixed with air.

The diagram is also useful as an indicator of the effect of changes in composition on explosibility. If a mixture is diluted with air, such as in the re-opening of a sealed area, its point on the diagram will move along a

* Excess nitrogen results from oxygen depletion of original air. This has a dampening effect on the explosibility of methane and is called an inert gas. The other inert gas commonly found in mine fire gases is carbon dioxide, from combustion. If this is present, the position of R will be different.

Table 2. *Lower and upper explosive limits of some combustible gases in air*[a]

Combustible gas	Limits of explosibility, volume (%)	
	Lower	Upper
Methane	5.0	15.0
Hydrogen	4.0	75.0
Carbon monoxide	12.5	74.0

[a] Gas concentration units are percentage by volume.

straight line towards point A (direction a, Fig. 5)*. This particular aspect is very important in the case of samples initially in a potentially explosive condition, because air dilution means that they must pass through an explosive state. If nitrogen is added, the sample point will move toward the origin, i.e. direction o. Increasing the methane content of the mixture causes a shift in direction b. Increasing the combustible content of a mine atmosphere may result from not only the presence of increasing volumes of combustible gas but also from reducing the air supply to an area by sealing. This, under some conditions, could take the atmosphere through an explosive state.

When a mixture contains a number of combustible and inert gases, the following procedure is followed to draw the Coward triangle.

1. Coordinates of points P and Q

The following equation is used to calculate the lower and upper limits of explosibility, i.e. the x-coordinates of P and Q:

$$L_{L,U} = \frac{100}{\Sigma\,(p_j/L_j)}$$

where $L_{L,U}$ = lower, upper limits of explosibility of a mixture, %;
p_j = concentration of combustible gas, j, as a percentage of total combustibles; and
L_j = lower, upper limits of explosibility for combustible gas, j, %.

The total concentration of higher hydrocarbons present is added to that of hydrogen. The lower and upper explosive limits of some combustible gases found in a mine fire are given in Table 2.

* If air is added to the mixture represented by X, the composition will change along the line X to A. The explosive range of the atmosphere represented by X is from P′ to Q′.

Table 3. *Nose limit contents*[a]

	Combustible		Oxygen	
	Nitrogen	Carbon dioxide	Nitrogen	Carbon dioxide
Methane	5.93	6.66	12.24	15.07
Hydrogen	4.30	5.73	5.12	7.61
Carbon monoxide	13.78	18.00	6.07	8.80

[a] Gas concentration units are percentage by volume.

The y-coordinates are:

for P, $20.93 - 0.209L_L$

for Q, $20.93 - 0.209L_U$.

2. Coordinates of nose point R

The combustible content at the nose point (i.e. x-coordinate) is calculated from the following equation:

$$C_R = \frac{N_2}{(N_2 + CO_2)}\left[\frac{100}{\Sigma(p_j/n_{1,j})}\right] + \frac{CO_2}{(N_2 + CO_2)}\left[\frac{100}{\Sigma(p_j/n_{2,j})}\right]$$

where C_R = combustible content at the nose point for the mixture;

$n_{1,j}$ = combustible content at the nose point for gas, j, in air and excess nitrogen;

$n_{2,j}$ = combustible content at the nose point for gas, j, in air and excess carbon dioxide; and

p_j = concentration of combustible gas, j, as a percentage of total combustibles.

The oxygen content at the nose point (i.e. y-coordinate) is calculated from the following equation*:

$$O_R = 0.2093(100 - X_L - C_R)$$

where O_R = oxygen content at the nose point for the mixture; and

X_L = excess inert gas at the nose point

$$= \frac{N_2}{(N_2 + CO_2)}\left[\frac{\Sigma v_{1,j}p_j}{\Sigma(p_j/n_{1,j})}\right] + \frac{CO_2}{(N_2 + CO_2)}\left[\frac{\Sigma v_{2,j}p_j}{\Sigma(p_j/n_{2,j})}\right]$$

where $v_{1,j}$ = volume of nitrogen required to render a unit volume of combustible gas, j, extinctive;

$v_{2,j}$ = volume of carbon dioxide required to render a unit volume of combustible gas, j, extinctive;

$n_{1,j}$ = oxygen content at the nose point for gas, j, in air and excess nitrogen; and

$n_{2,j}$ = oxygen content at the nose point for gas, j, in air and excess carbon dioxide.

Table 3 gives combustible and oxygen contents at the nose limits for methane, hydrogen and carbon monoxide in air and both excess nitrogen and carbon dioxide.

Table 4 gives volumes of inert gas required to render a unit volume of combustible gas extinctive.

* This is based on the quantity of inert gases – N_2 and CO_2 – required to render the atmosphere extinctive.

Table 4. *Requirements of inert gas for extinction of combustible gases*

	Inert gas	
Combustible gas	Nitrogen	Carbon dioxide
Methane	6.00	3.20
Hydrogen	16.55	10.20
Carbon monoxide	4.15	2.16

Gas concentration units are percentage by volume.

Fig. 6. (*a*) The Coward diagram; (*b*) corresponding modified diagram (after Ellicott, 1981).

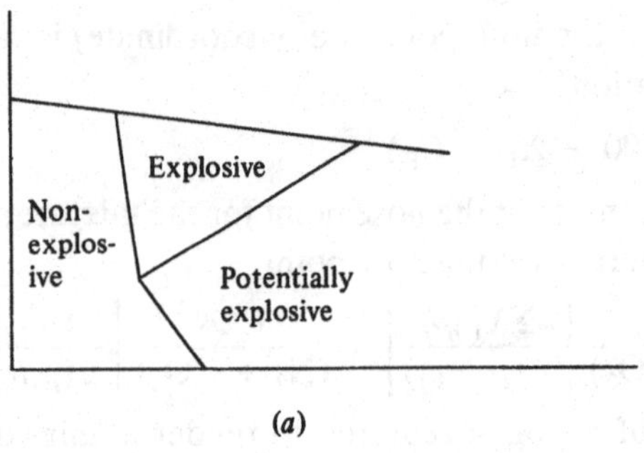

(*a*)

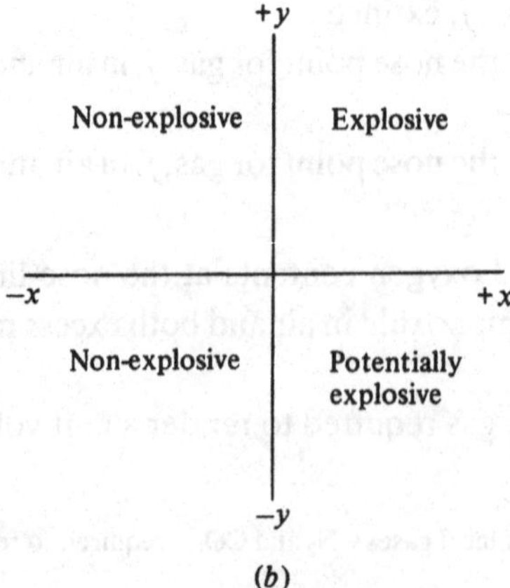

(*b*)

3. Coordinates of point S

$$\text{x-coordinate} = \frac{-20.93C_R}{O_R - 20.93}$$

$$\text{y-coordinate} = 0.$$

4. Coordinates of sample point X

x-coordinate = $\Sigma(CH_4 + H_2 + CO)$, i.e. total combustibles in the sample

y-coordinate = O_2 in the sample.

Ellicott's extension

Ellicott (1981) has transformed the Coward diagram (Fig. 6*a*) into a modified diagram (Fig. 6*b*) and the modified coordinates (x_m, y_m) of the sample then convey the explosive state of the analysed gas mixture according to

State	x_m	y_m
Explosive	+	+
Potentially explosive	+	−
Non-explosive	−	−
	−	+

They may be plotted on simple rectangular, x–y-axes to allow time trends to be observed.

Directional information of the Coward diagram with regard to air or inert dilution or increasing combustible content is retained in the modified diagram. In general, air dilution causes a shift of the sample point toward the top left of the modified diagram, inert dilution toward the bottom left and increasing combustible content toward the bottom right.

The details of and equations used in the transformation are as follows:

(1) The origin is moved to the nose point by the following coordinate transformation:

$$(x, y) \rightarrow [(x - C_R), (y - O_R)].$$

(2) The new coordinates of points P, Q, S and X are converted to polar form with the origin at the nose point as follows:

$$P \rightarrow r_P, \theta_P$$
$$Q \rightarrow r_Q, \theta_Q$$
$$S \rightarrow r_S, \theta_S$$
$$X \rightarrow r_X, \theta_X$$

(3) For the sample point X, the modified angular coordinates r_m, θ_m are calculated as follows (assuming $0 < \theta < 360$)*:

(*a*) Explosive case ($\theta_P > \theta_X > \theta_Q$)

$$r_m = r_X$$

$$\theta_m = \left[\frac{(\theta_X - \theta_Q)}{(\theta_X - \theta_Q) + (\theta_P - \theta_X)}\right] \times 90.$$

(*b*) Potentially explosive case ($\theta_Q > \theta_X$ or $\theta_X > \theta_S$)

$$r_m = r_X$$

$$\theta_m = \left\{\left[\frac{(\theta_X - \theta_S)}{(\theta_X - \theta_S) + (\theta_Q - \theta_X)}\right] \times 90\right\} + 270.$$

(*c*) Non-explosive case

$$r_m = r_X$$

$$\theta_m = \left\{\left[\frac{(\theta_X - \theta_P)}{(\theta_X - \theta_P) + (\theta_S - \theta_X)}\right] \times 180\right\} + 90.$$

(4) The modified angular coordinates are converted to rectangular (Cartesian) coordinates for the sample point as follows:

$r_m, \theta_m \rightarrow x_m, y_m.$

(5) The coordinates (x_m, y_m) then convey the explosive state of the analysed gas mixture as mentioned earlier.

The advantages of this method are the speed and reliability of processing sample analysis data and the reduction of the risk of misinterpreting graphical information when working under pressure.

A program (in BASIC) for the Casio FX 702P is given in Table 5 for determining the explosibility of mine atmospheres.

Table 6 gives details of compositions and other parameters for three mixtures of gases.

* If θ is between -180 and $+180$, equations for θ_m require modification.

Table 5. *Program for determining the explosibility of mine atmospheres*

```
5 WAIT 0
10 SET F2
20 INP "CH4=",A,"CO=",B,"H2=",C,"CO2=",D,"O2=",E
25 MODE 7
30 PRT "CH4=";A,"CO=";B,"H2=";C,"CO2=";D,"O2=";E
35 MODE 8
110 F=A+B+C:G=A/F*100:H=B/F*100:I=C/F*100
120 J=100-(F+E+D)
130 K=100/((G/5)+(H/12.5)+(I/4))
140 L=100/((G/15)+(H/74)+(I/75))
150 M=(100/((G/5.93)+(H/13.78)+(I/4.3)))*J/(J+D)
160 N=(100/((G/6.66)+(H/18)+(I/5.73)))*D/(J+D)+M
170 O=((6*G)+(4.15*H)+(16.55*I))/((G/5.93)+(H/13.78)+(I/4.3))
180 O=O*J/(J+D)
190 P=((3.2*G)+(2.16*H)+(10.2*I))/(((G/6.66)+(H/18)+(I/5.73)))
200 P=P*D/(J+D)+O
210 Q=.2093*(100-P-N)
220 R=-.209*K+20.93
230 S=-.209*L+20.93
240 T=-20.93*N/(Q-20.93)
250 U=0
255 MODE 7
260 PRT "X OF P=";K,"X OF Q=";L
270 PRT "X OF R=";N,"Y OF R=";Q,"X OF S=";T
280 PRT "X OF X=";F,"Y OF X=";E
285 MODE 8
290 V=(E-20.93)*(K-N)+F*(Q-R)
300 W=((K-N)*(-F*20.93)-(-F)*(K*Q-N*R))/V
301 X=F/W-1
305 MODE 7
310 PRT "X OF P1=";W
315 MODE 8
340 V=(E-20.93)*(L-N)+F*(Q-S)
350 W=((L-N)*(-F*20.93)+F*(L*Q-N*S))/V
351 Y=F/W-1
355 MODE 7
360 PRT "X OF Q1=";W
381 PRT "AIR REQD FOR LL MIX=";X
382 PRT "AIR REQD FOR UL MIX=";Y
383 MODE 8
390 K=K-N:L=L-N:T=T-N:F=F-N:N=N-N:R=R-Q:S=S-Q:U=U-Q:E=E-Q
400 Q=Q-Q:RPC K,R:K=Y:R=X:RPC L,S:L=Y:S=X:RPC T,U:T=Y:U=X
410 RPC F,E:F=Y:E=X
420 IF K>F THEN 440
430 GOTO 450
440 IF F>L THEN 540
450 IF L>F THEN 470
460 GOTO 480
470 IF F>T THEN 560
480 GOTO 490
490 IF F<0 THEN 520
500 Z=(F-K)/(T+360-K)*180+90
510 GOTO 580
520 Z=(F-K+360)/(T+360-K)*180+90
530 GOTO 580
540 Z=(F-L)/(K-L)*90
550 GOTO 580
560 Z=(F-T)/(L-T)*90+270
570 GOTO 580
580 V=E*COS Z
590 W= E*SIN Z
595 MODE 7
600 PRT "XM=";V,"YM=";W
605 MODE8
```

Table 6. *Details of composition and other parameters for three mixtures of gases*

Input	Mixture		
	1	2	3
Methane	1.93	3.30	3.18
Carbon monoxide	1.40	4.30	0.83
Hydrogen	4.00	4.90	—
Carbon dioxide	5.32	13.80	2.62
Oxygen	11.89	2.80	13.46
Output			
	1	2	3
x of P	4.89	5.61	5.71
x of Q	36.48	36.40	17.96
x of R	5.50	6.49	6.75
y of R	7.09	7.71	11.71
x of S	8.32	10.28	15.33
x of X	7.33	12.50	4.01
y of X	11.89	2.80	13.46
x of P′	5.14	6.18	
x of Q′	10.42	8.81	
Air required for LL mix	0.42	1.02	
Air required for UL mix	−0.30	0.42	
x_m	2.26	3.84	−2.22
y_m	4.61	−6.74	2.37

Gas concentration units are percentage by volume.
For an explanation of P, Q, R, S, X, P′, Q′, see Fig. 5.
x_m, y_m – modified x- and y-coordinates of mixture.
LL – lower limit; UL – upper limit.

References

Ellicott, C. W. (1981). Assessment of the explosibility of gas mixtures and monitoring of sample-time trends. In *Proceedings of the symposium on initions, explosions and fires*, University of Woolongong, Wollongong, ed. A. J. Hargraves, pp. 20.1–20.10. Melbourne: The Australasian Institute of Mining and Metallurgy.

Hughes, A. J. and Raybould, W. E. (1960). The rapid determination of the explosibility of mine fire gases. *Min. Engr.*, **20**, 37–53.

Smith, G. E. and Nugent, J. W. (1977). *The use of a handheld programmable calculator to determine the explosibility of mine atmospheres and gas mixtures.* Mine Enforcement and Safety Administration Inf. Rept 1069. 13 pp.

Zabetakis, M. G., Stahl, R. W. and Watson, H. A. (1959). *Determining the explosibility of mine atmospheres*, USBM IC 7901. Washington: US Bureau of Mines. 11 pp.

INDEX